Student Solutions Manual

Introductory and Intermediate Algebra
An Applied Approach

SIXTH EDITION

Richard N. Aufmann
Palomar Community College

Joanne S. Lockwood
Nashua Community College

Prepared by

Pat Foard
South Plains College

Ellena Reda
Dutchess Community College

CENGAGE Learning

Australia • Brazil • Mexico • Singapore • United Kingdom • United States

CENGAGE
Learning

For product information and technology assistance, contact us at **Cengage Learning Customer & Sales Support, 1-800-354-9706**.

For permission to use material from this text or product, submit all requests online at **www.cengage.com/permissions**
Further permissions questions can be emailed to **permissionrequest@cengage.com**.

ISBN-13: 978-1-285-41762-2
ISBN-10: 1-285-41762-3

Cengage Learning
200 First Stamford Place, 4th Floor
Stamford, CT 06902
USA

Cengage Learning is a leading provider of customized learning solutions with office locations around the globe, including Singapore, the United Kingdom, Australia, Mexico, Brazil, and Japan. Locate your local office at: **www.cengage.com/global**.

Cengage Learning products are represented in Canada by Nelson Education, Ltd.

To learn more about Cengage Learning Solutions, visit **www.cengage.com**.

Purchase any of our products at your local college store or at our preferred online store **www.cengagebrain.com**.

Printed in the United States of America
1 2 3 4 5 19 18 17 16 15

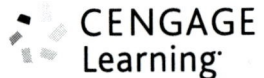

Contents

Chapter 1: Real Numbers and Variable Expressions

Prep Test

1. 127.16

2. $3416 + 42{,}561 + 537 = 46{,}514$

3. $5004 - 487 = 4517$

4. $407 \times 28 = 11{,}396$

5. $11{,}684 \div 23 = 508$

6. 24

7. 4

8. $3 \cdot 7$

9. $\frac{4}{10} = \frac{2}{5}$

10. **iv** Division by 0 is undefined.

Section 1.1

Concept Check

1. **a.** left

b. right

3. absolute

5. Find the absolute value of each number. Subtract the smaller number from the larger one. The sign of the final answer is the sign of the number with the larger absolute value.

7. Add the opposite of the second integer to the first integer.

Objective A Exercises

9. $8 > -6$

11. $-12 < 1$

13. $42 > 19$

15. $0 > -31$

17. $53 > -46$

19. $-23 < -8$

$-18 < -8$

$-8 = -8$

$0 > -8$

The elements -23 and -18 are less than -8.

21. $-33 < -10$

$-13 < -10$

$21 > -10$

$37 > 10$

The elements 21 and 37 are greater than -10.

23. (i) n is positive.

Objective B Exercises

25. -4

27. 9

29. 28

31. 14

33. -77

35. 0

37. 74

39. -82

41. -81

43. $|-83| > |58|$

45. $|43| < |-52|$

47. $|-68| > |-42|$

49. $|-45| < |-61|$

51. $p = -19; -p = 19$

$p = 0; -p = 0$

$p = 28; -p = -28$

53. $x = -45; -x = 45$

$x = 0; -x = 0$

$x = 17; -x = -17$

55. True

Objective C Exercises

57. $-3 + (-8) = -11$

59. $-8 + 3 = -5$

61. $-3 + (-80) = -83$

63. $-23 + (-23) = -46$

65. $16+(-16)=0$

67. $48+(-53)=-5$

69. $-17+(-3)+29=-20+29=9$

71. $-3+(-8)+12=-11+12=1$

73. $16-8=16+(-8)=8$

75. $7-14=7+(-14)=-7$

77. $-7-2=-7+(-2)=-9$

79. $7-(-2)=7+2=9$

81. $-6-(-3)=-6+3=-3$

83. $6-(-12)=6+12=18$

85. $13+(-22)+4+(-5)=-9+4+(-5)$
$=-5+(-5)=-10$

87. $-16+(-17)+(-18)+10=-33+(-18)+10$
$=-51+10=-41$

89. $26+(-15)+(-11)+(-12)=11+(-11)+(-12)$
$=0+(-12)=-12$

91. $-14+(-15)+(-11)+40=-29+(-11)+40$
$=-40+40=0$

93. $-4-3-2=-4+(-3)+(-2)$
$=-7+(-2)=-9$

95. $12-(-7)-8=12+7+(-8)$
$=19+(-8)=11$

97. $-19-(-19)-18=-19+19+(-18)$
$=0+(-18)=-18$

99. $-17-(-8)-(-9)=-17+8+9$
$=-9+9=0$

101. $-30-(-65)-29-4=-30+65+(-29)+(-4)$
$=35+(-29)+(-4)$
$=6+(-4)=2$

103. $-16-47-63-12=-63-63-12$
$=-126-12=-138$

105. $-47-(-67)-13-15=-47+(67)+(-13)+(-15)$
$=20+(-13)+(-15)$
$=7+(-15)=-8$

107. $-19-17-(-36)-12=-19+(-17)+36+(-12)$
$=-36+36+(-12)$
$=0+(-12)=-12$

109. Negative

111. Positive

Objective D Exercises

113. $14(3)=42$

115. $-7\cdot4=-28$

117. $(-12)(-5)=60$

119. $-11(23)=-253$

121. $(-17)(14)=-238$

123. $6(-19)=-114$

125. $12\div(-6)=-2$

127. $(-72)\div(-9)=8$

129. $-42\div6=-7$

131. $(-144)\div12=-12$

133. $48\div(-8)=-6$

135. $\frac{-49}{7}=-7$

137. $\frac{-44}{-4}=11$

139. $\frac{98}{-7}=-14$

141. $-\frac{-120}{8}=-(-15)=15$

143. $-\frac{-80}{-5}=-16$

145. $0\div(-9)=0$

147. $\frac{-261}{9}=-29$

149. $9\div 0$ is undefined.

151. $\frac{132}{-12}=-11$

153. $\frac{0}{0}$ is undefined

155. $7(5)(-3)=35(-3)=-105$

157. $9(-7)(-4)=-63(-4)=252$

159. $7(-2)(5)(-6)=-14(5)(-6)=-70(-6)=420$

161. $(-14)9(-11)0=-126(-11)0=(1386)0=0$

163. Negative

Objective E Exercises

165. **Strategy** To find the difference, subtract record low (–36°) from the record high (117°).

Solution $117-(-36)=117+36=153$

The difference is 153° F.

167. **Strategy** To find new temperature, add the rise (5°) to the original temperature (–19°).

Solution $-19+5=-14$

The temperature is –14°.

169. **Strategy** To find the difference, subtract the depth of the Mariana Trench (−11,520 m) from the height of Mt. Everest (8850 m).

Solution

$8850-(-11{,}520)=8850+11{,}520=20{,}370$

The difference is 20,370 m.

171. **a. Strategy** To find the score for each day relative to par, subtract par (72) from each day's scores.

Solution Day 1: $72-72=0$

Day 2: $68-72=-4$

Day 3: $70-72=-2$

Ken Duke's scores for the first three days were 0, –4, –2.

b. Strategy To find the score for the first three days, add the three scores.

Solution $0+(-4)+(-2)=-4+(-2)=-6$

Ken Duke's score for the first three days was −6.

c. Strategy To find the score for the first four days, find the score for the fourth day and add the fourth day's score to the first three day's scores.

Solution

Score for the fourth day: $68-72=-4$

Score for the first four days: $-6+(-4)=-10$

Ken Duke's score for the first four days was −10.

173. **Strategy** To find the average daily temperature:

- Add the seven temperature readings.
- Divide by the total by 7.

Solution

$-6+(-11)+1+5+(-3)+(-9)+(-5)$

$=-17+1+5+(-3)+(-9)+(-5)$

$=-16+5+(-3)+(-9)+(-5)$

$=-11+(-3)+(-9)+(-5)$

$=-14+(-9)+(-5)$

$=-23+(-5)=-28$

$-28\div 7=-4$

The average daily low temperature was −4° F.

175. False

177. **Strategy** To find average score, divide the total of the scores (–20) by ten.

Solution

$-20\div 10=-2$

The average score is –2.

179. **Strategy** To find the grade:
- Multiply the number of correct answers (20) by 5.
- Multiply the number of incorrect answers (5) by –5.
- Multiply the number of blank questions (2) by –2.
- Add the products.

Solution

$20 \times 5 = 100$

$5 \times (-5) = -25$

$2 \times (-2) = -4$

$100 + (-25) + (-4) = 75 + (-4) = 71$

The grade is 71.

Critical Thinking

181. **Strategy** To find largest difference, subtract the smallest number (–10) from the largest number (15).

Solution $15 - (-10) = 15 + 10 = 25$

The largest difference is 25.

183. **a.** True

b. True

Projects or Group Activities

185. Answers will vary. For example:

$-12 - (-6) = 12 + 6 = -6$.

Strategy: Write any negative number. Then subtract a negative number whose absolute value is 6 less than the absolute value of the first number.

187. –2, 4, –8, 16, . . .

To get each successive number, multiply by –2.

$16(-2) = -32$

$-32(-2) = 64$

$64(-2) = -128$

–32, 64, –128

Section 1.2

Concept Check

1. 3; 4; terminating

3. 0.01

5. equivalent, common denominator

7. reciprocal

9. $(-5)^6$

Objective A

11. $\frac{1}{8} = 1 \div 8 = 0.125$

13. $\frac{2}{9} = 2 \div 9 = 0.\overline{2}$

15. $\frac{1}{6} = 1 \div 6 = 0.1\overline{6}$

17. $\frac{9}{16} = 9 \div 16 = 0.5625$

19. $\frac{7}{12} = 7 \div 12 = 0.58\overline{3}$

21. $\frac{21}{40} = 21 \div 40 = 0.525$

Objective B Exercises

23. 100% = 100(0.01) = 1, multiplying by 1 does not change the value of the number.

25. $40\% = 40\left(\frac{1}{100}\right) = \frac{40}{100} = \frac{2}{5}$

$40\% = 40(0.01) = 0.40$

27. $88\% = 88\left(\frac{1}{100}\right) = \frac{88}{100} = \frac{22}{25}$

$88\% = 88(0.01) = 0.88$

29. $160\% = 160\left(\frac{1}{100}\right) = \frac{160}{100} = \frac{8}{5}$

$160\% = 160(0.01) = 1.60$

31. $87\% = 87\left(\frac{1}{100}\right) = \frac{87}{100}$

$87\% = 87(0.01) = 0.87$

33. $450\% = 450\left(\frac{1}{100}\right) = \frac{450}{100} = \frac{9}{2}$

$450\% = 450(0.01) = 4.50$

35. $4\frac{2}{7}\% = 4\frac{2}{7}\left(\frac{1}{100}\right) = \frac{30}{7}\left(\frac{1}{100}\right) = \frac{3}{70}$

37. $37\frac{1}{2}\% = 37\frac{1}{2}\left(\frac{1}{100}\right) = \frac{75}{2}\left(\frac{1}{100}\right) = \frac{3}{8}$

39. $\frac{1}{4}\% = \frac{1}{4}\left(\frac{1}{100}\right) = \frac{1}{400}$

41. $6\frac{1}{4}\% = 6\frac{1}{4}\left(\frac{1}{100}\right) = \frac{25}{4}\left(\frac{1}{100}\right) = \frac{1}{16}$

43. $5\frac{3}{4}\% = 5\frac{3}{4}\left(\frac{1}{100}\right) = \frac{23}{4}\left(\frac{1}{100}\right) = \frac{23}{400}$

45. $9.1\% = 9.1(0.01) = 0.091$

47. $16.7\% = 16.7(0.01) = 0.167$

49. $0.9\% = 0.9(0.01) = 0.009$

51. $9.15\% = 9.15(0.01) = 0.0915$

53. $18.23\% = 18.23(0.01) = 0.1823$

55. $0.37 = 0.37(100\%) = 37\%$

57. $0.02 = 0.02(100\%) = 2\%$

59. $0.125 = 0.125(100\%) = 12.5\%$

61. $1.36 = 1.36(100\%) = 136\%$

63. $0.004 = 0.004(100\%) = 0.4\%$

65. $\frac{83}{100} = \frac{83}{100}(100\%) = \frac{8300}{100}\% = 83\%$

67. $\frac{3}{8} = \frac{3}{8}(100\%) = \frac{300}{8}\% = 37\frac{1}{2}\%$

69. $\frac{4}{9} = \frac{4}{9}(100\%) = \frac{400}{9}\% = 44\frac{4}{9}\%$

71. $\frac{9}{20} = \frac{9}{20}(100\%) = \frac{900}{20}\% = 45\%$

73. $2\frac{1}{2} = 2\frac{1}{2}(100\%) = \frac{5}{2}(100\%) = \frac{500}{2}\%$

$= 250\%$

75. Greater than 1%.

Objective C Exercises

77. $-\frac{6}{13} + \frac{17}{26} = \frac{-12}{26} + \frac{17}{36} = \frac{-12+17}{26} = \frac{5}{26}$

79. $\frac{5}{8} - \left(-\frac{3}{4}\right) = \frac{5}{8} + \frac{6}{8} = \frac{5+6}{8} = \frac{11}{8}$

81. $\frac{11}{12} - \frac{5}{6} = \frac{11}{12} - \frac{10}{12} = \frac{11-10}{12} = \frac{1}{12}$

83. $-\frac{5}{8} - \left(-\frac{11}{12}\right) = \frac{-15}{24} + \frac{22}{24} = \frac{-15+22}{24} = \frac{7}{24}$

85. $\frac{1}{2} - \frac{2}{3} + \frac{1}{6} = \frac{3}{6} - \frac{4}{6} + \frac{1}{6} = \frac{3-4+1}{6} = \frac{0}{6} = 0$

87. $\frac{1}{2} - \frac{3}{8} - \left(-\frac{1}{4}\right) = \frac{4}{8} - \frac{3}{8} + \frac{2}{8} = \frac{4-3+2}{8} = \frac{3}{8}$

89. $2.54 - 3.6 = -1.06$

91. $-16.92 - 6.925 = -23.845$

93. $6.9027 - 17.692 = -10.7893$

95. $-3.09 - 4.6 - (-27.3) = -7.69 + 27.3 = 19.61$

97. $2.66 - (-4.66) - 8.2 = 2.66 + 4.66 - 8.2$

$= 7.32 - 8.2 = -0.88$

99. $-0.125 + 1.25 \approx 1$

Objective D Exercises

101. $\left(-\frac{3}{4}\right)\left(-\frac{8}{27}\right) = \frac{\cancel{3}\cdot\cancel{2}\cdot\cancel{2}\cdot 2}{\cancel{2}\cdot\cancel{2}\cdot\cancel{3}\cdot 3\cdot 3} = \frac{2}{9}$

103. $\left(\frac{5}{12}\right)\left(-\frac{8}{15}\right) = -\frac{\cancel{5}\cdot\cancel{2}\cdot\cancel{2}\cdot 2}{\cancel{2}\cdot\cancel{2}\cdot 3\cdot 3\cdot\cancel{5}} = -\frac{2}{9}$

105. $\frac{5}{12}\left(-\frac{8}{15}\right)\frac{1}{3} = -\frac{\cancel{5}\cdot\cancel{2}\cdot\cancel{2}\cdot 2}{\cancel{2}\cdot\cancel{2}\cdot 3\cdot 3\cdot\cancel{5}\cdot 3} = -\frac{2}{27}$

107. $\frac{3}{8} \div \frac{1}{4} = \frac{3}{8}\cdot\frac{4}{1} = \frac{3\cdot\cancel{2}\cdot\cancel{2}}{\cancel{2}\cdot\cancel{2}\cdot 2} = \frac{3}{2}$

109. $-\frac{5}{12} \div \frac{15}{32} = -\frac{5}{12}\cdot\frac{32}{15} = -\frac{\cancel{5}\cdot\cancel{2}\cdot\cancel{2}\cdot 2\cdot 2\cdot 2}{\cancel{2}\cdot\cancel{2}\cdot 3\cdot 3\cdot\cancel{5}} = -\frac{8}{9}$

111. $-\frac{4}{9} \div \left(-\frac{2}{3}\right) = -\frac{4}{9}\cdot\left(-\frac{3}{2}\right) = \frac{\cancel{2}\cdot 2\cdot\cancel{3}}{\cancel{3}\cdot 3\cdot\cancel{2}} = \frac{2}{3}$

113. $1.2(3.47) = 4.164$

115. $(-1.89)(-2.3) = 4.347$

117. $1.2(-0.5)(3.7) = (-0.6)(3.7) = -2.22$

119. $-1.27 \div (-1.7) \approx 0.75$

121. $0.0976 \div 0.042 \approx 2.32$

123. $-7.894 \div (-2.06) \approx 3.83$

125. **a.** Less than 1

b. Greater than 1

Objective E Exercises

127. $7^4 = 7 \cdot 7 \cdot 7 \cdot 7 = 2401$

129. $-4^3 = -(4 \cdot 4 \cdot 4) = -64$

131. $(-2)^3 = (-2)(-2)(-2) = -8$

133. $(-5)^3 = (-5)(-5)(-5) = -125$

135. $\left(-\frac{3}{4}\right)^3 = \left(-\frac{3}{4}\right)\left(-\frac{3}{4}\right)\left(-\frac{3}{4}\right) = -\frac{3 \cdot 3 \cdot 3}{4 \cdot 4 \cdot 4} = -\frac{27}{64}$

137. $(1.5)^3 = (1.5)(1.5)(1.5) = 3.375$

139. $\left(-\frac{1}{2}\right)^3 \cdot 8 = \left(-\frac{1}{2}\right)\left(-\frac{1}{2}\right)\left(-\frac{1}{2}\right) \cdot 2 \cdot 2 \cdot 2$

$= -\frac{\overset{1}{\cancel{2}} \cdot \overset{1}{\cancel{2}} \cdot \overset{1}{\cancel{2}}}{\underset{1}{\cancel{2}} \cdot \underset{1}{\cancel{2}} \cdot \underset{1}{\cancel{2}}} = -1$

141. $(-2) \cdot (-2)^2 = (-2)(-2)(-2) = -8$

143. $(-3)^3 \cdot 5^2 \cdot 10 = (-3)(-3)(-3) \cdot 5 \cdot 5 \cdot 10$

$= -27 \cdot 25 \cdot 10 = -675 \cdot 10 = -6750$

145. Negative

147. Positive

Objective F Exercises

149. $\sqrt{16} = 4$

151. $\sqrt{49} = 7$

153. $\sqrt{32} = \sqrt{16 \cdot 2} = \sqrt{16} \cdot \sqrt{2} = 4\sqrt{2}$

155. $\sqrt{8} = \sqrt{4 \cdot 2} = \sqrt{4} \cdot \sqrt{2} = 2\sqrt{2}$

157. $6\sqrt{18} = 6\sqrt{9 \cdot 2} = 6\sqrt{9} \cdot \sqrt{2}$
$= 6 \cdot 3\sqrt{2} = 18\sqrt{2}$

159. $5\sqrt{40} = 5\sqrt{4 \cdot 10} = 5\sqrt{4} \cdot \sqrt{10}$
$= 5 \cdot 2\sqrt{10} = 10\sqrt{10}$

161. $\sqrt{15} = \sqrt{3 \cdot 5} = \sqrt{15}$

163. $\sqrt{29}$

165. $-9\sqrt{72} = -9\sqrt{4 \cdot 9 \cdot 2} = -9\sqrt{4} \cdot \sqrt{9} \cdot \sqrt{2}$
$= -9 \cdot 2 \cdot 3\sqrt{2} = -54\sqrt{2}$

167. $\sqrt{45} = \sqrt{9 \cdot 5} = \sqrt{9} \cdot \sqrt{5} = 3\sqrt{5}$

169. $\sqrt{0} = 0$

171. $6\sqrt{128} = 6\sqrt{64 \cdot 2} = 6\sqrt{64} \cdot \sqrt{2}$
$= 6 \cdot 8\sqrt{2} = 48\sqrt{2}$

173. $\sqrt{240} \approx 15.492$

175. $\sqrt{288} \approx 16.971$

177. $\sqrt{256} = 16$

179. Between −11 and −10

181. Between 2 and 3

Objective G Exercises

183. **Strategy** To find the difference, subtract the low temperature (−48.9°) from the high temperature (6.67°).

Solution $6.67 - (-48.9) = 6.67 + 48.9 = 55.57$

The difference between the record high and record low temperature in Browing is 55.57° C.

185. **Strategy** To find the difference, subtract the melting point (−218.4°) from the boiling point (−182.962°).

Solution
$-182.962 - (-218.4) = -182.962 + 218.4 = 35.438$

The difference between the boiling point and melting point is 35.438° C.

187. **a. Strategy** To find the difference, subtract the oil production in 2008 (4.9 million) from the oil production in 1973 (9.2 million).

Solution
$9.2 - 4.9 = 4.3$
The difference in oil production is 4.3 million barrels per day.

b. Strategy To find the increase, subtract the oil production in 2008 (4.9) from the predicted production in 2020 (6.0).

Solution $6.0 - 4.9 = 1.1$
The increase in oil production from 2008 to 2020 is 1.1 million barrels per day.

189. **Strategy** To find how much butter the chef should use, add $\frac{1}{2}$ of $\frac{3}{4}$ c to $\frac{3}{4}$ c.

Solution $\frac{3}{4}\cdot\frac{1}{2}+\frac{3}{4}=\frac{3}{8}+\frac{3}{4}=\frac{3}{8}+\frac{6}{8}=\frac{9}{8}=1\frac{1}{8}$

The chef should use $1\frac{1}{8}$ c of butter.

191. **Strategy** To find number of servings, divide the total weight (24 oz.) by the number of ounces per serving $\left(1\frac{1}{2}\right)$.

Solution

$$24 \div 1\frac{1}{2} = \frac{24}{1} \div \frac{3}{2} = \frac{\overset{8}{\cancel{24}}}{1}\cdot\frac{2}{\underset{1}{\cancel{3}}} = \frac{8\cdot 2}{1} = 16$$

There are 16 servings in 1 box.

Critical Thinking

193. Answers will vary. For example:
a. 0.15
b. 1.05
c. 0.001

195. Yes, it is always possible to find a rational number between two given numbers. Explanations will vary. One method is to add the two numbers and divide the sum by 2

Projects or Group Activities

197. $a = 2, b = 3, c = 6$

Section 1.3

Concept Check

1. We need an Order of Operations Agreement to prevent there being more than one answer for a numerical expression.

Objective A Exercises

3. $4 - 8 \div 2 = 4 - 4 = 0$

5. $$\begin{aligned} 2(3-4)-(-3)^2 &= 2(-1)-(-3)^2 \\ &= 2(-1)-9 \\ &= -2-9 \\ &= -11 \end{aligned}$$

7. $24 - 18 \div 3 + 2 = 24 - 6 + 2 = 18 + 2 = 20$

9. $$\begin{aligned} 8-2(3)^2 &= 8-2(9) \\ &= 8-18 \\ &= -10 \end{aligned}$$

11. $$\begin{aligned} 12+16\div 4\cdot 2 &= 12+4\cdot 2 \\ &= 12+8 \\ &= 20 \end{aligned}$$

13. $$\begin{aligned} 27-18\div\left(-3^2\right) &= 27-18\div(-9) \\ &= 27+2 \\ &= 29 \end{aligned}$$

15. $$\begin{aligned} 16+15\div(-5)-2 &= 16+(-3)-2 \\ &= 13-2 \\ &= 11 \end{aligned}$$

17. $$\begin{aligned} 14-2^2-|4-7| &= 14-2^2-|-3| = 14-2^2-3 \\ &= 14-4-3 = 10-3 = 7 \end{aligned}$$

19. $$\begin{aligned} 3-2[8-(3-2)] &= 3-2[8-(1)] \\ &= 3-2[7] \\ &= 3-14 \\ &= -11 \end{aligned}$$

21. $$\begin{aligned} 6+\frac{16-4}{2^2+2}-2 &= 6+\frac{12}{4+2}-2 \\ &= 6+\frac{12}{6}-2 \\ &= 6+2-2 \\ &= 8-2 \\ &= 6 \end{aligned}$$

23. $18 \div |9-2^3| + (-3) = 18 \div |9-8| + (-3)$
$= 18 \div 1 + (-3)$
$= 18 + (-3)$
$= 15$

25. $4[16-(7-1)] \div 10 = 4[16-6] \div 10$
$= 4[10] \div 10$
$= 40 \div 10$
$= 4$

27. $20 \div (10-2^3) + (-5) = 20 \div (10-8) + (-5)$
$= 20 \div 2 + (-5) = 10 + (-5) = 5$

29. $4(-8) \div [2(7-3)^2] = 4(-8) \div [2(4)^2]$
$= 4(-8) \div [2(16)] = 4(-8) \div 32$
$= -32 \div 32 = -1$

31. $16 - 4 \cdot \frac{3^3-7}{2^3+2} - (-2)^2 = 16 - 4 \cdot \frac{27-7}{8+2} - (4)$
$= 16 - 4 \cdot \frac{20}{10} - 4$
$= 16 - 4 \cdot 2 - 4$
$= 16 - 8 - 4 = 8 - 4 = 4$

33. $0.3(1.7-4.8) + (1.2)^2 = 0.3(-3.1) + 1.44$
$= -0.93 + 1.44 = 0.51$

35. $(1.65-1.05)^2 \div 0.4 + 0.8 = (0.6)^2 \div 0.4 + 0.8$
$= 0.36 \div 0.4 + 0.8$
$= 0.9 + 0.8 = 1.7$

Critical Thinking

37. Answers will vary. For example, $\frac{17}{24}$ and $\frac{33}{48}$.

39. Answers will vary. For example:

a. $\frac{1}{2}$

b. 1

c. 2

Projects and Group Activities

41. $1{,}000{,}000 = 100^3$

A	B	C		A	B	C
1	8	27	is	1^3	2^3	3^3
64	125	216		4^3	5^3	6^3

A: $(1+3n)^3$
B: $(2+3n)^3$
C: $(3+3n)^3$
$100^3 = [1+3(33)]^3$
1,000,000 is in Column A.

Check Your Progress: Chapter 1

1. $\{1, 2, 3, 4, 5, 6, 7, 8\}$

2. $-7 < 1$
$0 < 1$
$2 > 1$
$5 > 1$
-7 and 0 are less than 1.

3. 13

4. $|-44| = 44$
$-|-18| = -18$

5. $|31| > |-13|$

6. $-47 + 23 = -24$

7. $-11 - (-27) = -11 + 27 = 16$

8. $-32 + 40 + (-9) = 8 + (-9) = -1$

9. $42 - (-82) - 65 - 7 = 42 + 82 - 65 - 7$
$= 124 - 65 - 7 = 59 - 7 = 52$

10. $16(-2) = -32$

11. $-9(7)(-5) = -63(-5) = 315$

12. $250 \div (-25) = -10$

13. $-\frac{-80}{-5} = -16$

14. $\frac{-58}{0}$ is undefined

15. $\frac{11}{16} = 11 \div 16 = 0.6875$

16. $\frac{7}{11} = 7 \div 11 = 0.\overline{63}$

17. $45\% = 45\left(\frac{1}{100}\right) = \frac{45}{100} = \frac{9}{20}$

$45\% = 45(0.01) = 0.45$

18. $14\frac{1}{2}\% = 14\frac{1}{2}\left(\frac{1}{100}\right) = \frac{29}{2}\left(\frac{1}{100}\right) = \frac{29}{200}$

19. $\frac{7}{8} = \frac{7}{8} \times 100\% = \frac{700}{8}\% = 87.5\%$

20. $0.08 = 0.08(100\%) = 8\%$

21. $\frac{5}{6} + \frac{3}{18} = \frac{15}{18} + \frac{3}{18} = \frac{18}{18} = 1$

22. $\frac{3}{24} - \frac{1}{6} = \frac{3}{24} - \frac{4}{24} = -\frac{1}{24}$

23. $-18.39 + 4.9 - 23.7 = -13.49 - 23.7 = -37.19$

24. $\frac{5}{8}\left(-\frac{9}{12}\right)\left(\frac{16}{25}\right) = -\frac{\overset{1}{\cancel{5}} \cdot \overset{3}{\cancel{9}} \cdot \overset{\overset{1}{\cancel{2}}}{\cancel{16}}}{\underset{1}{\cancel{8}} \cdot \underset{\underset{2}{\cancel{4}}}{\cancel{12}} \cdot \underset{5}{\cancel{25}}} = -\frac{3}{10}$

25. $-\frac{6}{11} \div \frac{9}{4} = -\frac{\overset{2}{\cancel{6}}}{11} \cdot \frac{4}{\underset{3}{\cancel{9}}} = -\frac{8}{33}$

26. $-1.6(0.2) = -0.32$

27. $3\sqrt{18} = 3\sqrt{9 \cdot 2} = 3\sqrt{3} \cdot \sqrt{2} = 3 \cdot 3\sqrt{2} = 6\sqrt{2}$

28. $\sqrt{27} = \sqrt{9 \cdot 3} = \sqrt{9} \cdot \sqrt{3} = 3\sqrt{3}$

29. $-3^2 \cdot (-2)^4 = -9(16) = -144$

30.
$$\begin{aligned} 5 - 4[3 - 2(7-1)] \div 9 &= 5 - 4[3 - 2(6)] \div 9 \\ &= 5 - 4[3 - 12] \div 9 \\ &= 5 - 4[-9] \div 9 \\ &= 5 + 36 \div 9 \\ &= 5 + 4 \\ &= 9 \end{aligned}$$

31.
$$\begin{aligned} -4 \cdot 2^3 - \frac{1-13}{2^2 \cdot 3} &= -4 \cdot 2^3 - \frac{-12}{4 \cdot 3} \\ &= -4 \cdot 2^3 - (-1) \\ &= -4 \cdot 8 - (-1) \\ &= -32 - (-1) \\ &= -32 + 1 \\ &= -31 \end{aligned}$$

32.
$$\begin{aligned} (8 - 3^2)^6 + (2 \cdot 3 - 7)^9 &= (8-9)^6 + (6-7)^9 \\ &= (-1)^6 + (-1)^9 \\ &= 1 - 1 \\ &= 0 \end{aligned}$$

33. **Strategy** To find the temperature, add the rise (8°) to the previous temperature (–3°).

Solution $-3° + 8° = 5°$
The temperature is 5° C.

34. **Strategy** To find the average low temperature:
- add the temperatures (–8°, –12°, 0°, –4°, 5°, –7°, –9°).
- add divide by the number of days in a week (7).

Solution
$$\begin{aligned} &-8 + (-12) + 0 + (-4) + 5 + (-7) + (-9) \\ &= -20 + 0 + (-4) + 5 + (-7) + (-9) \\ &= -20 + (-4) + 5 + (-7) + (-9) \\ &= -24 + 5 + (-7) + (-9) \\ &= -19 + (-7) + (-9) \\ &= -26 + (-9) \\ &= -35 \end{aligned}$$
$-35 \div 7 = -5$
The average low temperature is –5° C.

35. **Strategy** To find the temperature, subtract the rise (20.3°) from the high temperature (15.7°).

Solution $15.7° - 20.3° = -4.6°$
The temperature was –4.6° C.

Section 1.4

Concept Check

1. $2x^2$, $5x$, $\underline{-8}$

3. $-a^4$, $\underline{6}$

5. coefficient of $12a^2$: 12
coefficient of $-8ab$: -8
coefficient of $-b^2$: -1

7. reciprocal (or multiplicative inverse)

9. Like terms are variable terms with the same variable part. Constant terms are also like terms. Examples of like terms are $4x$ and $-9x$.

Examples of terms that are not alike are $4x^2$ and $-9x$. The terms 4 and 9 are also like terms; 4 and $4x$ are not.

11. less than, quotient

13. $25-x$

Objective A Exercises

15. $6b \div (-a)$

$6(3) \div (-2) = 18 \div (-2) = -9$

17. $b^2 - 4ac$

$$(3)^2 - 4(2)(-4) = 9 - 4(2)(-4) = 9 - (-32) = 9 + 32 = 41$$

19. $b^2 - c^2$

$3^2 - (-4)^2 = 9 - 16 = -7$

21. $a^2 + b^2$

$2^2 + 3^2 = 4 + 9 = 13$

23. $\dfrac{5ab}{6} - 3cb$

$$\frac{5(2)(3)}{6} - 3(-4)(3) = \frac{30}{6} - (-36) = 5 - (-36) = 41$$

25. $\dfrac{2d+b}{-a}$

$$\frac{2(3)+4}{-(-2)} = \frac{6+4}{2} = \frac{10}{2} = 5$$

27. $\dfrac{b-d}{c-a}$

$$\frac{4-3}{-1-(-2)} = \frac{1}{1} = 1$$

29. $(b+d)^2 - 4a$

$$(4+3)^2 - 4(-2) = 7^2 - 4(-2) = 49 - (-8) = 57$$

31. $(d-a)^2 \div 5$

$$\left[3-(-2)\right]^2 \div 5 = 5^2 \div 5 = 25 \div 5 = 5$$

33. $\dfrac{b-2a}{bc^2-d}$

$$\frac{4-2(-2)}{4(-1)^2-3} = \frac{4-(-4)}{4(1)-3} = \frac{8}{4-3} = \frac{8}{1} = 8$$

35. $\dfrac{1}{3}d^2 - \dfrac{3}{8}b^2$

$$\frac{1}{3}(3)^2 - \frac{3}{8}(4)^2 = \frac{1}{3}(9) - \frac{3}{8}(16) = 3 - 6 = -3$$

37. $\dfrac{-4bc}{2a-b}$

$$\frac{-4(4)(-1)}{2(-2)-4} = \frac{16}{-4-4} = \frac{16}{-8} = -2$$

39. $-\dfrac{2}{3}d - \dfrac{1}{5}(bd - ac)$

$$-\frac{2}{3}(3) - \frac{1}{5}\left[4(3) - (-2)(-1)\right] = -\frac{2}{3}(3) - \frac{1}{5}\left[12 - 2\right] = -\frac{2}{3}(3) - \frac{1}{5}(10) = -2 - 2 = -4$$

41. Positive

43. Negative

Objective B Exercises

45. $6x + 8x = 14x$

47. $9a - 4a = 5a$

49. $7 - 3b = 7 - 3b$

51. $-12a + 17a = 5a$

53. $-12xy + 17xy = 5xy$

55. $-3ab + 3ab = 0$

57. $-\dfrac{1}{2}x - \dfrac{1}{3}x = -\dfrac{3}{6}x - \dfrac{2}{6}x = -\dfrac{5}{6}x$

59. $2.3x + 4.2x = 6.5x$

61. $x - 0.55x = 0.45x$

63. $5a - 3a + 5a = 7a$

65. $-5x^2 - 12x^2 + 3x^2 = -14x^2$

67. $\frac{3}{4}x-\frac{1}{3}x-\frac{7}{8}x=\frac{18}{24}x-\frac{8}{24}x-\frac{21}{24}x=-\frac{11}{24}x$

69. $7x-3y+10x=17x-3y$

71. $3a+(-7b)-5a+b=-2a-6b$

73. $3x+(-8y)-10x+4x=-3x-8y$

75. $x^2-7x+(-5x^2)+5x=-4x^2-2x$

77. $-10x-10y-10y-10x=-20x-20y$

i. 0 No

ii. -20 No

iii. $-20y$ No

iv. $-20x-20y$ Yes

v. $-20y-20x$ Yes

(iv) and (v)

Objective C Exercises

79. $12(5x)=60x$

81. $-2(5a)=-10a$

83. $-5(-6y)=30y$

85. $(6x)12=72x$

87. $(7a)(-4)=-28a$

89. $(-12b)(-9)=108b$

91. $-8(7x^2)=-56x^2$

93. $\frac{1}{6}(6x^2)=x^2$

95. $\frac{1}{8}(8x)=x$

97. $-\frac{1}{4}(-4a)=a$

99. $-\frac{1}{9}(-9b)=b$

101. $(12x)\left(\frac{1}{12}\right)=x$

103. $(-10n)\left(-\frac{1}{10}\right)=n$

105. $\frac{1}{7}(14x)=2x$

107. $-0.25(8x)=-2x$

109. $-\frac{5}{8}(24a^2)=-15a^2$

111. $-0.75(-8y)=6y$

113. $(33y)\left(\frac{1}{11}\right)=3y$

115. $(-10x)\left(\frac{1}{5}\right)=-2x$

117. $(21y)\left(-\frac{3}{7}\right)=-9y$

Objective D Exercises

119. $2(4x-3)=8x-6$

121. $-2(a+7)=-2a-14$

123. $-3(2y-8)=-6y+24$

125. $-(x+2)=-x-2$

127. $(5-3b)7=35-21b$

129. $\frac{1}{3}(6-15y)=2-5y$

131. $3(5x^2+2x)=15x^2+6x$

133. $-2(-y+9)=2y-18$

135. $(-3x-6)5=-15x-30$

137. $2(-3x^2-14)=-6x^2-28$

139. $-3(2y^2-7)=-6y^2+21$

141. $3(x^2-y^2)=3x^2-3y^2$

143. $-\frac{2}{3}(6x-18y)=-4x+12y$

145. $-(6a^2-7b^2)=-6a^2+7b^2$

147. $4(x^2-3x+5)=4x^2-12x+20$

149. $\frac{3}{4}(2x-6y+8)=\frac{3}{2}x-\frac{9}{2}y+6$

151. $4(-3a^2 - 5a + 7) = -12a^2 - 20a + 28$

153. $-3(-4x^2 + 3x - 4) = 12x^2 - 9x + 12$

155. $5(2x^2 - 4xy - y^2) = 10x^2 - 20xy - 5y^2$

157. $-(8b^2 - 6b + 9) = -8b^2 + 6b - 9$

159. $12 - 7(y - 9) = 12 - 7y + 63 = -7y + 75$

i. $5(y - 9) = 5y - 45$ No

ii. $12 - 7y - 63 = -7y - 51$ No

iii. $12 - 7y + 63 = -7y + 75$ Yes

iv. $12 - 7y - 9 = -7y + 3$ No

161. $6a - (5a + 7) = 6a - 5a - 7 = a - 7$

163. $10 - (11x - 3) = 10 - 11x + 3 = -11x + 13$

165. $8 - (12 + 4y) = 8 - 12 - 4y = -4y - 4$

167. $2(x - 4) - 4(x + 2) = 2x - 8 - 4x - 8$

$= -2x - 16$

169. $6(2y - 7) - (3 - 2y) = 12y - 42 - 3 + 2y$

$= 14y - 45$

171. $2(a + 2b) - (a - 3b) = 2a + 4b - a + 3b = a + 7b$

173. $2[x + 2(x + 7)] = 2[x + 2x + 14] = 2[3x + 14]$

$= 6x + 28$

175. $-5[2x + 3(5 - x)] = -5[2x + 15 - 3x]$

$= -5[-x + 15] = 5x - 75$

177. $-2[3x - (5x - 2)] = -2[3x - 5x + 2]$

$= -2[-2x + 2] = 4x - 4$

179. $-7x + 3[x - (3 - 2x)] = -7x + 3[x - 3 + 2x]$

$= -7x + 3[3x - 3]$

$= -7x + 9x - 9 = 2x - 9$

181. $0.12(2x + 3) + x = 0.24x + 0.36 + x$

$= 1.24x + 0.36$

183. $0.03x + 0.04(1000 - x) = 0.03x + 40 - 0.04x$

$= -0.01x + 40$

Objective E Exercises

185. the unknown number: x

$\frac{x}{18}$

187. the unknown number: x

$x + 20$

189. the unknown number: x

the product of eleven and the number: $11x$

$11x - 8$

191. the unknown number: x

the quotient of the number and twenty: $\frac{x}{20}$

$40 - \frac{x}{20}$

193. the unknown number: x

the square of the number: x^2

twice the number: $2x$

$x^2 + 2x$

195. the unknown number: x

the difference between the number and 50: $x - 50$

$10(x - 50) = 10x - 500$

197. the unknown number: x

three more than the number: $x + 3$

$x - (x + 3) = x - x - 3 = -3$

199. the unknown number: x

twice the number: $2x$

the difference between twice the number and four: $2x - 4$

$(2x - 4) + x = 2x - 4 + x = 3x - 4$

201. the unknown number: x

the product of three and the number: $3x$

$x + 3x = 4x$

203. the unknown number: x

the sum of the number and six: $x + 6$

$(x + 6) + 5 = x + 6 + 5 = x + 11$

205. the unknown number: x

the sum of the number and ten: $x + 10$

$x - (x + 10) = x - x - 10 = -10$

207. number of visitors to the Metropolitan Museum of Art: M
number of visitors to the Louvre: $M +$ 3,800,000

209. noise level of a car horn: d
noise level of an ambulance siren: $d + 10$

211. U2's concert ticket sales: T
E Street Band's concert ticket sales: $T -$ 28,500,000

213. number of bones in your body: N
number of bones in your foot: $\frac{1}{4}N$

215. attendance at major league basketball games: B
attendance at major league baseball games: $B +$ 50,000,000

217. number of U.S. undergraduate students: N
number who attend a two-year college: $0.46N$

219. measure of the largest angle: L
measure of the smallest angle: $\frac{1}{2}L - 10$

Critical Thinking

221. The number of students enrolled in fall-term science classes.

223. length of wire: x
length of side of square: $\frac{1}{4}x$

225. Two examples of translation of $5x + 8$ are "eight more than the product of five and a number" and "the sum of five times a number and eight." Two examples of the translation of $5(x + 8)$ are "five times the sum of a number and eight" and "the product of five and eight more than a number.

Projects or Group Activities

227. a. Yes; $7 \otimes 5 = 5 \otimes 7$

$7 \otimes 5 = 23 \qquad 5 \otimes 7$

$= (5 \cdot 7) - (5 + 7) = 35 - 12 = 23$

b. No; $(7 \otimes 5) \otimes 2 \neq 7 \otimes (5 \otimes 2)$

$$\begin{aligned} &(7 \otimes 5) \otimes 2 \\ &= 23 \otimes 2 \\ &= (23 \cdot 2) - (23 + 2) \\ &= 46 - 25 \\ &= 21 \end{aligned}$$

$$\begin{aligned} &7 \otimes (5 \otimes 2) \\ &= 7 \otimes [(5 \cdot 2) - (5 + 2)] \\ &= 7 \otimes [10 - 7] \\ &= 7 \otimes 3 \\ &= (7 \cdot 3) - (7 + 3) \\ &= 21 - 10 \\ &= 11 \end{aligned}$$

229. (i) $2x + 4(2x + 1) = 2x + 8x + 4 = 10x + 4$

(ii) $x - (4 - 9x) + 8 = x - 4 + 9x + 8 = 10x + 4$

(iii) $7(x - 4) - 3(2x + 6) = 7x - 28 - 6x - 18$
$= x - 46$

(iv) $3(2x + 8) + 4(x - 5) = 6x + 24 + 4x - 20$
$= 10x + 4$

(v) $6 - 2[x + (3x - 4)] + 2(9x - 5)$
$= 6 - 2[x + 3x - 4] + 2(9x - 5)$
$= 6 - 2[4x - 4] + 2(9x - 5)$
$= 6 - 8x + 8 + 18x - 10$
$= 10x + 4$

i, ii, iv, and v are equivalent; they are equal to $10x + 4$.

Section 1.5

Concept Check

1. roster, set-builder, interval

Objective A Exercises

3. $A = \{16, 17, 18, 19, 20, 21\}$

5. $A = \{9, 11, 13, 15, 17\}$

7. $A \cup B = \{3, 4, 5, 6\}$

9. $A \cup B = \{-10, -9, -8, 8, 9, 10\}$

11. $A \cup B = \{1, 3, 7, 9, 11, 13\}$

13. $A \cap B = \{4, 5\}$

15. $A \cap B = \varnothing$

17. $A \cap B = \{\text{c, d, e}\}$

Objective B

19. $\{x | x > -5,\ x \in \text{negative integers}\}$

21. $\{x | x > 30,\ x \in \text{integers}\}$

23. $\{x | x > 8,\ x \in \text{real numbers}\}$

25. $(1, 2)$

27. $(3, \infty)$

29. $[-4, 5)$

31. $(-\infty, 2]$

33. $[-3, 1]$

35. $\{x | -5 < x < -3\}$

37. $\{x | x \le -2\}$

39. $\{x | -3 \le x \le -2\}$

41. $\{x | x \le 6\}$

43. $[-5, 4]$

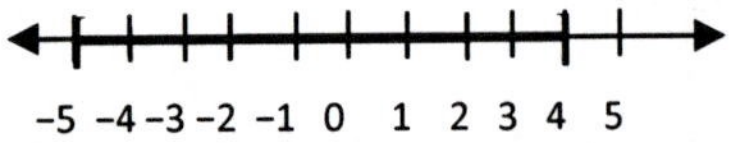

45. $\{x | x < 4\}$

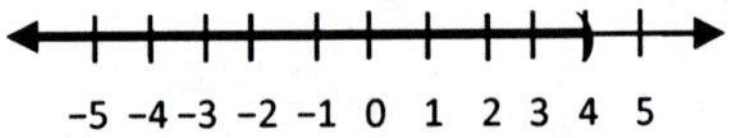

47. $\{x | x \le -4\}$

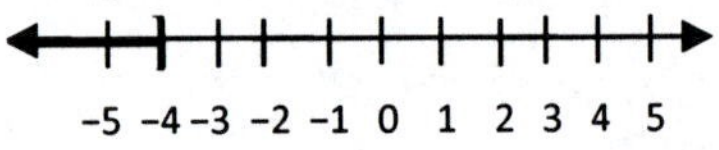

49. $(-\infty, 3]$

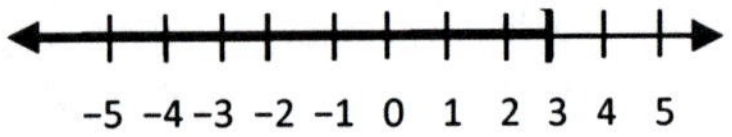

51. $[-1, 3)$

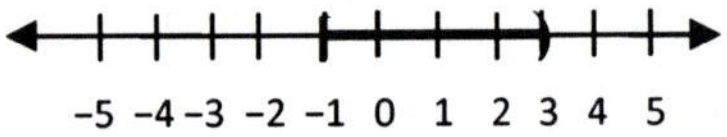

53. $\{x | -3 < x < 3\}$

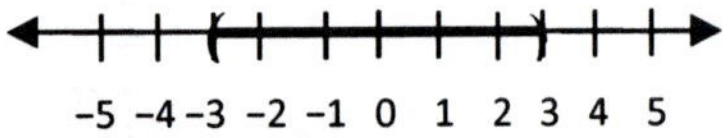

55. $\{x | 2 \le x \le 4\}$

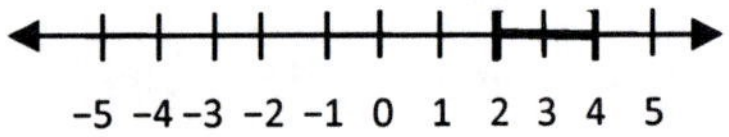

57. $\{x | -\infty < x < \infty\}$

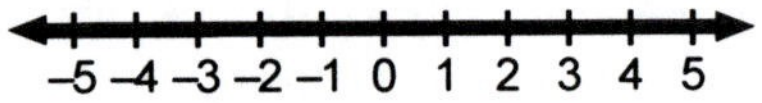

59. None

Critical Thinking

61. $m \ge 250$

63. True

Projects of Group Activities

65. Answers will vary. For example, $A = \{1, 2, 3, 4\}$ and $B = \{1, 2, 3, 4\}$.

Chapter 1 Review Exercises

1. $-4 < 1$ True
$0 < 1$ True
$11 < 1$ False
$x < 1$ for the values -4 and 0.

2. 4

3. $-|-5| = -(5) = -5$

4. $-3 + (-12) + 6 + (-4) = -15 + 6 + (-4)$
$= -9 + (-4) = -13$

5. $16 - (-3) - 18 = 16 + 3 - 18 = 19 - 18 = 1$

6. $-6(7) = -42$

7. $-100 \div 5 = -20$

8. $$\begin{array}{r} 0.28 \\ 25\overline{)7.00} \\ \underline{50} \\ 200 \\ \underline{200} \\ 0 \end{array}$$

$\frac{7}{25} = 0.28$

9. $6.2\% = 6.2(0.01) = 0.062$

10. $\frac{5}{8} = \frac{5}{8}(100\%) = \frac{500}{8}\% = 62.5\%$

11. $\frac{1}{3} - \frac{1}{6} + \frac{5}{12} = \frac{4}{12} - \frac{2}{12} + \frac{5}{12} = \frac{4-2+5}{12} = \frac{7}{12}$

12. $5.17 - 6.238 = -1.068$

13. $-\frac{18}{35} \div \frac{17}{28} = -\frac{18}{35} \cdot \frac{28}{17} = -\frac{2 \cdot 3 \cdot 3 \cdot 2 \cdot 2 \cdot \cancel{7}^{1}}{5 \cdot \cancel{7}_{1} \cdot 17} = -\frac{72}{85}$

14. $4.32(-1.07) = -4.6224$

15. $\left(-\frac{2}{3}\right)^4 = \left(-\frac{2}{3}\right)\left(-\frac{2}{3}\right)\left(-\frac{2}{3}\right)\left(-\frac{2}{3}\right) = \frac{16}{81}$

16. $2\sqrt{36} = 2 \cdot 6 = 12$

17. $-3\sqrt{120} = -3\sqrt{4 \cdot 30} = -3 \cdot 2\sqrt{30} = -6\sqrt{30}$

18. $$\begin{aligned} -3^2 + 4[18 + (12-20)] &= -3^2 + 4[18 + (-8)] \\ &= -3^2 + 4[10] \\ &= -9 + 40 = 31 \end{aligned}$$

19. $(b-a)^2 + c$

$$\begin{aligned} [3-(-2)]^2 + 4 = [3+2]^2 + 4 &= [5]^2 + 4 \\ &= 25 + 4 = 29 \end{aligned}$$

20. $$\begin{aligned} 6a - 4b + 2a &= 6a + 2a - 4b \\ &= (6+2)a - 4b \\ &= 8a - 4b \end{aligned}$$

21. $-3(-12y) = -3(-12)y = 36y$

22. $5(2x-7) = 5(2x) + 5(-7) = 10x - 35$

23. $$\begin{aligned} &-4(2x-9) + 5(3x+2) \\ &= -4(2x) - 4(-9) + 5(3x) + 5(2) \\ &= -8x + 36 + 15x + 10 \\ &= -8x + 15x + 36 + 10 \\ &= 7x + 46 \end{aligned}$$

24. $$\begin{aligned} 5[2 - 3(6x-1)] &= 5[2 - 18x + 3] \\ &= 5[5 - 18x] \\ &= 25 - 90x \\ &= -90x + 25 \end{aligned}$$

25. $\{1, 3, 5, 7\}$

26. $A \cap B = \{1, 5, 9\}$

27. $\{x \mid x > 3\}$

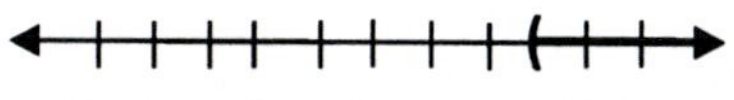

28. $[1, 4]$

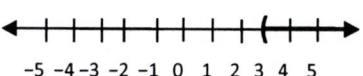

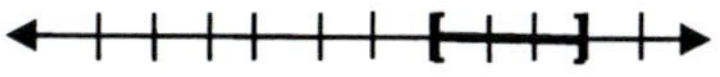

29. $(-4, \infty)$

30. **Strategy** To find the score:

- Multiply the number of correct answers by 6.
- Multiply the number of incorrect answers by −4.
- Multiply the number of blank answers by −2.
- Add the results.

Solution

$21(6) = 126$

$5(-4) = -20$

$4(-2) = -8$

$126 + (-20) + (-8) = 98$

The student's score was 98.

31. **Strategy** To find the percent

- Find the total number by adding the numbers in the three categories together.
- Divide the number opposing (1260) by the total number and multiply by 100%.

Solution $491 + 385 + 1260 = 2136$

$\left(\frac{1260}{2136}\right)100\% = 59.0\%$

59.0% oppose abolishing the penny.

32. the unknown number: x
twice the number: $2x$
one-half the number: $\frac{1}{2}x$

$$2x-\frac{1}{2}x=\left(2-\frac{1}{2}\right)x=\left(\frac{4}{2}-\frac{1}{2}\right)x=\frac{3}{2}x$$

33. number of American League cards: A
number of National League cards: $5A$

Chapter 1Test

1. $-2>-40$

2. 7

3. $-|-4|=-(4)=-4$

4. $16-30=-14$

5. $-22+14+(-8)=-8+(-8)=-16$

6. $16-(-30)-42=16+30-42=46-42=4$

7. $-561\div(-33)=17$

8. $\frac{7}{9}=0.\overline{7}$

9. $45\%=45\left(\frac{1}{100}\right)=\frac{45}{100}=\frac{9}{20}$

$45\%=45(0.01)=0.45$

10. $-\frac{2}{5}+\frac{7}{15}=-\frac{6}{15}+\frac{7}{15}=\frac{-6+7}{15}=\frac{1}{15}$

11. $6.02(-0.89)=-5.3578$

12. $\frac{5}{12}\div\left(-\frac{5}{6}\right)=\frac{5}{12}\cdot\left(-\frac{6}{5}\right)=-\frac{5\cdot 2\cdot 3}{2\cdot 2\cdot 3\cdot 5}=-\frac{1}{2}$

13. $\frac{3}{4}\cdot(4)^2=\frac{3}{4}\cdot 16=\frac{3\cdot 2\cdot 2\cdot 2\cdot 2}{2\cdot 2}=12$

14. $-2\sqrt{45}=-2\sqrt{9\cdot 5}=-2\sqrt{9}\cdot\sqrt{5}=-2\cdot 3\sqrt{5}=-6\sqrt{5}$

15. $16\div 2[8-3(4-2)]+1=16\div 2[8-3(2)]+1$
$=16\div 2[8-6]+1$
$=16\div 2[2]+1$
$=8[2]+1$
$=16+1$
$=17$

16. b^2-3ab

$(-2)^2-3(3)(-2)=4+18=22$

17. $3x-5x+7x=(3-5+7)x=5x$

18. $\frac{1}{5}(10x)=\frac{1}{5}(10)x=2x$

19. $-3(2x^2-7y^2)=-3(2x^2)-3(-7y^2)$
$=-6x^2+21y^2$

20. $2x-3(x-2)=2x-3(x)-3(-2)$
$=2x-3x+6$
$=-x+6$

21. $2x+3[4-(3x-7)]=2x+3[4-3x+7]$
$=2x+3[11-3x]$
$=2x+33-9x$
$=2x-9x+33$
$=-7x+33$

22. $\{-2, -1, 0, 1, 2, 3\}$

23. $\{x \mid x<-3, x\in \text{ real numbers}\}$

24. $A\cup B=\{1, 2, 3, 4, 5, 6, 7, 8\}$

25. $\{x \mid x<1\}$

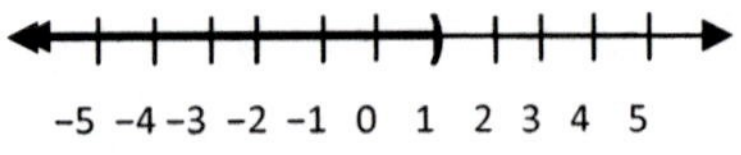

26. (0, 5)

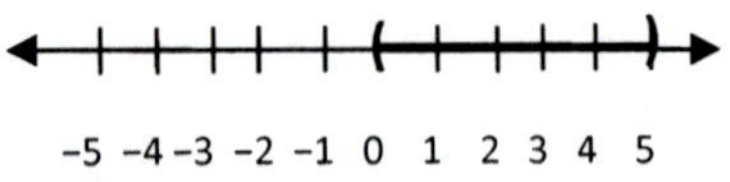

27. the number: x
the difference between a number and 3: $x-3$
$10(x-3)=10x-30$

28. catcher's throw: s
pitcher's fastball: $2s$

29. **a.** 1981, 1988, 1989, 1990, 1991, 1995

b. $-369.7-(-81.1)=-369.7+81.1=-288.6$

The difference between the trade balance in 1990 and 2000 was –\$288.6 billion.

c. The difference in trade was greatest from 1999 to 2000.

d. $\frac{-81.1}{-19.4}=4.18\approx 4$ times greater

e. $\frac{-369.7}{4}=-\$92.425$ billion

30. **Strategy** To find the difference between the highest temperature and the lowest temperature, subtract the lowest temperature (–81.4°) from the highest temperature (134.0°).

Solution $134.0-(-81.4)=134.0+81.4=215.4$

The difference between the highest temperature recorded in North America and the lowest temperature recorded is 215.4° F.

Chapter 2: First-Degree Equations and Inequalities

Prep Test

1. $\frac{9}{100} = 0.09$

2. $\frac{3}{4} = \frac{3}{4}(100\%) = \frac{300}{4}\% = 75\%$

3. $3x^2 - 4x - 1$

$3(-4)^2 - 4(-4) - 1$
$= 3(16) - 4(-4) - 1$
$= 48 + 16 - 1$
$= 63$

4. $R - 0.35R = (1 - 0.35)R = 0.65R$

5. $\frac{1}{2}x + \frac{2}{3}x = \left(\frac{1}{2} + \frac{2}{3}\right)x$

$= \left(\frac{3}{6} + \frac{4}{6}\right)x$

$= \frac{7}{6}x$

6. $6x - 3(6 - x) = 6x - 3(6) - 3(-x)$

$= 6x - 18 + 3x$
$= 9x - 18$

7. $0.22(3x + 6) + x = 0.66x + 1.32 + x = 1.66x + 1.32$

8. the unknown number: n
twice a number: $2n$
$5 - 2n$

9. speed of old card: s
speed of new card: $5s$

10. length of longer piece: x
length of shorter piece: $5 - x$

Section 2.1

Concept Check

1. **a.** equation
b. expression
c. expression
d. equation
e. expression

3. i, ii, and iv are equations in the form $x + a = b$.
You would subtract a from both sides.

5. Amount: 30; base: 40

7. unknown; 30; 24

9. Keith

Objective A Exercises

11.
$$\begin{array}{c|c} 2x & = 8 \\ \hline 2(4) & 8 \end{array}$$
$8 = 8$

Yes, 4 is a solution.

13.
$$\begin{array}{c|c} 2b - 1 & = 3 \\ \hline 2(-1) - 1 & 3 \\ -2 - 1 & 3 \end{array}$$
$-3 \neq 3$

No, -1 is not a solution.

15.
$$\begin{array}{c|c} 4 - 2m & = 3 \\ \hline 4 - 2(1) & 3 \\ 4 - 2 & 3 \end{array}$$
$2 \neq 3$

No, 1 is not a solution.

17.
$$\begin{array}{c|c} 2x + 5 & = 3x \\ \hline 2(5) + 5 & 3(5) \\ 10 + 5 & 15 \end{array}$$
$15 = 15$

Yes, 5 is a solution.

19.
$$\begin{array}{c|c} 3a + 2 & = 2 - a \\ \hline 3(-2) + 2 & 2 - (-2) \\ -6 + 2 & 2 + 2 \end{array}$$
$-4 \neq 4$

No, -2 is not a solution.

21.
$$\begin{array}{c|c} 2x^2 - 1 & = 4x - 1 \\ \hline 2(2)^2 - 1 & 4(2) - 1 \\ 2(4) - 1 & 8 - 1 \\ 8 - 1 & 7 \end{array}$$
$7 = 7$

Yes, 2 is a solution.

23.

$4y + 1$	$=$	3
$4(1/2) + 1$	$\vert$	3
$2 + 1$	$\vert$	3
	$3 = 3$	

Yes, $\dfrac{1}{2}$ is a solution.

25.

$8x - 1$	$=$	$12x + 3$
$8(3/4) - 1$	$\vert$	$12(3/4) + 3$
$6 - 1$	$\vert$	$9 + 3$
	$5 \neq 12$	

No, $\dfrac{3}{4}$ is not a solution.

Objective B Exercises

27. x will be greater than $\dfrac{19}{24}$ because you will

add $\dfrac{11}{16}$ to solve the equation.

29.
$$x + 5 = 7$$
$$x + 5 - 5 = 7 - 5$$
$$x = 2$$
The solution is 2.

31.
$$b - 4 = 11$$
$$b - 4 + 4 = 11 + 4$$
$$b = 15$$
The solution is 15.

33.
$$2 + a = 8$$
$$2 - 2 + a = 8 - 2$$
$$a = 6$$
The solution is 6.

35.
$$n - 5 = -2$$
$$n - 5 + 5 = -2 + 5$$
$$n = 3$$
The solution is 3.

37.
$$b + 7 = 7$$
$$b + 7 - 7 = 7 - 7$$
$$b = 0$$
The solution is 0.

39.
$$z + 9 = 2$$
$$z + 9 - 9 = 2 - 9$$
$$z = -7$$
The solution is -7.

41.
$$10 + m = 3$$
$$10 - 10 + m = 3 - 10$$
$$m = -7$$
The solution is -7.

43.
$$9 + x = -3$$
$$9 - 9 + x = -3 - 9$$
$$x = -12$$
The solution is -12.

45.
$$2 = x + 7$$
$$2 - 7 = x + 7 - 7$$
$$-5 = x$$
The solution is -5.

47.
$$4 = m - 11$$
$$4 + 11 = m - 11 + 11$$
$$15 = m$$
The solution is 15.

49.
$$12 = 3 + w$$
$$12 - 3 = 3 - 3 + w$$
$$9 = w$$
The solution is 9.

51.
$$4 = -10 + b$$
$$4 + 10 = -10 + 10 + b$$
$$14 = b$$
The solution is 14.

53.
$$m + \frac{2}{3} = -\frac{1}{3}$$
$$m + \frac{2}{3} - \frac{2}{3} = -\frac{1}{3} - \frac{2}{3}$$
$$m = -1$$
The solution is -1.

55.
$$x - \frac{1}{2} = \frac{1}{2}$$
$$x - \frac{1}{2} + \frac{1}{2} = \frac{1}{2} + \frac{1}{2}$$
$$x = 1$$
The solution is 1.

57.
$$\frac{5}{8} + y = \frac{1}{8}$$
$$\frac{5}{8} - \frac{5}{8} + y = \frac{1}{8} - \frac{5}{8}$$
$$y = -\frac{4}{8}$$
$$y = -\frac{1}{2}$$

The solution is $-\dfrac{1}{2}$.

59.
$$-\frac{5}{6}=x-\frac{1}{4}$$
$$-\frac{5}{6}+\frac{1}{4}=x-\frac{1}{4}+\frac{1}{4}$$
$$-\frac{10}{12}+\frac{3}{12}=x$$
$$-\frac{7}{12}=x$$
The solution is $-\frac{7}{12}$.

61.
$$d+1.3619=2.0148$$
$$d+1.3619-1.3619=2.0148-1.3619$$
$$d=0.6529$$
The solution is 0.6529.

63.
$$6.149=-3.108+z$$
$$6.149+3.108=-3.108+3.108+z$$
$$9.257=z$$
The solution is 9.257.

Objective C Exercises

65.
$$5x=-15$$
$$\frac{5x}{5}=\frac{-15}{5}$$
$$x=-3$$
The solution is –3.

67.
$$3b=0$$
$$\frac{3b}{3}=\frac{0}{3}$$
$$b=0$$
The solution is 0.

69.
$$-3x=6$$
$$\frac{-3x}{-3}=\frac{6}{-3}$$
$$x=-2$$
The solution is –2.

71.
$$-\frac{1}{6}n=-30$$
$$-6\left(-\frac{1}{6}n\right)=-6(-30)$$
$$n=180$$
The solution is 180.

73.
$$0=-5x$$
$$\frac{0}{-5}=\frac{-5x}{-5}$$
$$0=x$$
The solution is 0.

75.
$$\frac{x}{3}=2$$
$$3\left(\frac{x}{3}\right)=3(2)$$
$$x=6$$
The solution is 6.

77.
$$-\frac{y}{2}=5$$
$$-2\left(-\frac{1}{2}y\right)=-2(5)$$
$$y=-10$$
The solution is –10.

79.
$$\frac{3}{4}y=9$$
$$\frac{4}{3}\left(\frac{3}{4}y\right)=\frac{4}{3}(9)$$
$$y=12$$
The solution is 12.

81.
$$-\frac{2}{3}d=8$$
$$-\frac{3}{2}\left(-\frac{2}{3}d\right)=-\frac{3}{2}(8)$$
$$d=-12$$
The solution is –12.

83.
$$\frac{2n}{3}=0$$
$$\frac{3}{2}\left(\frac{2}{3}n\right)=\frac{3}{2}(0)$$
$$n=0$$
The solution is 0.

85.
$$\frac{-3z}{8}=9$$
$$-\frac{8}{3}\left(-\frac{3}{8}z\right)=-\frac{8}{3}(9)$$
$$z=-24$$
The solution is –24.

87.
$$\frac{2}{9}=\frac{2}{3}y$$
$$\frac{3}{2}\left(\frac{2}{9}\right)=\frac{3}{2}\left(\frac{2}{3}y\right)$$
$$\frac{1}{3}=y$$
The solution is $\frac{1}{3}$.

89.
$$\frac{x}{1.46} = 3.25$$
$$1.46\left(\frac{1}{1.46}x\right) = 1.46(3.25)$$
$$x = 4.745$$
The solution is 4.745.

91.
$$3.47a = 7.1482$$
$$\frac{3.47a}{3.47} = \frac{7.1482}{3.47}$$
$$a = 2.06$$
The solution is 2.06.

93.
$$2m + 5m = 49$$
$$7m = 49$$
$$\frac{7m}{7} = \frac{49}{7}$$
$$m = 7$$
The solution is 7.

95.
$$3n + 2n = 20$$
$$5n = 20$$
$$\frac{5n}{5} = \frac{20}{5}$$
$$n = 4$$
The solution is 4.

97.
$$10y - 3y = 21$$
$$7y = 21$$
$$\frac{7y}{7} = \frac{21}{7}$$
$$y = 3$$
The solution is 3.

99. Positive

101. Negative

Objective D Exercises

103.
$$P \cdot B = A$$
$$0.35(80) = A$$
$$A = 28$$
35% of 80 is 28.

105.
$$P \cdot B = A$$
$$0.012(60) = A$$
$$A = 0.72$$
1.2% of 60 is 0.72.

107.
$$P \cdot B = A$$
$$(1.25)B = 80$$
$$\frac{(1.25)B}{1.25} = \frac{80}{1.25}$$
$$B = 64$$
The number is 64.

109.
$$P \cdot B = A$$
$$P(50) = 12$$
$$\frac{P(50)}{50} = \frac{12}{50}$$
$$P = 0.24$$
$$P = 24\%$$
The percent is 24%.

111.
$$P \cdot B = A$$
$$0.18(40) = A$$
$$A = 7.2$$
18% of 40 is 7.2.

113.
$$P \cdot B = A$$
$$0.12(B) = 48$$
$$\frac{0.12(B)}{0.12} = \frac{48}{0.12}$$
$$B = 400$$
The number is 400.

115.
$$\frac{1}{3}(27) = A \quad \left(33\frac{1}{3}\% = \frac{1}{3}\right)$$
$$9 = A$$
$33\frac{1}{3}\%$ of 27 is 9.

117.
$$P(12) = 3$$
$$\frac{12P}{12} = \frac{3}{12}$$
$$P = 0.25$$
The percent is 25%.

119.
$$P \cdot B = A$$
$$P(6) = 12$$
$$\frac{P(6)}{6} = \frac{12}{6}$$
$$P = 2$$
$$P = 200\%$$
The percent is 200%.

121.
$$P \cdot B = A$$
$$0.0525B = 21$$
$$\frac{0.0525B}{0.0525} = \frac{21}{0.0525}$$
$$B = 400$$
The number is 400.

123.
$$P \cdot B = A$$
$$0.154(50) = A$$
$$A = 7.7$$
15.4% of 50 is 7.7.

125.
$$P \cdot B = A$$
$$0.005B = 1$$
$$\frac{0.005B}{0.005} = \frac{1}{0.005}$$
$$B = 200$$
The number is 200.

127.
$$P \cdot B = A$$
$$0.0075B = 3$$
$$\frac{0.0075B}{0.0075} = \frac{3}{0.0075}$$
$$B = 400$$
The number is 400.

129.
$$P \cdot B = A$$
$$2.5(12) = A$$
$$A = 30$$
250% of 12 is 30.

131. Less than

133. Strategy To find the percent, solve the basic percent equation $P \cdot B = A$ using $B = 26735$ and $A = 23126$.

Solution
$$P \cdot B = A$$
$$P \cdot 26735 = 23126$$
$$P = \frac{23126}{26735}$$
$$P = 0.979$$
97.9% of those that started, finished.

135. Strategy To find the percent:
- Add the deaths to get the total number.
- Add the deaths from a fall (30), fire (47), and drowning (200).
- Solve the basic percent equation $P \cdot B = A$ using B = total deaths and A = total deaths from a fall, fire, and drowning

Solution
Total deaths: $30 + 47 + 200 + 1950 = 2227$
Deaths from a fall, fire, or drowning:
$30 + 47 + 200 = 277$
$$P \cdot B = A$$
$$P2227 = 277$$
$$P \approx .12$$
12% of accidental deaths are not car accidents.

137. Strategy To find the percent, solve the basic percent equation $P \cdot B = A$ using $B = 2252$ and $A = 1850$.

Solution
$$P \cdot B = A$$
$$P2252 = 1850$$
$$P \approx .821$$
The percent of the vacation costs that are charged is 82.1%.

139. Strategy To find the simple interest rate, solve the simple interest equation using $I = \$72$, P = \$1200, and $t = 8$ months $= \frac{8}{12}$ years, for r.

Solution
$$I = Prt$$
$$72 = (1200)r\left(\frac{8}{12}\right)$$
$$72 = 800r$$
$$\frac{72}{800} = \frac{800r}{800}$$
$$0.09 = r$$
The annual simple interest rate is 9%.

141. Strategy To find the interest, solve the simple interest equation for each account:
First, using $P = \$1000$, $r = 7.5\% = 0.075$, and $t = 1$ year, for I.
Second, using $P = 3000 - 1000 = \$2000$, $r = 8.25\% = 0.0825$, and $t = 1$ year, for I.
Finally, find the total interest by adding the interest earned in each account.

Solution

$$I = Prt$$
$$I = (1000)(0.075)(1)$$
$$I = 75$$
$$I = Prt$$
$$I = (2000)(0.0825)(1)$$
$$I = 165$$

$75 + 165 = \$240$

Sal earned \$240 after one year.

143. Strategy To find the amount of interest earned by Makana:
First, find the interest rate of Marlys by solving simple interest equation with $I = 51$, $P = \$850$, and $t = 1$ year, for r.
Second, find Makana's interest rate by increasing Marlys' interest rate by 1%.
Finally, using the rate found in the previous step, $P = \$900$, $t = 1$ year, and solve for I.

Solution

$$I = Prt$$
$$51 = (850)(r)(1)$$
$$\frac{51}{850} = r$$
$$0.06 = r$$

Marlys' rate is 6%: $0.06 + 0.01 = 0.07$

$$I = Prt$$
$$I = (900)(0.07)(1)$$
$$I = 63$$

Makana would earn \$63.

145. Strategy The principal for each investment is the same amount. The time the interest accrued is the same for each account. If one account earns 6% and the other earns 9%, the combined interest earned is between 6% and 9%. To find simple interest rate on the combined accounts, solve the simple interest equation for each account:
First, using $P = \$1000$, $r = 9\% = 0.09$, and $t = 1$ year, for I.
Second, using $P = \$1000$, $r = 6\% = 0.06$, and $t = 1$ year, for I.
Finally, to find the combined interest rate, add the value of $P = 1000 + 1000 = \$2000$, and total the interest earned in both accounts, using the simple interest equation to find r.

Solution

$$I = Prt$$
$$I = (1000)(0.09)(1)$$
$$I = 90$$

$$I = Prt$$
$$I = (1000)(0.06)(1)$$
$$I = 60$$

$$90 + 60 = (2000)r(1)$$
$$\frac{150}{2000} = \frac{2000r}{2000}$$
$$0.075 = r$$

The interest rate earned on the combined accounts is between 6% and 9%.

147. Strategy To find the percent, solve the basic percent equation using $B = 250$ and $A = 5$. The percent is the unknown.

Solution

$$PB = A$$
$$P(250) = 5$$
$$\frac{250P}{250} = \frac{5}{250}$$
$$P = 0.02$$

There is a 2% concentration of hydrogen peroxide.

149. Strategy To find which brand has the greater concentration, solve the basic percent equation for Apple Dan's using B = 32 and A = 8. The percent is the unknown. Then solve the basic percent equation for the generic brand using B = 40 and A = 9. The percent is the unknown. Compare the percent of concentration.

Solution

$$PB = A$$
$$P(32) = 8$$
$$\frac{32P}{32} = \frac{8}{32}$$
$$P = 0.25 \text{ Apple Dan's}$$

$$PB = A$$
$$P(40) = 9$$
$$\frac{40P}{40} = \frac{9}{40}$$
$$P = 0.225 \text{ generic}$$

25% > 22.5%

Apple Dan's concentration is 25%. The generic's concentration is 22.5%.
Apple Dan's has the greater concentration.

151. Strategy To find the amount that is not glycerin, solve the basic percent equation, find the percent that is not glycerin using P = 100 – 75% = 25% = 0.25 and B = 50 g. The amount is unknown.

Solution

$$PB = A$$
$$0.25(50) = A$$
$$12.5 = A$$

There is 12.5 g of cream that is not glycerin.

153. Strategy To find the percent, solve the basic percent equation using B = 500 – 100 = 400 and A = 50. The percent is unknown.

Solution

$$PB = A$$
$$P(400) = 50$$
$$\frac{400P}{400} = \frac{50}{400}$$
$$P = 0.125$$

The percent concentration is 12.5%.

Objective E Exercises

155. (a) equal to
(b) less than

157. Strategy To find the number of miles per hour, solve $d = rt$ for d using d = 20 mi and $t = \frac{40}{60} = \frac{2}{3}$ h.

Solution

$$d = rt$$
$$20 = r\left(\frac{2}{3}\right)$$
$$\frac{3}{2}(20) = r\left(\frac{2}{3}\right)\left(\frac{3}{2}\right)$$
$$30 = r$$

The dietician's average rate of speed is 30 mph.

159. Strategy To find the number of miles traveled, solve $d = rt$ for d using d = 27 mi and $t = \frac{45}{60} = \frac{3}{4}$ h.

Solution

$$d = rt$$
$$27 = r\left(\frac{3}{4}\right)$$
$$36 = r$$

Marcella's average rate of speed is 36 mph.

161. Strategy To find the number of hours to walk the course:
Find the rate to run the course by solving $d = rt$ for r using d = 30 km and $t = 2$ h.

Decrease the rate by 3 km/h to find his walking rate.

Solve for $d = rt$ for t using d = 30 km and r equal to his walking rate.

Solution

$$d = rt$$
$$30 = r(2)$$
$$\frac{30}{2} = r$$
$$15 = r \quad \text{His running rate}$$
$$15 - 3 = 12 \quad \text{His walking rate}$$
$$d = rt$$
$$30 = 12t$$
$$\frac{30}{12} = t$$
$$2.5 = t$$

It would take Palmer 2.5 h to walk the course.

163. Strategy The distance is 8 mi. Therefore d = 8. The joggers are running toward each other, one at 5 mph and one at 7 mph. The rate is the sum of the two rates, or 12 mph. So, r = 12. To find the time solve $d = rt$ for t. Convert the answer to minutes.

Solution

$$d = rt$$
$$8 = 12t$$
$$\frac{8}{12} = t$$
$$\frac{2}{3} = t$$
$$\frac{2}{3}\text{ h} = \frac{2}{3}\cdot 60\text{ min} = 40\text{ min}$$

The two joggers will meet 40 min after they start.

165. Strategy The two cyclists are traveling in opposite directions, one at 8 mph and one at 9 mph. The rate is the sum of the two rates, or 17 mph. So, $r = 17$. The time traveled is $30\text{ min} = \frac{1}{2}\text{ h}$. So, $t = \frac{1}{2}$. To find the distance, solve $d = rt$ for d.

Solution

$$d = rt$$
$$d = 17\cdot\frac{1}{2}$$
$$d = 8.5$$

The two cyclists are 8.5 mi apart.

167. Strategy To find the number of miles apart:
Find the distance the first train travels by solving $d = rt$ for d using r = 45 and t = 2.
Find the distance the second train travels by solving $d = rt$ for d using r = 60 and t = 1.
Find the difference between these distances.

Solution

First train:	Second train:
$d = rt$	$d = rt$
$d = 45(2)$	$d = 60(1)$
$d = 90$	$d = 60$

$90 - 60 = 30$

The trains are 30 mi apart.

Critical Thinking

169.
$$\frac{2m + m}{5} = -9$$
$$\frac{3m}{5} = -9$$
$$\frac{5}{3}\cdot\frac{3m}{5} = -9\cdot\frac{5}{3}$$
$$m = -15$$

The solution is –15.

171. $\dfrac{1}{\frac{1}{x}}=5$

$\dfrac{1}{x}\cdot\dfrac{1}{\frac{1}{x}}=\dfrac{1}{x}\cdot 5$

$1=\dfrac{5}{x}$

$x\cdot 1=x\cdot\dfrac{5}{x}$

$x=5$

The solution is 5.

173. $\dfrac{4}{\frac{3}{b}}=8$

$\dfrac{3}{b}\cdot\dfrac{4}{\frac{3}{b}}=\dfrac{3}{b}\cdot 8$

$4=\dfrac{24}{b}$

$b\cdot 4=b\cdot\dfrac{24}{b}$

$4b=24$

$\dfrac{4b}{4}=\dfrac{24}{4}$

$b=6$

The solution is 6.

175. Lower
After the increase, the cost is now $1.1C$ $(C+0.1C=1.1C)$. After the decrease, the new price is $0.99C$ $[1.1C-0.1(1.1C)=1.1C-0.11C$ $=0.99C]$.

177. Employee B. If Employee B earned more before the raise and they got the same percent raise, then Employee B will get more after the raise.

Projects and Group Activities

179. Answers will vary. One example is $x+7=9$.

181. $\dfrac{3}{7}+\dfrac{1}{b}=2$

$7b\cdot\dfrac{3}{7}+7b\cdot\dfrac{1}{b}=7b\cdot 2$

$3b+7=14b$

$3b-3b+7=14b-3b$

$7=11b$

$\dfrac{7}{11}=\dfrac{11b}{11}$

$\dfrac{7}{11}=b$

The solution is $\dfrac{7}{11}$.

183. **a. Strategy** To find the percent for each region:
- find the total population by adding the number in each region (67.4, 113.6, 72.2, and 55.8)
- find the percent by solving $BP = A$ for P using A = the total population and B as the population for each region.

Solution

Total: 67.4 + 113.6 + 72.2 + 55.8 = 309

Midwest: 21.8%

$BP=A$

$309P=67.4$

$P=\dfrac{67.4}{309}$

$P=0.218$

South: 36.8%

$BP=A$

$309P=113.6$

$P=\dfrac{113.6}{309}$

$P=0.368$

West: 23.4%

$BP=A$

$309P=72.2$

$P=\dfrac{72.2}{309}$

$P=0.234$

Northwest: 18.1%

$BP=A$

$309P=55.8$

$P=\dfrac{55.8}{309}$

$P=0.181$

b. South, South

c. Strategy To find the percent California, solve the formula $BP = A$ for P using P = the total population and $A = 38$.

Solution

$$BP = A$$
$$309P = 38$$
$$P = \frac{38}{309}$$
$$P = 0.123$$

12.3% of the population lives in California.

d. Strategy To find the population for Wyoming, solve the formula $BP = A$ for A using B = the total population and $P = 0.00168$.

Solution

$$BP = A$$
$$309(0.00168) = A$$
$$0.52 = P$$

0.52 million = 520,000

The population of Wyoming is 520,000.

e. Answers will vary.

Section 2.2

Concept Check

1. a and i, b and iii, c and ii, d and iv

3. 5; 8

5. True

7. Subtract $2x$ from each side.

Objective A Exercises

9.
$$3x + 1 = 10$$
$$3x + 1 - 1 = 10 - 1$$
$$3x = 9$$
$$\frac{3x}{3} = \frac{9}{3}$$
$$x = 3$$
The solution is 3.

11.
$$2a - 5 = 7$$
$$2a - 5 + 5 = 7 + 5$$
$$2a = 12$$
$$\frac{2a}{2} = \frac{12}{2}$$
$$a = 6$$
The solution is 6.

13.
$$5 = 4x + 9$$
$$5 - 9 = 4x + 9 - 9$$
$$-4 = 4x$$
$$\frac{-4}{4} = \frac{4x}{4}$$
$$-1 = x$$
The solution is −1.

15.
$$2x - 5 = -11$$
$$2x - 5 + 5 = -11 + 5$$
$$2x = -6$$
$$\frac{2x}{2} = \frac{-6}{2}$$
$$x = -3$$
The solution is −3.

17.
$$4 - 3w = -2$$
$$4 - 4 - 3w = -2 - 4$$
$$-3w = -6$$
$$\frac{-3w}{-3} = \frac{-6}{-3}$$
$$w = 2$$
The solution is 2.

19.
$$8 - 3t = 2$$
$$8 - 8 - 3t = 2 - 8$$
$$-3t = -6$$
$$\frac{-3t}{-3} = \frac{-6}{-3}$$
$$t = 2$$
The solution is 2.

21.
$$4a - 20 = 0$$
$$4a - 20 + 20 = 0 + 20$$
$$4a = 20$$
$$\frac{4a}{4} = \frac{20}{4}$$
$$a = 5$$
The solution is 5.

23.
$$6 + 2b = 0$$
$$6 - 6 + 2b = 0 - 6$$
$$2b = -6$$
$$\frac{2b}{2} = \frac{-6}{2}$$
$$b = -3$$
The solution is −3.

25.
$$\begin{aligned} -2x+5&=-7\\ -2x+5-5&=-7-5\\ -2x&=-12\\ \frac{-2x}{-2}&=\frac{-12}{-2}\\ x&=6 \end{aligned}$$
The solution is 6.

27.
$$\begin{aligned} -1.2x+3&=-0.6\\ -1.2x+3-3&=-0.6-3\\ -1.2x&=-3.6\\ \frac{-1.2x}{-1.2}&=\frac{-3.6}{-1.2}\\ x&=3 \end{aligned}$$
The solution is 3.

29.
$$\begin{aligned} 2&=7-5a\\ 2-7&=7-7-5a\\ -5&=-5a\\ \frac{-5}{-5}&=\frac{-5a}{-5}\\ 1&=a \end{aligned}$$
The solution is 1.

31.
$$\begin{aligned} -35&=-6b+1\\ -35-1&=-6b+1-1\\ -36&=-6b\\ \frac{-36}{-6}&=\frac{-6b}{-6}\\ 6&=b \end{aligned}$$
The solution is 6.

33.
$$\begin{aligned} -3m-21&=0\\ -3m-21+21&=0+21\\ -3m&=21\\ \frac{-3m}{-3}&=\frac{21}{-3}\\ m&=-7 \end{aligned}$$
The solution is −7.

35.
$$\begin{aligned} -4y+15&=15\\ -4y+15-15&=15-15\\ -4y&=0\\ \frac{-4y}{-4}&=\frac{0}{-4}\\ y&=0 \end{aligned}$$
The solution is 0.

37.
$$\begin{aligned} 9-4x&=6\\ 9-9-4x&=6-9\\ -4x&=-3\\ \frac{-4x}{-4}&=\frac{-3}{-4}\\ x&=\frac{3}{4} \end{aligned}$$
The solution is $\frac{3}{4}$.

39.
$$\begin{aligned} 9x-4&=0\\ 9x-4+4&=0+4\\ 9x&=4\\ \frac{9x}{9}&=\frac{4}{9}\\ x&=\frac{4}{9} \end{aligned}$$
The solution is $\frac{4}{9}$.

41.
$$\begin{aligned} 1-3x&=0\\ 1-1-3x&=0-1\\ -3x&=-1\\ \frac{-3x}{-3}&=\frac{-1}{-3}\\ x&=\frac{1}{3} \end{aligned}$$
The solution is $\frac{1}{3}$.

43.
$$\begin{aligned} 12w+11&=5\\ 12w+11-11&=5-11\\ 12w&=-6\\ \frac{12w}{12}&=\frac{-6}{12}\\ w&=-\frac{6}{12}\\ w&=-\frac{1}{2} \end{aligned}$$
The solution is $-\frac{1}{2}$.

45.
$$\begin{aligned} 8b-3&=-9\\ 8b-3+3&=-9+3\\ 8b&=-6\\ \frac{8b}{8}&=\frac{-6}{8}\\ b&=-\frac{6}{8}\\ b&=-\frac{3}{4} \end{aligned}$$
The solution is $-\frac{3}{4}$.

47.
$$\begin{aligned} 7-9a&=4\\ 7-7-9a&=4-7\\ -9a&=-3\\ \frac{-9a}{-9}&=\frac{-3}{-9}\\ a&=\frac{3}{9}\\ a&=\frac{1}{3} \end{aligned}$$
The solution is $\frac{1}{3}$.

49.
$$\begin{aligned} 10&=-18x+7\\ 10-7&=-18x+7-7\\ 3&=-18x\\ \frac{3}{-18}&=\frac{-18x}{-18}\\ -\frac{3}{18}&=x\\ -\frac{1}{6}&=x \end{aligned}$$
The solution is $-\frac{1}{6}$.

51.
$$\begin{aligned} 9x+\frac{4}{5}&=\frac{4}{5}\\ 9x+\frac{4}{5}-\frac{4}{5}&=\frac{4}{5}-\frac{4}{5}\\ 9x&=0\\ \frac{9x}{9}&=\frac{0}{9}\\ x&=0 \end{aligned}$$
The solution is 0.

53.
$$\begin{aligned} 0.9&=10x-0.6\\ 0.9+0.6&=10x-0.6+0.6\\ 1.5&=10x\\ \frac{1.5}{10}&=\frac{10x}{10}\\ 0.15&=x \end{aligned}$$
The solution is 0.15.

55.
$$\begin{aligned} -4x+3&=9\\ -4x+3-3&=9-3\\ -4x&=6\\ \frac{-4x}{-4}&=\frac{6}{-4}\\ x&=-\frac{6}{4}\\ x&=-\frac{3}{2} \end{aligned}$$
The solution is $-\frac{3}{2}$.

57.
$$\begin{aligned} \frac{1}{3}m-1&=5\\ \frac{1}{3}m-1+1&=5+1\\ \frac{1}{3}m&=6\\ 3\left(\frac{1}{3}m\right)&=3\cdot 6\\ m&=18 \end{aligned}$$
The solution is 18.

59.
$$\begin{aligned} \frac{3}{4}n+7&=13\\ \frac{3}{4}n+7-7&=13-7\\ \frac{3}{4}n&=6\\ \frac{4}{3}\left(\frac{3}{4}n\right)&=\frac{4}{3}(6)\\ n&=8 \end{aligned}$$
The solution is 8.

61.
$$\begin{aligned} -\frac{3}{8}b+4&=10\\ -\frac{3}{8}b+4-4&=10-4\\ -\frac{3}{8}b&=6\\ -\frac{8}{3}\left(-\frac{3}{8}b\right)&=-\frac{8}{3}(6)\\ b&=-16 \end{aligned}$$
The solution is −16.

63.
$$\frac{y}{5}-2=3$$
$$\frac{y}{5}-2+2=3+2$$
$$\frac{y}{5}=5$$
$$5\left(\frac{1}{5}y\right)=5\cdot 5$$
$$y=25$$
The solution is 25.

65.
$$\frac{2}{3}x-\frac{5}{6}=-\frac{1}{3}$$
$$6\left(\frac{2}{3}x-\frac{5}{6}\right)=6\left(-\frac{1}{3}\right)$$
$$4x-5=-2$$
$$4x=3$$
$$x=\frac{3}{4}$$
The solution is $\frac{3}{4}$.

67.
$$\frac{1}{2}-\frac{2}{3}x=\frac{1}{4}$$
$$12\left(\frac{1}{2}-\frac{2}{3}x\right)=12\left(\frac{1}{4}\right)$$
$$6-8x=3$$
$$-8x=-3$$
$$x=\frac{3}{8}$$
The solution is $\frac{3}{8}$.

69.
$$\frac{3}{2}=\frac{5}{6}+\frac{3x}{8}$$
$$\frac{3}{2}-\frac{5}{6}=\frac{5}{6}-\frac{5}{6}+\frac{3x}{8}$$
$$\frac{2}{3}=\frac{3x}{8}$$
$$\frac{8}{3}\left(\frac{2}{3}\right)=\frac{8}{3}\left(\frac{3x}{8}\right)$$
$$\frac{16}{9}=x$$
The solution is $\frac{16}{9}$.

71.
$$\frac{11}{27}=\frac{4}{9}-\frac{2x}{3}$$
$$\frac{11}{27}-\frac{4}{9}=\frac{4}{9}-\frac{4}{9}-\frac{2x}{3}$$
$$-\frac{1}{27}=-\frac{2x}{3}$$
$$-\frac{3}{2}\left(-\frac{1}{27}\right)=-\frac{3}{2}\left(-\frac{2x}{3}\right)$$
$$\frac{1}{18}=x$$
The solution is $\frac{1}{18}$.

73.
$$7=\frac{2x}{5}+4$$
$$7-4=\frac{2x}{5}+4-4$$
$$3=\frac{2x}{5}$$
$$\frac{5}{2}(3)=\frac{5}{2}\left(\frac{2}{5}x\right)$$
$$\frac{15}{2}=x$$
The solution is $\frac{15}{2}$.

75.
$$7-\frac{5}{9}y=9$$
$$7-7-\frac{5}{9}y=9-7$$
$$-\frac{5}{9}y=2$$
$$-\frac{9}{5}\left(-\frac{5}{9}y\right)=-\frac{9}{5}(2)$$
$$y=-\frac{18}{5}$$
The solution is $-\frac{18}{5}$.

77.
$$5y+9+2y=23$$
$$7y+9=23$$
$$7y+9-9=23-9$$
$$7y=14$$
$$\frac{7y}{7}=\frac{14}{7}$$
$$y=2$$
The solution is 2.

79. $$\begin{aligned} 11z-3-7z&=9\\ 4z-3&=9\\ 4z-3+3&=9+3\\ 4z&=12\\ \frac{4z}{4}&=\frac{12}{4}\\ z&=3 \end{aligned}$$
The solution is 3.

81. Negative

83. Negative

85. $$\begin{aligned} 3x+4y&=13 \text{ when } y=-2\\ 3x+4(-2)&=13\\ 3x-8&=13\\ 3x-8+8&=13+8\\ 3x&=21\\ \frac{3x}{3}&=\frac{21}{3}\\ x&=7 \end{aligned}$$
The solution is 7.

87. $$\begin{aligned} 4-5x&=-1\\ 4-4-5x&=-1-4\\ -5x&=-5\\ \frac{-5x}{-5}&=\frac{-5}{-5}\\ x&=1 \end{aligned}$$

$x^2-3x+1;\ x=1$

$(1)^2-3(1)+1$

$1-3+1$

-1

Objective B Exercises

89. $$\begin{aligned} 6y+2&=y+17\\ 6y-y+2&=y-y+17\\ 5y+2&=17\\ 5y+2-2&=17-2\\ 5y&=15\\ \frac{5y}{5}&=\frac{15}{5}\\ y&=3 \end{aligned}$$
The solution is 3.

91. $$\begin{aligned} 13b-1&=4b-19\\ 13b-4b-1&=4b-4b-19\\ 9b-1&=-19\\ 9b-1+1&=-19+1\\ 9b&=-18\\ \frac{9b}{9}&=\frac{-18}{9}\\ b&=-2 \end{aligned}$$
The solution is –2.

93. $$\begin{aligned} 7a-5&=2a-20\\ 7a-2a-5&=2a-2a-20\\ 5a-5&=-20\\ 5a-5+5&=-20+5\\ 5a&=-15\\ \frac{5a}{5}&=\frac{-15}{5}\\ a&=-3 \end{aligned}$$
The solution is –3.

95. $$\begin{aligned} n-2&=6-3n\\ n+3n-2&=6+3n+3n\\ 4n-2&=6\\ 4n-2+2&=6+2\\ 4n&=8\\ \frac{4n}{4}&=\frac{8}{4}\\ n&=2 \end{aligned}$$
The solution is 2.

97. $$\begin{aligned} 4y-2&=-16-3y\\ 4y+3y-2&=-16-3y+3y\\ 7y-2&=-16\\ 7y-2+2&=-16+2\\ 7y&=-14\\ \frac{7y}{7}&=\frac{-14}{7}\\ y&=-2 \end{aligned}$$
The solution is –2.

99. $$\begin{aligned} m+0.4&=3m+0.8\\ m-3m+0.4&=3m-3m+0.8\\ -2m+0.4&=0.8\\ -2m+0.4-0.4&=0.8-0.4\\ -2m&=0.4\\ \frac{-2m}{-2}&=\frac{0.4}{-2}\\ m&=-0.2 \end{aligned}$$
The solution is –0.2.

101.
$$\begin{aligned} 5a+7&=2a+7\\ 5a-2a+7&=2a-2a+7\\ 3a+7&=7\\ 3a+7-7&=7-7\\ 3a&=0\\ \frac{3a}{3}&=\frac{0}{3}\\ a&=0 \end{aligned}$$
The solution is 0.

103.
$$\begin{aligned} 10-4n&=16-n\\ 10-4n+n&=16-n+n\\ 10-3n&=16\\ 10-10-3n&=16-10\\ -3n&=6\\ \frac{-3n}{-3}&=\frac{6}{-3}\\ n&=-2 \end{aligned}$$
The solution is –2.

105.
$$\begin{aligned} 3-2y&=15+4y\\ 3-2y-4y&=15+4y-4y\\ 3-6y&=15\\ 3-3-6y&=15-3\\ -6y&=12\\ \frac{-6y}{-6}&=\frac{12}{-6}\\ y&=-2 \end{aligned}$$
The solution is –2.

107.
$$\begin{aligned} 2b-10&=7b\\ 2b-2b-10&=7b-2b\\ -10&=5b\\ \frac{-10}{5}&=\frac{5b}{5}\\ -2&=b \end{aligned}$$
The solution is –2.

109.
$$\begin{aligned} 9y&=5y+16\\ 9y-5y&=5y-5y+16\\ 4y&=16\\ \frac{4y}{4}&=\frac{16}{4}\\ y&=4 \end{aligned}$$
The solution is 4.

111.
$$\begin{aligned} 6y-1&=2y+2\\ 6y-2y-1&=2y-2y+2\\ 4y-1&=2\\ 4y-1+1&=2+1\\ 4y&=3\\ \frac{4y}{4}&=\frac{3}{4}\\ y&=\frac{3}{4} \end{aligned}$$
The solution is $\frac{3}{4}$.

113.
$$\begin{aligned} 2y-7&=-1-2y\\ 2y+2y-7&=-1-2y+2y\\ 4y-7&=-1\\ 4y-7+7&=-1+7\\ 4y&=6\\ \frac{4y}{4}&=\frac{6}{4}\\ y&=\frac{3}{2} \end{aligned}$$
The solution is $\frac{3}{2}$.

115.
$$\begin{aligned} 5x&=3x-8\\ 5x-3x&=3x-3x-8\\ 2x&=-8\\ \frac{2x}{2}&=\frac{-8}{2}\\ x&=-4 \end{aligned}$$

$$\begin{aligned} &4x+2\\ &=4(-4)+2\\ &=-16+2\\ &=-14 \end{aligned}$$
The answer is –14.

117.
$$\begin{aligned} 2-6a&=5-3a\\ 2-6a+3a&=5-3a+3a\\ 2-3a&=5\\ 2-2-3a&=5-2\\ -3a&=3\\ \frac{-3a}{-3}&=\frac{3}{-3}\\ a&=-1 \end{aligned}$$

$$\begin{aligned} &4a^2-2a+1\\ &=4(-1)^2-2(-1)+1\\ &=4(1)-2(-1)+1\\ &=4+2+1\\ &=6+1\\ &=7 \end{aligned}$$
The answer is 7.

Objective C Exercises

119. (ii)

121. $$\begin{aligned} 6y+2(2y+3)&=16\\ 6y+4y+6&=16\\ 10y+6&=16\\ 10y+6-6&=16-6\\ 10y&=10\\ \frac{10y}{10}&=\frac{10}{10}\\ y&=1 \end{aligned}$$

The solution is 1.

123. $$\begin{aligned} 12x-2(4x-6)&=28\\ 12x-8x+12&=28\\ 4x+12&=28\\ 4x+12-12&=28-12\\ 4x&=16\\ \frac{4x}{4}&=\frac{16}{4}\\ x&=4 \end{aligned}$$

The solution is 4.

125. $$\begin{aligned} 9m-4(2m-3)&=11\\ 9m-8m+12&=11\\ m+12&=11\\ m+12-12&=11-12\\ m&=-1 \end{aligned}$$

The solution is –1.

127. $$\begin{aligned} 4(1-3x)+7x&=9\\ 4-12x+7x&=9\\ 4-5x&=9\\ 4-4-5x&=9-4\\ -5x&=5\\ \frac{-5x}{-5}&=\frac{5}{-5}\\ x&=-1 \end{aligned}$$

The solution is –1.

129. $$\begin{aligned} 0.22(x+6)&=0.2x+1.8\\ 0.22x+1.32&=0.2x+1.8\\ 0.22x-0.2x+1.32&=0.2x-0.2x+1.8\\ 0.02x+1.32&=1.8\\ 0.02x+1.32-1.32&=1.8-1.32\\ 0.02x&=0.48\\ \frac{0.02x}{0.02}&=\frac{0.48}{0.02}\\ x&=24 \end{aligned}$$

The solution is 24.

131. $$\begin{aligned} 0.3x+0.3(x+10)&=300\\ 0.3x+0.3x+3&=300\\ 0.6x+3&=300\\ 0.6x+3-3&=300-3\\ 0.6x&=297\\ \frac{0.6x}{0.6}&=\frac{297}{0.6}\\ x&=495 \end{aligned}$$

The solution is 495.

133. $$\begin{aligned} 5-(9-6x)&=2x-2\\ 5-9+6x&=2x-2\\ -4+6x&=2x-2\\ -4+6x-2x&=2x-2x-2\\ -4+4x&=-2\\ -4+4+4x&=-2+4\\ 4x&=2\\ \frac{4x}{4}&=\frac{2}{4}\\ x&=\frac{1}{2} \end{aligned}$$

The solution is $\frac{1}{2}$.

135. $$\begin{aligned} 3[2-4(y-1)]&=3(2y+8)\\ 3[2-4y+4]&=6y+24\\ 3[6-4y]&=6y+24\\ 18-12y&=6y+24\\ 18-12y-6y&=6y-6y+24\\ 18-18y&=24\\ 18-18-18y&=24-18\\ -18y&=6\\ \frac{-18y}{-18}&=\frac{6}{-18}\\ y&=-\frac{1}{3} \end{aligned}$$

The solution is $-\frac{1}{3}$.

137.
$$3a+2[2+3(a-1)]=2(3a+4)$$
$$3a+2[2+3a-3]=6a+8$$
$$3a+2[-1+3a]=6a+8$$
$$3a-2+6a=6a+8$$
$$9a-2=6a+8$$
$$9a-6a-2=6a-6a+8$$
$$3a-2=8$$
$$3a-2+2=8+2$$
$$3a=10$$
$$\frac{3a}{3}=\frac{10}{3}$$
$$a=\frac{10}{3}$$
The solution is $\frac{10}{3}$.

139.
$$-2[4-(3b+2)]=5-2(3b+6)$$
$$-2[4-3b-2]=5-6b-12$$
$$-2[2-3b]=-7-6b$$
$$-4+6b=-7-6b$$
$$-4+6b+6b=-7-6b+6b$$
$$-4+12b=-7$$
$$-4+4+12b=-7+4$$
$$12b=-3$$
$$\frac{12b}{12}=\frac{-3}{12}$$
$$b=-\frac{1}{4}$$
The solution is $-\frac{1}{4}$.

141.
$$4-3a=7-2(2a+5)$$
$$4-3a=7-4a-10$$
$$4-3a=-3-4a$$
$$4-3a+4a=-3-4a+4a$$
$$4+a=-3$$
$$4-4+a=-3-4$$
$$a=-7$$

$$a^2+7a$$
$$=(-7)^2+(7)(-7)$$
$$=49+(7)(-7)$$
$$=49-49$$
$$=0$$
The answer is 0.

Objective D Exercises

143. Strategy $F = 14$
Unknown: m

Solution
$$F=2.5+2.3(m-1)$$
$$14=2.5+2.3m-2.3$$
$$14=0.2+2.3m$$
$$14-0.2=0.2-0.2+2.3m$$
$$13.98=2.3m$$
$$\frac{13.98}{2.3}=\frac{2.3m}{2.3}$$
$$6.08\approx m$$
The customer drove 6 mi.

145. (a) $8-3=5$ ft

(b) The person who is 3 ft away.

(c) No

147. Solution To find the location of the fulcrum when the system balances, replace the variables F_1, F_2, and d in the lever system equation by the given values and solve for x.

Solution
$$F_1x=F_2(d-x)$$
$$70x=175(14-x)$$
$$70x=2450-175x$$
$$70x+175x=2450-175x+175x$$
$$245x=2450$$
$$\frac{245x}{245}=\frac{2450}{245}$$
$$x=10$$
The fulcrum is 10 ft from the child.

149. Strategy To find the location of the fulcrum when the system balances, replace the variables F_1, F_2, and d in the lever system equation by the given values and solve for x.

Solution
$$F_1x=F_2(d-x)$$
$$90x=60(12-x)$$
$$90x=720-60x$$
$$90x+60x=720$$
$$150x=720$$
$$\frac{150x}{150}=\frac{720}{150}$$
$$x=4.8$$
The fulcrum is 4.8 ft from the 90-lb child.

151. Strategy To find the force when the system balances, replace the variables F_2, x, and d in the lever system equation by the given values and solve for F_1.

Solution

$$\begin{aligned} F_1 x &= F_2(d-x) \\ F_1 \cdot 0.15 &= 30(9-0.15) \\ F_1 \cdot 0.15 &= 30(8.85) \\ 0.15F_1 &= 265.5 \\ \frac{0.15F_1}{0.15} &= \frac{265.5}{0.15} \\ F_1 &= 1770 \end{aligned}$$

A 1770-lb force is applied to the other end.

153. Strategy To find the break-even point, replace the variables P, C, and F in the cost equation by the given values and solve for x.

Solution

$$\begin{aligned} Px &= Cx + F \\ 325x &= 175x + 39{,}000 \\ 325x - 175x &= 39{,}000 \\ 150x &= 39{,}000 \\ \frac{150x}{150} &= \frac{39{,}000}{150} \\ x &= 260 \end{aligned}$$

The break-even point is 260 barbecues.

155. Strategy To find the break-even point, replace the variables P, C, and F in the cost equation by the given values and solve for x.

Solution

$$\begin{aligned} Px &= Cx + F \\ 49x &= 12x + 19{,}240 \\ 49x - 12x &= 19{,}240 \\ 37x &= 19{,}240 \\ \frac{37x}{37} &= \frac{19{,}240}{37} \\ x &= 520 \end{aligned}$$

The break-even point is 520 recorders.

157. Strategy $m = 8.3$ Unknown: C

Solution

$$\begin{aligned} m &= \frac{1}{6}(C-5) \\ 8.3 &= \frac{1}{6}(C-5) \\ 8.3 &= \frac{1}{6}C - \frac{5}{6} \\ 6 \cdot 8.3 &= 6 \cdot \frac{1}{6}C - 6 \cdot \frac{5}{6} \\ 49.8 &= C - 5 \\ 49.8 + 5 &= C - 5 + 5 \\ 54.8 &= C \end{aligned}$$

The mammal consumes 54.8 ml/min.

Critical Thinking

159. $\frac{1}{5}(25-10b)+4=\frac{1}{3}(9b-15)-6$

$$\begin{aligned} 5-2b+4 &= 3b-5-6 \\ 9-2b &= 3b-11 \\ 9-2b-3b &= 3b-3b-11 \\ 9-5b &= -11 \\ 9-9-5b &= -11-9 \\ -5b &= -20 \\ \frac{-5b}{-5} &= \frac{-20}{-5} \\ b &= 4 \end{aligned}$$

The solution is 4.

161. $\frac{2(5x-6)-3(x-4)}{7}=x+2$

$$\begin{aligned} \frac{10x-12-3x+12}{7} &= x+2 \\ \frac{7x}{7} &= x+2 \\ x &= x+2 \\ x-x &= x-x+2 \\ 0 &= 2 \end{aligned}$$

No solution

163. $3x-4(x-1)$ is an expression, not an equation. There must be a equals sign to have an equation. You cannot solve an expression.

Projects and Group Activities

165. **Strategy** Let x be the number.
Subtract 4 from the number: $x - 4$
300% of the result: $3(x - 4)$

Solution

$$3(x-4)=x$$
$$3x-12=x$$
$$3x-3x-12=x-3x$$
$$-12=-2x$$
$$\frac{-12}{-2}=\frac{-2x}{-2}$$
$$6=x$$

The number is 6.

167. **Strategy** Let x be the population in 1990.
Population after 10,000 increase: $x + 10,000$
Population after 10% decrease:
$x+10,000-0.1(x+10,000)$
$x+10,000-0.1x-1000$
$0.9x+9000$
6000 more than the beginning:
$0.9x+9000=x+6000$

Solution

$$0.9x+9000=x+6000$$
$$0.9x-0.9x+9000=x-0.9x+6000$$
$$9000=0.1x+6000$$
$$9000-6000=0.1x+6000-6000$$
$$3000=0.1x$$
$$\frac{3000}{0.1}=\frac{0.1x}{0.1}$$
$$30,000=x$$

The population in 1990 was 30,000.

Section 2.3

Concept Check

1. True

3. Ttue

5. equals

7. 1; 2; 2

Objective A Exercises

9. the unknown number: x

The difference between a number and 15	is	seven

$$x-15=7$$
$$x-15+15=7+15$$
$$x=22$$

The number is 22.

11. the unknown number: x

The difference between nine and a number	is	seven

$$9-x=7$$
$$9-9-x=7-9$$
$$-x=-2$$
$$\frac{-1x}{-1}=\frac{-2}{-1}$$
$$x=2$$

The number is 2.

13. the unknown number: x

The difference between five and twice a number	is	one

$$5-2x=1$$
$$5-5-2x=1-5$$
$$-2x=-4$$
$$\frac{-2x}{-2}=\frac{-4}{-2}$$
$$x=2$$

The number is 2.

15. the unknown number: x

The sum of twice a number and five	is	fifteen

$$2x+5=15$$
$$2x+5-5=15-5$$
$$2x=10$$
$$\frac{2x}{2}=\frac{10}{2}$$
$$x=5$$

The number is 5.

17. the unknown number: x

Six less than four times a number	is	twenty-two

$$
\begin{aligned}
4x - 6 &= 22 \\
4x - 6 + 6 &= 22 + 6 \\
4x &= 28 \\
\frac{4x}{4} &= \frac{28}{4} \\
x &= 7
\end{aligned}
$$

The number is 7.

19. the unknown number: x

Three times the difference between four times a number and seven	is	fifteen

$$
\begin{aligned}
3(4x - 7) &= 15 \\
12x - 21 &= 15 \\
12x - 21 + 21 &= 15 + 21 \\
12x &= 36 \\
\frac{12x}{12} &= \frac{36}{12} \\
x &= 3
\end{aligned}
$$

The number is 3.

21. the smaller number: x
the larger number: $20 - x$

Three times the smaller	is equal to	two times the larger

$$
\begin{aligned}
3x &= 2(20 - x) \\
3x &= 40 - 2x \\
3x + 2x &= 40 - 2x + 2x \\
5x &= 40 \\
\frac{5x}{5} &= \frac{40}{5} \\
x &= 8
\end{aligned}
$$

$20 - x = 20 - 8 = 12$

The smaller number is 8.

The larger number is 12.

23. the smaller number: x

the larger number: $14 - x$

The difference between two times the smaller and the larger	is	one

$$
\begin{aligned}
2x - (14 - x) &= 1 \\
2x - 14 + x &= 1 \\
3x - 14 &= 1 \\
3x &= 15 \\
\frac{3x}{3} &= \frac{15}{3} \\
x &= 5
\end{aligned}
$$

$14 - x = 14 - 5 = 9$

The smaller number is 5.

The larger number is 9.

25. First odd integer: n

Second odd integer: $n + 2$

Third odd integer: $n + 4$

The sum of the three integers is 51.

$$
\begin{aligned}
n + (n + 2) + (n + 4) &= 51 \\
3n + 6 &= 51 \\
3n &= 45 \\
n &= 15 \\
n + 2 &= 15 + 2 = 17 \\
n + 4 &= 15 + 4 = 19
\end{aligned}
$$

The three integers are 15, 17, and 19.

27. First odd integer: n

Second odd integer: $n + 2$

Third odd integer: $n + 4$

Three times the second number is one more than the sum of the first and third numbers.

$$
\begin{aligned}
3(n + 2) &= 1 + n + (n + 4) \\
3n + 6 &= 5 + 2n \\
n + 6 &= 5 \\
n &= -1 \\
n + 2 &= -1 + 2 = 1 \\
n + 4 &= -1 + 4 = 3
\end{aligned}
$$

The three integers are −1, 1, and 3.

29. First even integer: n

Second even integer: $n + 2$

Three times the first integer equals twice the second integer.

$$
\begin{aligned}
3n &= 2(n + 2) \\
3n &= 2n + 4 \\
n &= 4 \\
n + 2 &= 4 + 2 = 6
\end{aligned}
$$

The integers are 4 and 6.

31. (iii)

Objective B Exercises

33.

320	is	one-fourth of the length of the Golden Gate Bridge

$$320 = \frac{1}{4}x$$
$$4 \cdot 320 = 4 \cdot \frac{1}{4}x$$
$$1280 = x$$

The Golden Gate Bridge is 1280 ft.

35.

\$1400 million	was	one-sixty-fifth of gross national product

$$1400 = \frac{1}{65}x$$
$$65 \cdot 1400 = 65 \cdot \frac{1}{65}x$$
$$91{,}000$$

91,000 million = 91 billion

The gross national product was \$91 billion.

37. Strategy To find the length of the sides of the triangle, write and solve an equation using x to represent the length of an equal side.

Solution

Perimeter of 23 ft	is	x ft + x ft + $(2x-1)$ ft

$$23 = x + x + (2x-1)$$
$$23 = 4x - 1$$
$$24 = 4x$$
$$\frac{24}{4} = x$$
$$6 = x$$
$$2x - 1 = 12 - 1 = 11$$

The length of the sides are 6 ft, 6 ft and 11 ft.

39. **Strategy** Level of tv = 70
Blender = 70 + 20 = 90
Jet engine = 2(90) – 40

Solution
2(90) – 40 = 180 – 40 = 140
The jet engine is 140 decibels.

41. the area of Iceland: x
the area of Greenland: $21x$

The combined area	is	880,000 mi^2

$$x + 21x = 880{,}000$$
$$22x = 880{,}000$$
$$\frac{22x}{22} = \frac{880{,}000}{22}$$
$$x = 40{,}000$$

21(40,000) = 840,000
The area of Greenland is 840,000 mi^2.

43. **Strategy** To find the number of kilowatt hours, write and solve an equation using x to represent the number of kilowatt hours over 300.

Solution

The total cost	is	\$51.95

$$0.08(300) + 0.13x = 51.95$$
$$24 + 0.13x = 51.95$$
$$0.13x = 51.95 - 24$$
$$0.13x = 27.95$$
$$x = \frac{27.95}{0.13}$$
$$x = 215$$

The total number of kilowatt hours is 300 + 215 = 515.
The family used 515 kWh.

45. **Strategy** To find the amount of time that the phone was used, write and solve an equation using x to represent the amount of time.

Solution

\$80 plus \$0.40 per minute	is	\$100.40

$$80 + 0.40x = 100.40$$
$$0.40x = 100.40 - 80$$
$$0.40x = 20.40$$
$$x = \frac{20.40}{0.40}$$
$$x = 51$$

900 + 51 = 951
The business executive used the phone for 951 min.

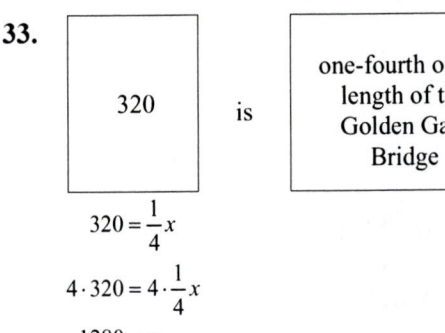

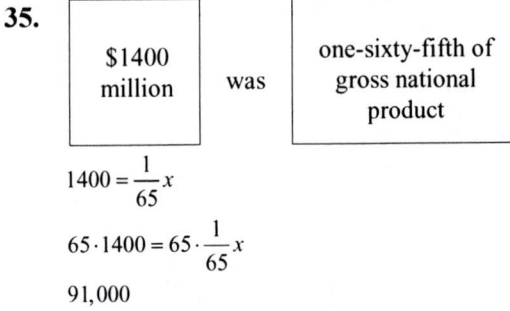

47. \$.15

Critical Thinking

49. Strategy length of the shorter piece: x
perimeter of the shorter square: x
length of the longer piece: $12-x$
perimeter of the longer piece: $12-x$

Solution

perimeter of the larger square	is	twice the perimeter of the shorter square

$$12-x=2x$$
$$12-x+x=2x+x$$
$$12=3x$$
$$\frac{12}{3}=\frac{3x}{3}$$
$$4=x \qquad 12-4=8$$

The perimeter of the larger square is 8 ft.

51. Strategy To find the time remaining:
- find the time to complete the whole trip
- subtract the time completed $\frac{1}{2}$ h from the whole trip to get the time remaining.

Solution

three-fifths of the trip	in	one-half hour

$$\frac{3}{5}t=\frac{1}{2}$$
$$\frac{5}{3}\cdot\frac{3}{5}t=\frac{5}{3}\cdot\frac{1}{2}$$
$$t=\frac{5}{6}$$
$$\frac{5}{6}-\frac{1}{2}=\frac{5}{6}-\frac{3}{6}=\frac{2}{6}=\frac{1}{3}$$

$\frac{1}{3}$ h is remaining

53. first number: n
second consecutive even number: $n+2$
third consecutive even number: $n+4$
fourth consecutive even number: $n+6$
The sum of the four number is −36.

$$n+n+2+n+4+n+6=-36$$
$$4n+12=-36$$
$$4n+12-12=-36-12$$
$$4n=-48$$
$$\frac{4n}{4}=\frac{-48}{4}$$
$$n=-12$$
$$n+2=-12+2=-10$$
$$n+4=-12+4=-8$$
$$n+6=-12+6=-6$$

The integers are −12, −10, −8, and −6.

55. first number: n
second consecutive odd number: $n+2$
third consecutive odd number: $n+4$
The sum of the first and the third is twice the second.

$$n+n+4=2(n+2)$$
$$2n+4=2n+4$$

Since an identity is true for all values, any three consecutive odd integers will make this true.

Projects or Group Activities

57. even

59. even

61. even

63. even

65. odd

Check Your Progress: Chapter 2

1. $2a(a-1)=3a+3$

$2(3)(3-1)$	$3(3)+3$
$6(2)$	$9+3$
12	12

Yes

2.
$$x+7=-4$$
$$x+7-7=-4-7$$
$$x=-11$$

The solution is −11.

3. $-3y = -27$

$\dfrac{-3y}{-3} = \dfrac{-27}{-3}$

$y = 9$

The solution is 9.

4. $P \cdot B = A$

$0.45 \cdot 160 = A$

$72 = A$

The 72 is 45% of 160.

5. $6 - 4a = -10$

$6 - 6 - 4a = -10 - 6$

$-4a = -16$

$\dfrac{-4a}{-4} = \dfrac{-16}{-4}$

$a = 4$

The solution is 4.

6. $8t + 1 = -1$

$8\left(-\dfrac{1}{4}\right) + 1$	-1
$-2 + 1$	-1
-1	-1

Yes

7. $\dfrac{1}{6} + b = -\dfrac{1}{3}$

$\dfrac{6}{1} \cdot \dfrac{1}{6} + 6 \cdot b = \dfrac{6}{1}\left(-\dfrac{1}{3}\right)$

$1 + 6b = -2$

$6b = -3$

$b = -\dfrac{1}{2}$

The solution is $-\dfrac{1}{2}$.

8. $5x - 4(3 - x) = 2(x - 1) - 3$

$5x - 12 + 4x = 2x - 2 - 3$

$9x - 12 = 2x - 5$

$9x - 2x - 12 = 2x - 2x - 5$

$7x - 12 = -5$

$7x - 12 + 12 = -5 + 12$

$7x = 7$

$\dfrac{7x}{7} = \dfrac{7}{7}$

$x = 1$

The solution is 1.

9. **Strategy** Solve the equation $P \cdot B = A$ for B using $P = 0.18$ and $A = 27$.

Solution

$P \cdot B = A$

$0.18B = 27$

$\dfrac{0.18B}{0.18} = \dfrac{27}{0.18}$

$B = 150$

18% of 150 is 27.

10. $6y + 5 - 8y = 3 - 4y$

$-2y + 5 = 3 - 4y$

$-2y + 4y + 5 = 3 - 4y + 4y$

$2y + 5 = 3$

$2y + 5 - 5 = 3 - 5$

$2y = -2$

$\dfrac{2y}{2} = \dfrac{-2}{2}$

$y = -1$

The solution is −1.

11. $x(x + 1) = x^2 + 5$

$4(4 + 1)$	$4^2 + 5$
$4(5)$	$16 + 5$
20	21

No

12. $84 = -16 + t$

$84 + 16 = -16 + 16 + t$

$100 = t$

The solution is 100.

13. $\dfrac{3}{4}c = \dfrac{3}{5}$

$\dfrac{4}{3} \cdot \dfrac{3}{4}c = \dfrac{4}{3} \cdot \dfrac{3}{5}$

$c = \dfrac{4}{5}$

The solution is $\dfrac{4}{5}$.

14.
$$9 = \frac{1}{2}d - 5$$
$$9 + 5 = \frac{1}{2}d - 5 + 5$$
$$14 = \frac{1}{2}d$$
$$2 \cdot 14 = 2 \cdot \frac{1}{2}d$$
$$28 = d$$

The solution is 28.

15. Strategy Solve $P \cdot B = A$ for P using 170 for B and 42.5 for A.

Solution
$$P \cdot B = A$$
$$P \cdot 170 = 42.5$$
$$\frac{P170}{170} = \frac{42.5}{170}$$
$$P = 0.25$$

42.5 is 25% of 170.

16.
$$-\frac{8}{9} = -\frac{2}{3}y$$
$$-\frac{3}{2}\left(-\frac{8}{9}\right) = -\frac{3}{2}\left(-\frac{2}{3}y\right)$$
$$\frac{4}{3} = y$$

The solution is $\frac{4}{3}$.

17. $3n + 2(n - 4) = 7$
$$3n + 2n - 8 = 7$$
$$5n - 8 = 7$$
$$5n - 8 + 8 = 7 + 8$$
$$5n = 15$$
$$\frac{5n}{5} = \frac{15}{5}$$
$$n = 3$$

The solution is 3.

18.
$$3x - 8 = 5x + 6$$
$$3x - 3x - 8 = 5x - 3x + 6$$
$$-8 = 2x + 6$$
$$-8 - 6 = 2x + 6 - 6$$
$$-14 = 2x$$
$$\frac{-14}{2} = \frac{2x}{2}$$
$$-7 = x$$

The solution is –7.

19. $2[3 - 5(x - 1)] = 7x - 1$
$$2[3 - 5x + 5] = 7x - 1$$
$$2[8 - 5x] = 7x - 1$$
$$16 - 10x = 7x - 1$$
$$16 - 10x - 7x = 7x - 7x - 1$$
$$16 - 17x = -1$$
$$16 - 16 - 17x = -1 - 16$$
$$-17x = -17$$
$$\frac{-17x}{-17} = \frac{-17}{-17}$$
$$x = 1$$

The solution is 1.

20. $18 = 2t$
$$\frac{18}{2} = \frac{2t}{2}$$
$$9 = t$$

The solution is 9.

21. the unknown number: x

The quotient of fifteen and an unknown number	is	–3

$$\frac{15}{x} = -3$$
$$x\left(\frac{15}{x}\right) = x(-3)$$
$$15 = -3x$$
$$\frac{15}{-3} = \frac{-3x}{-3}$$
$$-5 = x$$

The number is –5.

22. First odd integer: x
Second odd integer: $x + 2$
Third odd integer: $x + 4$
Fourth odd integer: $x + 6$
The sum of the integers is 24.
$$x + x + 2 + x + 4 + x + 6 = 24$$
$$4x + 12 = 24$$
$$4x + 12 - 12 = 24 - 12$$
$$4x = 12$$
$$\frac{4x}{4} = \frac{12}{4}$$
$$x = 3$$

$x + 2 = 5$
$x + 4 = 7$
$x + 6 = 9$
The integers are 3, 5, 7, and 9.

23. **Strategy** Solve the equation $B + BP = A$ for A using 1970 for B and 0.116 for P.

Solution

$$B + BP = A$$
$$1970 + 1970 \cdot 0.116 = A$$
$$1970 + 229 = A$$
$$2199 = A$$

The average consumption is 2199 calories.

24. **Strategy** To find the total time:
- find the time with the current to travel 24 mi (r = 10 + 2 = 12)
- find the time without the current to travel 24 mi (r = 10 – 2 = 8)
- add the two times and 1 hour to get the total trip

Solution

with the current:
$$rt = d$$
$$12t = 24$$
$$\frac{12t}{12} = \frac{24}{12}$$
$$t = 2$$

without the current:
$$rt = d$$
$$8t = 24$$
$$\frac{8t}{8} = \frac{24}{8}$$
$$t = 3$$

$2 + 3 + 1 = 6$

The total trip was 6 h.

25. **Strategy** Check the equation $F_1x = F_2(d - x)$ when $F_1 = 60$, $x = 3.5$, $F_2 = 50$, and $d = 8$.

Solution

F_1x	$F_2(d-x)$
$60 \cdot 3.5$	$50(8-3.5)$
210	$50(4.5)$
210	225

No

Section 2.4

Concept Check

1. \$10.50

3. \$.76

5. 100

7. True

9. False

11. In the formula $Q = Ar$, Q represents the quantity, A represents the amount, and r represents the percent. For example: there are 2 ml of a 25% acid solution. To find the value (Q), multiply the amount (2) by the percent (0.25).

$$Q = Ar$$
$$Q = 2 \cdot 0.25$$
$$Q = 0.5$$

There is 0.5 ml of acid in the solution.

Objective A Exercises

13. **Strategy**
- Amount of high-protein supplement: x
  Amount of vitamin supplement: $5 - x$

	Amount	Cost	Value
High-protein	x	6.75	$6.75x$
Vitamin	$5 - x$	3.25	$3.25(5 - x)$
Mixture	5	4.65	$4.65(5)$

- The sum of the values before mixing equals the value after mixing.

Solution

$$6.75x + 3.25(5 - x) = 4.65(5)$$
$$6.75x + 16.25 - 3.25x = 23.25$$
$$3.50x + 16.25 = 23.25$$
$$3.50x + 16.25 - 16.25 = 23.25 - 16.25$$
$$3.50x = 7.00$$
$$x = \frac{7.00}{3.50}$$
$$x = 2$$
$$5 - x = 3$$

To make the mixture, 2 lb of the high protein supplement and 3 lb of the vitamin supplement were used.

15. Strategy

- Amount of chamomile tea: x
  Amount of orange tea: 12

	Amount	Cost	Value
Chamomile	x	18.20	$18.20(x)$
Orange tea	12	12.25	12(12.25)
Mixture	$x + 12$	14.63	$14.63(x + 12)$

- The sum of the values before mixing equals the value after mixing.

Solution

$$18.20x + 12(12.25) = 14.63(x + 12)$$
$$18.20x + 147 = 14.63x + 175.56$$
$$18.20x - 14.63x + 147 = 14.63x - 14.63x + 175.56$$
$$3.57x + 147 = 175.56$$
$$3.57x + 147 - 147 = 175.56 - 147$$
$$3.57x = 28.56$$
$$\frac{3.57x}{3.57} = \frac{28.56}{3.57}$$
$$x = 8$$

The amount of chamomile tea needed is 8 lb.

17. Strategy

- Cost of mixture: x

	Amount	Cost	Value
Expensive coffee	8	9.20	8(9.20)
Cheaper coffee	12	5.50	12(5.50)
Mixture	20	x	$20(x)$

- The sum of the values before mixing equals the value after mixing.

Solution

$$8(9.20) + 12(5.50) = 20x$$
$$73.60 + 66 = 20x$$
$$139.60 = 20x$$
$$\frac{139.60}{20} = \frac{20x}{20}$$
$$6.98 = x$$

The cost of the coffee mixture is \$6.98.

19. Strategy

- Amount of \$1 herb: x

	Amount	Cost	Value
\$2 herb	30	2	2(30)
\$1 herb	x	1	$1x$
Mixture	$30 + x$	1.60	$1.6(30 + x)$

- The sum of the values before mixing equals the value after mixing.

Solution

$$2(30) + x = 1.6(30 + x)$$
$$60 + x = 48 + 1.6x$$
$$60 + x - x = 48 + 1.6x - x$$
$$60 = 48 + 0.6x$$
$$60 - 48 = 48 - 48 + 0.6x$$
$$12 = 0.6x$$
$$\frac{12}{0.6} = \frac{0.6x}{0.6}$$
$$20 = x$$

The amount of the \$1 herb is 20 oz.

21. Strategy

- Amount of pepper cheddar cheese: x
  Amount of Pennsylvania Jack: $5 - x$

	Amount	Cost	Value
Pepper cheddar	x	16	$16x$
Jack	$5 - x$	12	$12(5 - x)$
Mixture	5	13.20	13.20(5)

- The sum of the values before mixing equals the value after mixing.

Solution

$$16x + 12(5 - x) = 13.20(5)$$
$$16x + 60 - 12x = 66$$
$$4x + 60 - 60 = 66 - 60$$
$$4x = 6$$
$$\frac{4x}{4} = \frac{6}{4}$$
$$x = 1.5$$

$5 - 1.5 = 3.5$

The mixture needs 1.5 kg of pepper cheese and 3.5 kg of Pennsylvania Jack.

23. **Strategy**

- Amount of grain: 500
  Amount of meal: x

	Amount	Cost	Value
Grain	500	1.2	1.2(500)
Meal	x	0.8	$0.8x$
Mixture	$500+x$	1.05	$1.05(500+x)$

- The sum of the values before mixing equals the value after mixing.

Solution

$$1.2(500)+0.8x=1.05(500+x)$$
$$600+0.8x=525+1.05x$$
$$600+0.8x-0.8x=525+1.05x-0.8x$$
$$600=525+0.25x$$
$$600-525=525-525+0.25x$$
$$75=0.25x$$
$$\frac{75}{0.25}=\frac{0.25x}{0.25}$$
$$300=x$$

The mixture needs 300 lb of meal.

25. **Strategy**

- Amount of almonds: x
  Amount of walnuts: $100-x$

	Amount	Cost	Value
Almonds	x	6.50	$6.50(x)$
Walnuts	$100-x$	5.50	$5.50(100-x)$
Mixture	100	5.87	5.87(100)

- The sum of the values before mixing equals the value after mixing.

Solution

$$6.50x+5.50(100-x)=5.87(100)$$
$$6.50x+550-5.50x=587$$
$$x+550=587$$
$$x+550-550=587-550$$
$$x=37$$
$$100-x=63$$

The amount of almonds is 37 lb.
The amount of walnuts is 63 lb.

27. **Strategy**

- Cost of mixture: x

	Amount	Cost	Value
Sugar	40	2.00	40(2.00)
Flakes	120	1.20	120(1.20)
Mixture	160	x	$160x$

- The sum of the values before mixing equals the value after mixing.

Solution

$$40(2.00)+120(1.20)=160x$$
$$80+144=160x$$
$$224=160x$$
$$\frac{224}{160}=\frac{160x}{160}$$
$$1.40=x$$

The cost per pound of the sugar-coated cereal is \$1.40.

29. **Strategy**

- Number of bundles of seedlings: x
  Number of bundles of container-grown plants: $1720-x$

	Amount	Cost	Value
Seedlings	x	17	$17(x)$
Contain-grown	$14-x$	45	$45(14-x)$
Mixture	14		406

- The sum of the values of the seedlings and container-grown plants must equal the total spent.

Solution

$$17x+45(14-x)=406$$
$$17x+630-45x=406$$
$$-28x=-224$$
$$\frac{-28x}{-28}=\frac{-224}{-28}$$
$$x=8$$
$$14-x=14-8=6$$

The Park's Department bought 8 bundles of seedlings and 6 bundles of container-grown plants.

31. Strategy

- Amount of expensive lotion: 50
  Amount of supplement lotion: 100

	Amount	Cost	Value
Expensive lotion	50	4.00	4(50)
Supplement lotion	100	2.50	2.5(100)
Mixture	150	x	$150x$

- The sum of the values before mixing equals the value after mixing.

Solution

$$\begin{aligned} 200+250&=150x \\ 450&=150x \\ \frac{450}{150}&=\frac{150x}{150} \\ 3&=x \end{aligned}$$

The sunscreen mixture will cost $3.00.

33. iv

Objective B Exercises

35. Strategy

- The percent concentration of tomato juice in the mixture: x

	Amount	Percent	Quantity
50% juice	100	0.50	0.50(100)
25% juice	200	0.25	0.25(200)
Mixture	300	x	$300x$

- The sum of the quantities before mixing is equal to the quantity after mixing.

Solution

$$\begin{aligned} 0.50(100)+0.25(200)&=300x \\ 50+50&=300x \\ 100&=300x \\ \frac{1}{3}&=x \end{aligned}$$

The percent concentration of tomato juice in the mixture as $33\frac{1}{3}\%$.

37. Strategy

- Amount of 50% corn: x
- Amount of mixture: $x + 400$

	Amount	Percent	Quantity
50% corn	x	0.50	$0.50x$
80% corn	400	0.80	0.80(400)
Mixture	$x + 400$	0.75	$0.75(x + 400)$

- The sum of the quantities before mixing is equal to the quantity after mixing.

Solution

$$\begin{aligned} 0.50x+0.80(400)&=0.75(x+400) \\ 0.50x+320&=0.75x+300 \\ -0.25x&=-20 \\ x&=80 \end{aligned}$$

80 lbs of 50% corn must be used.

39. Strategy

- Amount of dark green paint: x
- Amount of mixture: $x + 5$

	Amount	Percent	Quantity
Light green paint	x	0.40	$0.40x$
Dark green paint	5	0.20	0.20(5)
25% yellow paint	$x + 5$	0.25	$0.25(x + 5)$

- The sum of the quantities before mixing is equal to the quantity after mixing.

Solution

$$\begin{aligned} 0.40x+0.20(5)&=0.25(x+5) \\ 0.40x+1&=0.25x+1.25 \\ 0.15x&=0.25 \\ x&=1\frac{2}{3} \end{aligned}$$

$1\frac{2}{3}$ gal of light green latex paint must be used.

41. Strategy

- Amount of 13% acid solution: x
- Amount of 18% acid solution: $50 - x$

	Amount	Percent	Quantity
13% acid	x	0.13	$0.13x$
18% acid	$50 - x$	0.18	$0.18(50 - x)$
16% acid mixture	50	0.16	0.16(50)

- The sum of the quantities before mixing is equal to the quantity after mixing.

Solution

$$0.13x + 0.18(50 - x) = 0.16(50)$$
$$0.13x + 9.00 - 0.18x = 8.00$$
$$-0.05x + 9.00 = 8.00$$
$$-0.05x = -1.00$$
$$x = 20$$
$$50 - x = 50 - 20 = 30$$

The amount of 13% solution is 20 ml.
The amount of 18% solution is 30 ml.

43. Strategy

- Percent concentration of the resulting alloy: x

	Amount	Percent	Quantity
Pure silver	30	1.00	30
20% silver	50	0.20	0.20(50)
Resulting mixture	80	x	$80x$

- The sum of the quantities before mixing is equal to the quantity after mixing.

Solution

$$30 + 10 = 80x$$
$$40 = 80x$$
$$0.50 = x$$

The percent concentration is 50%.

45. Strategy

- Amount of 40% mixture: x

	Amount	Percent	Quantity
Grass seed 1	x	0.40	$0.40x$
Grass seed 2	40	0.60	0.60(40)
60% mixture	$x + 40$	0.56	$0.56(x + 40)$

- The sum of the quantities before mixing is equal to the quantity after mixing.

Solution

$$0.4x + 0.6(40) = 0.56(x + 40)$$
$$0.4x + 24 = 0.56x + 22.4$$
$$0.4x - 0.4x + 24 = 0.56x - 0.4x + 22.4$$
$$24 = 0.16x + 22.4$$
$$24 - 22.4 = 0.16x + 22.4 - 22.4$$
$$1.6 = 0.16x$$
$$\frac{1.6}{0.16} = \frac{0.16x}{0.16}$$
$$10 = x$$

10 lb of the 40% mixture must be used.

47. Strategy

- Amount of pure silk: x
- Amount of 85% silk: $75 - x$

	Amount	Percent	Quantity
Pure silk	x	1.00	x
85% silk	$75 - x$	0.84	$0.85(75 - x)$
Mixture	75	0.96	0.96(75)

- The sum of the quantities before mixing is equal to the quantity after mixing.

Solution

$$x + 0.85(75 - x) = 0.96(75)$$
$$x + 63.75 - 0.85x = 72$$
$$0.15x + 63.75 - 63.75 = 72 - 63.75$$
$$0.15x = 8.25$$
$$x = 55$$

$75 - 55 = 20$

55 kg of pure silk and 20 kg of 85% silk are needed.

49. Strategy

- amount of pure ethanol: x

	Amount	Percent	Quantity
Pure ethanol	x	1.00	$1x$
E10	100	0.10	0.10(100)
mixture	$100 + x$	0.20	$0.20(100 + x)$

- The sum of the quantities before mixing is equal to the quantity after mixing.

Solution

$$x + 0.1(100) = 0.2(100 + x)$$
$$x + 10 = 20 + 0.2x$$
$$x - 0.20x + 10 = 20 + 0.2x - 0.2x$$
$$0.8x + 10 = 20$$
$$0.8x + 10 - 10 = 20 - 10$$
$$0.8x = 10$$
$$\frac{0.8x}{0.8} = \frac{10}{0.8}$$
$$x = 12.5$$

12.5 gal of ethanol need to be added.

51. Strategy • Amount of pure chocolate: x

	Amount	Percent	Quantity
50% chocolate	150	0.50	0.50(150)
Pure chocolate	x	1.00	$1.00x$
Mixture	$x + 150$	0.75	$0.75(x + 150)$

- The sum of the quantities before mixing is equal to the quantity after mixing.

Solution

$$0.50(150) + 1.00x = 0.75(150 + x)$$
$$75 + x = 112.5 + 0.75x$$
$$0.25x = 37.5$$
$$x = 150$$

150 oz of pure chocolate must be added.

53. False

Objective C Exercises

55. Strategy

- Speed of first plane: r
- Speed of second plane: $r + 25$

	Rate	Time	Distance
First plane	r	2	$2r$
Second plane	$r + 25$	2	$2(r + 25)$

- In 2 h, the planes are 470 miles apart.

Solution

$$2r + 2(r + 25) = 470$$
$$2r + 2r + 50 = 470$$
$$4r = 420$$
$$r = 105$$
$$r + 25 = 130$$

The first plane is flying at 105 mph and the second plane is flying at 130 mph.

57. Strategy • Time for first skater: t

• Time for second skater: $t - 10$

	Rate	Time	Distance
First skater	8	t	$8t$
Second skater	10	$t - 10$	$10(t - 10)$

- The skaters travel the same distance.

Solution

$$8t = 10(t - 10)$$
$$8t = 10t - 100$$
$$8t - 10t = 10t - 10t - 100$$
$$-2t = -100$$
$$\frac{-2t}{-2} = \frac{-100}{-2}$$
$$t = 50$$

Time for second skater = 50 – 10 = 40
The second skater overtakes the first 40 s after the second skater starts.

59. Strategy

- Time the motorboat travels: t
- Time the cabin cruiser travels: $t - 2$

	Rate	Time	Distance
Motorboat	9	t	$9t$
Cabin Cruiser	18	$t-2$	$18(t-2)$

- How many hours after the cabin cruiser leaves will the cabin cruiser meet up with the motorboat?

Solution

$$\begin{aligned} 9t &= 18(t-2) \\ 9t &= 18t - 36 \\ -9t &= -36 \\ t &= 4 \\ t - 2 &= 2 \end{aligned}$$

The cabin cruiser will overtake the motorboat in 2 h.

61. Strategy

- Time to airport: t
- Time in flight: $3 - t$

	Rate	Time	Distance
To airport	30	t	$30t$
In flight	60	$3-t$	$60(3-t)$

- The total trip is 150 mi.

Solution

$$\begin{aligned} 30t + 60(3-t) &= 150 \\ 30t + 180 - 60t &= 150 \\ 180 - 30t &= 150 \\ -30t &= -30 \\ t &= 1 \\ \text{Distance} &= 60(3-t) = 60(3-1) \\ &= 60(2) = 120 \end{aligned}$$

The corporate offices are 120 mi from the airport.

63. Strategy

- Speed for first 3 h: r
- Speed for second 3 h: $r - 5$

	Rate	Time	Distance
First 3h	r	3	$3r$
Second 3h	$r-5$	3	$3(r-5)$

- The total trip is 57 mi.

Solution

$$\begin{aligned} 3r + 3(r-5) &= 57 \\ 3r + 3r - 15 &= 57 \\ 6r - 15 &= 57 \\ 6r &= 72 \\ r &= 12 \\ \text{Distance} = 3r &= 3(12) = 36 \end{aligned}$$

The sailboat traveled 36 mi in the first 3 h.

65. Strategy

- Rate for freight train: r
- Rate for passenger train: $r + 20$

	Rate	Time	Distance
Freight	r	5	$5r$
Passenger	$r+20$	3	$3(r+20)$

- The trains travel the same distance.

Solution

$$\begin{aligned} 5r &= 3(r+20) \\ 5r &= 3r + 60 \\ 2r &= 60 \\ r &= 30 \\ r + 20 &= 30 + 20 = 50 \end{aligned}$$

The freight train travels at 30 mph.

The passenger train travels at 50 mph.

67. Strategy

- The time the first ship traveled: t
- The time the second ship traveled: $t-\frac{10}{25}$

	Rate	Time	Distance
First ship	25	t	$25t$
Second ship	35	$t-\frac{10}{25}$	$35\left(t-\frac{10}{25}\right)$

- The second ship catches up to the first ship.

Solution

$$25t = 35\left(t-\frac{10}{25}\right)$$
$$25t = 35t - 14$$
$$-10t = -14$$
$$t = 1.4$$
$$t-\frac{10}{25} = 1.4 - 0.4 = 1$$

The second ship catches up to the first ship in 1 h.

69. Strategy

- Rate of the second car: r
- Rate of the first car: $r + 10$
- $12 \text{ min} \div 60 = \frac{1}{5}$ hr

	Rate	Time	Distance
First car	$r + 10$	$\frac{1}{5}$	$\frac{1}{5}(r+10)$
Second car	r	$\frac{1}{5}$	$\frac{1}{5}r$

- The total distance traveled by the two cars is 36.

Solution

$$\frac{1}{5}(r+10)+\frac{1}{5}r = 36$$
$$\frac{1}{5}r+2+\frac{1}{5}r = 36$$
$$\frac{5}{1}\cdot\frac{1}{5}r+5\cdot 2+\frac{5}{1}\cdot\frac{1}{5}r = 5\cdot 36$$
$$r+10+r = 180$$
$$2r+10-10 = 180-10$$
$$2r = 170$$
$$\frac{2r}{2} = \frac{170}{2}$$
$$r = 85$$

$85 + 10 = 95$

The faster car is traveling 95 km/h.

71. Strategy

- Time for first car: t
- Time for second car: $t-\frac{1}{4}$

	Rate	Time	Distance
First driver	90	t	$90t$
Second driver	120	$t-\frac{1}{4}$	$120\left(t-\frac{1}{4}\right)$

- The second car will overtake the first when the distances are equal.

Solution

$$90t = 120\left(t-\frac{1}{4}\right)$$
$$90t = 120t - 30$$
$$90t - 120t = 120t - 120t - 30$$
$$-30t = -30$$
$$\frac{-30t}{-30} = \frac{-30}{-30}$$
$$t = 1$$

Distance $= 90t = 90(1) = 90$ mi

The track is on 50 mi, so the second will not overtake the first.

73. Strategy

- Time the car traveled: t
- Time the bus traveled: $t - 1$

	Rate	Time	Distance
Car	45	t	$45t$
Bus	60	$t - 1$	$60(t - 1)$

- The bus overtakes the car.

Solution

$$45t = 60(t-1)$$
$$45t = 60t - 60$$
$$-15t = -60$$
$$t = 4$$
$$45t = 180$$

The bus overtakes the car 180 mi from the starting point.

75. **Strategy**

- Time for the first part of the trip: t
- Time for the second part of the trip: $5 - t$

	Rate	Time	Distance
First part of trip	115	t	$115t$
Remainder of trip	125	$5-t$	$125(5-t)$

- The total distance traveled was 605 mi.

Solution

$$115t + 125(5-t) = 605$$
$$115t + 625 - 125t = 605$$
$$-10t + 625 - 625 = 605 - 625$$
$$-10t = -20$$
$$\frac{-10t}{-10} = \frac{-20}{-10}$$
$$t = 2$$

$5 - 2 = 3$

The plane traveled 2 h at 115 mph and 3 h at 125 mph.

Critical Thinking

77. **Strategy**

- Amount of Walnuts: x
- Amount of Cashews: $50 - 20 - x = 30 - x$

	Amount	Cost	Value
Walnuts	x	5.60	$5.60(x)$
Cashews	$30-x$	7.50	$7.50(30-x)$
Peanuts	20	4.00	4.00(20)
Mixture	50	5.72	5.72(50)

- The sum of the values before mixing equals the value after mixing.

Solution

$$5.60(x) + 7.50(30-x) + 4.00(20) = 5.72(50)$$
$$5.6x + 225 - 7.5x + 80 = 286$$
$$-1.9x + 305 = 286$$
$$-1.9x + 305 - 305 = 286 - 305$$
$$-1.9x = -19$$
$$\frac{-1.9x}{-1.9} = \frac{-19}{-1.9}$$
$$x = 10$$

$30 - 10 = 20$

The amount of walnuts is 10 lb.

The amount of cashews is 20 lb.

79. **Strategy**

- Amount of pure acid: x
- Amount of water: $10 - x$

	Amount	Percent	Quantity
Pure acid	x	1.00	$1.00x$
Water	$10-x$	0	$0(10-x)$
Mixture	10	0.30	0.30(10)

- The sum of the quantities before mixing is equal to the quantity after mixing.

Solution

$$1.00x + 0(10-x) = 0.30(10)$$
$$x = 3$$

$10 - 3 = 7$

3 L of pure acid and 7 L of water are mixed.

81. **Strategy**

- Number of adult tickets: x
- Number of child tickets: $120 - x$

	Amount	Cost	Value
Adult tickets	x	5.50	$5.50(x)$
Child tickets	$120-x$	2.75	$2.75(120-x)$

- The sum of the values must equal $563.75.

Solution

$$5.50x + 2.75(120-x) = 563.75$$
$$5.50x + 330 - 2.75x = 563.75$$
$$2.75x + 330 = 563.75$$
$$2.75x + 330 - 330 = 563.75 - 330$$
$$2.75x = 233.75$$
$$\frac{2.75x}{2.75} = \frac{233.75}{2.75}$$
$$x = 85$$

$120 - 85 = 35$

85 adult tickets and 35 child tickets were sold.

83. **Strategy**

- Time downstream: t
- Time upstream: $1 - t$

	Rate	Time	Distance
Downstream	12	t	$12t$
Upstream	4	$1 - t$	$4(1 - t)$

- The distance downstream is equal to the distance upstream.

Solution

$$12t = 4(1-t)$$
$$12t = 4 - 4t$$
$$12t + 4t = 4 - 4t + 4t$$
$$16t = 4$$
$$\frac{16t}{16} = \frac{4}{16}$$
$$t = \frac{1}{4}$$

$\frac{1}{4} \cdot 60$ min = 15 min

10:00 + 0:15 = 10:15

The campers turned around at 10:15 A.M.

Projects and Group Activities

85. **Strategy**

- Amount to be drained and pure to be added: x
- Amount of 20%: $15 - x$

	Amount	Percent	Quantity
Pure antifreeze	x	1.00	$1.00x$
20%	$15 - x$	0.20	$0.20(15 - x)$
Mixture	15	0.40	0.40(15)

- The sum of the quantities before mixing is equal to the quantity after mixing.

Solution

$$1.00x + 0.20(15 - x) = 0.40(15)$$
$$x + 3 - 0.2x = 6$$
$$0.8x + 3 = 6$$
$$0.8x + 3 - 3 = 6 - 3$$
$$0.8x = 3$$
$$\frac{0.8x}{0.8} = \frac{3}{0.8}$$
$$x = 3.75$$

3 .75 gal should be drained and replaced with antifreeze.

87. **Strategy**

- Find the total distance traveled and the total time.
- Divide the total distance by the total time to determine the average speed.

	Rate	Time	Distance
Leaving	10	2	10(2) = 20
Returning	20	$\frac{20}{20} = 1$	20(1) = 20
Total		3	40

Solution

Average speed $= \frac{40}{3} = 13\frac{1}{3}$

The bicyclist's average speed is $13\frac{1}{3}$ mph.

89. **Strategy**

- Time to ascend: t
- Time to descend: $12 - t$

	Rate	Time	Distance
Ascend	0.5	t	$0.5t$
Descend	1	$12 - t$	$1(12 - t)$

- The distances are the same.

Solution

$$0.5t = 1(12-t)$$
$$0.5t = 12 - t$$
$$0.5t + t = 12 - t + t$$
$$1.5t = 12$$
$$\frac{1.5t}{1.5} = \frac{12}{1.5}$$
$$t = 8$$

Distance: $0.5t = 0.5(8) = 4$ mi each way

The total distance was 8 mi.

Section 2.5

Concept Check

1. The Addition Property of Inequalities states that the same number can be added to each side of an inequality without changing the solution set of the inequality.
Examples will vary. For instance:

$$8 > 6 \qquad\qquad -5 < -1$$
$$8+7 > 6+7 \text{ and } -5+(-2) < -1+(-2)$$
$$15 > 13 \qquad\qquad -7 < -3$$

3. Replace x with each value to determine if the inequality holds.
i) $-17+7 \le -3$; $-10 \le -3$; solution
ii) $8+7 \le -3$; $15 \le -3$; not a solution
iii) $-10+7 \le -3$; $-3 \le -3$; solution
iv) $0+7 \le -3$; $7 \le -3$; not a solution

5. <

Objective A Exercises

7. $x-3<2$
$x<5$
$\{x|\, x<5\}$

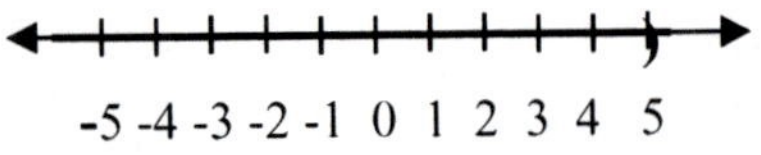

9. $4x \le 8$
$$\frac{4x}{4} \le \frac{8}{4}$$
$x \le 2$
$\{x|\, x \le 2\}$

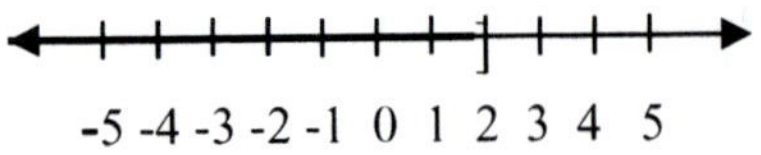

11. $-2x > 8$
$$\frac{-2x}{-2} < \frac{8}{-2}$$
$x < -4$
$\{x|\, x < -4\}$

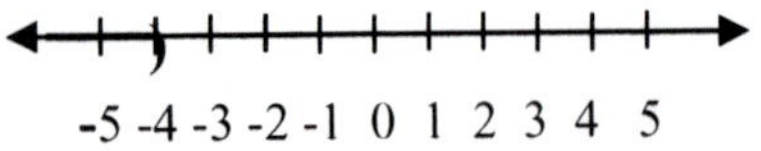

13. $3x-1 > 2x+2$
$x-1>2$
$x>3$
The solution set is $\{x|\, x>3\}$.

15. $2x-1>7$
$2x>8$
$$\frac{2x}{2} > \frac{8}{2}$$
$x>4$
The solution set is $\{x|\, x>4\}$.

17. $5x-2 \le 8$
$5x \le 10$
$$\frac{5x}{5} \le \frac{10}{5}$$
$x \le 2$
The solution set is $\{x|\, x \le 2\}$.

19. $6x+3 > 4x-1$
$6x > 4x-4$
$2x > -4$
$$\frac{2x}{2} > \frac{-4}{2}$$
$x > -2$
The solution set is $\{x|\, x > -2\}$.

21. $8x+1 \ge 2x+13$
$6x+1 \ge 13$
$6x \ge 12$
$$\frac{6x}{6} \ge \frac{12}{6}$$
$x \ge 2$
The solution set is $\{x|\, x \ge 2\}$.

23. $4-3x<10$
$-3x < 6$
$$\frac{-3x}{-3} > \frac{6}{-3}$$
$x > -2$
The solution set is $\{x|\, x > -2\}$.

25. $7-2x \ge 1$
$-2x \ge -6$
$$\frac{-2x}{-2} \le \frac{-6}{-2}$$
$x \le 3$
The solution set is $\{x|\, x \le 3\}$.

27. $-3-4x > -11$

$$-4x > -8$$

$$\frac{-4x}{-4} < \frac{-8}{-4}$$

$$x < 2$$

The solution set is $\{x|\, x < 2\}$.

29. $4x-2 < x-11$

$$3x-2 < -11$$

$$3x < -9$$

$$\frac{3x}{3} < \frac{-9}{3}$$

$$x < -3$$

The solution set is $\{x|\, x < -3\}$.

31. $x+7 \geq 4x-8$

$$-3x+7 \geq -8$$

$$-3x \geq -15$$

$$\frac{-3x}{-3} \leq \frac{-15}{-3}$$

$$x \leq 5$$

The solution set is $\{x|\, x \leq 5\}$.

33. $3x+2 \leq 7x+4$

$$-4x+2 \leq 4$$

$$-4x \leq 2$$

$$\frac{-4x}{-4} \geq \frac{2}{-4}$$

$$x \geq -\frac{1}{2}$$

The solution set is $\{x|\, x \geq -\frac{1}{2}\}$.

35. The solution to the inequality $nx > a$, where both n and a are negative contains both positive and negative numbers.

37. The solution to the inequality $x - n > -a$, where both n and a are positive and $n < a$ contains both positive and negative numbers.

39. $7x+3 < 4x+1$

$$7x+3-3 < 4x+1-3$$

$$7x < 4x-2$$

$$7x-4x < 4x-4x-2$$

$$3x < -2$$

$$\frac{3x}{3} < \frac{-2}{3}$$

$$x < -\frac{2}{3}$$

The solution is $\left(-\infty, -\frac{2}{3}\right)$.

41. $\frac{2}{3}x - \frac{3}{2} < \frac{7}{6} - \frac{1}{3}x$

$$6\left(\frac{2}{3}x - \frac{3}{2}\right) < 6\left(\frac{7}{6} - \frac{1}{3}x\right)$$

$$4x-9 < 7-2x$$

$$6x-9 < 7$$

$$6x < 16$$

$$\frac{6x}{6} < \frac{16}{6}$$

$$x < \frac{8}{3}$$

The solution is $\left(-\infty, \frac{8}{3}\right)$.

43. $\frac{1}{2}x - \frac{3}{4} < \frac{7}{4}x - 2$

$$4\left(\frac{1}{2}x - \frac{3}{4}\right) < 4\left(\frac{7}{4}x - 2\right)$$

$$2x-3 < 7x-8$$

$$-5x-3 < -8$$

$$-5x < -5$$

$$\frac{-5x}{-5} > \frac{-5}{-5}$$

$$x > 1$$

The solution is $(1, \infty)$.

45. $4(2x-1) > 3x-2(3x-5)$

$$8x-4 > 3x-6x+10$$

$$8x - 4 > -3x + 10$$
$$11x - 4 > 10$$
$$11x > 14$$
$$\frac{11x}{11} > \frac{14}{11}$$
$$x > \frac{14}{11}$$

The solution is $\left(\frac{14}{11}, \infty\right)$.

47. $2 - 5(x + 1) \geq 3(x - 1) - 8$
$$2 - 5x - 5 \geq 3x - 3 - 8$$
$$-3 - 5x \geq 3x - 11$$
$$-5x \geq 3x - 8$$
$$-8x \geq -8$$
$$\frac{-8x}{-8} \leq \frac{-8}{-8}$$
$$x \leq 1$$

The solution is $(-\infty, 1]$.

49. $3 + 2(x + 5) \geq x + 5(x + 1) + 1$
$$3 + 2x + 10 \geq x + 5x + 5 + 1$$
$$2x + 13 \geq 6x + 6$$
$$-4x + 13 \geq 6$$
$$-4x \geq -7$$
$$\frac{-4x}{-4} \leq \frac{-7}{-4}$$
$$x \leq \frac{7}{4}$$

The solution is $\left(-\infty, \frac{7}{4}\right]$.

51. $3 - 4(x + 2) \leq 6 + 4(2x + 1)$
$$3 - 4x - 8 \leq 6 + 8x + 4$$
$$-4x - 5 \leq 10 + 8x$$
$$-12x - 5 \leq 10$$
$$-12x \leq 15$$
$$\frac{-12x}{-12} \geq \frac{15}{-12}$$
$$x \geq -\frac{5}{4}$$

The solution is $\left[-\frac{5}{4}, \infty\right)$.

53. $12 - 2(3x - 2) \geq 5x - 2(5 - x)$
$$12 - 6x + 4 \geq 5x - 10 + 2x$$
$$16 - 6x \geq 7x - 10$$
$$-6x \geq 7x - 26$$
$$-13x \geq -26$$
$$\frac{-13x}{-13} \leq \frac{-26}{-13}$$
$$x \leq 2$$

The solution is $(-\infty, 2]$.

Objective B Exercises

55. $x - 3 \leq 1$ and $2x \geq -4$
$x \leq 4$ $\qquad x \geq -2$
$\{x \mid x \leq 4\}$ $\qquad \{x \mid x \geq -2\}$
$\{x \mid x \leq 4\} \cap \{x \mid x \geq -2\} = [-2, 4]$

57. $2x < 6$ or $x - 4 > 1$
$x < 3$ $\qquad x > 5$
$\{x \mid x < 3\}$ $\qquad \{x \mid x > 5\}$
$\{x \mid x < 3\} \cup \{x \mid x > 5\} = (-\infty, 3) \cup (5, \infty)$

59. $\frac{1}{2}x > -2$ and $5x < 10$
$x > -4$ $\qquad x < 2$
$\{x \mid x > -4\}$ $\qquad \{x \mid x < 2\}$
$\{x \mid x > -4\} \cap \{x \mid x < 2\} = (-4, 2)$

61. $\frac{2}{3}x > 4$ or $2x < -8$
$x > 6$ $\qquad x < -4$
$\{x \mid x > 6\}$ $\qquad \{x \mid x < -4\}$
$\{x \mid x > 6\} \cup \{x \mid x < -4\} = (-\infty, -4) \cup (6, \infty)$

63. $3x < -9$ and $x - 2 < 2$
$x < -3$ $\qquad x < 4$
$\{x \mid x < -3\}$ $\qquad \{x \mid x < 4\}$
$\{x \mid x < -3\} \cap \{x \mid x < 4\} = (-\infty, -3)$

65. $2x - 3 > 1$ and $3x - 1 < 2$
$2x > 4$ $\qquad 3x < 3$
$x > 2$ $\qquad x < 1$
$\{x \mid x > 2\}$ $\qquad \{x \mid x < 1\}$
$\{x \mid x > 2\} \cap \{x \mid x < 1\} = \emptyset$

67. $4x + 1 < 5$ and $4x + 7 > -1$
$4x < 4$ $4x > -8$
$x < 1$ $x > -2$
$\{x \mid x < 1\}$ $\{x \mid x > -2\}$
$\{x \mid x < 1\} \cap \{x \mid x > -2\} = (-2, 1)$

69. The inequality $x > -3$ or $x < 2$ describes all real numbers.

71. The inequality $x < -3$ or $x > 2$ describes two intervals of real numbers.

73. $6x - 2 < -14$ or $5x + 1 > 11$
$\mathbf{6x < -12}$ $\mathbf{5x > 10}$
$x < -2$ $x > 2$
$\{x \mid x < -2\}$ $\{x \mid x > 2\}$
$\{x \mid x < -2\} \cup \{x \mid x > 2\} = \{x \mid x < -2 \text{ or } x > 2\}$

75. $5 < 4x - 3 < 21$
$5 + 3 < 4x - 3 + 3 < 21 + 3$
$8 < 4x < 24$
$$\frac{8}{4} < \frac{4x}{4} < \frac{24}{4}$$
$2 < x < 6$
$\{x \mid 2 < x < 6\}$

77. $-2 < 3x + 7 < 1$
$-2 + (-7) < 3x + 7 + (-7) < 1 + (-7)$
$-9 < 3x < -6$
$$\frac{-9}{3} < \frac{3x}{3} < \frac{-6}{3}$$
$-3 < x < -2$
$\{x \mid -3 < x < -2\}$

79. $3x - 5 > 10$ or $3x - 5 < -10$
$3x > 15$ $3x < -5$
$x > 5$ $x < -\frac{5}{3}$
$\{x \mid x > 5\}$ $\{x \mid x < -\frac{5}{3}\}$
$\{x \mid x > 5\} \cup \{x \mid x < -\frac{5}{3}\} = \{x \mid x > 5 \text{ or } x < -\frac{5}{3}\}$

81. $8x + 2 \le -14$ and $4x - 2 > 10$
$8x \le -16$ $4x > 12$
$x \le -2$ $x > 3$
$\{x \mid x \le -2\}$ $\{x \mid x > 3\}$
$\{x \mid x \le -2\} \cap \{x \mid x > 3\} = \emptyset$

83. $5x + 12 \ge 2$ or $7x - 1 \le 13$
$5x \ge -10$ $7x \le 14$
$x \ge -2$ $x \le 2$
$\{x \mid x \ge -2\}$ $\{x \mid x \le 2\}$
$\{x \mid x \ge -2\} \cup \{x \mid x \le 2\}$ = the set of real numbers

85. $3 \le 7x - 14 \le 31$
$3 + 14 \le 7x - 14 + 14 \le 31 + 14$
$17 \le 7x \le 45$
$$\frac{17}{7} \le \frac{7x}{7} \le \frac{45}{7}$$
$$\frac{17}{7} \le x \le \frac{45}{7}$$
$\{x \mid \frac{17}{7} \le x \le \frac{45}{7}\}$

87. $1 - 3x < 16$ and $1 - 3x > -16$
$-3x < 15$ $-3x > -17$
$x > -5$ $x < \frac{-17}{-3}$
$\{x \mid x > -5\}$ $\{x \mid x < \frac{17}{3}\}$
$\{x \mid x > -5\} \cap \{x \mid x < \frac{17}{3}\} = \{x \mid -5 < x < \frac{17}{3}\}$

89. $6x + 5 < -1$ or $1 - 2x < 7$
$6x < -6$ $-2x < 6$
$x < -1$ $x > -3$
$\{x \mid x < -1\}$ $\{x \mid x > -3\}$
$\{x \mid x < -1\} \cup \{x \mid x > -3\}$ = The set of real numbers.

91. $9 - x \ge 7$ and $9 - 2x < 3$
$-x \ge -2$ $-2x < -6$
$x \le 2$ $x > 3$
$\{x \mid x \le 2\}$ $\{x \mid x > 3\}$
$\{x \mid x \le 2\} \cap \{x \mid x > 3\} = \emptyset$

Objective C Exercises

93. The temperature did not go above 42°F can be written as $t \le 42$.

95. The high temperature was 42°F can be written as $t \le 42$.

97. Strategy: Let x represent the width of the rectangle.
The length of the rectangle is $2x - 5$.
To find the maximum width solve the inequality $2L + 2W < 60$.

Solution:
$$\begin{aligned} 2L + 2W &< 60 \\ 2(2x-5) + 2x &< 60 \\ 4x - 10 + 2x &< 60 \\ 6x - 10 &< 60 \\ 6x &< 70 \\ x &< \frac{70}{6} = 11\frac{2}{3} \end{aligned}$$

The maximum width of the rectangle is 11 cm.

99. Strategy: Let d represent the number of days to run advertisement.
To find the maximum number of days the advertisement can run on the website solve the inequality $250 + 12d \le 1500$.

Solution:
$$\begin{aligned} 250 + 12d &\le 1500 \\ 12d &\le 1250 \\ d &\le \frac{1250}{12} \\ d &\le 104\frac{1}{6} \end{aligned}$$

You can run the advertisement for 104 days.

101. Strategy: Let x represent the cost of a gallon of paint.
Since a gallon of paint covers 100 square feet and the room is 320 square feet the homeowner will need to buy 4 gallons of paint.
To find the maximum cost per gallon solve the inequality $24 + 4x \le 100$.

Solution:
$$\begin{aligned} 24 + 4x &\le 100 \\ 4x &\le 76 \\ x &\le 19 \end{aligned}$$

The maximum that the homeowner can pay for a gallon of paint is \$19.

103. Strategy: To find the temperature range in Fahrenheit degrees solve the compound inequality $0 < \frac{5(F-32)}{9} < 30$.

Solution:
$$0 < \frac{5(F-32)}{9} < 30$$
$$\frac{9}{5}(0) < \frac{9}{5}\left(\frac{5(F-32)}{9}\right) < \frac{9}{5}(30)$$
$$0 < F - 32 < 54$$
$$0 + 32 < F - 32 + 32 < 54 + 32$$
$$32^\circ < F < 86^\circ$$

105. Strategy: Let N represent the amount of sales.
To find the minimum amount of sales solve the inequality $1000 + 0.05N \ge 3200$.

Solution:
$$\begin{aligned} 1000 + 0.05N &\ge 3200 \\ 0.05N &\ge 2200 \\ N &\ge 44{,}000 \end{aligned}$$

George's amount of sales must be \$44,000 or more per month.

107. Let x represent the number of gallons needed in the first month. To find the minimum, solve the inequality $x + (x+400) + (x+800) + (x+1200) + (x+1600) \ge 8500$.

Solution
$$x + x + 400 + (x+800) + (x+1200) + (x+1600) \le 8500$$
$$5x + 4000 \le 8500$$
$$5x + 4000 - 4000 \le 8500 - 4000$$
$$5x \le 4500$$
$$x \le 900$$

The company must make a minimum of 900 gal the first month.

109. Strategy: Let n represent the score on the last test.
To find the range of scores solve the inequality
$70 \le \dfrac{56+91+83+62+n}{5} \le 79.$

Solution:

$$70 \le \frac{56+91+83+62+n}{5} \le 79$$
$$70 \le \frac{292+n}{5} \le 79$$
$$5(70) \le 5 \cdot \frac{292+n}{5} \le 5(79)$$
$$350 \le 292 + n \le 395$$
$$350 - 292 \le 292 - 292 + n \le 395 - 292$$
$$58 \le n \le 103$$

Since 100 is the maximum core, the range of scores needed to receive an C grade is $58 \le n \le 100$.

Critical Thinking

111. a) $a \le 2x + 1 \le b$
$a - 1 \le 2x \le b - 1$
Since $-2 \le x \le 4$ we have $-4 \le 2x \le 6$
$a - 1 \le 2x$
$a - 1 \le -4$
$a \le -3$
The largest possible value of a is -3.

b) $2x \le b - 1$
$6 \le b - 1$
$7 \le b$
The smallest possible value of b is 7.

113. True

115. True

Section 2.6

Concept Check

1. $|2 - 8| = 6$
$|-6| = 6$
$6 = 6$
Yes, 2 is a solution.

3. $|3(-1) - 4| = 7$
$|-3 - 4| = 7$
$|-7| = 7$
$7 = 7$
Yes, -1 is a solution.

5. $|x| = 7$
$x = 7$ or $x = -7$
The solutions are 7 and -7.

7. $|-y| = 6$
$-y = 6$ or $-y = -6$
$y = -6$ or $y = 6$
The solutions are 6 and -6.

9. $|x| = -4$
There is no solution to this equation because the absolute value of a number must be nonnegative.

11. $|-t| = -3$
There is no solution to this equation because the absolute value of a number must be nonnegative.

13. $|x| > 3$
$x > 3$ or $x < -3$
$\{x \mid x > 3\}$ $\{x \mid x < -3\}$
$\{x \mid x > 3\} \cup \{x \mid x < -3\} = \{x \mid x > 3 \text{ or } x < -3\}$

15. $|x - 2| < 5$

Objective A Exercises

17. $|x + 2| = 3$
$x + 2 = 3$ or $x + 2 = -3$
$x = 1$ $x = -5$
The solutions are 1 and -5.

19. $|y-5|=3$
$y-5=3$ or $y-5=-3$
$y=8$ $\quad y=2$
The solutions are 2 and 8.

21. $|a-2|=0$
$a-2=0$
$a=2$
The solution is 2.

23. $|x-2|=-4$
There is no solution to this equation because the absolute value of a number must be nonnegative.

25. $|3-4x|=9$
$3-4x=9$ or $3-4x=-9$
$-4x=6$ $\quad -4x=-12$
$x=-\frac{3}{2}$ $\quad x=3$
The solutions are 3 and $-\frac{3}{2}$.

27. $|2x-3|=0$
$2x-3=0$
$2x=3$
$x=\frac{3}{2}$
The solution is $\frac{3}{2}$.

29. $|3x-2|=-4$
There is no solution to this equation because the absolute value of a number must be nonnegative.

31. $|x-2|-2=3$
$|x-2|=5$
$x-2=5$ or $x-2=-5$
$x=7$ $\quad x=-3$
The solutions are 7 and – 3.

33. $|3a+2|-4=4$
$|3a+2|=8$
$3a+2=8$ or $3a+2=-8$
$3a=6$ $\quad 3a=-10$
$a=2$ $\quad a=-\frac{10}{3}$
The solutions are 2 and $-\frac{10}{3}$.

35. $|2-y|+3=4$
$|2-y|=1$
$2-y=1$ or $2-y=-1$
$-y=-1$ $\quad -y=-3$
$y=1$ $\quad y=3$
The solutions are 1 and 3.

37. $|2x-3|+3=3$
$|2x-3|=0$
$2x-3=0$
$2x=3$
$x=\frac{3}{2}$
The solution is $\frac{3}{2}$.

39. $|2x-3|+4=-4$
$|2x-3|=-8$
There is no solution to this equation because the absolute value of a number must be nonnegative.

41. $|6x-5|-2=4$
$|6x-5|=6$
$6x-5=6$ or $6x-5=-6$
$6x=11$ $\quad 6x=-1$
$x=\frac{11}{6}$ $\quad x=-\frac{1}{6}$
The solutions are $\frac{11}{6}$ and $-\frac{1}{6}$.

43. $|3t+2|+3=4$
$|3t+2|=1$
$3t+2=1$ or $3t+2=-1$
$3t=-1$ $\quad 3t=-3$
$t=-\frac{1}{3}$ $\quad t=-1$
he solutions are $-\frac{1}{3}$ and – 1.

45. $3 - |x - 4| = 5$
$-|x - 4| = 2$
$|x - 4| = -2$
There is no solution to this equation because the absolute value of a number must be nonnegative.

47. $8 - |2x - 3| = 5$
$-|2x - 3| = -3$
$|2x - 3| = 3$
$2x - 3 = 3$ or $2x - 3 = -3$
$2x = 6$ $\quad$ $2x = 0$
$x = 3$ $\quad$ $x = 0$
The solutions are 3 and 0.

49. $|2 - 3x| + 7 = 2$
$|2 - 3x| = -5$
There is no solution to this equation because the absolute value of a number must be nonnegative.

51. $|8 - 3x| - 3 = 2$
$|8 - 3x| = 5$
$8 - 3x = 5$ or $8 - 3x = -5$
$-3x = -3$ $\quad$ $-3x = -13$
$x = 1$ $\quad$ $x = \frac{13}{3}$
The solutions are 1 and $\frac{13}{3}$.

53. $|2x - 8| + 12 = 2$
$|2x - 8| = -10$
There is no solution to this equation because the absolute value of a number must be nonnegative.

55. $2 + |3x - 4| = 5$
$|3x - 4| = 3$
$3x - 4 = 3$ or $3x - 4 = -3$
$3x = 7$ $\quad$ $3x = 1$
$x = \frac{7}{3}$ $\quad$ $x = \frac{1}{3}$
The solutions are $\frac{7}{3}$ and $\frac{1}{3}$.

57. $5 - |2x + 1| = 5$
$-|2x + 1| = 0$
$2x + 1 = 0$
$2x = -1$
$x = -\frac{1}{2}$
The solution is $-\frac{1}{2}$.

59. $6 - |2x + 4| = 3$
$-|2x + 4| = -3$
$|2x + 4| = 3$
$2x + 4 = 3$ or $2x + 4 = -3$
$2x = -1$ $\quad$ $2x = -7$
$x = -\frac{1}{2}$ $\quad$ $x = -\frac{7}{2}$
The solutions are $-\frac{1}{2}$ and $-\frac{7}{2}$.

61. $8 - |1 - 3x| = -1$
$-|1 - 3x| = -9$
$|1 - 3x| = 9$
$1 - 3x = 9$ $\quad$ $1 - 3x = -9$
$-3x = 8$ or $-3x = -10$
$x = -\frac{8}{3}$ $\quad$ $x = \frac{10}{3}$
The solutions are $-\frac{8}{3}$ and $\frac{10}{3}$.

63. $5 + |2 - x| = 3$
$|2 - x| = -2$
There is no solution to this equation because the absolute value of a number must be nonnegative.

65. Two positive solutions.

67. Two negative solutions.

Objective B Exercises

69. $|x+1|>2$

$x+1>2$ or $x+1<-2$

$x>1$ $\quad x<-3$

$\{x \mid x>1\}$ $\quad \{x \mid x<-3\}$

$\{x \mid x>1\} \cup \{x \mid x<-3\} = \{x \mid x>1 \text{ or } x<-3\}$

71. $|x-5| \le 1$

$-1 \le x-5 \le 1$

$-1+5 \le x-5+5 \le 1+5$

$4 \le x \le 6$

$\{x \mid 4 \le x \le 6\}$

73. $|2-x| \ge 3$

$2-x \ge 3$ or $2-x \le -3$

$-x \ge 1$ $\quad -x \le -5$

$x \le -1$ $\quad x \ge 5$

$\{x \mid x \le -1\}$ $\quad \{x \mid x \ge 5\}$

$\{x \mid x \le -1\} \cup \{x \mid x \ge 5\} = \{x \mid x \le -1 \text{ or } x \ge 5\}$

75. $|2x+1|<5$

$-5<2x+1<5$

$-5-1<2x+1-1<5-1$

$-6<2x<4$

$-3<x<2$

$\{x \mid -3<x<2\}$

77. $|5x+2|>12$

$5x+2>12$ or $5x+2<-12$

$5x>10$ $\quad 5x<-14$

$x>2$ $\quad x<-\frac{14}{5}$

$\{x \mid x>2\}$ $\quad \{x \mid x<-\frac{14}{5}\}$

$\{x \mid x>2\} \cup \{x \mid x<-\frac{14}{5}\}$

$=\{x \mid x>2 \text{ or } x<-\frac{14}{5}\}$

79. $|4x-3| \le -2$

The absolute value of a number must be nonnegative. The solution set is the empty set $\emptyset$.

set $\emptyset$.

81. $|2x+7|>-5$

$2x+7>-5$ or $2x+7<5$

$2x>-12$ $\quad 2x<-2$

$x>-6$ $\quad x<-1$

$\{x \mid x>-6\}$ $\quad \{x \mid x<-1\}$

$\{x \mid x>-6\} \cup \{x \mid x<-1\}$= The set of all real numbers.

83. $|4-3x| \ge 5$

$4-3x \ge 5$ or $4-3x \le -5$

$-3x \ge 1$ $\quad -3x \le -9$

$x \le -\frac{1}{3}$ $\quad x \ge 3$

$\{x \mid x \le -\frac{1}{3}\}$ $\quad \{x \mid x \ge 3\}$

$\{x \mid x \le -\frac{1}{3}\} \cup \{x \mid x \ge 3\}$

$=\{x \mid x \le -\frac{1}{3} \text{ or } x \ge 3\}$

85. $|5-4x| \le 13$

$-13 \le 5-4x \le 13$

$-13+(-5) \le 5+(-5)-4x \le 13+(-5)$

$-18 \le -4x \le 8$

$\frac{18}{4} \ge x \ge -2$

$\{x \mid -2 \le x \le \frac{9}{2}\}$

87. $|6-3x| \le 0$

$0 \le 6-3x \le 0$

$-6 \le -3x \le -6$

$2 \le x \le 2$

$2 \le x \le 2 = \{x \mid x=2\}$

89. $|2-9x|>20$

$2-9x>20$ or $2-9x<-20$

$-9x>18$ $\quad -9x<-22$

$x<-2$ $\quad x>\frac{22}{9}$

$\{x \mid x<-2\}$ $\quad \{x \mid x>\frac{22}{9}\}$

$\{x \mid x<-2\} \cup \{x \mid x>\frac{22}{9}\}$

$=\{x \mid x<-2 \text{ or } x>\frac{22}{9}\}$

91. $|2x-3|+2<8$
$|2x-3|<6$
$-6<2x-3<6$
$-6+3<2x-3+3<6+3$
$-3<2x<9$
$-\frac{3}{2}<x<\frac{9}{2}$
$\{x\,|-\frac{3}{2}<x<\frac{9}{2}\}$

93. $|2-5x|-4>-2$
$|2-5x|>2$
$2-5x>2$ or $2-5x<-2$
$-5x>0$ $\quad -5x<-4$
$x<0$ $\quad x>\frac{4}{5}$
$\{x\,|\,x<0\}$ $\quad \{x\,|\,x>\frac{4}{5}\}$
$\{x\,|\,x<0\}\cup\{x\,|x>\frac{4}{5}\}=\{x\,|\,x<0 \text{ or } x>\frac{4}{5}\}$

95. $8-|2x-5|<3$
$-|2x-5|<-5$
$|2x-5|>5$
$2x-5<-5$ or $2x-5>5$
$2x<0$ $\quad 2x>10$
$x<0$ $\quad x>5$
$\{x\,|\,x<0\}$ $\quad \{x\,|\,x>5\}$
$\{x\,|\,x<0\}\cup\{x\,|x>5\}=\{x\,|\,x<0 \text{ or } x>5\}$

97. All negative solutions.

Objective C Exercises

99. The desired dosage is 3 ml. The tolerance is 0.2 ml.

101. Strategy: Let d represent the diameter of the bushing, T the tolerance and x the lower and upper limits of the diameter.
Solve the absolute value inequality $|x-d|\le T$.

Solution: $|x-d|\le T$
$|x-1.75|\le 0.008$
$-0.008\le x-1.75\le 0.008$
$-0.008+1.75\le x-1.75+1.75$
$\le 0.008+1.75$
$1.742\le x\le 1.758$
The lower and upper limits of the diameter of the bushing are 1.742 in. and 1.758 in.

103. Strategy: Let L represent the length of the piston.
Solve the absolute value inequality $|L-9\frac{5}{8}|\le\frac{1}{32}$.

Solution: $|L-9\frac{5}{8}|\le\frac{1}{32}$
$-\frac{1}{32}\le L-9\frac{5}{8}\le\frac{1}{32}$
$-\frac{1}{32}+9\frac{5}{8}\le L-9\frac{5}{8}+9\frac{5}{8}\le\frac{1}{32}+9\frac{5}{8}$
$9\frac{19}{32}\le L\le 9\frac{21}{32}$
The upper and lower limits of the length of the piston are $9\frac{19}{32}$ in. and $9\frac{21}{32}$ in.

105. Strategy: Let x represent the percent of American voters who felt the economy is an important issue.
Solve the absolute value inequality $|x-41|\le 3$.

Solution: $|x-41|\le 3$
$-3\le x-41\le 3$
$-3+41\le x-41+41\le 3+41$
$38\le x\le 44$
The lower and upper limits of American voters who felt the economy is an important issue 38% and 44%.

107. Strategy: Let M represent the range, in ohms, for a resistor.
Let T represent the tolerance of the resistor.
Solve the absolute value inequality $|M - 29{,}000| \le T$.

Solution: $T = (0.02)(29{,}000)$
$= 580$ ohm

$$|M - 29{,}000| \le 580$$
$$-580 \le M - 29{,}000 \le 580$$
$$-580 + 29{,}000 \le M - 29{,}000 + 29{,}000 \le 580 + 29{,}000$$
$$28{,}420 \le M \le 29{,}580$$

The upper and lower limits of the resistor are 28,420 ohms and 29,580 ohms.

Critical Thinking

109. a) The equation $|x + 3| = x + 3$ is true for all x for which $x + 3 \ge 0$.

$$x + 3 \ge 0$$
$$x \ge -3$$
$$\{x \mid x \ge -3\}$$

b) The equation $|a - 4| = 4 - a$ is true for all a for which $4 - a \ge 0$.

$$4 - a \ge 0$$
$$-a \ge -4$$
$$a \le 4$$
$$\{a \mid a \le 4\}$$

111. $-2 \le x \le 2$
$-a \le 3x - 2 \le a,\ a \ge 0$
For $x = 2$ we have $3x - 2 = 4$ and $3x - 2 < 4$ for $-2 \le x \le 2$.
For $3x - 2 \le a$ to be true a must be greater than or equal to 4. The smallest possible value of a is 4.

Projects or Group Activities

113. $|3x - 4| = 2x + 10$

$$3x - 4 = 2x + 10$$
$$3x - 2x - 4 = 2x - 2x + 10$$
$$x - 4 = 10$$
$$x - 4 + 4 = 10 + 4$$
$$x = 14$$

or

$$3x - 4 = -(2x + 10)$$
$$3x - 4 = -2x - 10$$
$$3x + 2x - 4 = -2x + 2x - 10$$
$$5x - 4 = -10$$
$$5x - 4 + 4 = -10 + 4$$
$$5x = -6$$
$$\frac{5x}{5} = \frac{-6}{5}$$
$$x = -\frac{6}{5}$$

The solutions are 14 and $-\frac{6}{5}$.

115. $|3x + 1| = 2x - 5$

$$3x + 1 = 2x - 5$$
$$3x - 2x + 1 = 2x - 2x - 5$$
$$x + 1 = -5$$
$$x + 1 - 1 = -5 - 1$$
$$x = -6$$

or

$$3x + 1 = -(2x - 5)$$
$$3x + 1 = -2x + 5$$
$$3x + 2x + 1 = -2x + 2x + 5$$
$$5x + 1 = 5$$
$$5x + 1 - 1 = 5 - 1$$
$$5x = 4$$
$$\frac{5x}{5} = \frac{4}{5}$$
$$x = \frac{4}{5}$$

Since $2x - 5 = 2(-6) - 5 = -12 - 5 = -17$

and $2x - 5 = 2\left(-\frac{6}{5}\right) - 5 = -\frac{12}{5} - \frac{25}{5} = -\frac{37}{5}$,

there is no solution.

Chapter 2 Review Exercises

1. $x+3=24$
$x=24-3$
$x=21$
The solution is 21.

2. $x+5(3x-20)=10(x-4)$
$x+15x-100=10x-40$
$16x-100=10x-40$
$6x=60$
$\frac{6x}{6}=\frac{60}{6}$
$x=10$
The solution is 10.

3. $5x-6=29$
$5x=29+6$
$5x=35$
$\frac{5x}{5}=\frac{35}{5}$
$x=7$
The solution is 7.

4. $5x-2=4x+5$

$5(3)-2$	$4(3)+5$
$15-2$	$12+5$

$13\neq 17$
No, 3 is not a solution.

5. $\frac{3}{5}a=12$
$a=12\cdot\frac{5}{3}$
$a=20$
The solution is 20.

6. $3x-7>-2$
$3x-7+7>-2+7$
$3x>5$
$\frac{3x}{3}>\frac{5}{3}$
$x>\frac{5}{3}$
The solution is $\left(\frac{5}{3},\infty\right)$.

7. $P(12)=30$
$\frac{P(12)}{12}=\frac{30}{12}$
$P=2.5$
The percent is 250%.

8. $5x+3=10x-17$
$3+17=10x-5x$
$20=5x$
$4=x$
The solution is 4.

9. $7-[4+2(x-3)]=11(x+2)$
$7-[4+2x-6]=11x+22$
$7-[-2+2x]=11x+22$
$7+2-2x=11x+22$
$9-2x=11x+22$
$9-22=11x+2x$
$-13=13x$
$-1=x$
The solution is -1.

10. $6+|3x-3|=2$
$6-6+|3x-3|=2-6$
$|3x-3|=-4$
There is no solution to this equation because the absolute value of a number must be non-negative.

11. $|2x-5|<3$
$-3<2x-5<3$
$-3+5<2x-5+5<3+5$
$2<2x<8$
$\frac{2}{2}<\frac{2x}{2}<\frac{8}{2}$
$1<x<4$
The solution set is $\{x\,|\,1<x<4\}$.

12. $3x<4$
$\frac{3x}{3}<\frac{4}{3}$
$x<\frac{4}{3}$ and

$x+2>-1$
$x+2-2>-1-2$
$x>-3$

$\left\{x\,|\,x<\frac{4}{3}\right\}\cup\{x\,|\,x>-3\}=\left\{x\,|\,-3<x<\frac{4}{3}\right\}$

The solution is set is $\left\{x\,|\,-3<x<\frac{4}{3}\right\}$.

13.

$$\begin{aligned} 3x-2 &> x-4 \\ 3x-x-2 &> x-x-4 \\ 2x-2 &> -4 \\ 2x-2+2 &> -4+2 \\ 2x &> -2 \\ \frac{2x}{2} &> \frac{-2}{2} \\ x &> -1 \end{aligned} \quad \text{or} \quad \begin{aligned} 7x-5 &< 3x+3 \\ 7x-3x-5 &< 3x-3x+3 \\ 4x-5 &< 3 \\ 4x-5+5 &< 3+5 \\ 4x &< 8 \\ \frac{4x}{4} &< \frac{8}{4} \\ x &< 2 \end{aligned}$$

$\{x \mid x > -1\} \cup \{x \mid x < 2\}$
$= \{x \mid x \text{ is any real number}\}$

The interval is $(-\infty, \infty)$.

14.

$$\begin{aligned} 4x-5 &\ge 3 \\ 4x-5+5 &\ge 3+5 \\ 4x &\ge 8 \\ \frac{4x}{4} &\ge \frac{8}{4} \\ x &\ge 2 \end{aligned} \quad \text{and} \quad \begin{aligned} 4x-5 &\le -3 \\ 4x-5+5 &\le -3+5 \\ 4x &\le 2 \\ \frac{4x}{4} &\le \frac{2}{4} \\ x &\le \frac{1}{2} \end{aligned}$$

The solution set is $\{x \mid x \ge 2\} \cup \left\{x \mid x \le \frac{1}{2}\right\}$.

15.

$$\begin{aligned} 3y-5 &= 3-2y \\ 3y+2y-5 &= 3-2y+2y \\ 5y-5 &= 3 \\ 5y-5+5 &= 3+5 \\ 5y &= 8 \\ \frac{5y}{5} &= \frac{8}{5} \\ y &= \frac{8}{5} \end{aligned}$$

The solution is $\frac{8}{5}$.

16.

$$\begin{aligned} 4x-5+x &= 6x-8 \\ 5x-5 &= 6x-8 \\ 5x-6x-5 &= 6x-6x-8 \\ -x-5 &= -8 \\ -x-5+5 &= -8+5 \\ -x &= -3 \\ \frac{-x}{-1} &= \frac{-3}{-1} \\ x &= 3 \end{aligned}$$

The solution is 3.

17.

$$\begin{aligned} 3(x-4) &= -5(6-x) \\ 3x-12 &= -30+5x \\ 3x-5x-12 &= -30+5x-5x \\ -2x-12 &= -30 \\ -2x-12+12 &= -30+12 \\ -2x &= -18 \\ \frac{-2x}{-2} &= \frac{-18}{-2} \\ x &= 9 \end{aligned}$$

The solution is 9.

18.

$$\begin{aligned} \frac{3x-2}{4}+1 &= \frac{2x-3}{2} \\ 4\left(\frac{3x-2}{4}+1\right) &= 4\left(\frac{2x-3}{2}\right) \\ 3x-2+4 &= 4x-6 \\ 3x+2 &= 4x-6 \\ 3x-4x+2 &= 4x-4x-6 \\ -x+2 &= -6 \\ -x+2-2 &= -6-2 \\ -x &= -8 \\ \frac{-x}{-1} &= \frac{-8}{-1} \\ x &= 8 \end{aligned}$$

The solution is 8.

19.

$$\begin{aligned} |5x+8| &= 0 \\ 5x+8 &= 0 \\ 5x+8-8 &= 0-8 \\ 5x &= -8 \\ \frac{5x}{5} &= \frac{-8}{5} \\ x &= -\frac{8}{5} \end{aligned}$$

The solution is $-\frac{8}{5}$.

20. $|5x-4| < -2$

There is no solution to this equation because the absolute value of a number must be non-negative.

21. Strategy Given: $F_1 = 120, x = 2,$
$d - x = 12 - 2 = 10$
Unknown: F_2

Solution

$$F_1 x = F_2(d - x)$$
$$120(2) = F_2(10)$$
$$240 = 10F_2$$
$$24 = F_2$$

The force is 24 lb.

22. Strategy

- Speed on winding road: r
- Speed on level road: $r + 20$

	Rate	Time	Distance
Winding road	r	3	$3r$
Level road	$r + 20$	2	$2(r + 20)$

- The total trip was 200 mi.

Solution

$$3r + 2(r + 20) = 200$$
$$3r + 2r + 40 = 200$$
$$5r + 40 = 200$$
$$5r = 160$$
$$r = 32$$

The average speed on the winding road was 32 mph.

23. Strategy

- Amount of cranberry juice: x
- Amount of apple juice: $10 - x$

	Amount	Cost	Value
Cranberry juice	x	1.79	$1.79(x)$
Apple juice	$10 - x$	1.19	$1.19(10 - x)$
Mixture	10	1.61	$1.61(10)$

- The sum of the quantities before mixing is equal to the quantity after mixing.

Solution

$$1.79x + 1.19(10 - x) = 1.61(10)$$
$$1.79x + 11.90 - 1.19x = 16.10$$
$$0.60x = 4.2$$
$$x = 7$$
$$10 - x = 10 - 3 = 7$$

The amount of cranberry juice was 7 qt. The amount of apple juice was 3 qt.

24. Strategy • First integer: n
- Second integer: $n + 1$
- Third integer: $n + 2$
- Four times the second integer equals the sum of the first and third integer.

Solution

$$4(n + 1) = n + n + 2$$
$$4n + 4 = 2n + 2$$
$$2n = -2$$
$$n = -1$$

The integers are −1, 0, and 1.

25. The unknown number is x.

Four less than five times a number	is	sixteen

$$5x - 4 = 16$$
$$5x - 4 + 4 = 16 + 4$$
$$5x = 20$$
$$\frac{5x}{5} = \frac{20}{5}$$
$$x = 4$$

The number is 4.

26. The height of the Eiffel Tower: x

1472	is	654 less than twice the height of the Eiffel Tower

$$1472 = 2x - 654$$
$$2126 = 2x$$
$$1063 = x$$

The Eiffel Tower is 1063 feet tall.

27. Strategy

- Time for jet plane: t
- Time for propeller-driven plane: $t + 2$

	Rate	Time	Distance
Jet	600	t	$600t$
Propeller	200	$t + 2$	$200(t + 2)$

- The two traveled the same distance.

Solution

$$600t = 200(t + 2)$$
$$600t = 200t + 400$$
$$400t = 400$$
$$t = 1$$

Distance = 600t = 600(1) = 600

The jet overtakes the propeller-driven plane 600 mi from the starting point.

28. **Strategy** • Let b represent the diameter of the bushing, T the tolerance, and d the lower and upper limits of diameter. Solve the absolute value inequality $|d-b| \le T$ for d.

Solution

$$|d-b| \le T$$
$$-0.003 < d - 2.75 \le 0.003$$
$$-0.003 < d - 2.75 \le 0.003$$
$$-0.003 + 2.75 \le d - 2.75 + 2.75 \le 0.003 + 2.75$$
$$2.747 \le d \le 2.753$$

The lower limit of the bushing is 2.747 in. and the upper limit is 2.753 in.

29. **Strategy**

- Amount of butter fat in the mixture: x

	Amount	Percent	Quantity
Cream	5	0.30	0.3(5)
Milk	8	0.04	0.04(8)
Mixture	13	x	$13x$

- The sum of the quantities before mixing is equal to the quantity after mixing.

Solution

$$0.30(5) + 0.04(8) = 13x$$
$$1.5 + 0.32 = 13x$$
$$1.82 = 13x$$
$$0.14 = x$$

The mixture is 14% butterfat.

30. **Strategy**

- Time to island: t
- Time to return: $2\frac{1}{3} - t = \frac{7}{3} - t$

	Rate	Time	Distance
To island	16	t	$16t$
Return	12	$\frac{7}{3} - t$	$12\left(\frac{7}{3} - t\right)$

- The distance to the island equals the distance to return.

Solution

$$16t = 12\left(\frac{7}{3} - t\right)$$
$$16t = 28 - 12t$$
$$16t + 12t = 28 - 12t + 12t$$
$$28t = 28$$
$$\frac{28t}{28} = \frac{28}{28}$$
$$t = 1$$
$$16t = 16(1) = 16$$

The distance from the island to the dock is 16 mi.

Chapter 2 Test

1.

$$3x - 2 = 5x + 8$$
$$3x - 3x - 2 = 5x - 3x + 8$$
$$-2 - 8 = 2x + 8 - 8$$
$$-10 = 2x$$
$$\frac{-10}{2} = \frac{2x}{2}$$
$$-5 = x$$

The solution is −5.

2.

$$x - 3 = -8$$
$$x - 3 + 3 = -8 + 3$$
$$x = -5$$

The solution is –5.

3.

$$3x - 5 = -14$$
$$3x - 5 + 5 = -14 + 5$$
$$3x = -9$$
$$\frac{3x}{3} = \frac{-9}{3}$$
$$x = -3$$

The solution is –3

4.
$$\begin{aligned}4-2(3-2x)&=2(5-x)\\4-6+4x&=10-2x\\-2+4x&=10-2x\\-2+2+4x+2x&=10+2-2x+2x\\6x&=12\\\frac{6x}{6}&=\frac{12}{6}\\x&=2\end{aligned}$$
The solution is 2.

5.
$$x^2-3x=2x-6$$

$(-2)^2-3(-2)$	$2(-2)-6$
$4-3(-2)$	$-4-6$
$4+6$	-10

$$10\neq-10$$
No, –2 is not a solution.

6.
$$\begin{aligned}7-4x&=-13\\7-7-4x&=-13-7\\-4x&=-20\\\frac{-4x}{-4}&=\frac{-20}{-4}\\x&=5\end{aligned}$$
The solution is 5.

7.
$$\begin{aligned}P\cdot B&=A\\0.005(8)&=A\\0.04&=A\end{aligned}$$
0.5% of 8 is 0.04.

8.
$$\begin{aligned}5x-2(4x-3)&=6x+9\\5x-8x+6&=6x+9\\-3x+6&=6x+9\\-3x+3x+6-9&=6x+3x+9-9\\-3&=9x\\\frac{-3}{9}&=\frac{9x}{9}\\-\frac{1}{3}&=x\end{aligned}$$
The solution is $-\frac{1}{3}$.

9.
$$\begin{aligned}5x+3-7x&=2x-5\\-2x+3&=2x-5\\-2x+2x+3+5&=2x+2x-5+5\\8&=4x\\\frac{8}{4}&=\frac{4x}{4}\\2&=x\end{aligned}$$
The solution is 2.

10.
$$\begin{aligned}\frac{3}{4}x&=-9\\\frac{4}{3}\left(\frac{3}{4}x\right)&=-9\left(\frac{4}{3}\right)\\x&=-12\end{aligned}$$
The solution is –12.

11.
$$\begin{aligned}4-3(x+2)&<2(2x+3)-1\\4-3x-6&<4x+6-1\\-2-3x&>4x+5\\-2+2-3x-4x&<4x-4x+5+2\\-7x&<7\\\frac{-7x}{-7}&>\frac{7}{-7}\\x&>-1\end{aligned}$$
The solution is $(-1,\infty)$.

12.
$$\begin{aligned}4x-1&>5\\4x-1+1&>5+1\\4x&>6\\\frac{4x}{4}&>\frac{6}{4}\\x&>\frac{3}{2}\end{aligned}\qquad\text{or}\qquad\begin{aligned}2-3x&<8\\2-2-3x&<8-2\\-3x&<6\\\frac{-3x}{-3}&>\frac{6}{-3}\\x&>-2\end{aligned}$$
$$\left\{x\mid x>\frac{3}{2}\right\}\cup\{x\mid x>-2\}=\{x\mid x>-2\}$$
The solution set is $\{x\mid x>-2\}$.

13.
$$\begin{aligned}4-3x&\geq7\\4-4-3x&\geq7-4\\-3x&\geq3\\\frac{-3x}{-3}&\leq\frac{3}{-3}\\x&\leq-1\end{aligned}\qquad\text{and}\qquad\begin{aligned}2x+3&\geq7\\2x+3-3&\geq7-3\\2x&\geq4\\\frac{2x}{2}&\geq\frac{4}{2}\\x&\geq2\end{aligned}$$
$$\{x\mid x\leq-1\}\cap\{x\mid x\geq2\}=\varnothing$$
There is no solution.

14. $|3-5x|=12$

$$3-5x=12 \qquad 3-5x=-12$$
$$3-3-5x=12-3 \qquad 3-3-5x=-12-3$$
$$-5x=9 \qquad -5x=-15$$
$$\frac{-5x}{-5}=\frac{9}{-5} \qquad \frac{-5x}{-5}=\frac{-15}{-5}$$
$$x=-\frac{9}{5} \qquad x=3$$

The solutions are $-\frac{9}{5}$ and 3.

15.
$$2-|2x-5|=-7$$
$$2-2-|2x-5|=-7-2$$
$$-|2x-5|=-9$$
$$\frac{-|2x-5|}{-1}=\frac{-9}{-1}$$
$$|2x-5|=9$$

$$2x-5=9 \qquad 2x-5=-9$$
$$2x-5+5=9+5 \qquad 2x-5+5=-9+5$$
$$2x=14 \qquad 2x=-4$$
$$\frac{2x}{2}=\frac{14}{2} \qquad \frac{2x}{2}=\frac{-4}{2}$$
$$x=7 \qquad x=-2$$

The solutions are –2 and 7.

16. $|3x-5|\le 4$
$$-4\le 3x-5\le 4$$
$$-4+5\le 3x-5+5\le 4+5$$
$$1\le 3x\le 9$$
$$\frac{1}{3}\le\frac{3x}{3}\le\frac{9}{3}$$
$$\frac{1}{3}\le x\le 3$$

The solution set is $\left\{x\,|\,\frac{1}{3}\le x\le 3\right\}$.

17. Strategy • Amount rye flour: x

• Amount wheat flour: $15-x$

	Amount	Percent	Quantity
Rye	x	0.70	$0.70(x)$
Wheat	$15-x$	0.40	$0.40(15-x)$
Mixture	15	0.60	$15(0.60)$

• The sum of the quantities before mixing is equal to the quantity after mixing.

Solution
$$0.70x+0.40(15-x)=0.60(5)$$
$$0.70x+6-0.40x=9$$
$$0.30x+6=9$$
$$0.30x=3$$
$$x=10$$
$$15-x=15-10=5$$
The amount of rye flour is 10 lb.
The amount of wheat flour is 5 lb.

18. Strategy • First even integer: n

• Second even integer: $n+2$

• Third even integer: $n+4$

• The sum of the integers is 36.

Solution
$$n+n+2+n+4=36$$
$$3n+6=36$$
$$3n+6-6=36-6$$
$$3n=30$$
$$\frac{3n}{3}=\frac{30}{3}$$
$$n=10$$
$$n+2=10+2=12$$
$$n+4=10+4=14$$
The integers are 10, 12, and 14.

19. Strategy • Amount pure water: x

	Amount	Percent	Quantity
Water	x	0.00	$0.00(x)$
20% solution	5	0.20	$0.20(5)$
Mixture	$x+5$	0.16	$0.16(x+5)$

• The sum of the quantities before mixing is equal to the quantity after mixing.

Solution

$$0.00x + 0.20(5) = 0.16(x+5)$$
$$1 = 0.16x + 0.8$$
$$0.2 = 0.16x$$
$$1.25 = x$$

1.25 gal of water must be added.

20. The number: x

The three times the number: $3x$

The difference between 3 times the number and 15	is	27

$$3x - 15 = 27$$
$$3x - 15 + 15 = 27 + 15$$
$$3x = 42$$
$$\frac{3x}{3} = \frac{42}{3}$$
$$x = 14$$

The number is 14.

21. Strategy

• Rate of the skier: x

• Rate of the snowmobile: $x + 4$

	Rate	Time	Distance
Skier	x	3	$3x$
Snowmobile	$x+4$	1	$1(x+4)$

• The two traveled the same distance.

Solution

$$3x = x + 4$$
$$2x = 4$$
$$x = 2$$
$$x + 4 = 2 + 4 = 6$$

The rate of the snowmobile is 6 mph.

22. Strategy Write and solve an equation letting x represent the number of LCD flat panel TVs and $140 - x$ represent the LCD rear projection TVs.

Solution

$$3(140 - x) = x - 20$$
$$420 - 3x = x - 20$$
$$440 = 4x$$
$$110 = x$$

The company makes 110 LCD flat panel TVs each day.

23. Strategy • The smaller number: x

• The larger number: $18 - x$

Solution

$$4x - 7 = 2(18 - x) + 5$$
$$4x - 7 = 36 - 2x + 5$$
$$4x - 7 = 41 - 2x$$
$$6x = 48$$
$$x = 8$$
$$18 - x = 18 - 8 = 10$$

The smaller number is 8.

The larger number is 10.

24. Strategy

• Time for flight out: t

• Time for flight in: $7 - t$

	Rate	Time	Distance
Flight out	90	t	$90t$
Flight in	120	$7-t$	$120(7-t)$

• The distance traveled is the same.

Solution

$$90t = 120(7 - t)$$
$$90t = 840 - 120t$$
$$210t = 840$$
$$t = 4$$

Distance $= 90t = 90(4) = 360$

The distance to the airport is 360 mi.

25. **Strategy** Given: $m_1 = 100$, $T_1 = 80$, $m_2 = 50$, and $T_2 = 20$
Unknown: T

Solution

$$\begin{aligned} m_1(T_1 - T) &= m_2(T - T_2) \\ 100(80 - T) &= 50(T - 20) \\ 8000 - 100T &= 50T - 1000 \\ -150T &= -9000 \\ T &= 60 \end{aligned}$$

The final temperature is 60° C.

26. **Strategy** To find the number of miles, write and solve an inequality using N to represent the number of miles.

Solution
cost of Gambelli < cost of McDougal

$$\begin{aligned} 40 + 0.25N &< 58 \\ 40 - 40 + 0.25N &< 58 - 40 \\ 0.25N &< 18 \\ \frac{0.18N}{0.18} &< \frac{18}{0.25} \\ N &< 72 \end{aligned}$$

Gambelli will cost less if you drive less the 72 mi.

27. **Strategy** • Let b represent the diameter of the bushing, T the tolerance, and d the lower and upper limits of diameter. Solve the absolute value inequality $|d - b| \le T$ for d.

Solution

$$\begin{aligned} |d - b| &\le T \\ |d - 2.65| &< 0.002 \\ -0.002 < d - 2.65 &\le 0.002 \\ -0.002 + 2.65 \le d - 2.65 + 2.65 &\le 0.002 + 2.65 \\ 2.648 \le d &\le 2.652 \end{aligned}$$

The lower limit of the bushing is 2.648 in. and the upper limit is 2.652 in.

Cumulative Review Exercises

1. $-6 - (-20) - 8 = -6 + 20 - 8 = 14 - 8 = 6$

2. $(-2)(-6)(-4) = 12(-4) = -48$

3. $-\frac{5}{6} - \left(-\frac{7}{16}\right) = -\frac{40}{48} - \left(-\frac{21}{48}\right) = \frac{-40 - (-21)}{48} = \frac{-40 + 21}{48}$
$= -\frac{19}{48}$

4. $-\frac{7}{3} \div \frac{7}{6} = -\frac{7}{3} \cdot \frac{6}{7} = -\frac{7 \cdot 6}{3 \cdot 7} = -2$

5. $-4^2 \cdot \left(-\frac{3}{2}\right)^3 = -(4)(4)\left(-\frac{3}{2}\right)\left(-\frac{3}{2}\right)\left(-\frac{3}{2}\right)$
$= -16\left(-\frac{27}{8}\right) = 54$

6. $25 - 3\frac{(5-2)^2}{2^3 + 1} + 2 = 25 - 3\frac{(3)^2}{8 + 1} + 2$
$= 25 - 3\frac{9}{9} + 2$
$= 25 - 3 + 2$
$= 22 + 2 = 24$

7. $3(a - c) - 2ab$
$= 3[2 - (-4)] - 2(2)(3) = 3[2 + 4] - 2(2)(3)$
$= 3[6] - 2(2)(3) = 18 - 2(2)(3) = 18 - 4(3)$
$= 18 - 12 = 6$

8. $3x - 8x + (-12x) = -5x + (-12x)$
$= -5x - 12x = -17x$

9. $2a - (-b) - 7a - 5b = 2a + b - 7a - 5b$
$= (2a - 7a) + (b - 5b)$
$= -5a + (-4b) = -5a - 4b$

10. $(16x)\left(\frac{1}{8}\right) = \frac{1}{8}(16x) = \left(\frac{1}{8} \cdot 16\right)x = 2x$

11. $-4(-9y) = 4(9y) = (4 \cdot 9)y = 36y$

12. $-2(-x^2 - 3x + 2) = -2(-x^2) + (-2)(-3x) + (-2)(2)$
$= 2x^2 + 6x - 4$

13. $-3[2x - 4(x - 3)] + 2 = -3[2x - 4x + 12] + 2$
$= -3[-2x + 12] + 2 = 6x - 36 + 2$
$= 6x - 34$

14. $A \cap B = \{-4, -2, 0, 2\} \cap \{-4, 0, 4, 8\} = \{-4, 0\}$

15. $\{x \mid x < 3\} \cap \{x \mid x > -2\}$

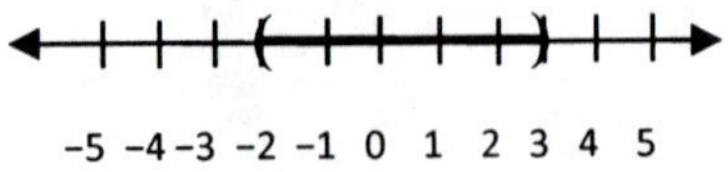

16.

x^2+6x+9	$=$	$x+3$
$(-3)^2+6(-3)+9$		$-3+3$
$9-18+9$		0
$-9+9$		0
	$0=0$	

Yes, –3 is a solution.

17. Percent · Base = Amount

$$32\% \cdot 60 = A$$
$$0.32 \cdot 60 = A$$
$$19.2 = A$$

32% of 60 is 19.2

18.

$$\frac{3}{5}x = -15$$
$$\frac{5}{3}\cdot\frac{3}{5}x = \frac{5}{3}\cdot(-15)$$
$$1\cdot x = -25$$
$$x = -25$$

The solution is −25.

19.

$$7x - 8 = -29$$
$$7x - 8 + 8 = -29 + 8$$
$$7x = -21$$
$$\frac{7x}{7} = \frac{-21}{7}$$
$$x = -3$$

The solution is −3.

20.

$$13 - 9x = -14$$
$$13 - 13 - 9x = -14 - 13$$
$$-9x = -27$$
$$\frac{-9x}{-9} = \frac{-27}{-9}$$
$$x = 3$$

The solution is 3.

21.

$$8x - 3(4x-5) = -2x - 11$$
$$8x - 12x + 15 = -2x - 11$$
$$-4x + 15 = -2x - 11$$
$$-2x = -26$$
$$x = 13$$

The solution is 13.

22. Percent · Base = Amount

$$25\% \cdot B = 30$$
$$0.25B = 30$$
$$\frac{0.25B}{0.25} = \frac{30}{0.25}$$
$$B = 120$$

25% of 120 is 30.

23.

$$5x - 8 = 12x + 13$$
$$5x - 12x - 8 = 5x - 5x + 13$$
$$-7x - 8 = 13$$
$$-7x - 8 + 8 = 13 + 8$$
$$-7x = 21$$
$$\frac{-7x}{-7} = \frac{21}{-7}$$
$$x = -3$$

The solution is –3.

24.

$$11 - 4x = 2x + 8$$
$$11 - 4x - 2x = 2x - 2x + 8$$
$$11 - 6x = 8$$
$$11 - 11 - 6x = 8 - 11$$
$$-6x = -3$$
$$\frac{-6x}{-6} = \frac{-3}{-6}$$
$$x = \frac{1}{2}$$

The solution is $\frac{1}{2}$.

25.

$$3 - 2(2x-1) \ge 3(2x-2) + 1$$
$$3 - 4x + 2 \ge 6x - 6 + 1$$
$$-4x + 5 \ge 6x - 5$$
$$-4x - 6x + 5 \ge -5$$
$$-10x + 5 \ge -5$$
$$-10x + 5 - 5 \ge -5 - 5$$
$$-10x \ge -10$$
$$\frac{-10x}{-10} \le \frac{-10}{-10}$$
$$x \le 1$$

The solution set is $\{x \mid x \le 1\}$.

26.

$$3x + 2 \le 5$$
$$3x + 2 - 2 \le 5 - 2$$
$$3x \le 3$$
$$\frac{3x}{3} \le \frac{3}{3}$$
$$x \le 1$$

and

$$x + 5 \ge 1$$
$$x + 5 - 5 \ge 1 - 5$$
$$x \ge -4$$

$$\{x \mid x \le 1\} \cap \{x \mid x \ge -4\} = \{x \mid -4 \le x \le 1\}$$

The solution set is $\{x \mid -4 \le x \le 1\}$.

27. $|3-2x|=5$

$$3-2x=5 \qquad 3-2x=-5$$
$$3-3-2x=5-3 \qquad 3-3-2x=-5-3$$
$$-2x=2 \qquad -2x=-8$$
$$\frac{-2x}{-2}=\frac{2}{-2} \qquad \frac{-2x}{-2}=\frac{-8}{-2}$$
$$x=-1 \qquad x=4$$

The solutions are –1 and 4.

28. $|3x-1|>5$

$$3x-1<-5 \qquad 3x-1>5$$
$$3x-1+1<-5+1 \qquad 3x-1+1>5+1$$
$$3x<-4 \quad \text{or} \quad 3x>6$$
$$\frac{3x}{3}<\frac{-4}{3} \qquad \frac{3x}{3}>\frac{6}{3}$$
$$x<-\frac{4}{3} \qquad x>2$$

$$\left\{x \mid x<-\frac{4}{3}\right\} \cup \{x \mid x>2\} = \left\{x \mid x>2 \text{ or } x<-\frac{4}{3}\right\}$$

The solution set is $\left\{x \mid x>2 \text{ or } x<-\frac{4}{3}\right\}$.

27. $55\% = 55\left(\frac{1}{100}\right)=\frac{55}{100}=\frac{11}{20}$

30. $1.03 = 1.03(100\%) = 103\%$

31. **Strategy** Given: $m_1 = 300$, $T_1 = 750$, $m_2 = 100$, and $T_2 = 15$

Unknown: T

Solution

$$m_1(T_1-T)=m_2(T-T_2)$$
$$300(75-T)=100(T-15)$$
$$22{,}500-300T=100T-1500$$
$$-400T=-24{,}000$$
$$T=60$$

The final temperature is 60° C.

32. The unknown number: x

The difference between 12and the product of 3 and a number	is	–18

$$12-5x=-18$$
$$-5x=-30$$
$$x=6$$

The number is 6.

33. To find the area of the garage, let x = the area.

200 ft² more than three times the area of the garage	is	2000 ft²

$$3x+200=2000$$
$$3x=1800$$
$$x=600$$

The area of th4e garage is 600 ft².

34. **Strategy** • Amount of oat flour: x

	Amount	Cost	Quantity
Oat	x	0.80	$0.80x$
Wheat	40	0.50	0.50(40)
Mixture	$x+40$	0.60	$0.60(x+40)$

Solution

$$0.80x+0.50(40)=0.60(x+40)$$
$$0.80x+20=0.60x+24$$
$$0.20x=4$$
$$x=20$$

20 lb of oat flour are needed for the mixture.

35. **Strategy** • Amount pure gold: x

	Amount	Percent	Quantity
Pure gold	x	1.00	$1.00x$
Alloy	100	0.20	0.20(100)
Mixture	$x+100$	0.36	$0.36(x+100)$

• The sum of the quantities before mixing is equal to the quantity after mixing.

Solution

$$1.00x+0.20(100)=0.36(x+100)$$
$$1.00x+20=0.36x+36$$
$$0.64x=16$$
$$x=25$$

25 g of pure gold must be added.

36. Strategy

- Time running: t
- Time jogging: $55 - t$

	Rate	Time	Distance
Running	8	t	$8t$
Jogging	3	$55 - t$	$3(55 - t)$

- The distance traveled is the same.

Solution

$$\begin{aligned} 8t &= 3(55 - t) \\ 8t &= 165 - 3t \\ 11t &= 165 \\ t &= 15 \end{aligned}$$

Distance = 8t = 8(15) = 120

The length of the track is 120 m.

Chapter 3: Geometry

Prep Test

1. $2(18)+2(10)=36+20=56$

2. $x+47=90$
$x=43$

3. $32+97+x=180$
$129+x=180$
$x=51$

4. abc
$(2)(3.14)(9)=6.28(9)=56.52$

5. xyz^3
$\left(\frac{4}{3}\right)(3.14)(3^3)=\frac{4}{3}(3.14)27=113.04$

6. $\frac{1}{2}a(b+c)$
$=\frac{1}{2}(6)(25+15)=\frac{1}{2}(6)(40)=3(40)=120$

Section 3.1

Concept Check

1. 12; 5; x; 4

3. 160°; 140°; 360°

5. a; b

7. c; d; 180°

9. **a.** $\angle a$, $\angle b$, and $\angle c$

b. $\angle y$ and $\angle z$

c. $\angle x$

Objective A Exercises

11. The measure of the given angle is 40°. The measure of the angle is between 0° and 90°, so the angle is acute.

13. The measure of the given angle is 115°. The measure of the angle is between 90° and 180°, so the angle is obtuse.

15. The measure of the given angle is 90°. The angle is right.

17. **Strategy** Complementary angles are two angles whose sum is 90°. To find the complement, let x represent the complement of a 62° angle. Write an equation and solve for x.

Solution
$x+62°=90°$
$x=28°$
The complement of a 62° angle is a 28° angle.

19. **Strategy** Supplementary angles are two angles whose sum is 180°. To find the supplement, let x represent the supplement of a 162° angle. Write an equation and solve for x.

Solution
$x+162°=180°$
$x=18°$
The supplement of a 162° angle is an 18° angle.

21. $AB+BC+CD=AD$
$12+BC+9=35$
$21+BC=35$
$BC=14$
$BC=14$ cm

23. $QR+RS=QS$
$QR+3(QR)=QS$
$7+3\cdot 7=QS$
$7+21=QS$
$28=QS$
$QS=28$ ft

25. $EF+FG=EG$
$EF+\frac{1}{2}(EF)=EG$
$20+\frac{1}{2}(20)=EG$
$20+10=EG$
$30=EG$
$EG=30$ m

27. $\angle LOM+\angle MON=\angle LON$
$53°+\angle MON=139°$
$\angle MON=139°-53°=86°$
The measure of $\angle MON$ is 86°.

29. **Strategy** To find the measure of $\angle x$,

write an equation using the fact that the sum of the measures of $\angle x$ and $\angle 2x$ is 90°. Solve for $\angle x$.

Solution

$$\begin{aligned} x + 2x &= 90^\circ \\ 3x &= 90^\circ \\ x &= 30^\circ \end{aligned}$$

The measure of $\angle x$ is 30°.

31. **Strategy** To find the measure of $\angle x$, write an equation using the fact that the sum of x and $x + 18^\circ$ is 90°. Solve for x.

Solution

$$\begin{aligned} x + x + 18^\circ &= 90^\circ \\ 2x + 18^\circ &= 90^\circ \\ 2x &= 72^\circ \\ x &= 36^\circ \end{aligned}$$

The measure of $\angle x$ is 36°.

33. **Strategy** To find the measure of $\angle a$, write an equation using the fact that the sum of $\angle a$ and 74° is 145°. Solve for $\angle a$.

Solution

$$\begin{aligned} \angle a + 74^\circ &= 145^\circ \\ \angle a &= 71^\circ \end{aligned}$$

The measure of $\angle a$ is 71°.

35. **Strategy** To find the measure of $\angle a$, write an equation using the fact that the sum of $\angle a$ and 53° is 180°. Solve for $\angle a$.

Solution

$$\begin{aligned} \angle a + 53^\circ &= 180^\circ \\ \angle a &= 127^\circ \end{aligned}$$

The measure of $\angle a$ is 127°.

37. **Strategy** To find the measure of $\angle a$, write an equation using the fact that the sum of $\angle a$ and 76° and 168° is 360°. Solve for $\angle a$.

Solution

$$\begin{aligned} \angle a + 76^\circ + 168 &= 360^\circ \\ \angle a + 244^\circ &= 360^\circ \\ \angle a &= 116^\circ \end{aligned}$$

The measure of $\angle a$ is 116°.

39. **Strategy** The sum of the measures of the three angles shown is 180°. To find x, write an equation and solve for x.

Solution

$$\begin{aligned} 3x + 4x + 2x &= 180^\circ \\ 9x &= 180^\circ \\ x &= 20^\circ \end{aligned}$$

The measure of x is 20°.

41. **Strategy** The sum of the measures of the three angles shown is 180°. To find x, write an equation and solve for x.

Solution

$$\begin{aligned} 5x + (x + 20^\circ) + 2x &= 180^\circ \\ 8x + 20^\circ &= 180^\circ \\ 8x &= 160^\circ \\ x &= 20^\circ \end{aligned}$$

The measure of x is 20°.

43. **Strategy** The sum of the measures of the four angles shown is 360°. To find x, write an equation and solve for x.

Solution

$$\begin{aligned} 3x + 4x + 6x + 5x &= 360^\circ \\ 18x &= 360^\circ \\ x &= 20^\circ \end{aligned}$$

The measure of x is 20°.

45. **Strategy**

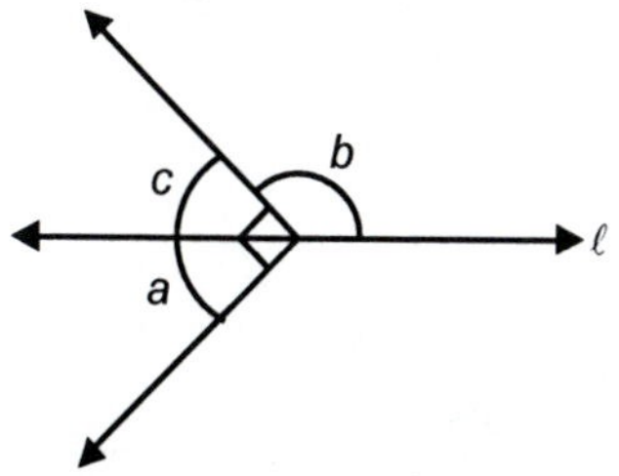

To find the measure of $\angle b$:

- Use the fact that $\angle a$ and $\angle c$ are complementary angles.
- Find $\angle b$ by using the fact that $\angle c$ and $\angle b$ are supplementary angles.

Solution

$$\begin{aligned} \angle a + \angle c &= 90^\circ \\ 51^\circ + \angle c &= 90^\circ \\ \angle c &= 39^\circ \end{aligned}$$

$$\begin{aligned} \angle b + \angle c &= 180^\circ \\ \angle b + 39^\circ &= 180^\circ \\ \angle b &= 141^\circ \end{aligned}$$

The measure of $\angle b$ is 141°.

47. **Strategy** To find the measure of $\angle BOC$, use the fact that the sum of the measures of the angles x, $\angle AOB$, and $\angle BOC$ is 180°. Since $\overline{AO} \perp \overline{OB}$, $\angle AOB = 90°$.

Solution

$$x + \angle AOB + \angle BOC = 180°$$
$$x + 90° + \angle BOC = 180°$$
$$\angle BOC = 90° - x$$

The measure of $\angle BOC$ is $90° - x$.

Objective B Exercises

49. **Strategy** The angles labeled are adjacent angles of intersecting lines and are, therefore, supplementary angles. To find x, write an equation and solve for x.

Solution

$$x + 74° = 180°$$
$$x = 106°$$

The measure of x is 106°.

51. **Strategy** The angles labeled are vertical angles and are, therefore, equal. To find x, write an equation and solve for x.

Solution

$$5x = 3x + 22°$$
$$2x = 22°$$
$$x = 11°$$

The measure of x is 11°.

53. **Strategy** • To find the measure of $\angle a$, use the fact that corresponding angles of parallel lines are equal.
• To find the measure of $\angle b$, use the fact that adjacent angles of intersecting lines are supplementary.

Solution

$$\angle a = 38°$$
$$\angle b + \angle a = 180°$$
$$\angle b + 38° = 180°$$
$$\angle b = 142°$$

The measure of $\angle a$ is 38°.
The measure of $\angle b$ is 142°.

55. **Strategy** • To find the measure of $\angle a$, use the fact that alternate interior angles of parallel lines are equal.
• To find the measure of $\angle b$, use the fact that adjacent angles of intersecting lines are supplementary.

Solution

$$\angle a = 47°$$
$$\angle a + \angle b = 180°$$
$$47° + \angle b = 180°$$
$$\angle b = 133°$$

The measure of $\angle a$ is 47°.
The measure of $\angle b$ is 133°.

57. **Strategy**

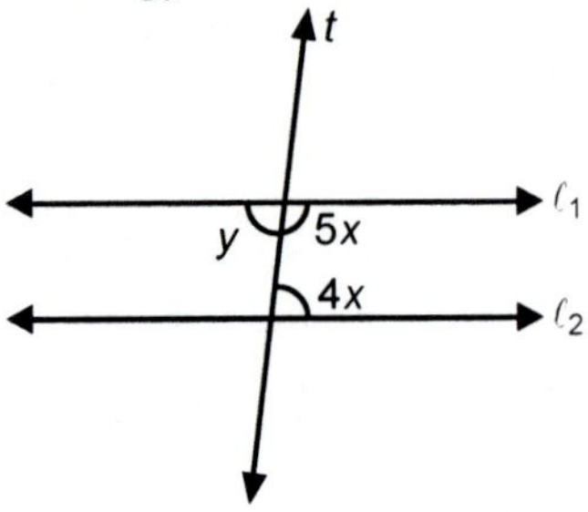

$4x = y$ because alternate interior angles have the same measure.
$y + 5x = 180°$ because adjacent angles of intersecting lines are supplementary.
Substitute $4x$ for y and solve for x.

Solution

$$4x + 5x = 180°$$
$$9x = 180°$$
$$x = 20°$$

The measure of x is 20°.

59. False

61. True, if l_1 and l_2 were parallel, then $\angle a = \angle c$ and $\angle a$ and $\angle c$ are supplementary.

Objective C Exercises

63. Strategy

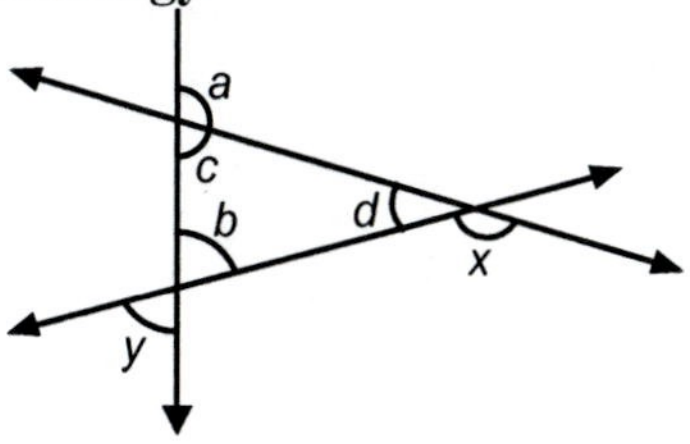

- To find the measure of angle y, use the fact that $\angle b$ and $\angle y$ are vertical angles.
- To find the measure of angle x:

Find the measure of angle c by using the fact that the sum of an interior and an exterior angle is 180°.
Find the measure of angle d by using the fact that the sum of the interior angles of triangle is 180°.
Find the measure of angle x by using the fact that the sum of an interior and an exterior angle is 180°.

Solution

$\angle y = \angle b = 70°$

$\angle a + \angle c = 180°$
$95° + \angle c = 180°$
$\angle c = 85°$

$\angle b + \angle c + \angle d = 180°$
$70° + 85° + \angle d = 180°$
$155° + \angle d = 180°$
$\angle d = 25°$

$\angle d + \angle x = 180°$
$25° + \angle x = 180°$
$\angle x = 155°$

The measure of $\angle x$ is 155°.
The measure of $\angle y$ is 70°.

65. Strategy

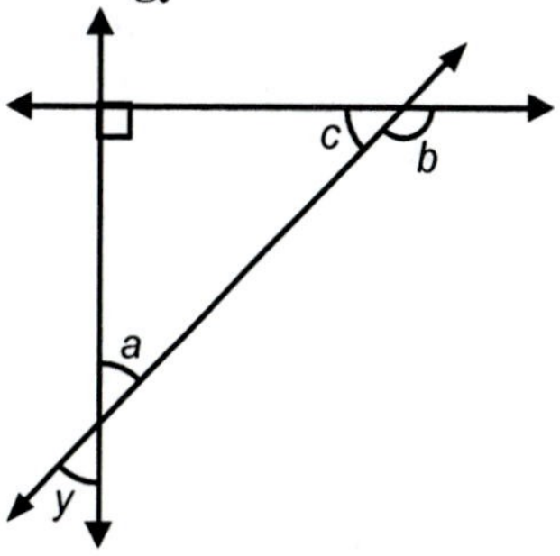

- To find the measure of angle a, use the fact that $\angle a$ and $\angle y$ are vertical angles.
- To find the measure of angle b:

Find the measure of angle c by using the fact that the sum of the interior angles of a triangle is 180°.
Find the measure of angle b by using the fact that the sum of an interior and an exterior angle is 180°.

Solution

$\angle a = \angle y = 45°$
$\angle a + \angle c + 90° = 180°$
$45° + \angle c + 90° = 180°$
$\angle c + 135° = 180°$
$\angle c = 45°$

$\angle c + \angle b = 180°$
$45° + \angle b = 180°$
$\angle b = 135°$

The measure of $\angle a$ is 45°.
The measure of $\angle b$ is 135°.

67. **Strategy** To find the measure of the third angle, use the fact that the sum of the measures of the interior angles of a triangle is 180°. Write and solve an equation using x to represent the measure of the third angle.

Solution

$$x + 90° + 30° = 180°$$
$$x + 120° = 180°$$
$$x = 60°$$

The measure of the third angle is 60°.

69. **Strategy** To find the measure of the third angle, use the fact that the sum of the measures of the interior angles of a triangle is 180°. Write an equation using x to represent the measure of the third angle. Solve the equation for x.

Solution

$$x + 42° + 103° = 180°$$
$$x + 145° = 180°$$
$$x = 35°$$

The measure of the third angle is 35°.

71. True

Critical Thinking

73. **Strategy** To find which point is the midpoint of two others, find the midpoint of each pair.

Solution

A and B: $\dfrac{-2.5+2}{2} = \dfrac{-.5}{2} = -.25$

A and C: $\dfrac{-2.5+5}{2} = \dfrac{2.5}{2} = 1.25$

A and D: $\dfrac{-2.5+3.5}{2} = \dfrac{1}{2} = .5$

B and C: $\dfrac{2+5}{2} = \dfrac{7}{2} = 3.5$

B and D: $\dfrac{2+3.5}{2} = \dfrac{5.5}{2} = 2.75$

C and D: $\dfrac{5+3.5}{2} = \dfrac{8.5}{2} = 4.25$

D (3.5) is the midpoint of B and C.

75. The students answers should include that the forest ranger should measure the diameter of the tree and use the formula $C = \pi d$ and solve for d.

Projects and Group Activities

77. $\angle a + \angle b + \angle c = 180°$

$\angle c + \angle x = 180°$ so

$$\angle a + \angle b + \angle c = \angle c + \angle x$$
$$\angle a + \angle b + \angle c - \angle c = \angle c - \angle c + \angle x$$
$$\angle a + \angle b = \angle x$$

Any exterior angle is equal to the sum of the opposite interior angles.

Section 3.2

Concept Check

1. The polygon has 6 sides so the polygon is a hexagon.

3. The polygon has 5 sides so the polygon is a pentagon.

5. The triangle has no sides equal so the triangle is a scalene triangle.

7. The triangle has three sides equal so the triangle is an equilateral triangle.

9. The triangle has one obtuse angle so the triangle is an obtuse triangle.

11. The triangle has three acute angle so the triangle is an acute triangle.

Objective A Exercises

13. **Strategy** To find the perimeter, use the formula for the perimeter of a triangle. Substitute 12 for a, 20 for b, and 24 for c. Solve for P.

Solution

$P = a + b + c = 12 + 20 + 24 = 56$

The perimeter is 56 in.

15. **Strategy** To find the perimeter, use the formula for the perimeter of a square. Substitute 3.5 for s and solve for P.

Solution

$P = 4s = 4 \cdot 3.5 = 14$

The perimeter is 14 ft.

17. **Strategy** To find the perimeter, use the formula for the perimeter of a rectangle. Substitute 13 for *L* and 10.5 for *W*. Solve for *P*.

Solution

$P = 2L + 2W = 2 \cdot 13 + 2 \cdot 10.5$

$= 26 + 21 = 47$

The perimeter is 47 mi.

19. **Strategy** To find the circumference, use the circumference formula that involves the radius. For the exact answer, leave the answer in terms of π. For an approximation, use the π key on a calculator. $r = 4$.

Solution

$C = 2\pi r = 2\pi(4) = 8\pi \approx 25.13$

The circumference is 8π cm. The circumference is approximately 25.13 cm.

21. **Strategy** To find the circumference, use the circumference formula that involves the radius. For the exact answer, leave the answer in terms of π. For an approximation, use the π key on a calculator. $r = 5.5$.

Solution

$C = 2\pi r = 2\pi(5.5) = 11\pi \approx 34.56$

The circumference is 11π mi. The circumference is approximately 34.56 mi.

23. **Strategy** To find the circumference, use the circumference formula that involves the diameter. For the exact answer, leave the answer in terms of π. For an approximation, use the π key on a calculator. $d = 17$.

Solution

$C = \pi d = \pi(17) = 17\pi \approx 53.41$

The circumference is 17π ft. The circumference is approximately 53.41 ft.

25. **Strategy** To find the perimeter, use the formula for the perimeter of a triangle. Substitute 3.8 for *a*, 5.2 for *b*, and 8.4 for *c*. Solve for *P*.

Solution

$P = a + b + c = 3.8 + 5.2 + 8.4 = 17.4$

The perimeter is 17.4 cm.

27. **Strategy** To find the perimeter, use the formula for the perimeter of a triangle. Substitute $2\frac{1}{2}$ for *a* and *b*, and 3 for *c*. Solve for *P*.

Solution

$P = a + b + c = 2\frac{1}{2} + 2\frac{1}{2} + 3 = 8$

The perimeter is 8 cm.

29. **Strategy** To find the perimeter, use the formula for the perimeter of a rectangle. Substitute 8.5 for *L* and 3.5 for *W*. Solve for *P*.

Solution

$P = 2L + 2W = 2(8.5) + 2(3.5) = 17 + 7 = 24$

The perimeter is 24 m.

31. **Strategy** To find the perimeter, multiply the measure of one of the equal sides (3.5) by 5.

Solution

$P = 5(3.5) = 17.5$

The perimeter is 17.5 in.

33. **Strategy** To find the perimeter, use the formula for the perimeter of a square. Substitute 12.2 for *s*. Solve for *P*.

Solution

$P = 4s = 4(12.2) = 48.8$

The perimeter is 48.8 cm.

35. **Strategy** To find the circumference, use the circumference formula that involves the diameter. Leave the answer in terms of π. $d = 1.5$.

Solution

$C = \pi d = \pi(1.5) = 1.5\pi$

The circumference is 1.5π in.

37. **Strategy** To find the circumference, use the circumference formula that involves the radius. An approximation is asked for; use the π key on a calculator. $r = 36$.

Solution

$C = 2\pi r = 2\pi(36) = 72\pi \approx 226.19$

The circumference is approximately 226.19 cm.

39. **Strategy** To find the amount of fencing, use the formula for the perimeter of a rectangle. Substitute 18 for L and 12 for W. Solve for P.

Solution

$P = 2L + 2W = 2(18) + 2(12)$

$= 36 + 24 = 60$

The 60 ft of fencing is needed.

41. **Strategy** To find the amount to be nailed down, use the formula for the perimeter of a rectangle. Substitute 12 for L and 10 for W. Solve for P.

Solution

$P = 2L + 2W = 2(12) + 2(10)$

$= 24 + 20 = 44$

44 ft of carpet must be nailed down.

43. **Strategy** To find the length, use the formula for the perimeter of a rectangle. Substitute 440 for P and 100 for W. Solve for L.

Solution

$$P = 2L + 2W$$
$$440 = 2L + 2(100)$$
$$440 = 2L + 200$$
$$240 = 2L$$
$$120 = L$$

The length is 120 ft.

45. **Strategy** To find the third side of the banner, use the formula for the perimeter of a triangle. Substitute 46 for P, 18 for a, and 18 for b. Solve for c.

Solution

$$P = a + b + c$$
$$46 = 18 + 18 + c$$
$$46 = 36 + c$$
$$10 = c$$

The third side of the banner is 10 in.

47. **Strategy** To find the length of each side, use the formula for the perimeter of a square. Substitute 48 for P. Solve for s.

Solution

$$P = 4s$$
$$48 = 4s$$
$$12 = s$$

The length of each side is 12 in.

49. **Strategy** To find the length of the diameter,use the circumference formula that involves the diameter. An approximation is asked for; use the π key on a calculator. $C = 8$.

Solution

$$C = \pi d$$
$$8 = \pi d$$
$$\frac{8}{\pi} = d$$
$$2.55 \approx d$$

The diameter is approximately 2.55 cm.

51. **Strategy** To find the length of molding, use the circumference formula that involves the diameter. An approximation is asked for; use the π key on a calculator. $d = 4.2$.

Solution

$C = \pi d = \pi(4.2) \approx 13.19$

The length of molding is approximately 13.19 ft.

53. **Strategy** To find the distance:

- Convert the diameter to feet.
- Multiply the circumference by 8.

An approximation is asked for; use the π key on a calculator.

Solution

24 in. = 2 ft

distance $= 8C = 8\pi d = 8\pi(2) = 16\pi \approx 50.27$

The bicycle travels approximately 50.27 ft.

55. **Strategy** To find the circumference of the earth, use the circumference formula that involves the radius. An approximation is asked for; use the π key on a calculator. $r = 6356$.

Solution

$C = 2\pi r = 2\pi(6356) = 12,712\pi \approx 39,935.93$

The circumference of the earth is approximately 39,935.93 km.

57. The perimeter of a square is $P = 4s$, and the circumference of a circle is $C = \pi d$. If the side of the square is equal to the diameter of the circle, then $s = d$. The perimeter of the square can be written $P = 4d$. $4d > \pi d$, since $4 > \pi$. The perimeter of the square is greater than the circumference of the circle.

Objective B Exercises

59. **Strategy** To find the area, use the formula for the area of a rectangle. Substitute 12 for L and 5 for W. Solve for A.

Solution

$A = LW = 12(5) = 60$

The area is 60 ft^2.

61. **Strategy** To find the area, use the formula for the area of a square. Substitute 4.5 for s. Solve for A.

Solution

$A = s^2 = (4.5)^2 = 20.25$

The area is 20.25 in^2.

63. **Strategy** To find the area, use the formula for the area of a triangle. Substitute 42 for b and 26 for h. Solve for A.

Solution

$A = \frac{1}{2}bh = \frac{1}{2}(42)(26) = 546$

The area is 546 ft^2.

65. **Strategy** To find the area, use the formula for the area of a circle. Substitute 4 for r. Solve for A. For the exact answer, leave the answer in terms of π. For an approximation, use the π key on a calculator.

Solution

$A = \pi r^2 = \pi(4)^2 = 16\pi \approx 50.27$

The area is $16\pi \text{ cm}^2$. The area is approximately 50.27 cm^2.

67. **Strategy** To find the area, use the formula for the area of a circle. Substitute 5.5 for r. Solve for A. For the exact answer, leave the answer in terms of π. For an approximation, use the π key on a calculator.

Solution

$A = \pi r^2 = \pi(5.5)^2 = 30.25\pi \approx 95.03$

The area is $30.25\pi \text{ mi}^2$. The area is approximately 95.03 mi^2.

69. **Strategy** To find the area:

- Find the radius of the circle.
- Use the formula for the area of a circle.

For the exact answer, leave the answer in terms of π. For an approximation, use the π key on a calculator.

Solution

$r = \frac{1}{2}d = \frac{1}{2}(17) = 8.5$

$A = \pi r^2 = \pi(8.5)^2 = 72.25\pi \approx 226.98$

The area is $72.25\pi \text{ ft}^2$. The area is approximately 226.98 ft^2.

71. **Strategy** To find the area, use the formula for the area of a square. Substitute 12.5 for s. Solve for A.

Solution

$A = s^2 = (12.5)^2 = 156.25$

The area is 156.25 cm^2.

73. **Strategy** To find the area, use the formula for the area of a rectangle. Substitute 38 for L and 15 for W. Solve for A.

Solution

$A = LW = 38(15) = 570$

The area is 570 in^2.

75. **Strategy** To find the area, use the formula for the area of a parallelogram. Substitute 16 for b and 12 for h. Solve for A.

Solution

$A = bh = 16(12) = 192$

The area is 192 in^2.

77. **Strategy** To find the area, use the formula for the area of a triangle. Substitute 6 for *b* and 4.5 for *h*. Solve for *A*.

Solution

$A = \frac{1}{2}bh = \frac{1}{2}(6)(4.5) = 13.5$

The area is 13.5 ft^2.

79. **Strategy** To find the area, use the formula for the area of a trapezoid. Substitute 35 for b_1, 20 for b_2, and 12 for *h*. Solve for *A*.

Solution

$A = \frac{1}{2}h(b_1 + b_2) = \frac{1}{2} \cdot 12(35 + 20) = 330$

The area is 330 cm^2.

81. **Strategy** To find the area, use the formula for the area of a circle. Leave the answer in terms of π. $r = 5$.

Solution

$A = \pi r^2 = \pi(5)^2 = 25\pi$

The area is $25\pi \text{ in}^2$.

83. **Strategy** To find the area, use the formula for the area of a rectangle where $l = 150$ and $w = 70$.

Solution

$A = lw$

$A = 150 \cdot 70$

$A = 10{,}500$

The area of the reserve is 10,500 mi^2.

85. **Strategy** To find the area, use the formula for the area of a circle. Leave the answer in terms of π. $r = 50$.

Solution

$A = \pi r^2 = \pi(50)^2 = 2500\pi$

The area is $2500\pi \text{ ft}^2$.

87. **Strategy** To find the area, use the formula for the area of a square. Substitute 8.5 for *s*. Solve for *A*.

Solution

$A = s^2 = (8.5)^2 = 72.25$

The area of the patio is 72.25 m^2.

89. **Strategy** To find the amount of turf, use the formula for the area of a rectangle. Substitute 100 for *L* and 75 for *W*. Solve for *A*.

Solution

$A = LW = 100(75) = 7500$

7500 yd^2 of artificial turf must be purchased.

91. **Strategy** To find the width, use the formula for the area of a rectangle. Substitute 300 for *A* and 30 for *L*. Solve for *W*.

Solution

$A = LW$

$300 = 30W$

$10 = W$

The width of the rectangle is 10 in.

93. **Strategy** To find the length of the base, use the formula for the area of a triangle. Substitute 50 for *A* and 5 for *h*. Solve for *b*.

Solution

$A = \frac{1}{2}bh$

$50 = \frac{1}{2}b(5)$

$50 = \frac{5}{2}b$

$20 = b$

The base of the triangle is 20 m.

95. **Strategy** To find the number of quarts of stain:

- Use the formula for the area of a rectangle to find the area of the deck.
- Divide the area of the deck by the area one quart will cover (50).

Solution

$A = LW$

$A = 10(8)$

$A = 80$

$80 \div 50 = 1.6$

Because a portion of a second quart is needed, 2 qt of stain should be purchased.

97. **Strategy** To find the cost of the wallpaper:

- Use the formula for the area of a rectangle to find the areas of the two walls.
- Add the areas of the two walls.
- Divide the total area by the area in one roll (40) to find the total number of rolls.
- Multiply the number of rolls by 37.

Solution

$$A_1 = LW = 9(8) = 72$$
$$A_2 = LW = 11(8) = 88$$
$$A = A_1 + A_2 = 72 + 88 = 160$$
$$160 \div 40 = 4$$
$$4 \cdot 37 = 148$$

The cost to wallpaper the two walls is \$148.

99. **Strategy** To find the storage unit needed, use the formula for the area of a rectangle to find the floor space for each unit. Then chose the smallest unit that has over 175 ft^2 of floor space.

Solution

$$A_1 = 10 \cdot 5 = 50$$
$$A_2 = 10 \cdot 10 = 100$$
$$A_3 = 10 \cdot 15 = 150$$
$$A_4 = 10 \cdot 20 = 200$$
$$A_5 = 10 \cdot 25 = 250$$
$$A_6 = 20 \cdot 30 = 300$$

The 10 × 20 unit should be selected.

101. **Strategy** To find the increase in area:

- Use the formula for the area of a circle to find the area of a circle with $r = 6$.
- Use the formula for the area of a circle to find the area of a circle with radius $r = 2(6) = 12$.
- Subtract the area of the smaller circle from the area of the larger circle. An approximation is asked for; use the π key on a calculator.

Solution

$$A_1 = \pi r^2 = \pi(6)^2 = 36\pi$$
$$A_2 = \pi(12)^2 = 144\pi$$
$$A_2 - A_1 = 144\pi - 36\pi = 108\pi \approx 339.29$$

The area is increased by 339.29 cm^2.

103. **Strategy** To find the cost of the paint:

- Use the formula for the area of a rectangle to find the area of the two walls that measure 15 by 9 and the two walls that measure 12 by 9.
- Add the areas to find the total area.
- Divide the total area by the area that one gallon will cover (400).
- Multiply the number of gallons by 29.98.

Solution

$$A_1 = 2(LW) = 2[15(9)] = 270$$
$$A_1 = 2(LW) = 2[12(9)] = 216$$
$$A_1 + A_2 = 270 + 216 = 486$$
$$486 \div 400 = 1.22$$

Because a portion of a second gallon is needed, 2 gal of paint should be purchased.

$$2(29.98) = 59.96$$

The paint will cost \$59.96.

105. **Strategy** To find the amount of material needed:

- add 1 ft to the length (4)
- multiply the width by 2
- find the area for each window by multiplying the length (4 + 1 = 5) by the width (2 × 2 = 4)
- multiply the area of each window by 4

Solution

$$A = lw$$
$$A = 5 \times 4 = 20$$
$$20 \times 4 = 80$$

80 ft^2 are needed for the drapes.

Critical Thinking

107. $A = LW$

Double the length and double the width:

$A = (2L)(2W) = 4LW$

$4LW$ is four times the quantity *LW*. The area of the resulting rectangle is 4 times larger.

Projects or Group Activities

109. If $2l + 2w = 20$, the $l + w = 10$. The possible whole number combinations are 1 and 9, 2 and 8, 3 and 7, 4 and 6, and 5 and 5. Find the area each combination to find the largest area.

$A = 9 \times 1 = 9$

$A = 8 \times 2 = 16$

$A = 7 \times 3 = 21$

$A = 6 \times 4 = 24$

$A = 5 \times 5 = 25$

The rectangle with the largest area 5 by 5.

111. Answers will vary.

Check Your Progress: Chapter 3

1. **Strategy** To find AC:

- find AB by taking $\frac{1}{3}(BC)(BC = 15)$
- add AB and BC.

Solution

$AB == \frac{1}{3}(BC) = \frac{1}{3}(15) = 5$

$AC = AB + BC = 5 + 15 = 20$

$AC = 20$ ft.

2. **Strategy** Supplementary angles are two angles whose sum is 180°. To find the supplement, let x represent the supplement of a 12° angle. Write an equation and solve for x.

Solution

$x + 12° = 180°$

$x = 168°$

The supplement of a 12° angle is a 168° angle.

3. **Strategy** • To find the measure of $\angle c$,use the fact that oppostie angles of intersecting lines are equal.
• To find the measure of $\angle b$, use the fact that adjacent angles of intersecting lines are supplementary.
• To find the measure of $\angle d$,use the fact that opposite angles of intersecting lines are equal.

Solution

$\angle c = 42°$

$\angle a + \angle b = 180°$

$42° + \angle b = 180°$

$\angle b = 138°$

$\angle d = \angle b = 138°$

The measure of $\angle c$ is 42°.
The measure of $\angle b$ is 138°.
The measure of $\angle d$ is 138°.

4. **Strategy** The sum of the three interior angles of a triangle is 180°. To find the third angle, let x represent the third angle where the other two angles are 23° and 90°. Write an equation and solve for x.

Solution

$x + 23° + 90° = 1800°$

$x + 113° = 180°$

$x = 67°$

The third angle is 67°.

5. **Strategy** To find the area, use the formula for the area of a square. Substitute 40 for s. Solve for A.

Solution

$A = s^2 = (40)^2 = 1600$

The area is 1600 in^2.

6. **Strategy** To find x, write an equation using the fact that the sum of the three angles is 180°. Solve for x.

Solution

$4x - 10 + 3x + 2x + 10 = 180$

$9x = 180$

$x = 20$

The measure if x is 20°.

7. **Strategy** To find the circumference, use the circumference formula that involves the diameter. An approximation is asked for π; use the π key on a calculator. $d = 12$

Solution

$C = \pi d = \pi(12) \approx 37.70$

The circumference is approximately 37.70 cm.

8. **Strategy** • To find the measure of $\angle a$,use the fact that alternate interior angles of parallel lines are equal.
• To find the measure of $\angle b$, use the fact that adjacent angles of intersecting lines are supplementary.

Solution

$\angle a = 135°$

$\angle a + \angle b = 180°$

$135° + \angle b = 180°$

$\angle b = 45°$

The measure of $\angle a$ is 135°.
The measure of $\angle b$ is 45°.

9. **Strategy** To find the base, use the formula for area of a triangle, where $A = 20$ and $h = 8$ and solve for b.

Solution

$A = \frac{1}{2}bh$

$20 = \frac{1}{2}b(8)$

$20 = 4b$

$5 = b$

The base is 5 m.

10. **Strategy** To find the area, use the formula for the area of a parallelogram where $h = 7$ and $b = 14$ and solve for A.

Solution

$A = bh$

$A = 14(7)$

$A = 98$

The area is 98 m^2.

11. **Strategy** To find the side, use the perimeter of a square where $P = 38$ and solve for s.

Solution

$A = 4s$

$38 = 4s$

$9.5 = x$

The sides are 9.5 in.

12. **Strategy** • To find $\angle x$, use the fact that any exterior angle is equal to the sum of the opposite two interior angles.
• To find the measure of $\angle y$, use the fact that $\angle x$ and $\angle y$ are supplementary angles.

Solution

$\angle b + \angle x = \angle a$

$48° + \angle x = 72°$

$\angle x = 24°$

$\angle x + \angle y = 180°$

$24° + \angle y = 180°$

$\angle y = 156°$

The measure of $\angle x = 24°$
The measure of $\angle y = 156°$.

13. **Strategy** To find the length, use the formula for the area of a rectangle where $A = 128$ and $w = 8$ and solve for l.

Solution

$A = lw$

$128 = l8$

$16 = l$

The length is 16 m.

14. **Strategy** To find the perimeter, add the length of the three sides.

Solution

$P = 15 + 9 + 12$

$P = 36$

The perimeter is 36 in.

15. **Strategy** To find the area, use the formula for the area of a circle. Substitute $\frac{1}{2}(2.8)=1.4$ for r. Solve for A. For the exact answer, leave the answer in terms of π. For an approximation, use the π key on a calculator.

Solution

$A=\pi r^2=\pi(1.4)^2=1.96\pi\approx 6.16$

The area is 1.96π m^2. The area is approximately 6.16 m^2.

16. **Strategy** To find the area, use the formula for the area of a trapezoid where $h = 6$, $b_1 = 14$ and $b_2 = 10$.

Solution

$A=\frac{1}{2}(b_1+b_2)h$

$A=\frac{1}{2}(14+10)6$

$A=\frac{1}{2}(24)6$

$A=12(6)$

$A=72$

The area is 72 cm^2.

17. **Strategy** To find the amount of molding, find the perimeter of the rectangle with $l = 10$ and $w = 8\frac{1}{2}$.

Solution

$P=2l+2w$

$P=2(10)+2\left(8\frac{1}{2}\right)$

$P=20+17$

$P=37$

37 ft of molding is needed.

Section 3.3

Concept Check

1. **a.** cone

b. cube

c. sphere

d. cylinder

3. s^2+2sl; l; s

Objective A Exercises

5. **Strategy** To find the volume, use the formula for the volume of a rectangular solid. $L=14$, $W=10$, $H=6$.

Solution $V=LWH=14(10)(6)=840$

The volume is 840 in^3.

7. **Strategy** To find the volume, use the formula for the volume of a pyramid. $s=3$, $h=5$.

Solution

$V=\frac{1}{3}s^2h=\frac{1}{3}(3^2)(5)=\frac{1}{3}(9)(5)=15$

The volume is 15 ft^3.

9. **Strategy** To find the volume:
- Find the radius of the sphere. $d=3$.
- Use the formula for the volume of a sphere.

Solution

$r=\frac{1}{2}d=\frac{1}{2}(3)=1.5$

$V=\frac{4}{3}\pi r^3=\frac{4}{3}\pi(1.5)^3=\frac{4}{3}\pi(3.375)$

$V=4.5\pi\approx 14.14$

The volume is 4.5π cm^3. The volume is approximately 14.14 cm^3.

11. **Strategy** To find the volume, use the formula for the volume of a rectangular solid. $L=6.8$, $W=2.5$, $H=2$.

Solution

$V=LWH=6.8(2.5)(2)=34$

The volume of the storage unit is 34 m^3.

13. **Strategy** To find the volume, use the formula for the volume of a cube. $s = 2.5$.

Solution

$V=s^3=(2.5)^3=15.625$

The volume of the cube is 15.625 in^3.

15. **Strategy** To find the volume:

- Find the radius of the sphere. $d = 6$.
- Use the formula for the volume of a sphere.

Solution

$r = \frac{1}{2}d = \frac{1}{2}(6) = 3$

$V = \frac{4}{3}\pi r^3 = \frac{4}{3}\pi(3)^3 = \frac{4}{3}\pi(27) = 36\pi$

The volume is 36π ft^3.

17. **Strategy** To find the volume:

- Find the radius of the cylinder. $d = 24$.
- Use the formula for the volume of a cylinder. $h = 18$.

Solution

$r = \frac{1}{2}d = \frac{1}{2}(24) = 12$

$V = \pi r^2 h = \pi(12^2)(18) = \pi(144)(18)$

$V = 2592\pi \approx 8143.01$

The volume of the cylinder is approximately 8143.01 cm^3.

19. **Strategy** To find the volume:

- Find the radius of the base of the cone. $d = 10$.
- Use the formula for the volume of a cone. $h = 15$.

Solution

$r = \frac{1}{2}d = \frac{1}{2}(10) = 5$

$V = \frac{1}{3}\pi r^2 h = \frac{1}{3}\pi(5)^2(15) = \frac{1}{3}\pi(25)(15)$

$V = 125\pi \approx 392.70$

The volume of the cone is 392.70 cm^3.

21. **Strategy** To find the volume, use the formula for the volume of a pyramid. $s = 9$, $h = 8$.

Solution

$V = \frac{1}{3}s^2 h = \frac{1}{3}(9^2)(8) = \frac{1}{3}(81)(8) = 216$

The volume of the pyramid is 216 m^3.

23. **Strategy** To find the height, use the formula for the volume of a rectangular solid. $V = 1836$, $L = 18$, $W = 12$.

Solution

$V = LWH$

$1836 = 18(12)(H)$

$1836 = 216H$

$8.5 = H$

The height of the aquarium is 8.5 in.

25. **Strategy** To find the height:

- Find the radius of the base of the cylinder. $d = 14$.
- Use the formula for the volume of a cylinder. $V = 2310$.

Solution

$r = \frac{1}{2}d = \frac{1}{2}(14) = 7$

$V = \pi r^2 h$

$2310 = \pi(7)^2 h$

$2310 = \pi(49)h$

$15.01 \approx h$

The height of the cone is about 15.01 cm.

27. **Strategy** To find the volume of the portion not used:

- Use the formula for the volume of a cylinder. $d = 16$, $r = 8$, $h = 30$
- Take $\frac{1}{4}$ of the volume to find the unused portion.

Solution

$V = \pi r^2 h$

$V = \pi(8)^2(30)$

$V \approx 6031.86$

$\frac{1}{4}V \approx \frac{1}{4}(6031.86) \approx 1507.96$

The volume of the unused portion would be 1507.96 ft^3.

29. **Strategy** To find the volume of the lock use the formula for the volume of a rectangular solid. $l = 1000$, $w = 110$, $h = 60$.

Solution

$V = lwh$

$V = 1000(110)(60)$

$V = 6,600,000$

The volume of lock is 6,600,000 ft^3.

31. Yes

33. No

35. **Strategy** To find the volume, use the formula for the volume of a rectangular solid. $L = 360$, $w = 160$, and $h = 3$.

Solution

$V = LWH$

$V = 360(160)(3)$

$V = 172,800$

The volume of the guacamole would be 172,800 ft^3.

Objective B Exercises

37. **Strategy** To find the surface area, use the formula for the surface area of a rectangular solid. $L = 5$, $W = 4$, $H = 3$.

Solution

$SA = 2LW + 2LH + 2WH$

$SA = 2(5)(4) + 2(5)(3) + 2(4)(3)$

$SA = 40 + 30 + 24 = 94$

The surface area of the rectangular solid is 94 m^2.

39. **Strategy** To find the surface area, use the formula for the surface area of a pyramid. $s = 4$, $l = 5$.

Solution

$SA = s^2 + 2sl$

$SA = 4^2 + 2(4)(5) = 16 + 40 = 56$

The surface area of the pyramid is 56 m^2.

41. **Strategy** To find the surface area, use the formula for the surface area of a cylinder. $r = 6$, $h = 2$.

Solution $SA = 2\pi r^2 + 2\pi rh$

$SA = 2\pi(6^2) + 2\pi(6)(2) = 2\pi(36) + 24\pi$

$SA = 72\pi + 24\pi = 96\pi \approx 301.59$

The surface area of the cylinder is 96π in^2. The surface area of the cylinder is approximately 301.59 in^2.

43. **Strategy** To find the surface area, use the formula for the surface area of a rectangular solid. $H = 5$, $L = 8$, $W = 4$.

Solution

$SA = 2LW + 2LH + 2WH$

$SA = 2(8)(4) + 2(8)(5) + 2(4)(5)$

$SA = 64 + 80 + 40 = 184$

The surface area of the rectangular solid is 184 ft^2.

45. **Strategy** To find the surface area, use the formula for the surface area of a cube. $s = 3.4$

Solution

$SA = 6s^2 = 6(3.4)^2 = 6(11.56) = 69.36$

The surface area of the cube is 69.36 m^2.

47. **Strategy** To find the surface area:

- Find the radius of the sphere. $d = 15$.
- Use the formula for the surface area of a sphere.

Solution

$r = \frac{1}{2}d = \frac{1}{2}(15) = 7.5$

$SA = 4\pi r^2 = 4\pi(7.5)^2 = 4\pi(56.25) = 225\pi$

The surface area of the sphere is 225π cm^2.

49. **Strategy** To find the surface area, use the formula for the surface area of a cylinder. $r = 4$, $h = 12$.

Solution

$SA = 2\pi r^2 + 2\pi rh = 2\pi(4)^2 + 2\pi(4)(12)$

$SA = 2\pi(16) + 96\pi = 32\pi + 96\pi = 128\pi$

$SA \approx 402.12$

The surface area of the cylinder is approximately 402.12 in^2.

51. **Strategy** To find the surface area, use the formula for the surface area of a cone. $r = 1.5$, $l = 2.5$.

Solution

$SA = \pi r^2 + \pi rl = \pi(1.5)^2 + \pi(1.5)(2.5)$

$SA = \pi(2.25) + 3.75\pi = 6\pi$

The surface area of the cone is $6\pi\ \text{ft}^2$.

53. **Strategy** To find the surface area, use the formula for the surface area of a pyramid. $s = 9$, $l = 12$.

Solution

$SA = s^2 + 2sl = 9^2 + 2(9)(12)$

$SA = 81 + 216 = 297$

The surface area of the pyramid is 297 in^2.

55. **Strategy** To find the width, use the formula for the surface area of a rectangular solid. $SA = 108$, $L = 6$, and $H = 4$.

Solution

$$\begin{aligned} SA &= 2LW + 2LH + 2WH \\ 108 &= 2(6)W + 2(6)(4) + 2W(4) \\ 108 &= 12W + 48 + 8W \\ 108 &= 20W + 48 \\ 60 &= 20W \\ 3 &= W \end{aligned}$$

The width of the rectangular solid is 3 cm.

57. **Strategy** To find the amount of fabric:

- Find the radius of the sphere. $d = 32$.
- Use the formula for the surface area of a sphere.

Solution

$$r = \frac{1}{2}d = \frac{1}{2}(32) = 16$$

$$SA = 4\pi r^2 = 4\pi(16)^2 = 4\pi(256)$$

$$SA = 1024\pi \approx 3217$$

Approximately 3217 ft^2 of fabric was used to construct the balloon.

59. **Strategy** To find the amount of glass, use the formula for the surface area of a rectangular solid. Omit the top of the fish tank. The formula becomes $SA = LW + 2LH + 2WH$. $L = 12$, $W = 8$, $H = 9$.

Solution

$SA = LW + 2LH + 2WH$

$SA = 12(8) + 2(12)(9) + 2(8)(9)$

$SA = 96 + 216 + 144 = 456$

The fish tank requires 456 in^2 of glass.

Critical Thinking

61. **a.** The distance from the edge of the base to the vertex of a regular pyramid is longer than the distance, perpendicular to the base, from the base to the vertex. The statement is always true.

b. The distance from the edge of the base of a cone to the vertex is longer than the distance, perpendicular to the base, from the base to the vertex. The statement is never true.

c. The four triangular faces of a regular pyramid could be equilateral triangles, but they could be isosceles triangles that are not equilateral. The statement is sometimes true.

63. Compare the volume of the cone to the volume of the cylinder.

$\text{Cone} = \frac{1}{3}\pi r^2 h \quad \text{Cylinder} = \pi r^2 h$

The volume of cylinder is 3 times the volume of the cone. The cone can be filled 3 times.

Projects or Group Activities

65. Answers will vary.

Chapter 3 Review Exercises

1. **Strategy** • To find the measure of angle c, use the fact that the sum of an interior and an exterior angle is 180°.

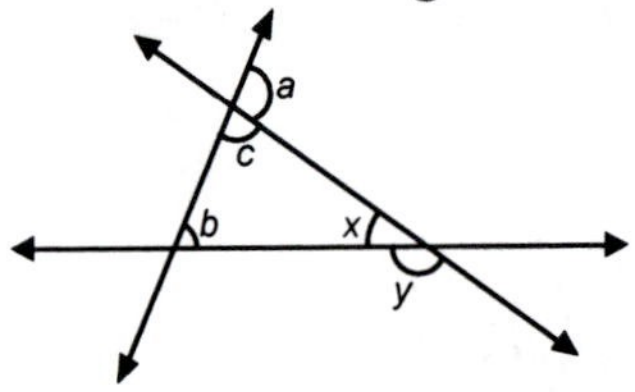

• To find the measure of angle x, use the fact that the sum of the measurements of the interior angles of a triangle is 180°.
• To find the measure of angle y, use the fact that the sum of an interior and an exterior angle is 180°.

Solution

$$\angle a + \angle c = 180° \quad 74° + \angle c = 180° \quad \angle c = 106°$$

$$\angle b + \angle c + \angle x = 180° \quad 52° + 106° + \angle x = 180° \quad 158° + \angle x = 180° \quad \angle x = 22°$$

$$\angle x + \angle y = 180°, \quad 22° + \angle y = 180° \quad \angle y = 158°$$

The measure of $\angle x$ is 22° and the measure of $\angle y$ is 158°.

2. **Strategy** To find the perimeter, use the formula for the perimeter of a rectangle. $l = 8$, $w = 5$

Solution

$P = 2l + 2w$

$P = 2(8) + 2(5)$

$P = 16 + 10$

$P = 26$

The perimeter is 26 ft.

3. **Strategy** To find the volume, use the formula for the volume of a rectangular solid. $l = 8$, $w = 7$, $h = 3$

Solution

$V = lwh$

$V = 8(7)3$

$V = 168$

The volume is 168 in^3.

4. **Strategy** To find x, use the fact that adjacent angles of intersecting lines are supplementary.

Solution

$112° + x = 180°$

$x = 68°$

The measure of x is 68°.

5. **Strategy** To find the area, use for formula for the area of a circle. $d = 9$, $r = 4.5$, $\pi = 3.14$

Solution

$A = \pi r^2$

$A = 3.14(4.5)^2$

$A = 63.585$

The area is 63.585 cm^2.

6. **Strategy** To find the surface area, use the formula for the surface area of a cylinder. $d = 4$, $r = 2$, $h = 8$

Solution

$SA = 2\pi r^2 + 2\pi rh$

$SA = 2\pi(2^2) + 2\pi(2)(8) = 2\pi(4) + 32\pi$

$SA = 8\pi + 32\pi = 40\pi \approx 125.66$

The surface area is 125.66 m^2.

7. $AC = AB + BC = 3(BC) + BC$

$= 4(BC) = 4(11) = 44$

The length of AC is 44 cm.

8. **Strategy** The sum of the measures of the three angles shown is 180°. To find x, write an equation and solve for x.

Solution

$4x + 3x + (x + 28°) = 180°$

$8x + 28° = 180°$

$8x = 152°$

$x = 19°$

The measure of x is 19°.

9. **Strategy** To find the area, use the formula for the area of a rectangle. $l = 8$, $w = 4$

Solution

$A = lw$

$A = 8(4)$

$A = 32$

The area is 32 in^2.

10. **Strategy** To find the volume, use the formula for the volume of a pyramid. $s = 6$, $h = 8$.

Solution

$V = \frac{1}{3}s^2h = \frac{1}{3}(6)^2(8) = \frac{1}{3}(36)(8) = 96 \text{ cm}^3$

The volume of the pyramid is 96 cm^3.

11. **Strategy** To find the perimeter, add the three sides.

Solution

$P = 10 + 16 + 16$

$P = 42$

The perimeter is 42 in.

12. **Strategy** $\angle a = 138°$ because alternate interior angles of parallel lines are equal. $\angle a + \angle b = 180°$ because adjacent angles of intersecting lines are supplementary.

Solution

$\angle a + \angle b = 180°$

$\angle a = 138°$

$138° + \angle b = 180°$

$\angle b = 42°$

The measure of $\angle a$ is 138°. The measure of $\angle b$ is 42°.

13. **Strategy** To find the surface area, use the formula for the surface area of a rectangular solid. $L = 10$, $W = 5$, $H = 4$.

Solution

$SA = 2LW + 2LH + 2WH$

$SA = 2(10)(5) + 2(10)(4) + 2(5)(4)$

$SA = 100 + 80 + 40 = 220$

The surface area of the solid is 220 ft^2.

14. **Strategy** To find *BC*, use the fact that the sum of *AB, BC,* and *CD* is equal to *AD*. Write an equation and solve for *BC*. *AB* = 15, *CD* = 6, *AD* = 24

Solution

$AD = AB + BC + CD$

$24 = 15 + BC + 6$

$24 = 21 + BC$

$3 = BC$

The length of *BC* is 3.

15. **Strategy** To find the volume use the formula of the volume of the cube. $S = 3.5$

Solution

$V = s^3$

$V = (3.5)^3$

$V = 42.875$

The volume is 42.875 in^3.

16. **Strategy** Supplementary angles are two angles whose sum is 180°. To find the supplement, let *x* represent the supplement of a 32° angle. Write and equation and solve for *x*.

Solution

$32° + x = 180°$

$x = 148°$

The supplement of a 32° angle is a 148° angle.

17. **Strategy** To find the volume, use the formula for the volume of a rectangular solid. $L = 6.5$, $W = 2$, $H = 3$.

Solution

$V = LWH = (6.5)(2)(3) = 39$

The area is 39 ft^3.

18. **Strategy** To find the third angle, use the fact that the sum of the measures of the interior angles of a triangle is 180°. Let x = the third angle.

Solution

$37° + 48° + x = 180°$

$85° + x = 180°$

$x = 95°$

The third angle is 95°.

19. **Strategy** To find the base, use the formula for the area of a triangle. Substitute 7 for *h* and 28 for *A*. Solve for *b*.

Solution

$A = \frac{1}{2}bh$

$28 = \frac{1}{2}b(7)$

$56 = 7b$

$8 = b$

The base of the triangle is 8 cm.

20. **Strategy** To find the volume, use the formula for the volume of a sphere. The radius of the sphere is one-half the diameter.

Solution

$r = \frac{1}{2}d = \frac{1}{2}(12) = 6$

$V = \frac{4}{3}\pi r^3 = \frac{4}{3}\pi(6^3) = \frac{4}{3}\pi(216) = 288\pi$

The volume is 288π mm^3.

21. **Strategy** To find the length of each side, use the formula for the perimeter of a square. $P = 86$.

Solution

$P = 4s$

$86 = 4s$

$21.5 = s$

The side of the square is 21.5 cm.

22. **Strategy** To find the number of cans of paint:
- Find the surface area by using the formula for the surface area of a cylinder.
- Divide the surface area by 200.

Solution

$SA = 2\pi r^2 + 2\pi rh$

$SA = 2\pi(6)^2 + 2\pi(6)(15)$

$SA = 2\pi(36) + 180\pi = 72\pi + 180\pi$

$SA = 252\pi$

$252\pi \div 200 \approx 3.96$

Because a portion of a fourth can is needed, 4 cans of paint should be purchased.

23. **Strategy** To find the amount of fencing, use the formula for the perimeter of a rectangle.

Solution

$P = 2L + 2W = 2(56) + 2(48) = 112 + 96$

$= 208$

208 yd of fencing are needed to fence the park.

24. **Strategy** To find the area, use the formula for the area of a square solid. $s = 9.5$

Solution

$A = s^2 = 9.5^2 = 90.25$

The area of the patio is 90.25 m^2.

25. **Strategy** To find the area of the walkway:
- Find the length and width of the total area.
- Find the total area.
- Subtract the area of the plot of grass from the total area.

Solution

$L = 40 + 4 = 44$

$W = 25 + 4 = 29$

$A = 44 \times 29 = 1276$

Area of the grass:

$A = 40 \times 25 = 1000$

Area of the walkway is $1276 - 1000 = 276$

The area of the walkway is 276 m^2.

Chapter 3 Test

1. **Strategy** To find the volume, use the formula for the volume of a rectangular solid. $l = 7$, $w = 6$, $h = 4$

Solution

$V = lwh$

$V = 7(6)(4)$

$V = 168$

The volume is 169 ft^2.

2. **Strategy** To find the measure of $\angle a$, use the fact that $\angle a$ and 37° add to 180°. Write an equation and solve for $\angle a$.

Solution

$\angle a + 37° = 180°$

$\angle a = 143°$

The measure of $\angle a$ is 143°.

3. **Strategy** To find the area, use the formula for the area of a rectangle. $l = 15$, $w = 7.4$

Solution

$A = lw$

$A = 15(7.4)$

$A = 111$

The area is 111 m^2.

4. **Strategy** To find the area, use the formula for the area of a triangle. $b = 7$, $h = 12$

Solution

$$A = \frac{1}{2}bh$$
$$A = \frac{1}{2}(7)(12)$$
$$A = 42$$

The area is 42 ft^2.

5. **Strategy** To find the volume, use the formula for the volume of a cone. $r = 7$, $h = 16$

Solution

$$V = \frac{1}{3}\pi r^2 h$$
$$V = \frac{1}{3}\pi (7)^2 (16)$$
$$V = \frac{784}{3}\pi$$

The volume is $\frac{784}{3}\pi$ cm^2.

6. **Strategy** To find the surface area, use the formula for the surface area of a pyramid. $s = 3$, $l = 11$

Solution

$$A = s^2 + 2sl$$
$$A = 3^2 + 2(3)(11)$$
$$A = 9 + 66$$
$$A = 75$$

The surface area is 75 m^2.

7. **Strategy** To find the volume, use the formula for the volume of a cylinder. $r = 7$, $h = 30$

Solution

$$V = \pi r^2 h$$
$$V = \pi (7)^2 30$$
$$V = 4618.14$$

The volume is 4618.14 cm^3.

8. **Strategy** To find the area, use the formula for the area of a trapezoid. $b_1 = 33$, $b_2 = 30$, $h = 6$

Solution

$$A = \frac{1}{2}h(b_1 + b_2)$$
$$A = \frac{1}{2}(6)(33 + 20)$$
$$A = 3(53)$$
$$A = 159$$

The area is 159 in^2.

9. The sum of the angles is 180°. Write and solve an equation.

$$3x + 6x = 180°$$
$$9x = 180°$$
$$x = 20°$$

The measure of x is 20°.

10. **Strategy** To find the surface area, use the formula for the surface area of a pyramid. $s = 5$, $l = 5$

Solution

$$A = s^2 + 2sl$$
$$A = 5^2 + 2(5)(5)$$
$$A = 25 + 50$$
$$A = 75$$

The surface area is 75 m^2.

11. The sum of the angles is 180°. Write and solve an equation.

$$4x + 10° + x = 180°$$
$$5x = 170°$$
$$x = 34°$$

The measure of x is 34°.

12. The figure has eight sides. The figure is an octagon.

13. **Strategy** To find the surface area, use the formula for the surface area of a cylinder. $h = 15, r = 10$

Solution

$S = 2\pi r^2 + 2\pi rh$

$S = 2\pi(10)^2 + 2\pi(10)(15)$

$S = 200\pi + 300\pi$

$S = 500\pi$

The surface area is 500π cm^2.

14. **Strategy** To find the measure of $\angle a$, use the fact that the sum of the opposite interior angles is equal to the exterior angle. Write an equation and solve.

Solution

$\angle a + 98° = 159°$

$\angle a = 61°$

The measure of $\angle a$ is 61°.

15. **Strategy** To find the perimeter, use the formula for the perimeter of a square. $s = 5$

Solution

$P = 4s$

$P = 4(5)$

$P = 20$

The perimeter is 20 m.

16. **Strategy** To find the perimeter, use the formula for the perimeter of a rectangle. $l = 8, w = 5$

Solution

$P = 2l + 2w$

$P = 2(8) + 2(5)$

$P = 16 + 10$

$P = 26$

The perimeter is 26 cm.

17. **Strategy** Supplementary angles are two angles whose sum is 180°. To find the supplement, write an equation and solve.

Solution

$x + 41° = 180°$

$x = 139°$

The supplement of a 41° angle is a 139° angle.

18. **Strategy** To find the third angle in a triangle, use the fact that the sum of all three angles in a triangle is 180°. Write an equation and solve for the third angle.

Solution

$41° + 37° + x = 180°$

$78° + x = 180°$

$x = 102°$

The measure of the third angle is 102°.

19. Right triangles have a 90° angle and two acute angles.

$32° + 90° + x = 180°$

$x = 58°$

The measure of the other two angles in the triangle are 90° and 58°.

20. Distance $= 10C$

$C = \pi D = 3.14 \cdot 28 \div 12 \text{ in} \approx 7.33 \text{ ft}$

Distance $= 10 \cdot 7.33 = 73.3$ ft

The bicycle traveled 73.3 ft.

21. **a.** **Strategy** To find the total area that must be cleaned, use the formula for the area of a rectangle to find the area of one side of the glass. Multiply the area by two both sides of the glass. $l = 40, w = 20$

Solution

$A = lw$

$A = 40(20)$

$A = 800$

$2A = 2(800) = 1600$

The total area to be cleaned is 1600 ft^2.

b. **Strategy** To find the volume of the window in cubic inches, change the length and width to inches by multiplying by 12. Use the formula for the volume of a rectangular solid. $l = 40$ ft, $w = 20$ ft, $h = 12.5$ in

Solution

$40 \times 12 = 480$

$20 \times 12 = 240$

$V = lwh$

$V = 480(240)12.5$

$V = 1{,}440{,}000$

The volume is 1,440,000 in^2.

22. **Strategy** To find the volume, use the formula for the volume of a cylinder. $d = 9, h = 18$

Solution

$r = \frac{1}{2}d = \frac{1}{2}(9) = 4.5$

$V = \pi r^2 h = \pi \cdot 4.5^2 \cdot 18 = 1145.11 \text{ ft}^3$

The volume of the silo is 1145.11 ft^2.

23. **Strategy** To find the area, use the formula for the area of a triangle. $b = 8, h = 2.75$

Solution

$A = \frac{1}{2}bh = \frac{1}{2}(8)(2.75) = 11 \text{ m}^2$

The area is 11 m^2.

24. **Strategy** To find the area, use the formula for the area of a circle. $d = 11$ ft 6 in $=$ 11.5 ft, $r = 5.75$, $\pi = 3.14$

Solution

$A = \pi r^2$

$A \approx 3.14(5.75)^2$

$A \approx 103.82$

The area is 103.82 ft^2.

25. **a.** **Strategy** To find the area, use the formula for the area of a rectangle. $l = 9, w = 5$

Solution

$A = lw$

$A = 9(5)$

$A = 45$

The area of a cell is 45 ft^2.

b. **Strategy** To find the volume, use the formula for the area of a rectangular solid. $l = 9, w = 5, h = 7$

Solution

$V = lwh$

$V = 9(5)(7)$

$V = 315$

The volume is 315 ft^3.

Cumulative Review Exercises

1.

$x \le 1$

$-3 \le 1$ True

$0 \le 1$ True

$1 \le 1$ True

The inequality is true for –3, 0, and 1.

2. $8.9\% = 8.9(0.01) = 0.089$

3. $\frac{7}{20} = \frac{7}{20}(100\%) = \frac{700}{20}\% = 35\%$

4. $-\frac{4}{9} \div \frac{2}{3} = -\frac{4}{9} \cdot \frac{3}{2} = -\left(\frac{4 \cdot 3}{9 \cdot 2}\right) = -\frac{2}{3}$

5. $5.7(-4.3) = -24.51$

6. $-\sqrt{125} = -\sqrt{25 \cdot 5} = -\sqrt{25}\sqrt{5} = -5\sqrt{5}$

7. $5 - 3[10 + (5-6)^2] = 5 - 3[10 + (-1)^2] = 5 - 3[10 + 1]$

$= 5 - 3[11] = 5 - 33 = -28$

8. $a(b-c)^3 = -1(-2-[-4])^3 = -1(-2+4)^3 = -1(2)^3$

$= -1(8) = -8$

9. $5m + 3n - 8m = (5m - 8m) + 3n$

$= -3m + 3n$

10. $-7(-3y) = [-7(-3)]y = 21y$

11. $4(3x+2) - (5x-1) = 12x + 8 - 5x + 1$

$= 7x + 9$

12. $\{-2, -1\}$

13. $C \cup D = \{-10, 0, 10, 20, 30\}$

14.

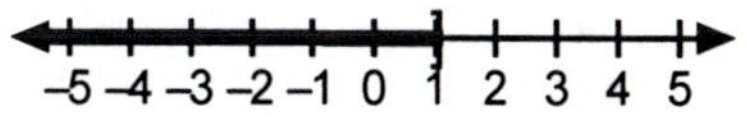

15.

$4x + 2 = 6x - 8$

$4x - 6x + 2 = 6x - 6x - 8$

$-2x + 2 = -8$

$-2x + 2 - 2 = -8 - 2$

$-2x = -10$

$\frac{-2x}{-2} = \frac{-10}{-2}$

$x = 5$

The solution is 5.

16. $$\begin{aligned} 3(2x+5) &= 18 \\ 6x+15 &= 18 \\ 6x+15-15 &= 18-15 \\ 6x &= 3 \\ \frac{6x}{6} &= \frac{3}{6} \\ x &= \frac{1}{2} \end{aligned}$$

The solution is $\frac{1}{2}$.

17. $$\begin{aligned} 4y-3 &\ge 6y+5 \\ -2y-3 &\ge 5 \\ -2y &\ge 8 \\ \frac{-2y}{-2} &\le \frac{8}{-2} \\ y &\le -4 \end{aligned}$$

The solution is $(-\infty,-4)$..

18. $$\begin{aligned} 8-4(3x+5) &\le 6(x-8) \\ 8-12x-20 &\le 6x-48 \\ -12-12x &\le 6x-48 \\ -12-18x &\le -48 \\ -18x &\le -36 \\ \frac{-18x}{-18} &\ge \frac{-36}{-18} \\ x &\ge 2 \end{aligned}$$

The solution set is $\{x|x \ge 2\}$.

19. $2x-3>5$ or $x+4<1$

$2x>8$ $\quad x<-3$

$x>4$

$\{x|x>4\}$ $\quad \{x|x<-3\}$

$\{x|x>4\} \cup \{x|x<-3\} = \{x|x<-3 \text{ or } x>4\}$

20. $$\begin{aligned} -3 \le 2x-7 &\le 5 \\ -3+7 \le 2x-7+7 &\le 5+7 \\ 4 \le 2x &\le 12 \\ \frac{4}{2} \le \frac{2x}{2} &\le \frac{12}{2} \\ 2 \le x &\le 6 \end{aligned}$$

$\{x|2 \le x \le 6\}$

21. $|3x-1|=2$

$3x-1=2$ or $3x-1=-2$

$3x=3$ $\quad 3x=-1$

$x=1$ $\quad x=-\frac{1}{3}$

The solutions are 1 and $-\frac{1}{3}$.

23. **Strategy** To find x, use the fact that adjacent angles of intersecting lines are supplementary.

Solution

$49°+x=180°$

$x=131°$

The measure of x is 131°.

24. **Strategy** The unknown number: x

the difference between four times a number and ten	is	two

$$\begin{aligned} 4x-10 &= 2 \\ 4x &= 12 \\ x &= 3 \end{aligned}$$

The number is 3.

25. **Strategy** To find the third angle, use the fact that the sum of the measures of the interior angles of a triangle is 180°. Let $x=$ the third angle.

Solution

$$\begin{aligned} 37°+21°+x &= 180° \\ 58°+x &= 180° \\ x &= 122° \end{aligned}$$

The third angle is 122°.

26. Strategy: Total amount of interest earned: x

	Principal	Rate	Interest
Amount at 4.5%	5000	0.045	0.045(5000)
Amount at 3.5 %	2500	0.035	0.035(2500)

Solution:

$$\begin{aligned} 0.045(5000)+0.035(2500) &= x \\ 225+87.5 &= x \\ 312.5 &= x \end{aligned}$$

The total amount of interest earned was $312.50.

27. **Strategy** To find the third side of the triangle, use the formula for the perimeter of a triangle. In an isosceles triangle, two sides are of equal measure.

Solution

$$\begin{aligned} P &= a+b+c \\ 19.5 &= 7.5+7.5+c \\ 19.5 &= 15+c \\ 4.5 &= c \end{aligned}$$

The third side measures 4.5 m.

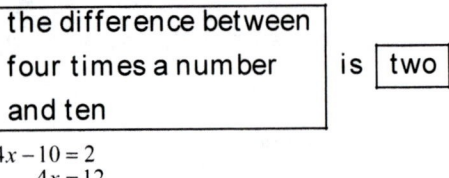

28. **Strategy** Use the formula for the volume of a rectangular solid. Substitute 144 for V, 12 for L, and 4 for W. Solve for H.

Solution

$$V = LWH$$
$$144 = 12(4)H$$
$$144 = 48H$$
$$3 = H$$

The height is 3 ft.

29. **Strategy** Solve the equation for D. $P = 35$

Solution

$$P = 15 + \frac{1}{2}D$$
$$35 = 15 + \frac{1}{2}D$$
$$20 = \frac{1}{2}D$$
$$2(20) = 2\left(\frac{1}{2}D\right)$$
$$40 = D$$

The depth is 40 ft.

30. **Strategy** Solve the equation $C = 79 + 0.35x$ for x. $C = 86.70$

Solution

$$C = 79 + 0.35x$$
$$86.70 = 79 + 0.35x$$
$$7.70 = 0.35x$$
$$22 = x$$

There were 22 text messages.

Chapter 4: Linear Functions and Inequalities in Two Variables

Prep Test

1. $-4(x-3) = -4x + 12$

2. $\sqrt{(-6)^2 + (-8)^2} = \sqrt{36+64} = \sqrt{100} = 10$

3. $\dfrac{3-(-5)}{2-6} = \dfrac{3+5}{2-6} = \dfrac{8}{-4} = -2$

4. $-2x + 5$
   $-2(-3) + 5 = 6 + 5 = 11$

5. $\dfrac{2r}{r-1}$
   $\dfrac{2(5)}{5-1} = \dfrac{10}{4} = 2.5$

6. $2p^3 - 3p + 4$
   $$\begin{aligned} 2(-1)^3 - 3(-1) + 4 &= 2(-1) - 3(-1) + 4 \\ &= -2 + 3 + 4 \\ &= 5 \end{aligned}$$

7. $\dfrac{x_1 + x_2}{2}$
   $\dfrac{7+(-5)}{2} = \dfrac{2}{2} = 1$

8. $$\begin{aligned} 3x - 4y &= 12 \\ 3x - 4(0) &= 12 \\ 3x &= 12 \\ x &= 4 \end{aligned}$$

Section 4.1

Concept Check

1. Quadrant II

3. y-axis

5. Answers will vary. For example, (–3, 2) and (5, 2)

7. Answers will vary. Any point with the y-coordinate lies on the line.

9. right; down

11. (6, –5)

Objective A Exercises

13.

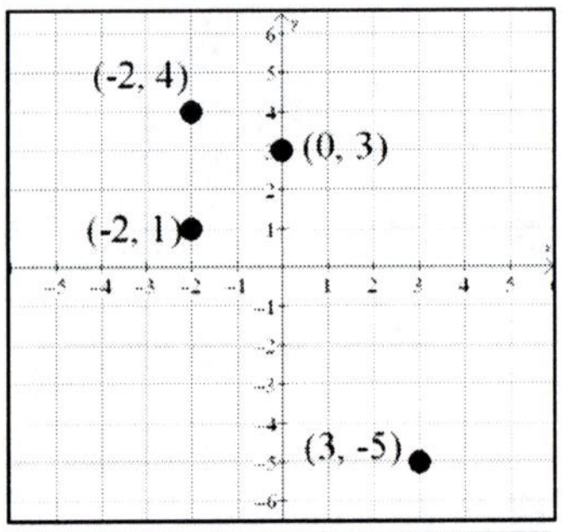

15.

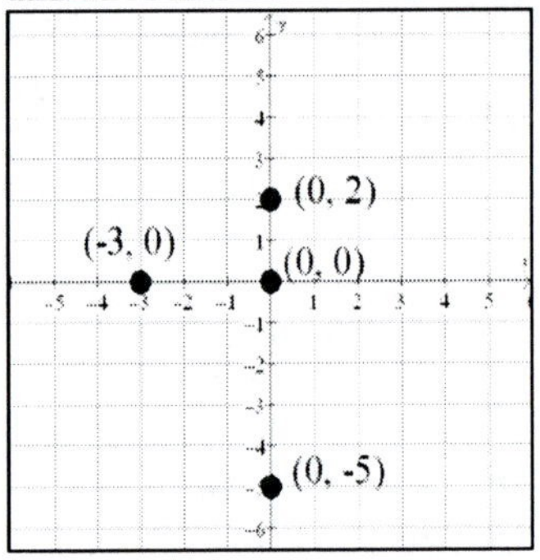

17.

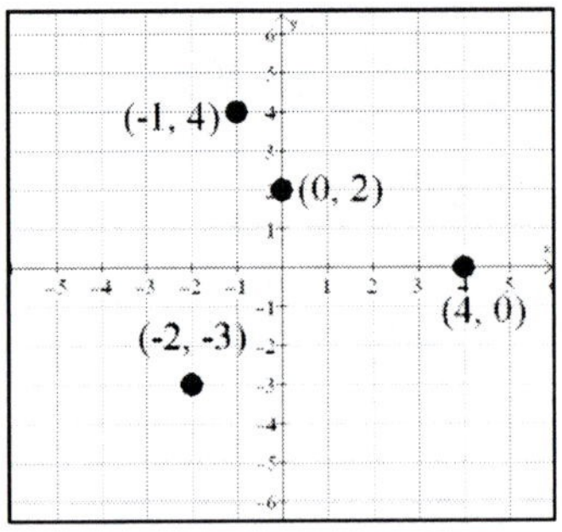

19. A (2, 3)
    B (4, 0)
    C (−4, 1)
    D (−2, −2)

21. A (−2, 5)
    B (3, 4)
    C (0, 0)
    D (−3, −2)

23. **a.** Abscissa of point A: 2
Abscissa of point C: -4

b. Ordinate of point B: 1
Ordinate of point D: -3

25. **a.** Quadrant I

b. Quadrant II

c. Quadrant IV

d. Quadrant III

Objective B Exercises

27.

	$y = -x + 7$
4	$-(3) + 7$
	$-3 + 7$
	4

$4 = 4$

Yes, (3, 4) is a solution of $y = -x + 7$.

29.

	$y = \frac{1}{2}x - 1$
2	$\frac{1}{2}(-1) - 1$
	$-\frac{1}{2} - 1$
	$-\frac{3}{2}$

$2 \neq -\frac{3}{2}$

No, (2, −3) is not a solution of $y = \frac{1}{2}x - 1$

31.

$2x - 5y$	$= 4$
$2(4) - 5(1)$	4
$8 - 5$	
3	

$3 \neq 4$

No, (4, 1) is a not solution of $2x - 5y = 4$

33. If $x > 2$, $-3x < -6$, then $-3x + 6 < 0$.
y is negative.

35.

x	$y = 2x$	(x, y)
-2	$2(-2) = -4$	$(-2, -4)$
-1	$2(-1) = -2$	$(-1, -2)$
0	$2(0) = 0$	$(0, 0)$
2	$2(2) = 4$	$(2, 4)$

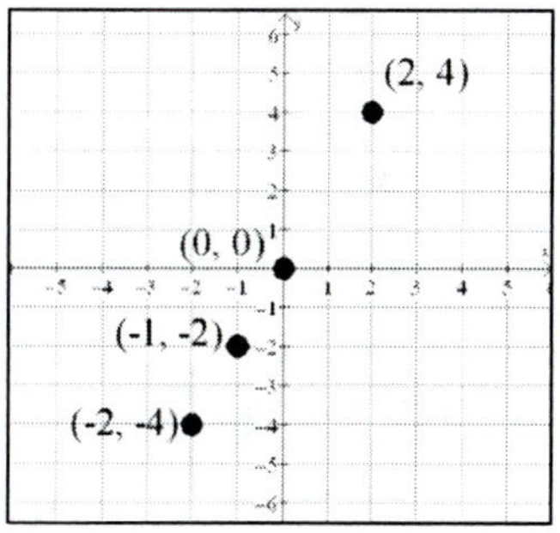

37.

x	$y = \frac{2}{3}x + 1$	(x, y)
-3	$\frac{2}{3}(-3) + 1 = -1$	$(-3, -1)$
0	$\frac{2}{3}(0) + 1 = 1$	$(0, 1)$
3	$\frac{2}{3}(3) + 1 = 3$	$(3, 3)$

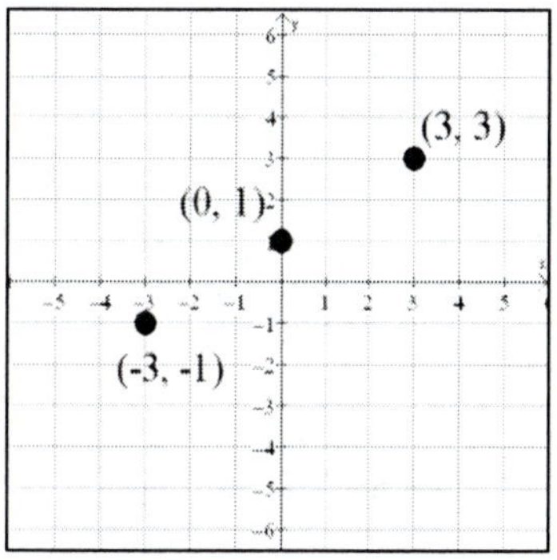

39. Solve $2x+3y=6$ for y.

$$2x+3y=6$$
$$3y=-2x+6$$
$$y=-\frac{2}{3}x+2$$

x	$y=-\frac{2}{3}x+2$	(x, y)
-3	$-\frac{2}{3}(-3)+2=4$	$(-3, 4)$
0	$-\frac{2}{3}(0)+2=2$	$(0, 2)$
3	$-\frac{2}{3}(3)+2=0$	$(3, 0)$

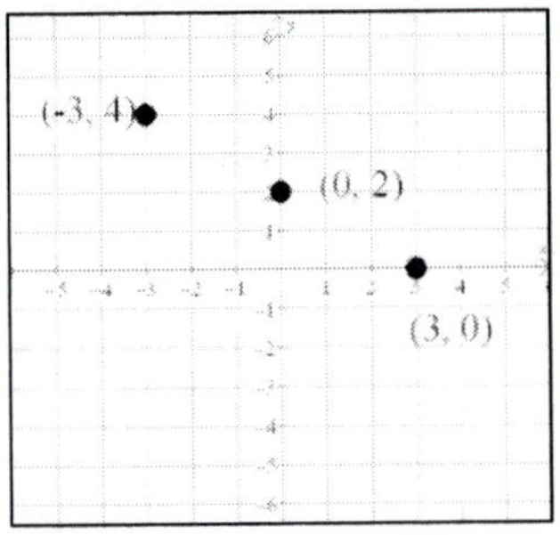

Objective C Exercises

41.

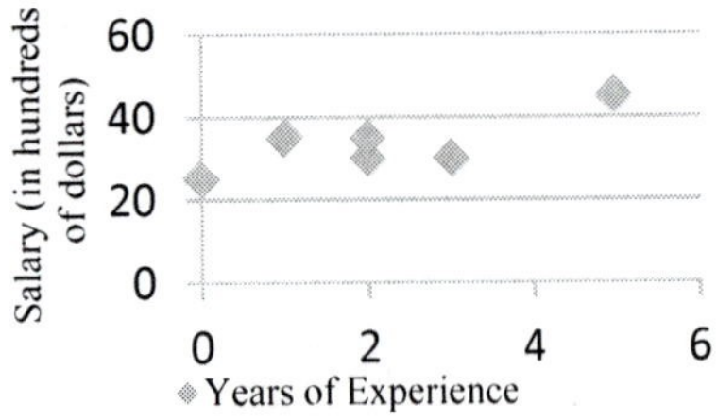

43.

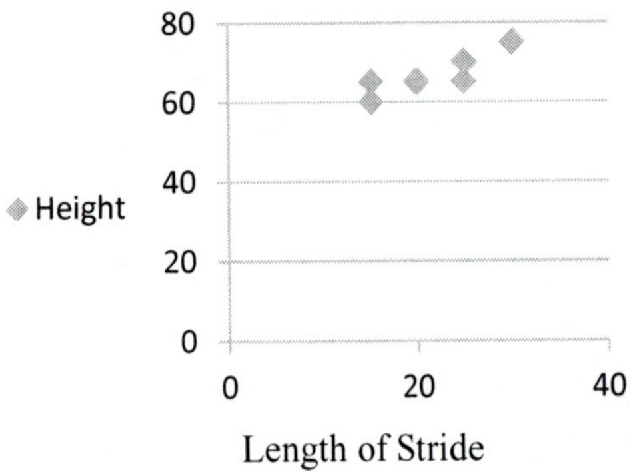

45. 200 s

47. The point will be graphed with an x-coordinate of 1200 and a y–coordinate for the 1200-meter race. The graph will have an additional point.

49. **a.** 15 mpg

b. 13 mpg

Critical Thinking

51.

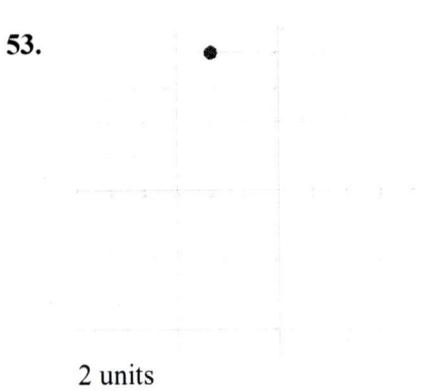

4 units

53.

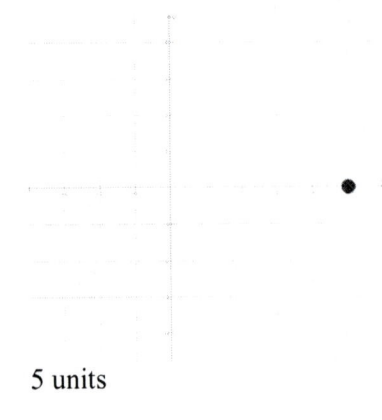

2 units

55.

5 units

57.

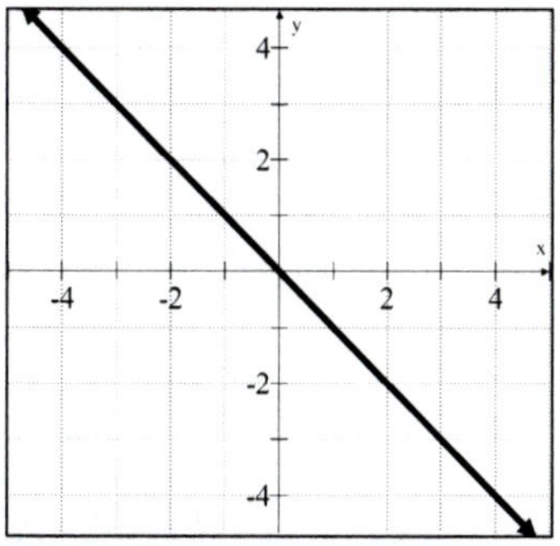

Projects and Group Activities

59. The graph will be a circle with the center at the origin and radius of 5 units.

Section 4.2

Concept Check

1. 5

3. Domain $\{-3, -2, -1, 1, 2, 3\}$
Range: $\{1, 4, 9\}$

5. Yes

Objective A Exercises

7. This is a function because each x-coordinate is paired with only one y-coordinate.
Domain: $\{0, 2, 3, 4, 5\}$
Range: $\{0, 4, 6, 8, 10\}$

9. This is a function because each x-coordinate is paired with only one y-coordinate.
Domain: $\{-4, -2, 0, 3\}$
Range: $\{-5, -1, 5\}$

11. This is a function because each x-coordinate is paired with only one y-coordinate.
Domain: $\{—2, -1, 0, 1, 2\}$
Range: $\{-3, 3\}$

13. This is not a function because there are x-coordinates paired with two different y-coordinates. They are $(1,1)$, $(1, -1)$ and $(4,2)$, $(4, -2)$.
Domain: $\{1, 4, 9\}$ Range: $\{-2, -1, 1, 2, 3\}$

15. a. Yes, this table defines a function because there is only one cost for any weight.
b. \$1.05
c. \$.65
d. \$.45

17. True

19. $f(x) = 5x - 4$
$f(3) = 5(3) - 4$
$f(3) = 15 - 4$
$f(3) = 11$

21. $f(x) = 5x - 4$
$f(0) = 5(0) - 4$
$f(0) = -4$

23. $G(t) = 4 - 3t$
$G(0) = 4 - 3(0)$
$G(0) = 4$

25. $G(t) = 4 - 3t$
$G(-2) = 4 - 3(-2)$
$G(-2) = 4 + 6$
$G(-2) = 10$

27. $q(r) = r^2 - 4$
$q(3) = 3^2 - 4$
$q(3) = 9 - 4$
$q(3) = 5$

29. $q(r) = r^2 - 4$
$q(-2) = (-2)^2 - 4$
$q(-2) = 4 - 4$
$q(-2) = 0$

31. $F(x) = x^2 + 3x - 4$
$F(4) = 4^2 + 3(4) - 4$
$F(4) = 16 + 12 - 4$
$F(4) = 24$

33. $F(x) = x^2 + 3x - 4$
$F(-3) = (-3)^2 + 3(-3) - 4$
$F(-3) = 9 - 9 - 4$
$F(-3) = -4$

35. $q\,H(p)=\dfrac{3p}{p+2}$

$H(1)=\dfrac{3(1)}{1+2}$

$H(1)=\dfrac{3}{3}$

$H(1)=1$

35. $H(p)=\dfrac{3p}{p+2}$

$H(-3)=\dfrac{3(-3)}{-3+2}$

$H(-3)=\dfrac{-9}{-1}$

$H(-3)=9$

37. $H(p)=\dfrac{3p}{p+2}$

$H(v)=\dfrac{3v}{v+2}$

39. $s(t)=t^3-3t+4$
$s(-1)=(-1)^3-3(-1)+4$
$s(-1)=-1+3+4$
$s(-1)=6$

41. $s(t)=t^3-3t+4$
$s(a)=a^3-3a+4$

43. $P(x)=4x+7$
$P(-2+h)-P(-2)$
$=4(-2+h)+7-[4(-2)+7]$
$=-8+4h+7+8-7$
$=4h$

45. Evaluate the function $s=f(v)=0.015v^3$ when $v=15$.
$s=f(v)=0.015(15)^3=50.625$
The windmill will produce 50.625 watts.

47. \$114.29

49. a) \$4.75 per game
b) \$4.00 per game

51. Values of x for which $x-1=0$ are excluded from the domain of the function.
$x-1=0$
$x=1$
$\{x\mid x\neq 1\}$

53. No values are excluded. $\{x\mid -\infty<x<\infty\}$

55. No values are excluded. $\{x\mid -\infty<x<\infty\}$

57. No values are excluded. $\{x\mid -\infty<x<\infty\}$

59. No values are excluded $\{x\mid -\infty<x<\infty\}$

61. Values of x for which $x+2=0$ are excluded from the domain of the function.
$x+2=0$
$x=-2$
$\{x\mid x\neq -2\}$

Critical Thinking

63. a. $\{(-2,-8), (-1,-1), (0,0), (1,1), (2,8)\}$
b. Yes, this set of ordered pairs defines a function because each member of the domain is assigned exactly one member of the range.

65. A relation and a function are similar in that both are sets of ordered pairs. A function is a specific type of relation. A function is a relation in which there are no two ordered pairs with the same first element.

67. a. The speed of the paratrooper 11.5 s after the beginning of the jump is 36.3 ft/s.
b. 30 ft/s

69. a. 110 beats/min
b. 75 beats/min

Section 4.3

Concept Check

1. If the three ordered pairs appear to lie on the same line, then it is more likely that your calculations are accurate.

3. 0; 0

Objective A Exercises

5. $P = f(d) = 0.097d + 1$

$P = f(500) = 0.097(500) + 1 = 49.5$

At 500 m below the surface the pressure is 49.5 atm.

7. $y = x - 3$

x	y
−1	−4
0	−3
3	0

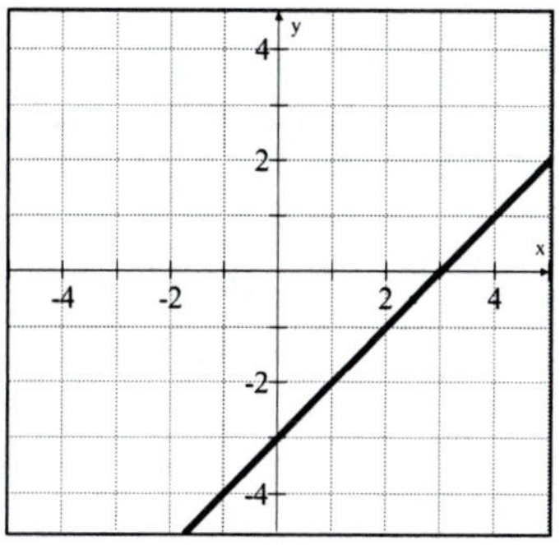

9. $y = -3x + 2$

x	y
0	2
1	−1
2	−4

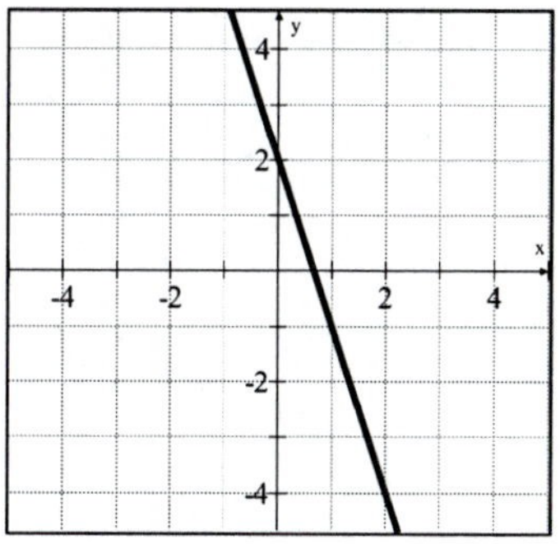

11. $f(x) = 3x - 4$

x	y
0	−4
1	−1
2	2

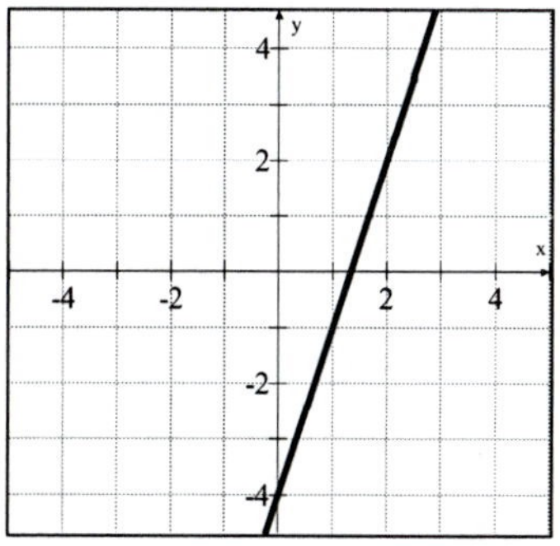

13.

$$f(x) = -\frac{2}{3}x$$

x	$f(x)$
−3	2
0	0
3	−2

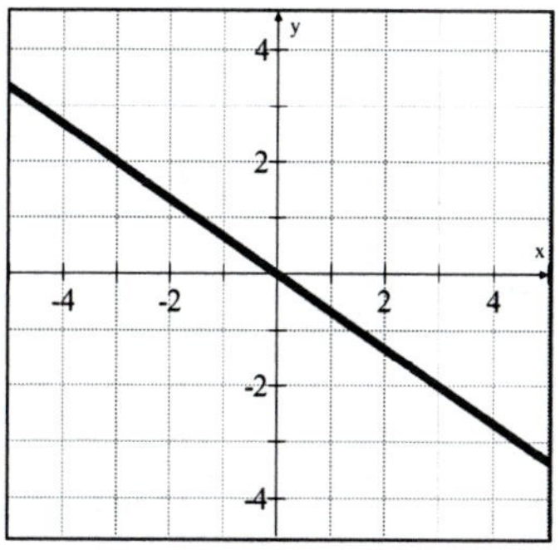

15. $y = \frac{2}{3}x - 4$

x	$f(x)$
0	−4
3	−2
6	0

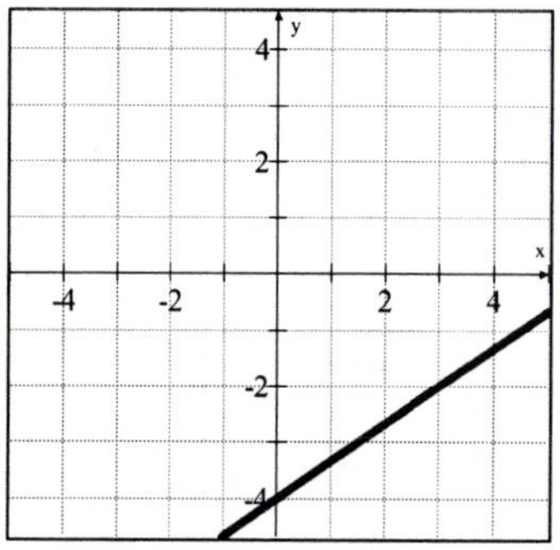

17. $f(x) = -\frac{1}{3}x + 2$

x	$f(x)$
−3	3
0	2
3	1

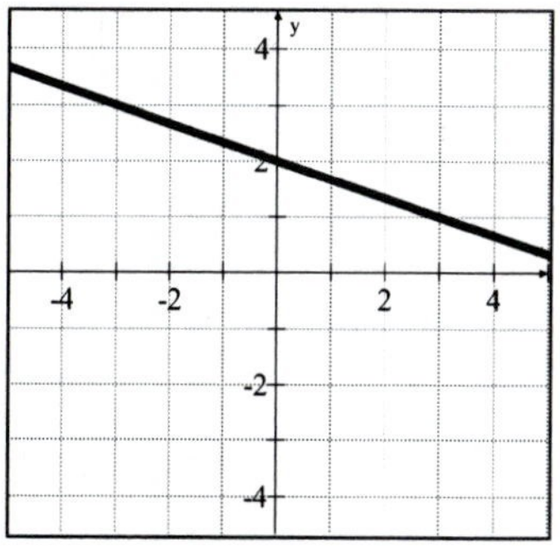

Objective B Exercises

19. $2x + y = -3$

$y = -2x - 3$

x	y
0	−3
1	−5
−1	−1

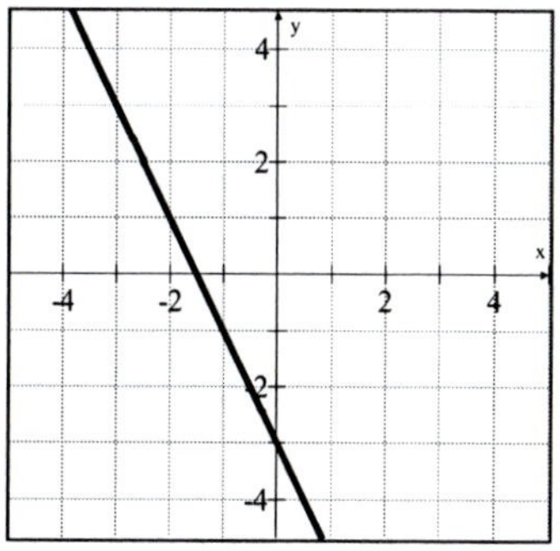

21. $x-4y=8$

$$-4y=-x+8$$

$$y=\frac{1}{4}x-2$$

x	y
0	−2
4	−1
8	0

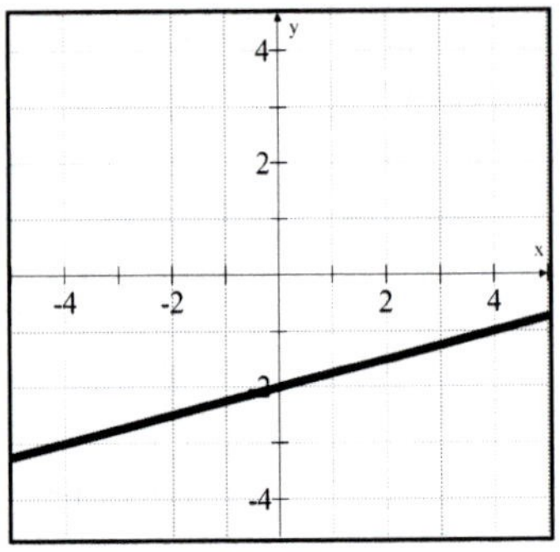

23. $4x+3y=12$

$$3y=-4x+12$$

$$y=-\frac{4}{3}x+4$$

x	y
0	4
3	0
6	−4

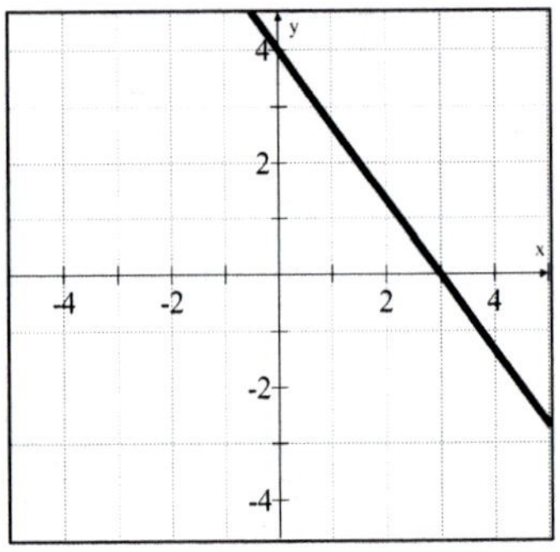

25. $x-3y=0$

$$-3y=-x$$

$$y=\frac{1}{3}x$$

x	y
0	0
3	1
−3	−1

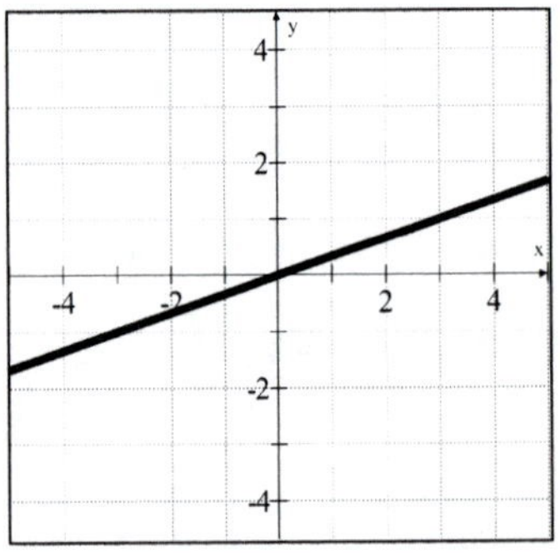

27. $y=-2$

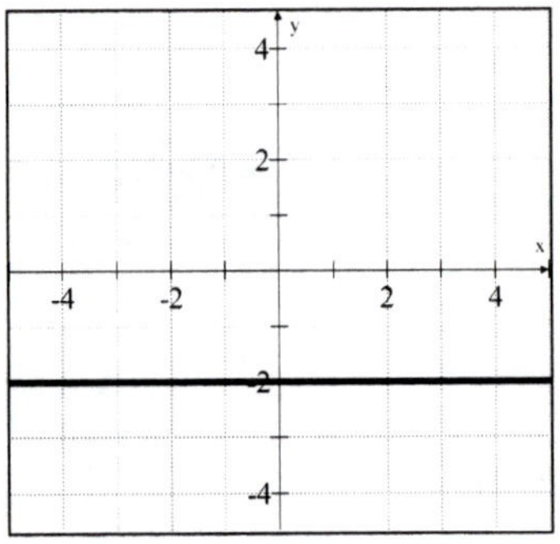

29. $x=-3$

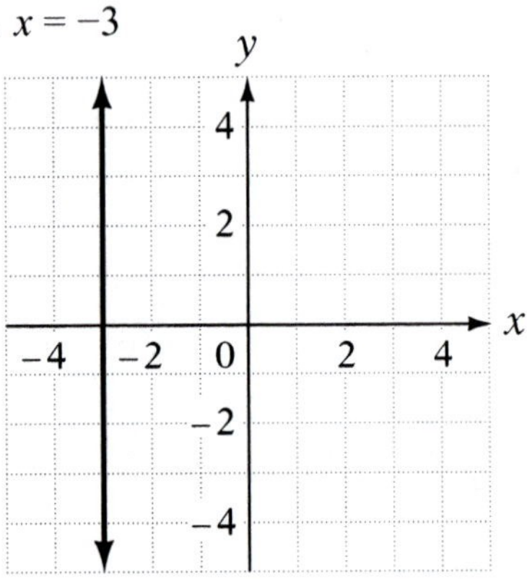

31. $3x-y=-2$

$$-y=-3x-2$$

$$y=3x+2$$

x	y
0	2
1	5
−1	−1

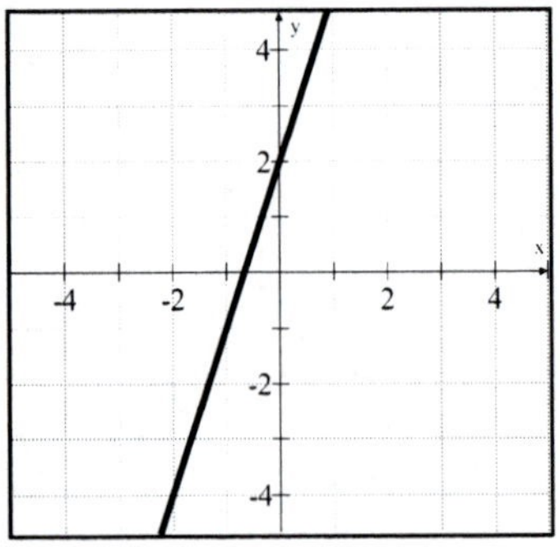

33. $3x - 2y = 8$

$$-2y = -3x + 8$$
$$y = \frac{3}{2}x - 4$$

x	y
0	−4
2	−1
4	2

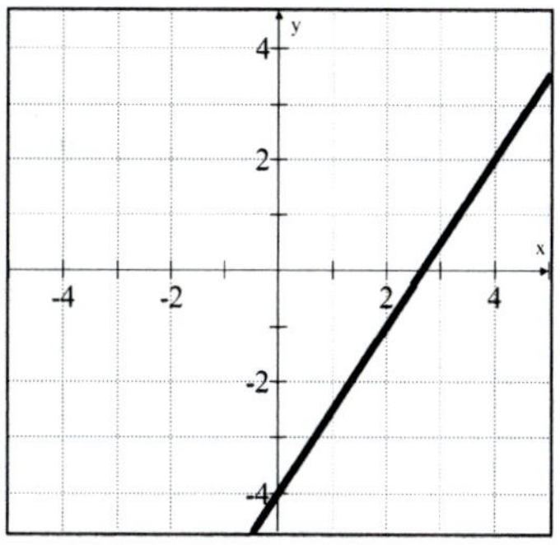

35. No. If $B = 0$, then it is not possible to solve $Ax + By = C$ for y.

Objective C Exercises

37. x- intercept:

$$x - 2y = -4$$
$$x - 2(0) = -4$$
$$x = -4$$

$(-4,0)$

y-intercept:

$$x - 2y = -4$$
$$0 - 2y = -4$$
$$-2y = -4$$
$$y = 2$$

$(0,2)$

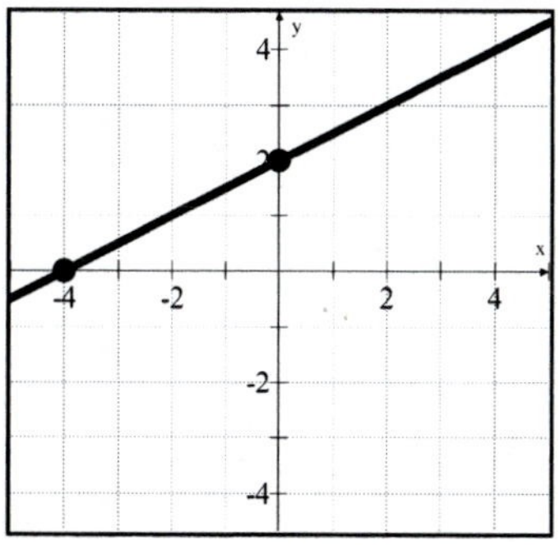

39. x- intercept:

$$2x - 3y = 9$$
$$2x - 3(0) = 9$$
$$2x = 9$$
$$x = \frac{9}{2}$$

$\left(\frac{9}{2},0\right)$

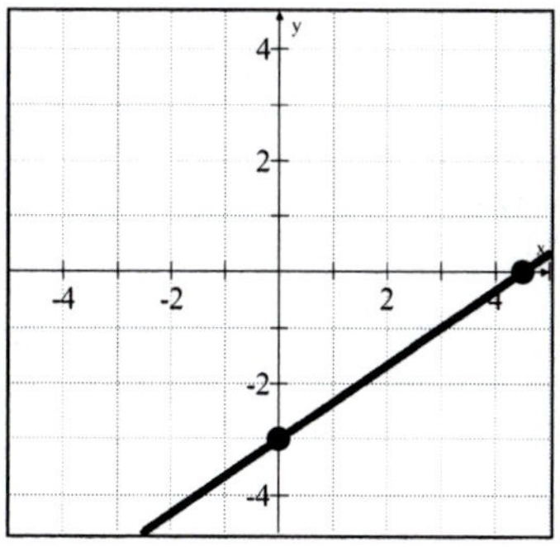

y-intercept:

$$2x - 3y = 9$$
$$2(0) - 3y = 9$$
$$-3y = 9$$
$$y = -3$$

$(0,-3)$

41. x- intercept:

$$2x + y = 3$$
$$2x + 0 = 3$$
$$2x = 3$$
$$x = \frac{3}{2}$$

$\left(\frac{3}{2},0\right)$

y-intercept:

$$2x + y = 3$$
$$2(0) + y = 3$$
$$y = 3$$

$(0,3)$

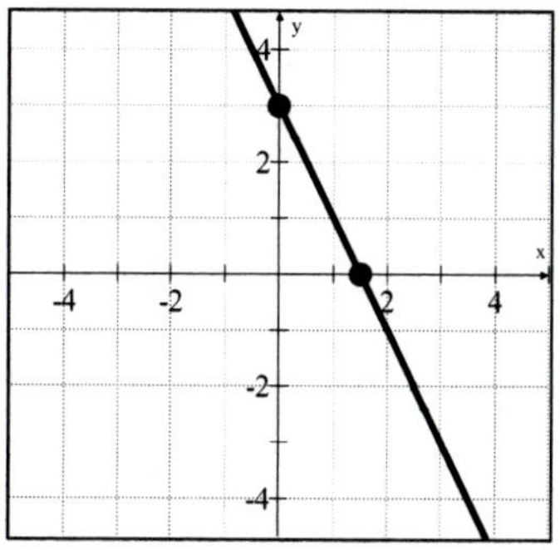

43. x- intercept:

$$3x + 2y = 4$$
$$3x + 2(0) = 4$$
$$3x = 4$$
$$x = \frac{4}{3}$$

$\left(\frac{4}{3},0\right)$

y-intercept:

$$3x + 2y = 4$$
$$3(0) + 2y = 4$$
$$2y = 4$$
$$y = 2$$

$(0,2)$

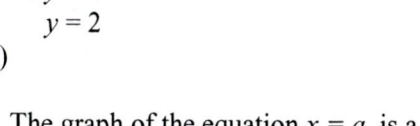

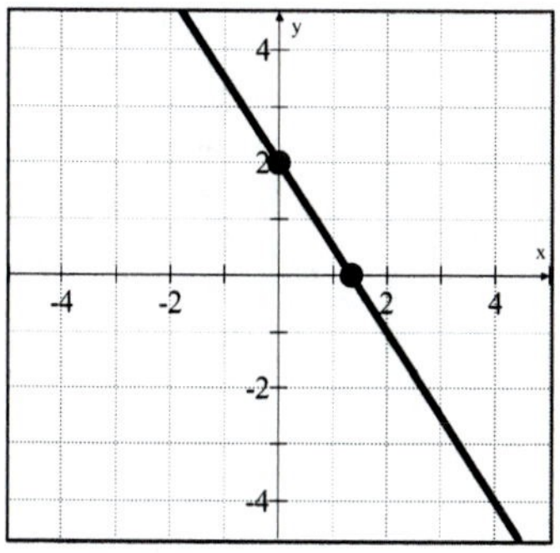

45. No. The graph of the equation $x = a$ is a vertical line and has no y-intercept.

47.

$$f(x) = 4x + 8$$
$$0 = 4x + 8$$
$$0 - 8 = 4x + 8 - 8$$
$$-8 = 4x$$
$$\frac{-8}{4} = \frac{4x}{4}$$
$$-2 = x$$

49.
$$s(t)=\frac{3}{4}t-9$$
$$0=\frac{3}{4}t-9$$
$$4(0)=4\left(\frac{3}{4}t\right)-9(4)$$
$$0=3t-36$$
$$0+36=3t-36+36$$
$$36=3t$$
$$\frac{36}{3}=\frac{3t}{3}$$
$$12=t$$

51.
$$f(x)=4x$$
$$0=4x$$
$$\frac{0}{4}=\frac{4x}{4}$$
$$0=4$$

53.
$$g(x)=\frac{3}{2}x-4$$
$$0=\frac{3}{2}x-4$$
$$2(0)=2\left(\frac{3}{2}x\right)-4(2)$$
$$0=3x-8$$
$$0+8=3x-8+8$$
$$8=3x$$
$$\frac{8}{3}=\frac{3x}{3}$$
$$\frac{8}{3}=x$$

Objective D Exercises

55. $B = 1200t$
$B = 1200(7)$
$B = 8400$
The heart of a hummingbird will beat 8400 times in 7 min.

57. $W = 11t$

t	W
0	0
5	55
15	165

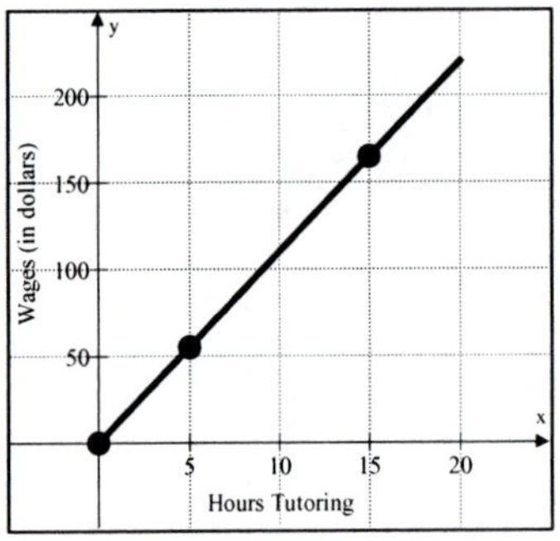

Marlys receives \$165 for tutoring 15 h.

59. $C = 80n + 5000$

n	C
0	5000
50	9000
100	13000

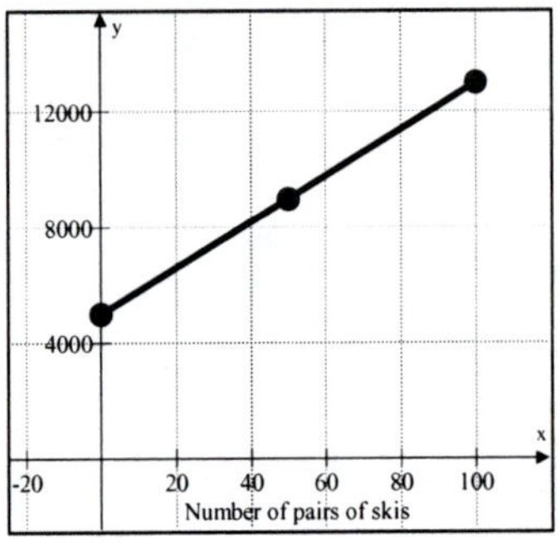

The cost of manufacturing 50 pairs of skis is \$9000.

61. a) $D = -30t$

t	D
0	0
20	−600
65	−1950

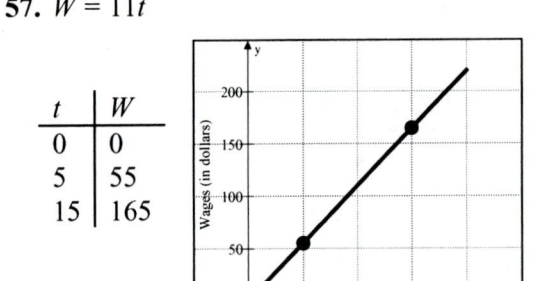

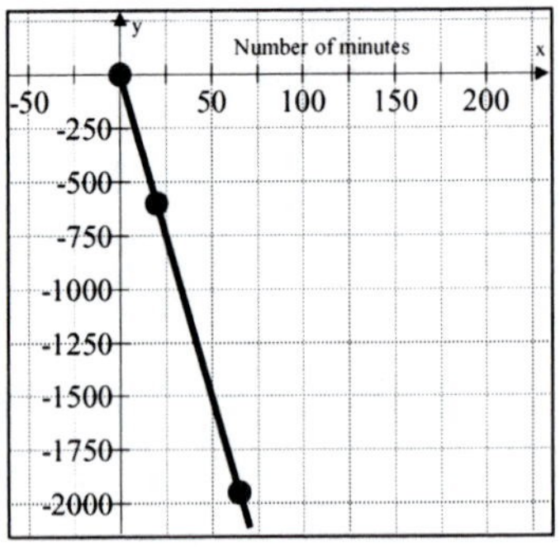

After 65 min, Alvin is 1950 m below sea level.

b) $D = -48t$

t	D
0	0
20	−960
65	−3120

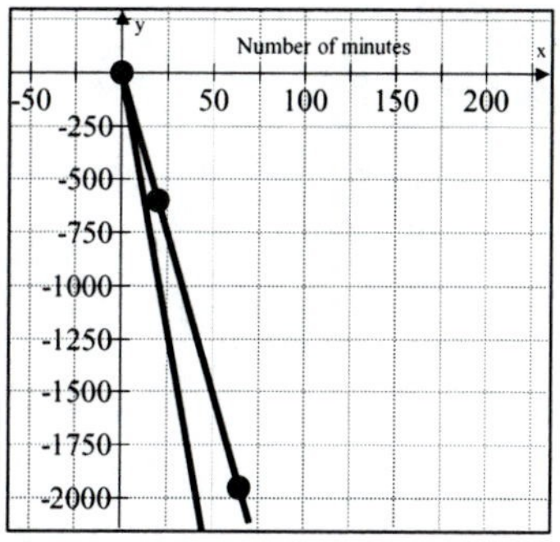

The point is below (65, −1950).

Critical Thinking

63. To graph the equation of a straight line by plotting points, find three ordered pair solutions of the equation. Plot these ordered pairs in a rectangular coordinate system. Draw a straight line through the points.

65. The x and y intercepts of the graph of the equation $4x + 3y = 0$ are both (0,0). A straight line is determined by two points, so we need to find another point on the line in order to graph this equation.

Projects or Group Activities

67. $\frac{x}{2} + \frac{y}{3} = 1$

x-intercept (2,0)
y-intercept (0,3)

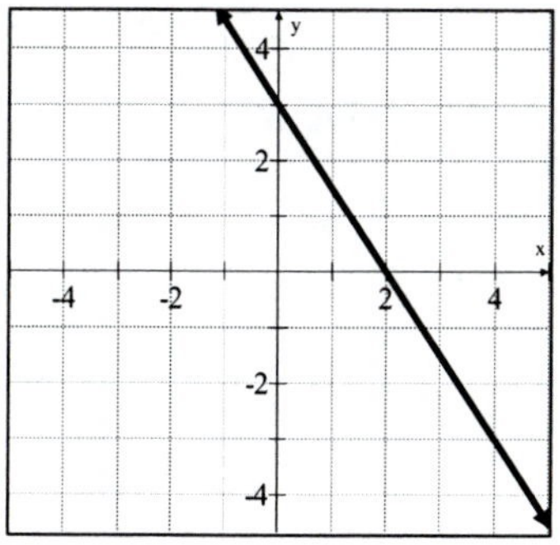

Check Your Progress: Chapter 4

1.

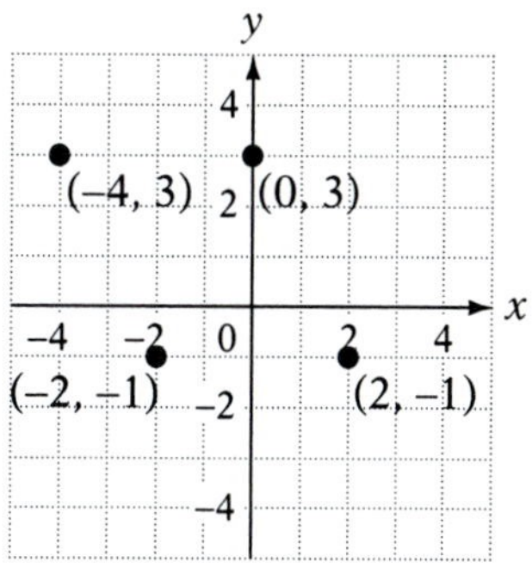

2. $f(x) = -\frac{4}{5}x + 4$

x	$f(x)$
0	4
5	0
−5	8

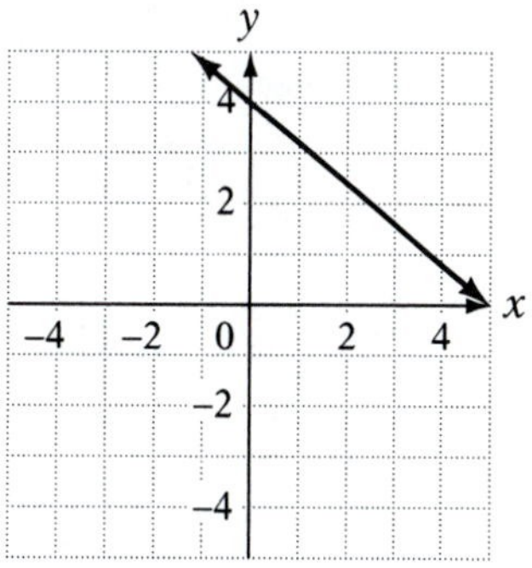

3. $3x - 4y = 12$

x-intercept

$3x - 4(0) = 12$

$3x = 12$

$x = 4$

(4, 0)

y-intercept

$3(0) - 4y = 12$

$-4y = 12$

$y = -3$

(0, –3)

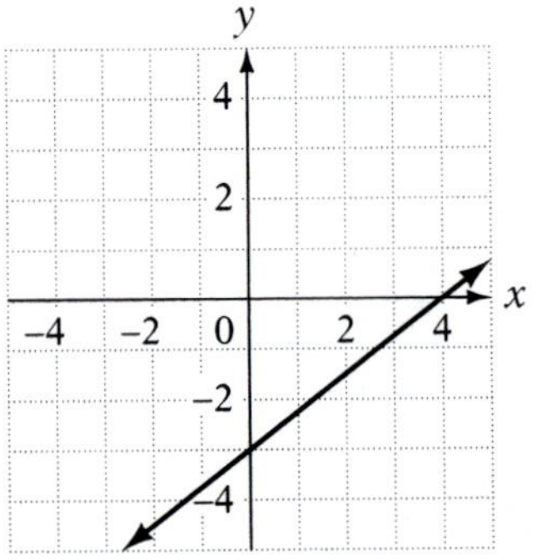

4. $y = -3x + 2$

$y = -3(-1) + 2$

$y = 3 + 2$

$y = 5$

$(-1, 5)$

$y = -3(0) + 2$

$y = 2$

$(0, 2)$

$y = -3(1) + 2$

$y = -3 + 2$

$y = -1$

$(1, -1)$

$y = -3(2) + 2$

$y = -6 + 2$

$y = -4$

$(2, -4)$

5. A vertical line passing through (–5, 0).

6. D: $\{-3, -2, -1\}$ R: $\{-2, -1, 0\}$

No x-value is repeated, so yes it is a function.

7. $h(s) = s^2 + 3s$

$h(-3) = (-3)^2 + 3(-3)$

$h(-3) = 9 - 9$

$h(-3) = 0$

8.

y	$-\frac{1}{3}x - 1$
0	$-\frac{1}{3}(-3) - 1$
	$1 - 1$
	0

Yes, (–3, 0) is a solution.

9. Yes

10. No x-value is repeated, so yes it is a function.

D: $\{-5, 0, 1, 2\}$ R: $\{1, 2, 3\}$

11. $2x \neq 0$

$x \neq 0$

$D = \{x \mid x \neq 0\}$

12. $s(t) = -3t^2 + 4t - 1$

$s(-3) = -3(-3)^2 + 4(-3) - 1$

$s(-3) = -3(9) - 12 - 1$

$s(-3) = -27 - 12 - 1$

$s(-3) = -40$

13. $f(x) = 2x - 3$

$f(2-a) = 2(2-a) - 3$

$f(2-a) = 4 - 2a - 3$

$f(2-a) = 1 - 2a$

14. $f(x) = -\frac{3}{4}x + 2$

x	$f(x)$
0	2
4	–1
–4	5

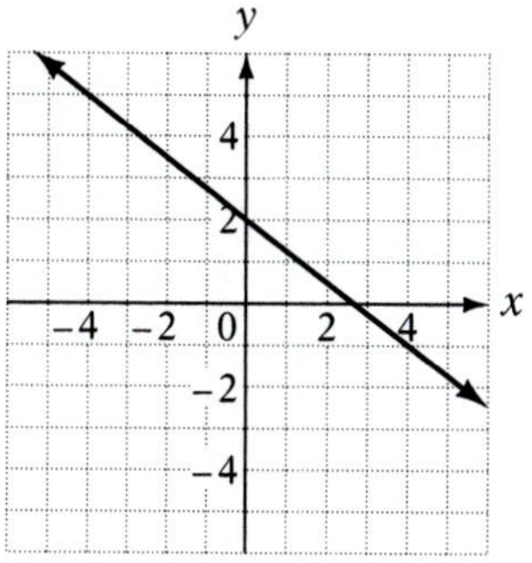

15. $y = 2x - 3$

x	y
0	-3
1	–1
2	1

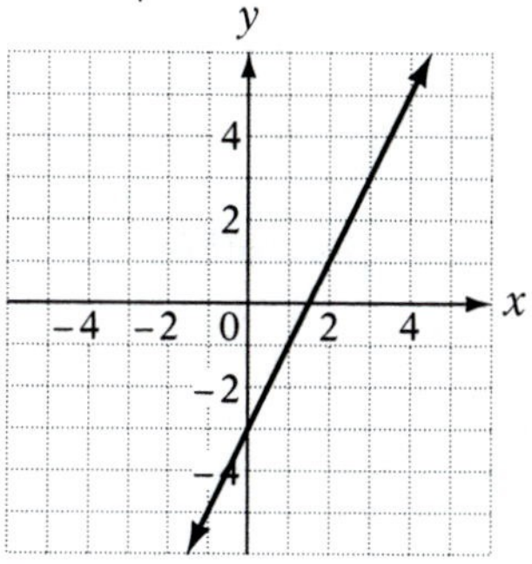

16. $y + 5 = 0$

$y = -5$

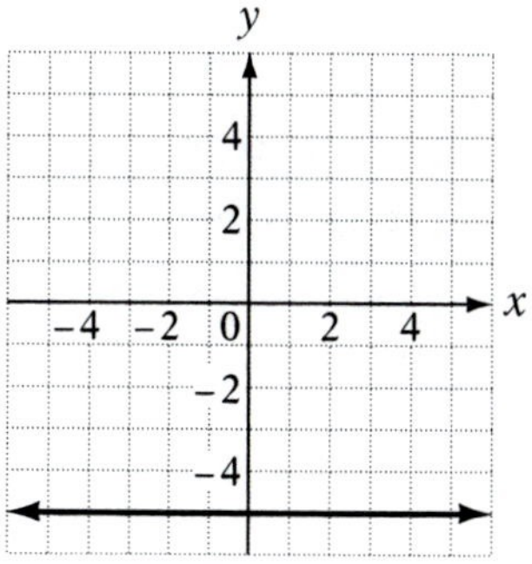

17. $x = 4$

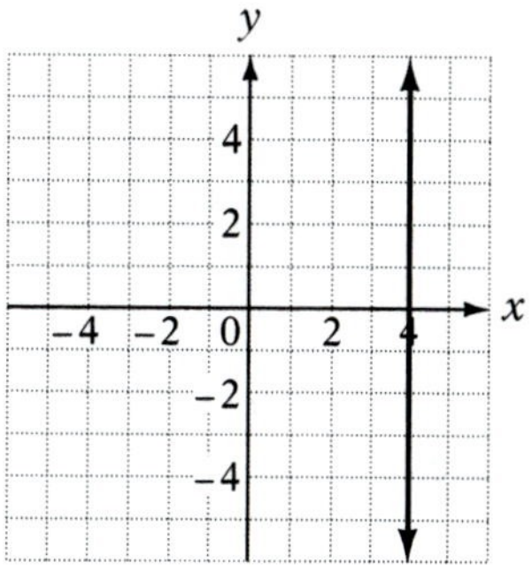

18. $4x - 5y = 20$

x-intercept

$4x - 5(0) = 20$

$4x = 20$

$x = 5$

(5, 0)

y-intercept

$4(0) - 5y = 20$

$-5y = 20$

$y = -4$

(0, –4)

19. $g(t) = 3t + 6$

$0 = 3t + 6$

$0 - 6 = 3t + 6 - 6$

$-6 = 3t$

$\frac{-6}{3} = \frac{3t}{3}$

$-2 = t$

20. $F = 2.80M + 2.20$

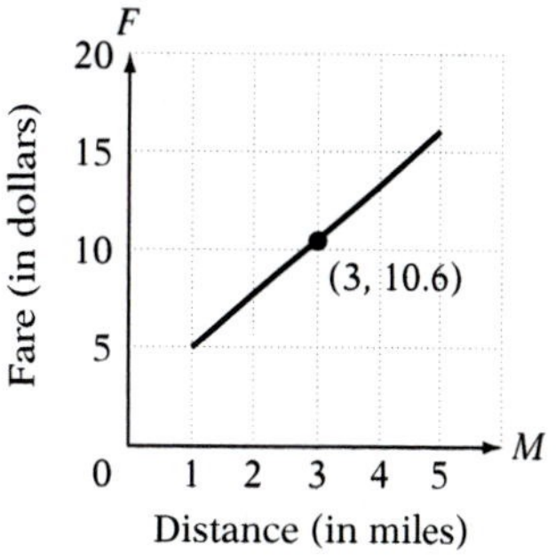

A 3-mile taxi ride costs $10.60.

Section 4.4

Concept Check

1. increases

3. 3; (0, 4)

5. −1; (0, 1)

Objective A Exercises

7. (1,3), (3,1)

$$m = \frac{y_2 - y_1}{x_2 - x_1} = \frac{1-3}{3-1} = \frac{-2}{2} = -1$$

The slope is −1.

9. (−1,4), (2,5)

$$m = \frac{y_2 - y_1}{x_2 - x_1} = \frac{5-4}{2-(-1)} = \frac{1}{3}$$

The slope is $\frac{1}{3}$.

11. (−1,3), (−4, 5)

$$m = \frac{y_2 - y_1}{x_2 - x_1} = \frac{5-3}{-4-(-1)} = \frac{2}{-3} = -\frac{2}{3}$$

The slope is $-\frac{2}{3}$.

13. (0,3), (4,0)

$$m = \frac{y_2 - y_1}{x_2 - x_1} = \frac{3-0}{0-4} = \frac{3}{-4} = -\frac{3}{4}$$

The slope is $-\frac{3}{4}$.

15. (2,4), (2,−2)

$$m = \frac{y_2 - y_1}{x_2 - x_1} = \frac{4-(-2)}{2-2} = \frac{6}{0}$$

The slope is undefined.

17. (2,5), (−3,−2)

$$m = \frac{y_2 - y_1}{x_2 - x_1} = \frac{5-(-2)}{2-(-3)} = \frac{7}{5}$$

The slope is $\frac{7}{5}$.

19. (2,3), (−1,3)

$$m = \frac{y_2 - y_1}{x_2 - x_1} = \frac{3-3}{2-(-1)} = \frac{0}{3} = 0$$

The slope is 0.

21. (0,4), (−2, 5)

$$m = \frac{y_2 - y_1}{x_2 - x_1} = \frac{5-4}{-2-0} = \frac{1}{-2} = -\frac{1}{2}$$

The slope is $-\frac{1}{2}$.

23. (−3,−1), (−3,4)

$$m = \frac{y_2 - y_1}{x_2 - x_1} = \frac{4-(-1)}{-3-(-3)} = \frac{5}{0}$$

The slope is undefined.

25. If *a* and *c* are equal the slope of *l* is undefined.

27. $m = \frac{240-80}{6-2} = \frac{160}{4} = 40$

The average speed of the motorist is 40 mph.

29. $m = \frac{275-125}{20-50} = \frac{150}{-30} = -5$

The temperature of the oven decreases 5°/min.

31. $m = \frac{13-6}{40-180} = \frac{7}{-140} = -0.05$

Approximately 0.05 gallon of fuel is used for each mile that the car is driven.

33. $m = \frac{5000}{14.19} = 352.4$

The average speed of the runner was 352.4m/min.

35. a) $\dfrac{6\ in}{5\ ft} = \dfrac{6\ in}{60\ in} = \dfrac{1}{10} > \dfrac{1}{12}$

No it does not meet the requirements for ANSI.

b) $\dfrac{12}{170} = \dfrac{6}{85} < \dfrac{1}{12}$

Yes, it does meet the requirements for ANSI.

37. $m = \dfrac{1.8 - 3.7}{5 - 2} = \dfrac{-1.9}{3} = -\dfrac{19}{30} \approx 0.6\overline{3}$

Each minute, the water in the lock decreases by $0.6\overline{3}$ million gallons.

Objective B Exercises

39. $y = \dfrac{1}{2}x + 2$

$m = \dfrac{1}{2}$

y – intercept (0,2)

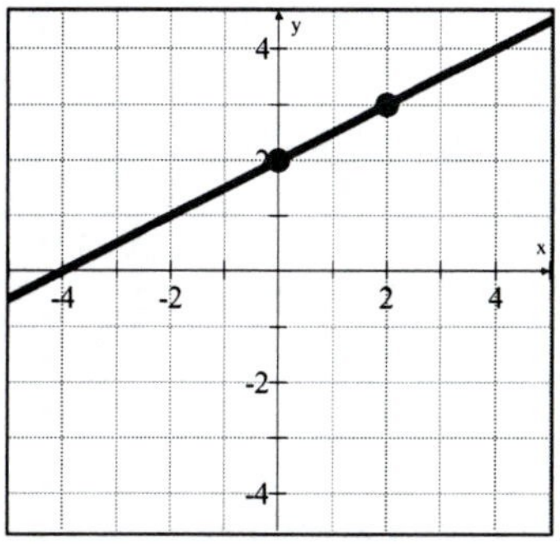

41. $y = -\dfrac{3}{2}x$

$m = -\dfrac{3}{2}$

y – intercept (0,0)

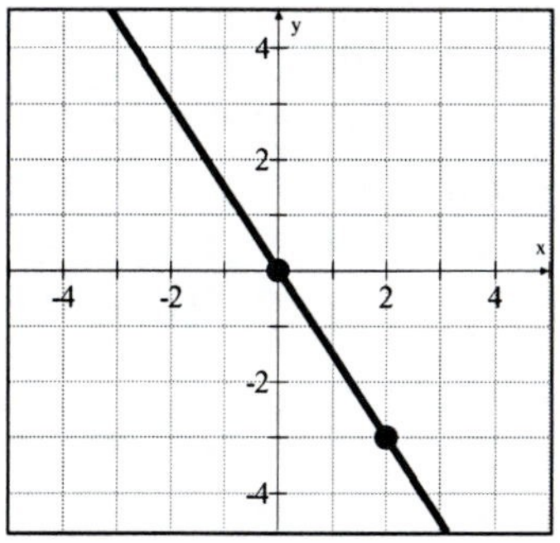

43. $y = -\dfrac{1}{2}x + 2$

$m = -\dfrac{1}{2}$

y – intercept (0,2)

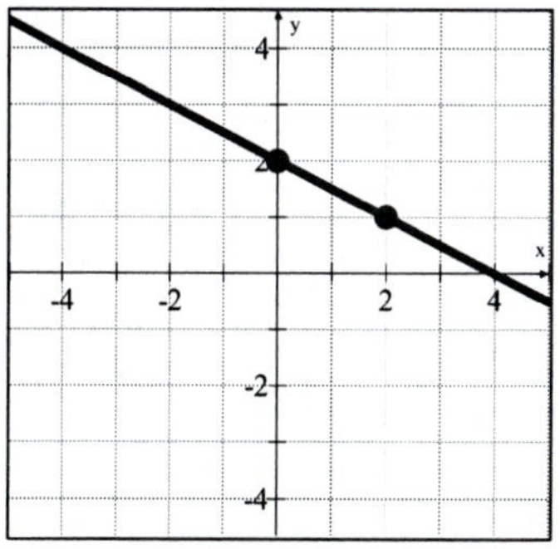

45. $y = 2x - 4$

$m = 2$

y – intercept (0,–4)

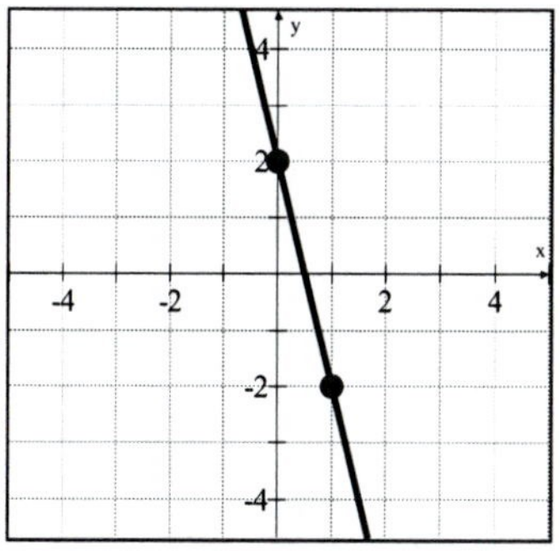

47. $4x - y = 1$

$-y = -4x + 1$

$y = 4x - 1$

$m = 4$

y – intercept (0,–1)

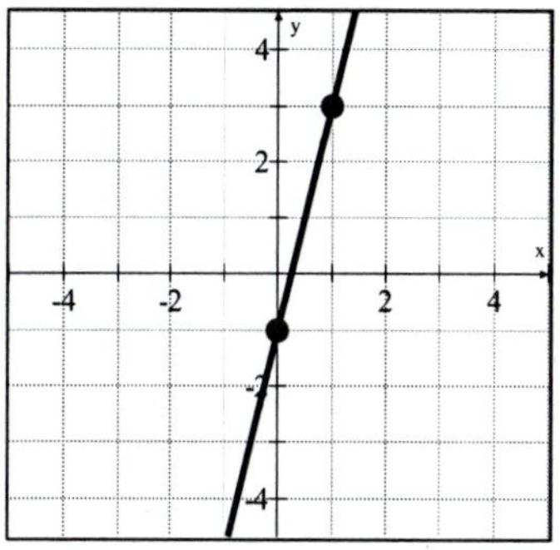

49. $x - 3y = 3$

$-3y = -x + 3$

$y = \dfrac{1}{3}x - 1$

$m = \dfrac{1}{3}$

y-intercept (0,–1)

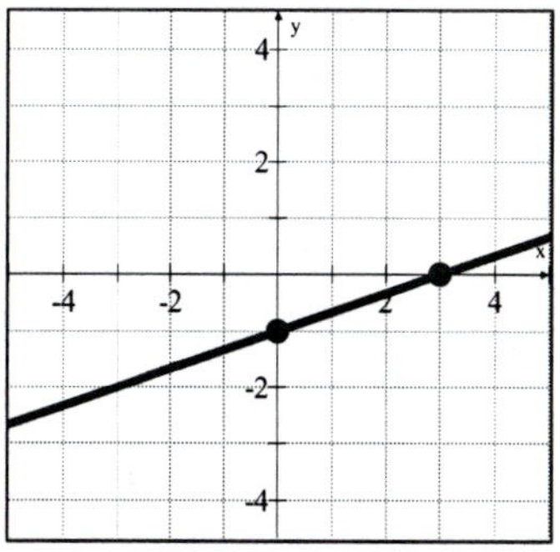

51.

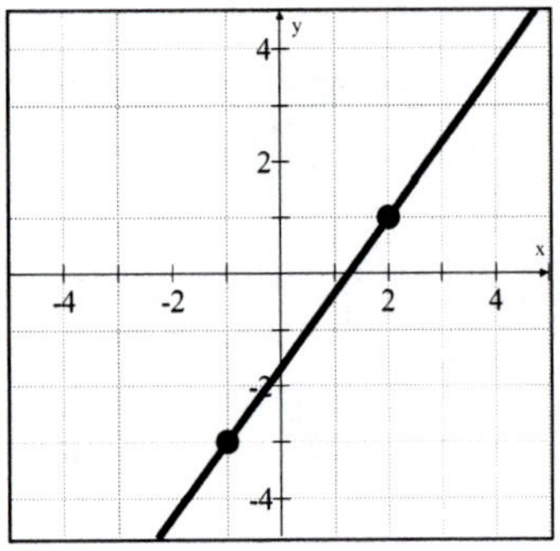

53.

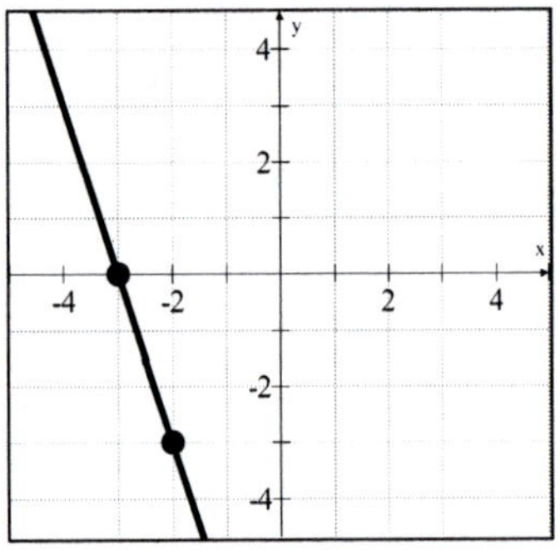

55.

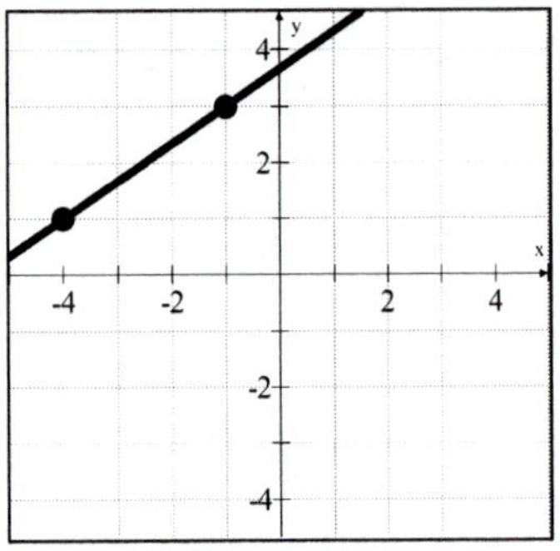

57. a) Below
b) Negative

Critical Thinking

59. increases by 2

61. increases by 2

Projects or Group Activities

63. The slope between each pair of points must be the same if all of the points line on the same line.

a) $P_1 = (2,5)$
$P_2 = (-1,-1)$
$P_3 = (3,7)$

P_1 to P_2: $m = \dfrac{5-(-1)}{2-(-1)} = \dfrac{6}{3} = 2$

P_2 to P_3: $m = \dfrac{7-(-1)}{3-(-1)} = \dfrac{8}{4} = 2$

P_3 to P_1: $m = \dfrac{7-5}{3-2} = \dfrac{2}{1} = 2$

Yes, the points all lie on the same line.

b) $P_1 = (-1,5)$
$P_2 = (0,3)$
$P_3 = (-3,4)$

P_1 to P_2: $m = \dfrac{5-3}{-1-0} = \dfrac{2}{-1} = -2$

P_2 to P_3: $m = \dfrac{4-3}{-3-0} = \dfrac{1}{-3} = -\dfrac{1}{3}$

P_3 to P_1: $m = \dfrac{5-4}{-1-(-3)} = \dfrac{1}{2}$

No, the points do not all lie on the same line.

65. $P_1 = (k, 1)$
$P_2 = (0,-1)$
$P_3 = (2, -2)$

P_2 to P_3: $m = -\dfrac{1}{2}$

The slope from P_1 to P_2 and from P_1 to P_3 must also be $-\dfrac{1}{2}$. Set the slope from P_1 to P_2 equal to $-\dfrac{1}{2}$ and solve for k.

$\dfrac{2}{k-0} = -\dfrac{1}{2}$

$k = -4$

Section 4.5

Concept Check

1. One

3. A point on the line

Objective A Exercises

5. When we know the slope and the y-intercept we can find the equation of the line using the slope-intercept form, $y = mx + b$. The value of the slope is substituted in for m, and the y-coordinate of the y-intercept is substituted in for b.

7. The line must pass through the origin (0,0).

9. $m = 2, b = 5$
$y = mx + b$
$y = 2x + 5$
The equation of the line is $y = 2x + 5$.

11. $m = \frac{1}{2}, (x_1, y_1) = (2, 3)$

$$y - y_1 = m(x - x_1)$$
$$y - 3 = \frac{1}{2}(x - 2)$$
$$y - 3 = \frac{1}{2}x - 1$$
$$y = \frac{1}{2}x + 2$$

The equation of the line is $y = \frac{1}{2}x + 2$.

13. $m = \frac{5}{4}, (x_1, y_1) = (-1, 4)$

$$y - y_1 = m(x - x_1)$$
$$y - 4 = \frac{5}{4}[x - (-1)]$$
$$y - 4 = \frac{5}{4}(x + 1)$$
$$y - 4 = \frac{5}{4}x + \frac{5}{4}$$
$$y = \frac{5}{4}x + \frac{21}{4}$$

The equation of the line is $y = \frac{5}{4}x + \frac{21}{4}$.

15. $m = -\frac{5}{3}, (x_1, y_1) = (3, 0)$

$$y - y_1 = m(x - x_1)$$
$$y - 0 = -\frac{5}{3}(x - 3)$$
$$y = -\frac{5}{3}(x - 3)$$
$$y = -\frac{5}{3}x + 5$$

The equation of the line is $y = -\frac{5}{3}x + 5$.

17. $m = -3, (x_1, y_1) = (2, 3)$

$$y - y_1 = m(x - x_1)$$
$$y - 3 = -3(x - 2)$$
$$y - 3 = -3x + 6$$
$$y = -3x + 9$$

The equation of the line is $y = -3x + 9$.

19. $m = -3$, $(x_1, y_1) = (-1, 7)$

$y - y_1 = m(x - x_1)$

$y - 7 = -3[x - (-1)]$

$y - 7 = -3(x + 1)$

$y - 7 = -3x - 3$

$y = -3x + 4$

The equation of the line is $y = -3x + 4$.

21. $m = \frac{2}{3}$, $(x_1, y_1) = (-1, -3)$

$y - y_1 = m(x - x_1)$

$y - (-3) = \frac{2}{3}[(x - (-1)]$

$y + 3 = \frac{2}{3}(x + 1)$

$y + 3 = \frac{2}{3}x + \frac{2}{3}$

$y = \frac{2}{3}x - \frac{7}{3}$

The equation of the line is $y = \frac{2}{3}x - \frac{7}{3}$.

23. $m = \frac{1}{2}$, $(x_1, y_1) = (0, 0)$

$y - y_1 = m(x - x_1)$

$y - 0 = \frac{1}{2}(x - 0)$

$y = \frac{1}{2}x$

The equation of the line is $y = \frac{1}{2}x$.

25. $m = 3$, $(x_1, y_1) = (2, -3)$

$y - y_1 = m(x - x_1)$

$y - (-3) = 3(x - 2)$

$y + 3 = 3x - 6$

$y = 3x - 9$

The equation of the line is $y = 3x - 9$.

27. $m = -\frac{2}{3}$, $(x_1, y_1) = (3, 5)$

$y - y_1 = m(x - x_1)$

$y - 5 = -\frac{2}{3}(x - 3)$

$y - 5 = -\frac{2}{3}x + 2$

$y = -\frac{2}{3}x + 7$

The equation of the line is $y = -\frac{2}{3}x + 7$.

29. $m = -1$, $b = -3$

$y = -x - 3$

The equation of the line is $y = -x - 3$.

31. $m = \frac{7}{5}$, $(x_1, y_1) = (1, -4)$

$y - y_1 = m(x - x_1)$

$y - (-4) = \frac{7}{5}(x - 1)$

$y + 4 = \frac{7}{5}x - \frac{7}{5}$

$y = \frac{7}{5}x - \frac{27}{5}$

The equation of the line is $y = \frac{7}{5}x - \frac{27}{5}$.

33. $m = -\frac{2}{5}$, $(x_1, y_1) = (4, -1)$

$y - y_1 = m(x - x_1)$

$y - (-1) = -\frac{2}{5}(x - 4)$

$y + 1 = -\frac{2}{5}x + \frac{8}{5}$

$y = -\frac{2}{5}x + \frac{3}{5}$

The equation of the line is $y = -\frac{2}{5}x + \frac{3}{5}$.

35. Slope is undefined, $(x_1, y_1) = (3, -4)$
The line is a vertical line. All points on the line have an abscissa of 3.
The equation of the line is $x = 3$.

37. $m = -\frac{5}{4}$, $(x_1, y_1) = (-2, -5)$

$$y - y_1 = m(x - x_1)$$
$$y - (-5) = -\frac{5}{4}[x - (-2)]$$
$$y + 5 = -\frac{5}{4}(x + 2)$$
$$y + 5 = -\frac{5}{4}x - \frac{10}{4}$$
$$y = -\frac{5}{4}x - \frac{15}{2}$$

The equation of the line is $y = -\frac{5}{4}x - \frac{15}{2}$.

39. $m = 0$, $(x_1, y_1) = (-2, -3)$

$$y - y_1 = m(x - x_1)$$
$$y - (-3) = 0[x - (-2)]$$
$$y + 3 = 0$$
$$y = -3$$

The equation of the line is $y = -3$.

41. $m = -2$, $(x_1, y_1) = (4, -5)$

$$y - y_1 = m(x - x_1)$$
$$y - (-5) = -2(x - 4)$$
$$y + 5 = -2x + 8$$
$$y = -2x + 3$$

The equation of the line is $y = -2x + 3$.

43. Slope is undefined, $(x_1, y_1) = (-5, -1)$
The line is a vertical line. All points on the line have an abscissa of -5.
The equation of the line is $x = -5$.

Objective B Exercises

45. Check that the coordinates of each given point are a solution of your equation.

47. (0,2), (3,5)

$$m = \frac{y_2 - y_1}{x_2 - x_1} = \frac{5 - 2}{3 - 0} = \frac{3}{3} = 1$$
$$y - y_1 = m(x - x_1)$$
$$y - 2 = 1(x - 0)$$
$$y - 2 = x$$
$$y = x + 2$$

The equation of the line is $y = x + 2$.

49. (0,−3), (−4,5)

$$m = \frac{y_2 - y_1}{x_2 - x_1} = \frac{5 - (-3)}{-4 - 0} = \frac{8}{-4} = -2$$
$$y - y_1 = m(x - x_1)$$
$$y - (-3) = -2(x - 0)$$
$$y + 3 = -2x$$
$$y = -2x - 3$$
$$y = -2x - 3$$

The equation of the line is $y = -2x - 3$.

51. (2,3), (5,5)

$$m = \frac{y_2 - y_1}{x_2 - x_1} = \frac{5 - 3}{5 - 2} = \frac{2}{3}$$
$$y - y_1 = m(x - x_1)$$
$$y - 3 = \frac{2}{3}(x - 2)$$
$$y - 3 = \frac{2}{3}x - \frac{4}{3}$$
$$y = \frac{2}{3}x + \frac{5}{3}$$

The equation of the line is $y = \frac{2}{3}x + \frac{5}{3}$.

53. (−1,3), (2,4)

$$m = \frac{y_2 - y_1}{x_2 - x_1} = \frac{4-3}{2-(-1)} = \frac{1}{3}$$

$$y - y_1 = m(x - x_1)$$

$$y - 3 = \frac{1}{3}[x - (-1)]$$

$$y - 3 = \frac{1}{3}(x+1)$$

$$y - 3 = \frac{1}{3}x + \frac{1}{3}$$

$$y = \frac{1}{3}x + \frac{10}{3}$$

The equation of the line is $y = \frac{1}{3}x + \frac{10}{3}$.

55. (−1,−2), (3,4)

$$m = \frac{y_2 - y_1}{x_2 - x_1} = \frac{4-(-2)}{3-(-1)} = \frac{6}{4} = \frac{3}{2}$$

$$y - y_1 = m(x - x_1)$$

$$y - 4 = \frac{3}{2}(x-3)$$

$$y - 4 = \frac{3}{2}x - \frac{9}{2}$$

$$y = \frac{3}{2}x - \frac{1}{2}$$

The equation of the line is $y = \frac{3}{2}x - \frac{1}{2}$.

57. (0,3), (2,0)

$$m = \frac{y_2 - y_1}{x_2 - x_1} = \frac{3-0}{0-2} = -\frac{3}{2}$$

$$y - y_1 = m(x - x_1)$$

$$y = -\frac{3}{2}(x-2)$$

$$y = -\frac{3}{2}x + 3$$

The equation of the line is $y = -\frac{3}{2}x + 3$.

59. (−3,−1), (2,−1)

$$m = \frac{y_2 - y_1}{x_2 - x_1} = \frac{-1-(-1)}{2-(-3)} = \frac{0}{5} = 0$$

$$y - y_1 = m(x - x_1)$$

$$y - (-1) = 0[(x - (-3)]$$

$$y + 1 = 0$$

$$y = -1$$

The equation of the line is $y = -1$.

61. (−2,−3), (−1,−2)

$$m = \frac{y_2 - y_1}{x_2 - x_1} = \frac{-3-(-2)}{-2-(-1)} = \frac{-1}{-1} = 1$$

$$y - y_1 = m(x - x_1)$$

$$y - (-3) = 1[x - (-2)]$$

$$y + 3 = 1(x+2)$$

$$y = x - 1$$

The equation of the line is $y = x - 1$.

63. (−2,3), (2,−1)

$$m = \frac{y_2 - y_1}{x_2 - x_1} = \frac{3-(-1)}{-2-2} = \frac{4}{-4} = -1$$

$$y - y_1 = m(x - x_1)$$

$$y - (-1) = -1(x-2)$$

$$y + 1 = -x + 2$$

$$y = -x + 1$$

The equation of the line is $y = -x + 1$.

65. (2,3), (5,−5)

$$m = \frac{y_2 - y_1}{x_2 - x_1} = \frac{3-(-5)}{2-5} = \frac{8}{-3} = -\frac{8}{3}$$

$$y - y_1 = m(x - x_1)$$

$$y - 3 = -\frac{8}{3}(x-2)$$

$$y - 3 = -\frac{8}{3}x + \frac{16}{3}$$

$$y = -\frac{8}{3}x + \frac{25}{3}$$

The equation of the line is $y = -\frac{8}{3}x + \frac{25}{3}$.

67. (2,0), (0,–1)

$$m = \frac{y_2 - y_1}{x_2 - x_1} = \frac{0-(-1)}{2-0} = \frac{1}{2}$$

$$y - y_1 = m(x - x_1)$$

$$y - 0 = \frac{1}{2}(x-2)$$

$$y = \frac{1}{2}x - 1$$

The equation of the line is $y = \frac{1}{2}x - 1$.

69. (3,–4), (–2,–4)

$$m = \frac{y_2 - y_1}{x_2 - x_1} = \frac{-4-(-4)}{3-(-2)} = \frac{0}{5} = 0$$

$$y - y_1 = m(x - x_1)$$

$$y - (-4) = 0(x-3)$$

$$y + 4 = 0$$

$$y = -4$$

The equation of the line is $y = -4$.

71. (0,0), (4,3)

$$m = \frac{y_2 - y_1}{x_2 - x_1} = \frac{3-0}{4-0} = \frac{3}{4}$$

$$y - y_1 = m(x - x_1)$$

$$y - 0 = \frac{3}{4}(x-0)$$

$$y = \frac{3}{4}x$$

The equation of the line is $y = \frac{3}{4}x$.

73. (2,–1), (–1, 3)

$$m = \frac{y_2 - y_1}{x_2 - x_1} = \frac{3-(-1)}{-1-2} = \frac{4}{-3} = -\frac{4}{3}$$

$$y - y_1 = m(x - x_1)$$

$$y - (-1) = -\frac{4}{3}(x-2)$$

$$y + 1 = -\frac{4}{3}x + \frac{8}{3}$$

$$y = -\frac{4}{3}x + \frac{5}{3}$$

The equation of the line is $y = -\frac{4}{3}x + \frac{5}{3}$.

75. (–2,5), (–2,–5)

$$m = \frac{y_2 - y_1}{x_2 - x_1} = \frac{5-(-5)}{-2-(-2)} = \frac{10}{0}$$

The slope is undefined. The line is a vertical line. All points on the line have an abscissa of –2. The equation of the line is $x = -2$.

77. (2,1), (–2,–3)

$$m = \frac{y_2 - y_1}{x_2 - x_1} = \frac{1-(-3)}{2-(-2)} = \frac{4}{4} = 1$$

$$y - y_1 = m(x - x_1)$$

$$y - 1 = 1(x-2)$$

$$y = x - 1$$

The equation of the line is $y = x - 1$.

79. (–4, –3), (2,5)

$$m = \frac{y_2 - y_1}{x_2 - x_1} = \frac{5-(-3)}{2-(-4)} = \frac{8}{6} = \frac{4}{3}$$

$$y - y_1 = m(x - x_1)$$

$$y - 5 = \frac{4}{3}(x-2)$$

$$y - 5 = \frac{4}{3}x - \frac{8}{3}$$

$$y = \frac{4}{3}x + \frac{7}{3}$$

The equation of the line is $y = \frac{4}{3}x + \frac{7}{3}$.

81. (0,3), (3,0)

$$m = \frac{y_2 - y_1}{x_2 - x_1} = \frac{3-0}{0-3} = \frac{3}{-3} = -1$$

$$y - y_1 = m(x - x_1)$$
$$y - 0 = -1(x - 3)$$
$$y = -x + 3$$

The equation of the line is $y = -x + 3$.

Objective C Exercises

83. Strategy: Let x represent the number of minutes after takeoff.
Let y represent the height of the plane in feet.
Use the slope-intercept form of an equation to find the equation of the line.

Solution:
a) y-intercept (0,0); slope is 1200
$y = mx + b$
$y = 1200x + 0$
The linear function is $f(x) = 1200x$.

$$0 \le x \le 26\frac{2}{3}$$

b) Find the height of the plane 11 minutes after takeoff.
$y = 1200(11) = 13{,}200$
Eleven minutes after takeoff the height of the plane will be 13,200 ft.

85. Strategy: Let x represent the year.
Let y represent the percent of trees that are hardwoods.
Use the point – slope formula to find the equation of the line.

Solution:
a) (1964, 57), (2004, 82)

$$m = \frac{82-57}{2004-1964} = \frac{25}{40} = 0.625$$

$$y - y_1 = m(x - x_1)$$
$$y - 57 = 0.625(x - 1964)$$
$$y - 57 = 0.625x - 1227.5$$
$$y = 0.625x - 1170.5$$

The linear equation is $f(x) = 0.625x - 1170.5$
b) Predict the percent of trees that will be hardwoods in 2020.
$y = 0.625(2020) - 1170.5 = 92$
In 2020 it is predicted that 92% of the trees will be hardwoods.

87. Strategy: Let x represent the number of miles driven.
Let y represent the number of gallons of gas in the tank.
Use the slope-intercept form of an equation to find the equation of the line.

Solution:
a) y-intercept (0,16); slope is -0.032
$y = -0.032x + 16$
Since $0 \le y \le 16$, we have

$$0 \le -0.032x + 16 \le 16$$
$$-16 \le -0.032x \le 0$$
$$500 \ge x \ge 0$$

The linear function is $f(x) = -0.032x + 16$, for $500 \ge x \ge 0$.
b) Find the number of gallons of gas left in the tank after driving 150 mi.
$y = -0.032(150) + 16 = 11.2$
After driving 150 mi there are 11.2 gal of gas left in the tank.

89. Strategy: Let x represent the price of a motorcycle.
Let y represent the number of motorcycles sold.
Use the point – slope formula to find the equation of the line.

Solution:
a) (9000, 50,000), (8750, 55,000)

$$m = \frac{55{,}000 - 50{,}000}{8750 - 9000} = \frac{5000}{-250} = -20$$
$$y - y_1 = m(x - x_1)$$
$$y - 50{,}000 = -20(x - 9000)$$
$$y - 50{,}000 = -20x + 180{,}000)$$
$$y = -20x + 230{,}000$$

The linear function is $f(x) = -20x + 230{,}000$.
b) Find the number of motorcycles sold when the price is \$8500.
$y = -20(8500) + 230{,}000 = 60{,}000$
When the price of a motorcycle is \$8500, 60,000 will be sold.

91. Strategy: Let x represent the number of ounces of lean hamburger.
Let y represent the number of calories.
Use the point – slope formula to find the equation of the line.

Solution:
a) (2, 126), (3, 189)

$$m = \frac{189 - 126}{3 - 2} = 63$$
$$y - y_1 = m(x - x_1)$$
$$y - 126 = 63(x - 2)$$
$$y - 126 = 63x - 126$$
$$y = 63x$$

The linear function is $f(x) = 63x$.
b) Find the number of calories in a 5 oz serving.
$y = 63(5) = 315$
There are 315 calories in a 5 oz serving of lean hamburger.

93. Substitute 15,000 in for $f(x)$ and solve the equation for x.

95. (2,5), (0,3)

$$m = \frac{5 - 3}{2 - 0} = \frac{2}{2} = 1$$
$$y - y_1 = m(x - x_1)$$
$$y - 3 = 1(x - 0)$$
$$y = x + 3$$
$$f(x) = x + 3$$

97. (1,3), (−1,5)

$$m = \frac{5 - 3}{-1 - 1} = \frac{2}{-2} = -1$$
$$y - y_1 = m(x - x_1)$$
$$y - 3 = -1(x - 1)$$
$$y - 3 = -x + 1$$
$$y = -x + 4$$
$$f(x) = -x + 4$$
$$f(4) = -4 + 4 = 0$$

99. Given $m = \frac{4}{3}$ and a point (3,2)

a) $y - y_1 = m(x - x_1)$

$$y - 2 = \frac{4}{3}(x - 3)$$
$$y - 2 = \frac{4}{3}x - 4$$
$$y = \frac{4}{3}x - 2$$

For $x = -6$ we have

$$y = \frac{4}{3}(-6) - 2 = -8 - 2 = -10$$

b) For $y = 6$ we have

$$6 = \frac{4}{3}x - 2$$
$$8 = \frac{4}{3}x$$
$$\frac{3}{4} \cdot 8 = \frac{3}{4} \cdot \frac{4}{3}x$$
$$6 = x$$

Critical Thinking

101. Student solutions will vary.
Find the equation of the line:

$$m = \frac{0-6}{6-(-3)} = \frac{-6}{9} = -\frac{2}{3}$$

$$y - 0 = -\frac{2}{3}(x-6)$$

$$y = -\frac{2}{3}x + 4$$

Possible answers are:

If x = 0, $y = -\frac{2}{3}(0) + 4 = 4$

x = 3, $y = -\frac{2}{3}(3) + 4 = 2$

x = 6, $y = -\frac{2}{3}(6) + 4 = 0$

(0,4), (3,2), (6,0)

103. Find the x and y coordinates for the midpoint of the line segment:

$$x_m = \frac{2+(-4)}{2} = \frac{-2}{2} = -1$$

$$y_m = \frac{5+1}{2} = \frac{6}{2} = 3$$

The midpoint is (-1, 3).
Use the point-slope formula to find the equation of the line.

$$y - y_1 = m(x - x_1)$$
$$y - 3 = -2[x-(-1)]$$
$$y - 3 = -2(x+1)$$

Projects or Group Activities

105. **Strategy** Using the points (5, 77) and (–2, 154), write the equation of the line and solve for x when y is 99.

Solution

$$m = \frac{154-77}{-2-5} = \frac{77}{-7} = -11$$

$$y - 77 = -11(x-5)$$
$$y - 77 = -11x + 55$$
$$y = -11x + 132$$

$$99 = -11x + 132$$
$$-33 = -11x$$
$$3 = x$$

When the maximum speed is 99 km/h, the slope is 3°. Since it is positive, the slope is up.

Section 4.6

Concept Check

1. Two lines are parallel if they have the same slope and different y-intercepts.

3. slope

5. $m = -5$

7. $m = -\frac{1}{3}$

9. $3x + 2y = 6$

$$2y = -3x + 6$$
$$y = -\frac{3}{2}x + 3$$
$$m_1 = -\frac{3}{2}$$
$$m_2 = \frac{2}{3}$$

11. Yes, the lines are parallel. $x = -4$ is a vertical line and $x = 4$ is vertical line.

Objective A Exercises

13. No, the lines are not parallel because their slopes are not equal.

15. Yes, the lines are perpendicular. Their slopes are negative reciprocals of each other.

17. $2x+3y=2$

$$3y=-2x+2$$
$$y=-\frac{2}{3}x+\frac{2}{3}$$
$$m_1=-\frac{2}{3}$$
$$2x+3y=-4$$
$$3y=-2x-4$$
$$y=-\frac{2}{3}x-\frac{4}{3}$$
$$m_2=-\frac{2}{3}$$

Since $m_1=m_2=-\frac{2}{3}$ the lines are parallel.

19. $x-4y=2$

$$-4y=-x+2$$
$$y=\frac{1}{4}x-\frac{1}{2}$$
$$m_1=\frac{1}{4}$$
$$4x+y=8$$
$$y=-4x+8$$
$$m_2=-4$$

Since $m_1\cdot m_2=\frac{1}{4}\cdot(-4)=-1$ the lines are perpendicular.

21. $m_1=\dfrac{6-2}{1-3}=\dfrac{4}{-2}=-2$

$$m_2=\frac{-1-3}{-1-(-1)}=\frac{-4}{0}$$
$$m_1\neq m_2$$

The lines are not parallel.

23. $m_1=\dfrac{-1-2}{4-(-3)}=\dfrac{-3}{7}=-\dfrac{3}{7}$

$$m_2=\frac{-4-3}{-2-1}=\frac{-7}{-3}=\frac{7}{3}$$
$$m_1\cdot m_2=-\frac{3}{7}\left(\frac{7}{3}\right)=-1$$

The lines are perpendicular.

25. Since the new line is parallel to $y=2x+1$ both lines will have the same slope. The slope of the new line is $m=2$.

Use the point-slope formula to find the equation of the line.
$m=2$ and $(3,-2)$

$$y-y_1=m(x-x_1)$$
$$y-(-2)=2(x-3)$$
$$y+2=2x-6$$
$$y=2x-8$$

The equation of the line is $y=2x-8$.

27. Since the new line is perpendicular to $y=-\frac{2}{3}x-2$ the slope of the new line must be the negative reciprocal of the slope of the given line. The slope of the new line is $m=\frac{3}{2}$.

Use the point-slope formula to find the equation of the line.

$m=\frac{3}{2}$ and $(-2,-1)$

$$y-y_1=m(x-x_1)$$
$$y-(-1)=\frac{3}{2}[x-(-2)]$$
$$y+1=\frac{3}{2}(x+2)$$
$$y+1=\frac{3}{2}x+3$$
$$y=\frac{3}{2}x+2$$

The equation of the line is $y=\frac{3}{2}x+2$.

29. Since the new line is parallel to $2x - 3y = 2$ both lines will have the same slope.

$2x - 3y = 2$

$-3y = -2x + 2$

$y = \frac{2}{3}x - \frac{2}{3}$

The slope of the new line is $m = \frac{2}{3}$.

Use the point-slope formula to find the equation of the line.

$m = \frac{2}{3}$ and $(-2, -4)$

$y - y_1 = m(x - x_1)$

$y - (-4) = \frac{2}{3}[x - (-2)]$

$y + 4 = \frac{2}{3}(x + 2)$

$y + 4 = \frac{2}{3}x + \frac{4}{3}$

$y = \frac{2}{3}x - \frac{8}{3}$

The equation of the line is $y = \frac{2}{3}x - \frac{8}{3}$.

31. Since the new line is perpendicular to $y = -3x + 4$ the slope of the new line must be the negative reciprocal of the slope of the given line.

$m_1 = -3$

$-3 \cdot m_2 = -1$ therefore $m_2 = \frac{1}{3}$

The slope of the new line is $m_2 = \frac{1}{3}$.

Use the point-slope formula to find the equation of the line.

$m_2 = \frac{1}{3}$ and $(4, 1)$

$y - y_1 = m(x - x_1)$

$y - 1 = \frac{1}{3}(x - 4)$

$y - 1 = \frac{1}{3}x - \frac{4}{3}$

$y = \frac{1}{3}x - \frac{1}{3}$

The equation of the line is $y = \frac{1}{3}x - \frac{1}{3}$.

33. Since the new line is perpendicular to $3x-5y=2$ the slope of the new line must be the negative reciprocal of the slope of the given line.

$$3x-5y=2$$
$$-5y=-3x+2$$
$$y=\frac{3}{5}x-\frac{2}{5}$$
$$m_1=\frac{3}{5}$$
$$\frac{3}{5}\cdot m_2=-1$$
$$m_2=-\frac{5}{3}$$

The slope of the new line is $m_2=-\frac{5}{3}$.

Use the point-slope formula to find the equation of the line.

$$m_2=-\frac{5}{3}\text{ and }(-1,-3)$$
$$y-y_1=m(x-x_1)$$
$$y-(-3)=-\frac{5}{3}[x-(-1)]$$
$$y+3=-\frac{5}{3}(x+1)$$
$$y+3=-\frac{5}{3}x-\frac{5}{3}$$
$$y=-\frac{5}{3}x-\frac{14}{3}$$

The equation of the line is $y=-\frac{5}{3}x-\frac{14}{3}$.

Critical Thinking

35. Since the new line is perpendicular to the line through the points (3, 4) and (–1, 2), find the slope of the line and the slope of the new line must be the negative reciprocal of the slope of the given line. To find a point on the line, find the midpoint of the points.

$$m_1=\frac{2-4}{-1-3}=\frac{-2}{-4}=\frac{1}{2}$$
$$\frac{1}{2}\cdot m_2=-1$$
$$m_2=-2$$

Midpoint: $\left(\frac{3+(-1)}{2},\frac{4+2}{2}\right)=(1,3)$

The slope of the new line is $m_2=-2$.

(1, 3) is a point on the line. Use the point-slope formula to find the equation of the line.

$$m_2=-2\text{ and }(1,3)$$
$$y-y_1=m(x-x_1)$$
$$y-3=-2(x-1)$$
$$y-3=-2x+2$$
$$y=-2x+5$$

The equation of the line is $y=-2x+5$.

Projects or Group Activities

37. Use the points (0,0) and (6, 3).

$$m_1=\frac{3-0}{6-0}=\frac{3}{6}=\frac{1}{2}$$
$$m_1\cdot m_2=-1$$
$$\frac{1}{2}\cdot m_2=-1$$
$$m_2=-2$$

Using the point (6, 3)

$$y-3=-2(x-6)$$
$$y-3=-2x+12$$
$$y=-2x+15$$

The equation of the line is $y=-2x+15$.

Section 4.7

Concept Check

1. A half-plane is the set of points on one side of a line in the plane.

3. $y > 2x - 7$
$0 > 2(0) - 7$
$0 > -7$
Yes, (0,0) is a solution.

5. $y \le -\dfrac{2}{3}x - 8$
$0 \le -\dfrac{2}{3}(0) - 8$
$0 \le -8$
No, (0,0) is not a solution.

Objective A Exercises

7. $y \le \dfrac{3}{2}x - 3$

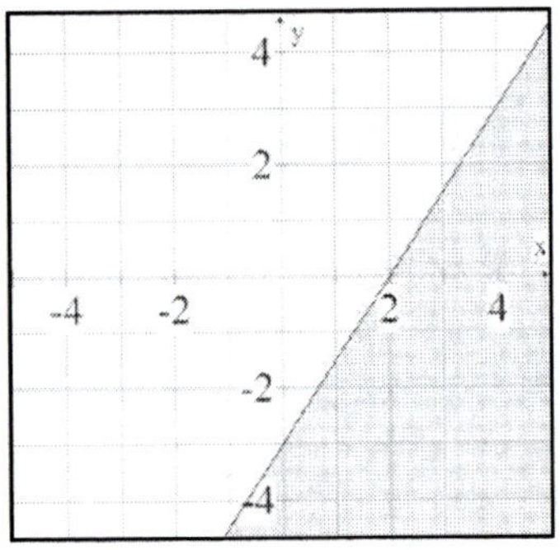

9. $y < -\dfrac{1}{3}x + 1$

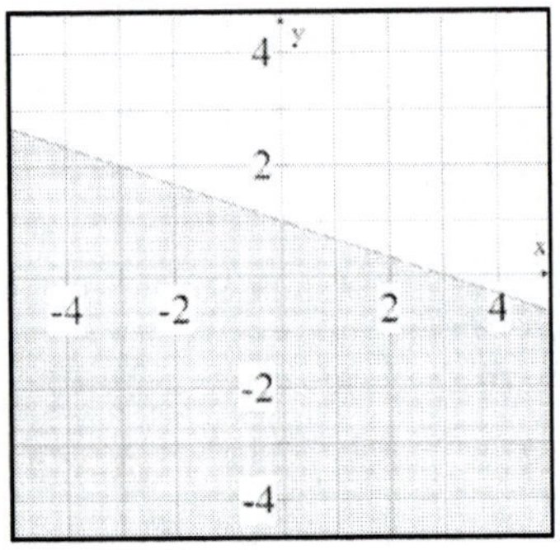

11. $4x - 5y > 10$
$-5y > -4x + 10$
$y < \dfrac{4}{5}x - 2$

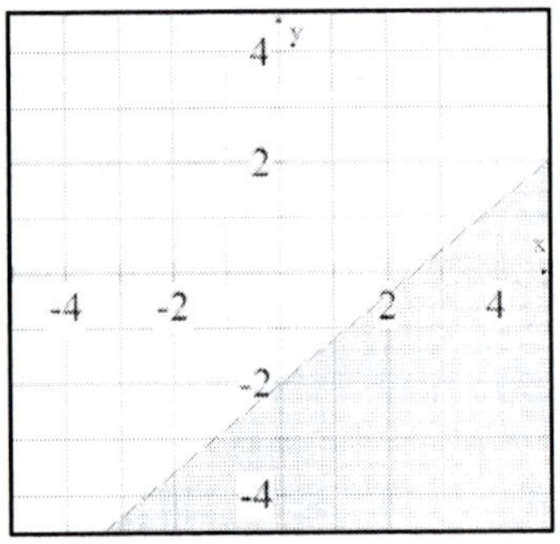

13. $x + 3y < 6$
$3y < -x + 6$
$y < -\dfrac{1}{3}x + 2$

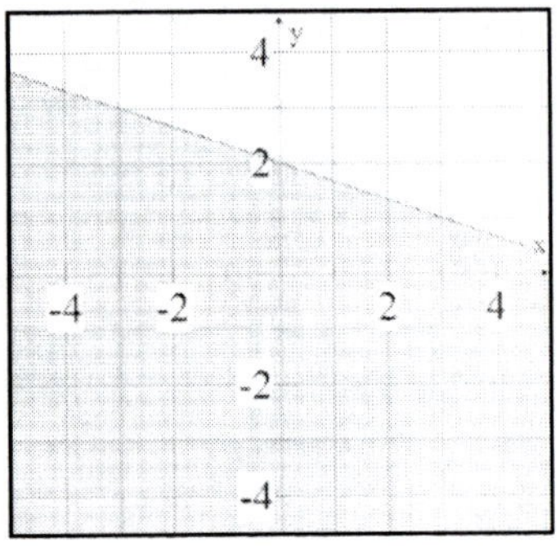

15. $2x + 3y \ge 6$
$3y \ge -2x + 6$
$y \ge -\dfrac{2}{3}x + 2$

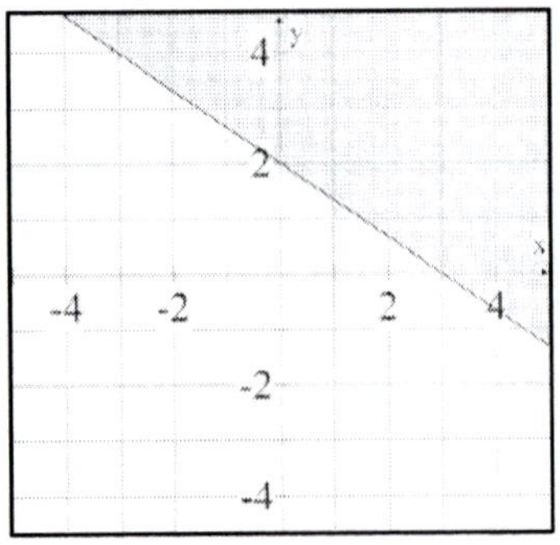

17. $-x+2y>-8$

$2y>x-8$

$y>\frac{1}{2}x-4$

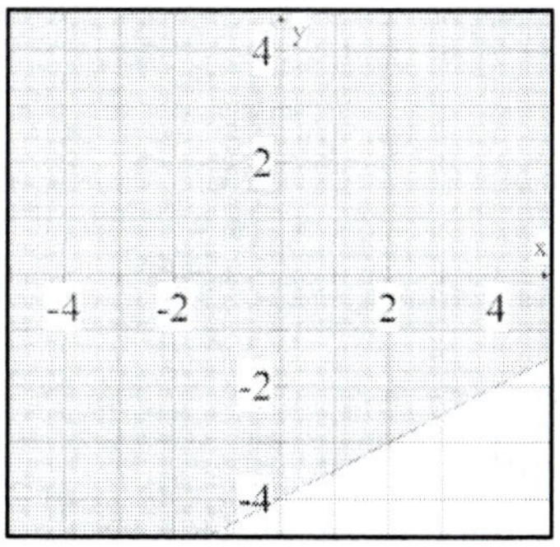

19. $y-4<0$

$y<4$

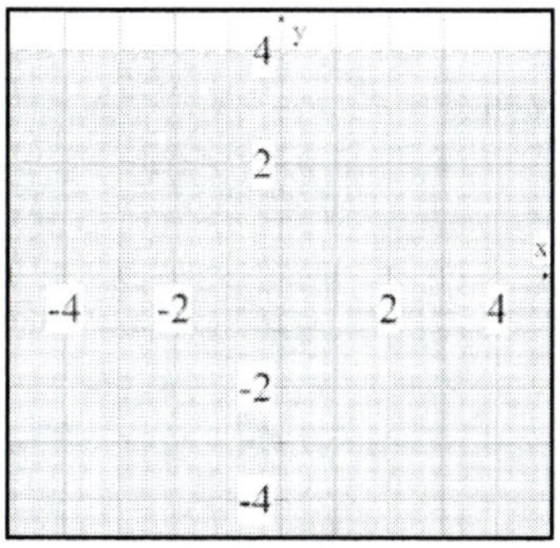

21. $6x+5y<15$

$5y<-6x+15$

$y<-\frac{6}{5}x+3$

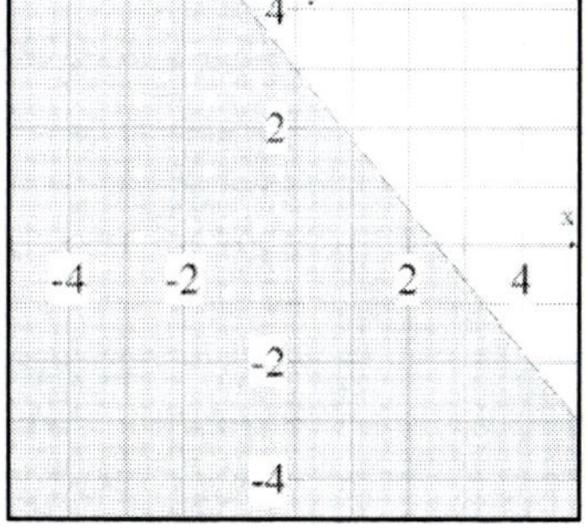

23. $-5x+3y\geq -12$

$3y\geq 5x-12$

$y\geq \frac{5}{3}x-4$

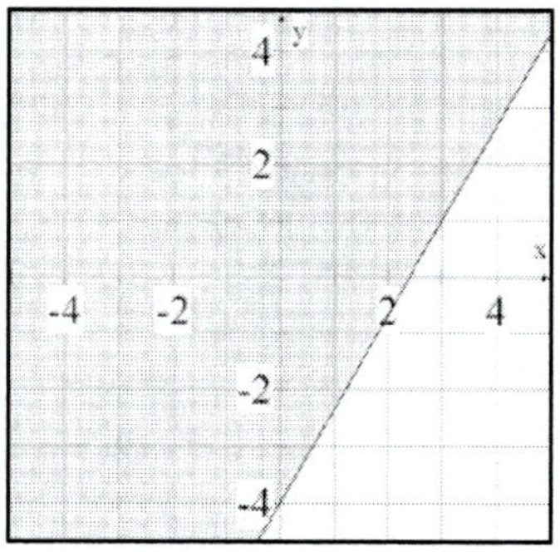

25. Quadrant I

Critical Thinking

27. The inequality $y<3x-1$ is not a function because given a value of x there is more than one corresponding value of y. For example, both (3,2) and (3,−1) are ordered pairs that satisfy the inequality. This contradicts the definition of a function because there are two ordered pairs with the same first coordinate and different second coordinates.

Projects or Group Activities

29. Since $|x|$ means $x=x$ or $x=-x$, and $|y|$ means $y=y$ and $y=-y$, graph the solution that satisfies all of the following equations.

$x+y\leq 5$

$x-y\leq 5$

$-x+y\leq 5$

$-x-y\leq 5$

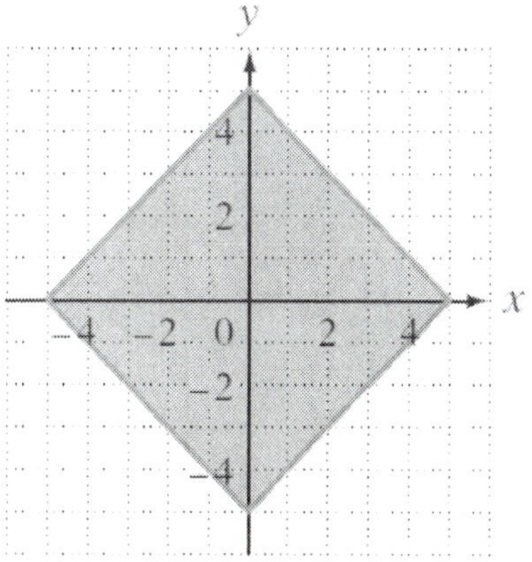

Chapter 4 Review Exercises

1. **a.**

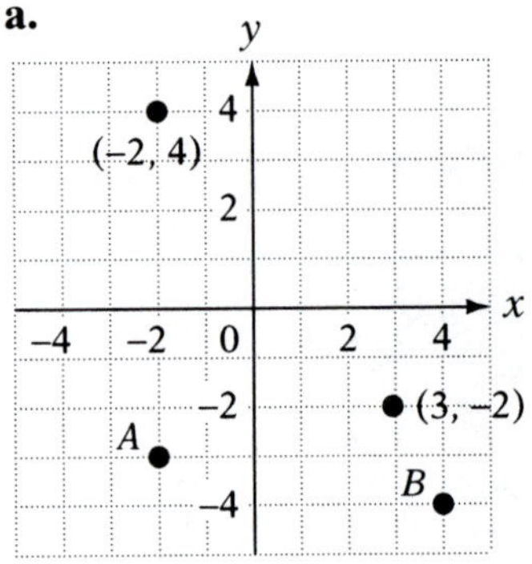

b. −2

c. −4

2.

x	y
−4	0
−2	−1
0	−2
2	−3

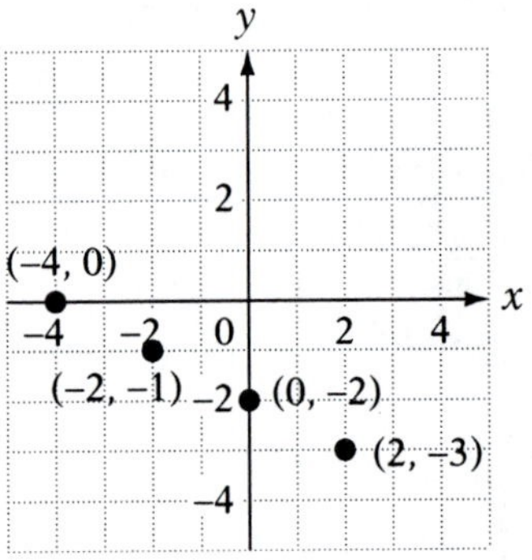

3. $y = \dfrac{x}{x-2}$

$y = \dfrac{4}{4-2} = \dfrac{4}{2} = 2$

The ordered pair is (4,2).

4. $P(x) = 3x + 4$

$P(-2) = 3(-2) + 4 = -6 + 4 = -2$

$P(a) = 3(a) + 4 = 3a + 4$

5. $f(x) = -\dfrac{2}{3}x + 4$

$0 = -\dfrac{2}{3}x + 4$

$3(0) = 3\left(-\dfrac{2}{3}x\right) + 3(4)$

$0 = -2x + 12$

$-12 = -2x$

$6 = x$

6. $f(x) = \dfrac{x}{x+4}$

The function is not defined for zero in the denominator.

$x + 4 = 0$

$x = -4$

$f(x)$ is not defined for $x = -4$

7. $m = -\dfrac{5}{2}$ and (6, 1)

$y - y_1 = m(x - x_1)$

$y - 1 = -\dfrac{5}{2}(x - 6)$

$y - 1 = -\dfrac{5}{2}x + 15$

$y = -\dfrac{5}{2}x + 16$

8. Domain = {−1, 0, 1, 5}
Range = {0, 2, 4}

9. $y = -\frac{2}{3}x - 2$

x-intercept:

$0 = -\frac{2}{3}x - 2$

$2 = -\frac{2}{3}x$

$x = -3$

(−3, 0)

y-intercept:

$y = -\frac{2}{3}(0) - 2$

$y = -2$

(0, −2)

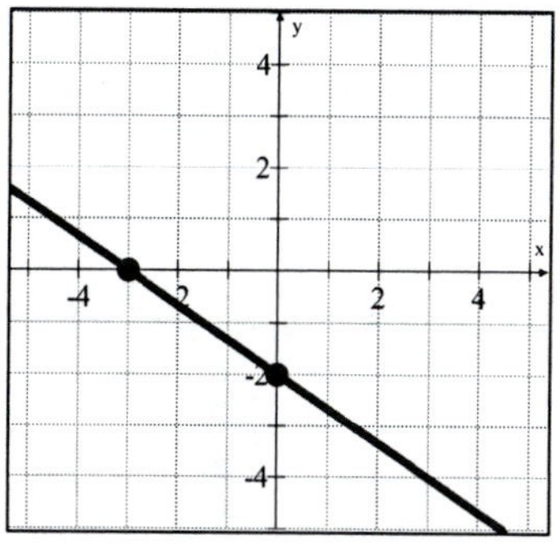

10. $3x + 2y = -6$

x-intercept:

$3x + 2(0) = -6$

$3x = -6$

$x = -2$

(−2, 0)

y-intercept:

$3(0) + 2y = -6$

$2y = -6$

$y = -3$

(0, −3)

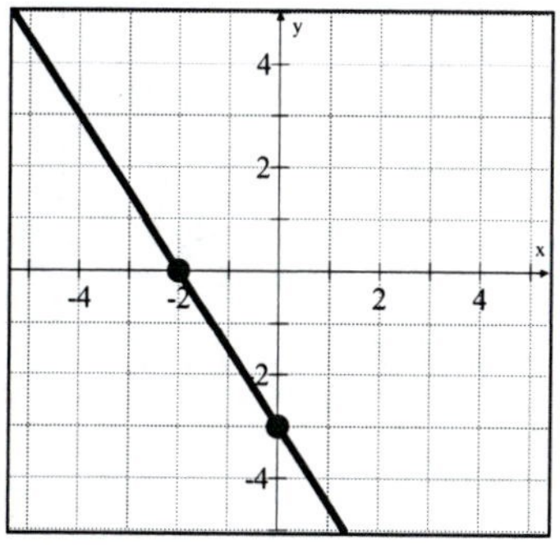

11. $y = -2x + 2$

x	y
0	2
1	0
-1	4

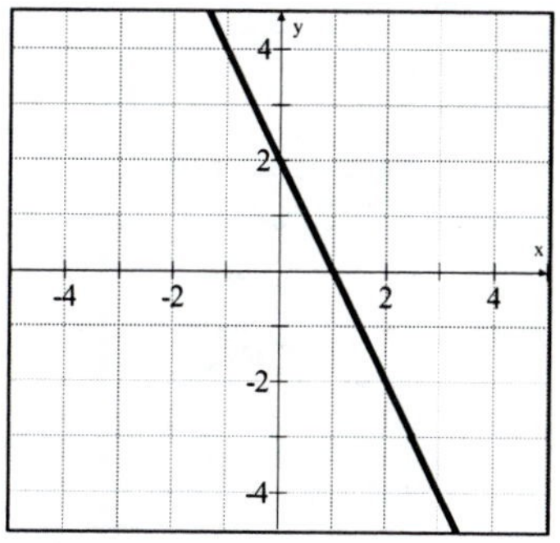

12. $4x - 3y = 12$

$-3y = -4x + 12$

$y = \frac{4}{3}x - 4$

x	y
0	−4
3	0
6	4

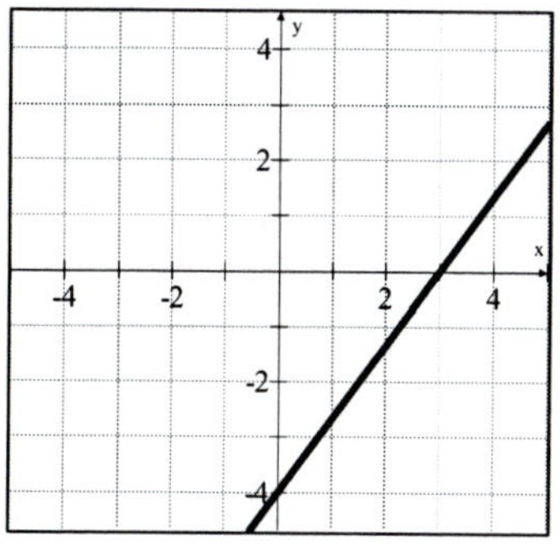

13. (3, −2) and (−1, 2)

$m = \frac{y_2 - y_1}{x_2 - x_1}$

$m = \frac{-2-2}{3-(-1)} = \frac{-4}{4} = -1$

14. Use the point-slope formula to find the equation of the line.

$m = \frac{5}{2}$ and (−3, 4)

$y - y_1 = m(x - x_1)$

$y - 4 = \frac{5}{2}[x - (-3)]$

$y - 4 = \frac{5}{2}(x + 3)$

$y - 4 = \frac{5}{2}x + \frac{15}{2}$

$y = \frac{5}{2}x + \frac{23}{2}$

15. $5x + 3y = 15$

x-intercept:

$5x + 3(0) = 15$

$5x = 15$

$x = 3$

(3, 0)

y-intercept

$5(0) + 3y = 15$

$3y = 15$

$y = 5$

(0, 5)

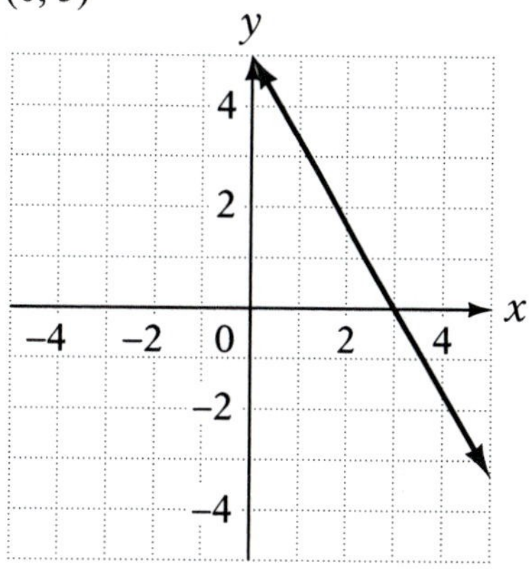

16. $3x-2y=-6$

x-intercept:

$3x-2(0)=-6$

$3x=-6$

$x=-2$

$(-2, 0)$

y-intercept:

$3(0)-2y=-6$

$-2y=-6$

$y=3$

$(0, 3)$

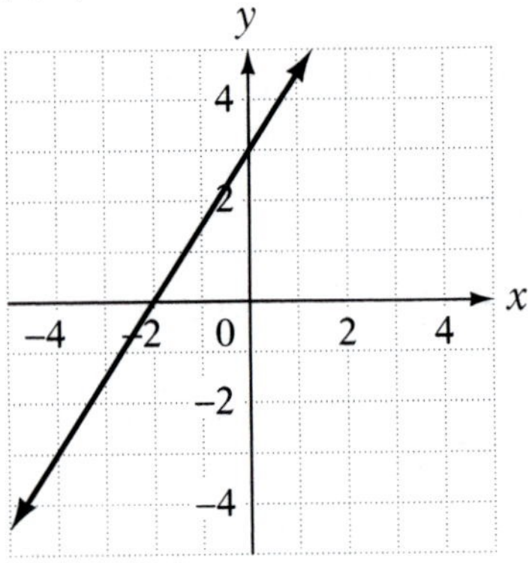

17. $m=-\dfrac{1}{4}$ and $(-2, 3)$

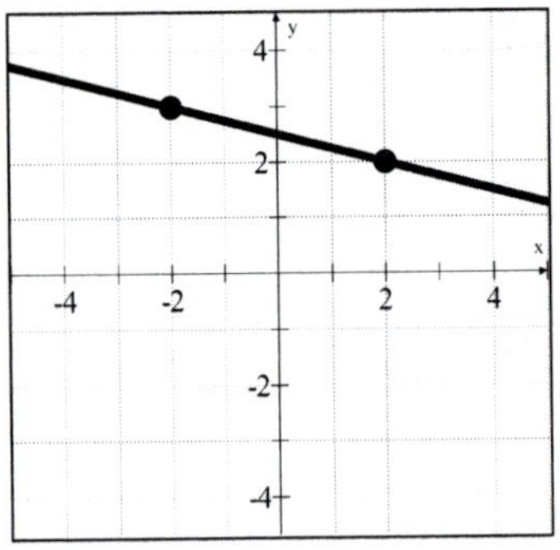

18. $m=-\dfrac{1}{3}$ and $(-1, 4)$

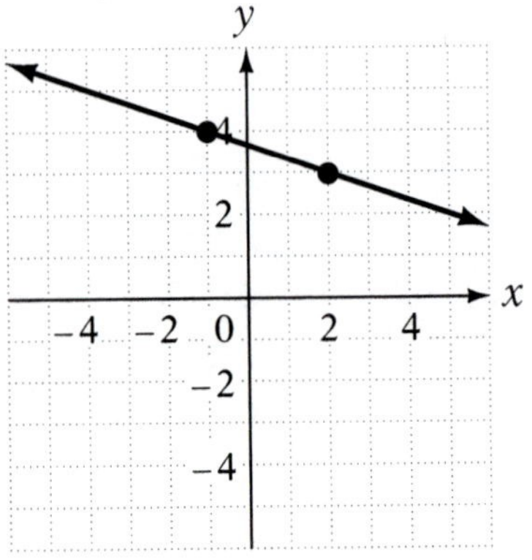

19. The slope for the parallel line is $m=-4$.

$m=-4$ and $(-2, 3)$

$y-y_1=m(x-x_1)$

$y-3=-4[(x-(-2)]$

$y-3=-4(x+2)$

$y-3=-4x-8$

$y=-4x-5$

The equation of the line is $y=-4x-5$.

20. The slope for the perpendicular line is $m=\dfrac{5}{2}$.

$m=\dfrac{5}{2}$ and $(-2, 3)$

$y-y_1=m(x-x_1)$

$y-3=\dfrac{5}{2}[(x-(-2)]$

$y-3=\dfrac{5}{2}(x+2)$

$y-3=\dfrac{5}{2}x+5$

$y=\dfrac{5}{2}x+8$

The equation of the line is $y=\dfrac{5}{2}x+8$.

21.

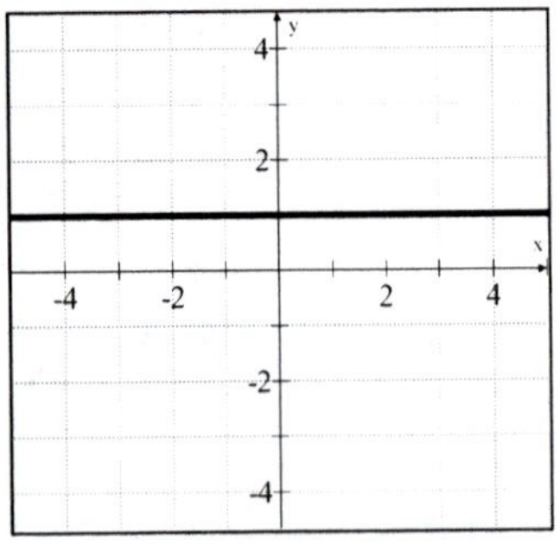

22.

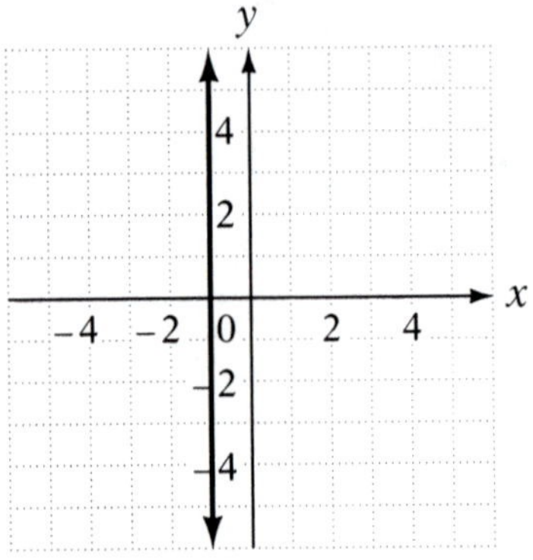

23. $m = -\frac{2}{3}$ and $(-3, 3)$

$$y - y_1 = m(x - x_1)$$
$$y - 3 = -\frac{2}{3}[(x - (-3)]$$
$$y - 3 = -\frac{2}{3}(x + 3)$$
$$y - 3 = -\frac{2}{3}x - 2$$
$$y = -\frac{2}{3}x + 1$$

The equation of the line is $y = -\frac{2}{3}x + 1$.

24. $(-8, 2)$ and $(4, 5)$

$$m = \frac{5 - 2}{4 - (-8)} = \frac{3}{12} = \frac{1}{4}$$
$$y - y_1 = m(x - x_1)$$
$$y - 5 = \frac{1}{4}(x - 4)$$
$$y - 5 = \frac{1}{4}x - 1$$
$$y = \frac{1}{4}x + 4$$

The equation of the line is $y = \frac{1}{4}x + 4$.

25. $y \geq 2x - 3$

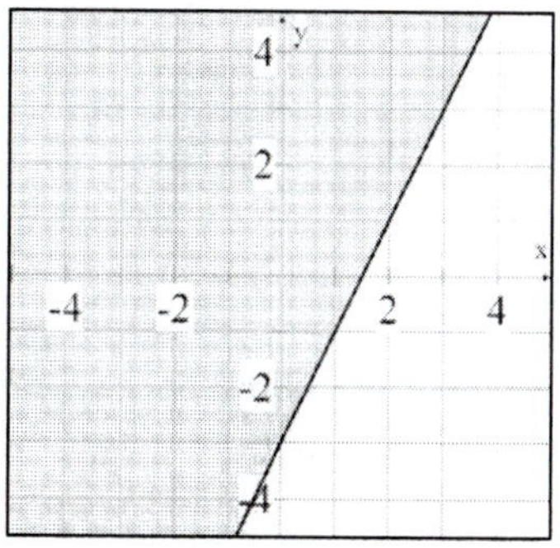

26. $3x - 2y < 6$

$$-2y < -3x + 6$$
$$y > \frac{3}{2}x - 3$$

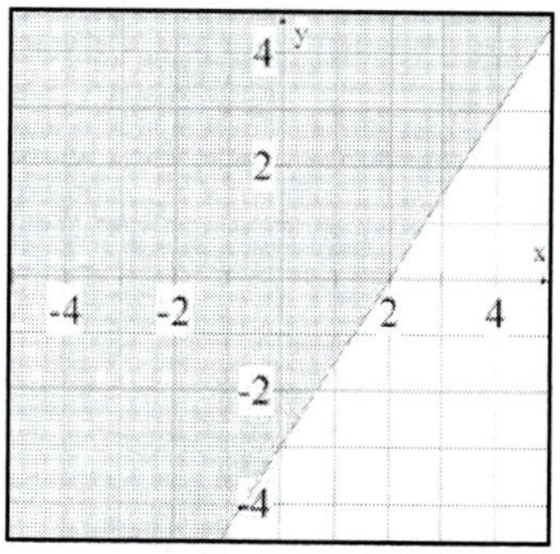

27. $(-2,4)$ and $(4,-3)$

$$m = \frac{4 - (-3)}{-2 - 4} = \frac{7}{-6} = -\frac{7}{6}$$
$$y - y_1 = m(x - x_1)$$
$$y - (-3) = -\frac{7}{6}(x - 4)$$
$$y + 3 = -\frac{7}{6}x + \frac{28}{6}$$
$$y = -\frac{7}{6}x + \frac{5}{3}$$

The equation of the line is $y = -\frac{7}{6}x + \frac{5}{3}$.

28. $4x - 2y = 7$

$$-2y = -4x + 7$$
$$y = 2x - \frac{7}{2}$$

The slope for the parallel line is $m = 2$.

$m = 2$ and $(-2, -4)$

$$y - y_1 = m(x - x_1)$$
$$y - (-4) = 2[(x - (-2)]$$
$$y + 4 = 2(x + 2)$$
$$y + 4 = 2x + 4$$
$$y = 2x$$

The equation of the line is $y = 2x$.

29. The slope for the parallel line is $m = -3$.
$m = -3$ and $(3, -2)$

$$y - y_1 = m(x - x_1)$$
$$y - (-2) = -3(x - 3)$$
$$y + 2 = -3x + 9$$
$$y = -3x + 7$$

The equation of the line is $y = -3x + 7$.

30. $y = -\frac{2}{3}x + 6$

$$m_1 = -\frac{2}{3}$$
$$m_1 \cdot m_2 = -1$$
$$-\frac{2}{3} \cdot m_2 = -1$$
$$m_2 = \frac{3}{2}$$

The slope for the perpendicular line is $m = \frac{3}{2}$.

$m = \frac{3}{2}$ and $(2, 5)$

$$y - y_1 = m(x - x_1)$$
$$y - 5 = \frac{3}{2}(x - 2)$$
$$y - 5 = \frac{3}{2}x - 3$$
$$y = \frac{3}{2}x + 2$$

The equation of the line is $y = \frac{3}{2}x + 2$.

31.

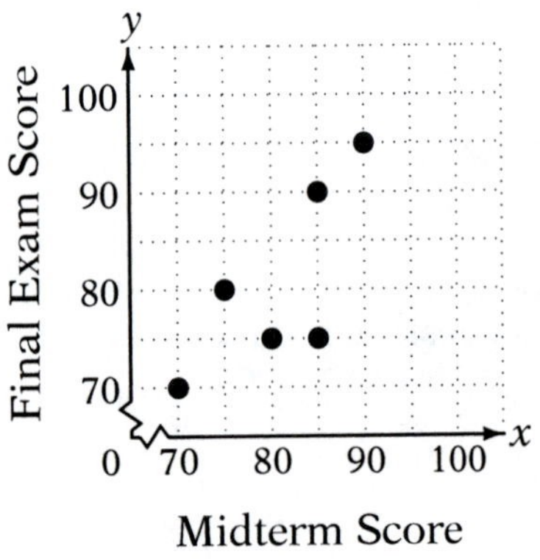

32. Strategy: Let x represent the room rate. Let y represent the number of rooms occupied.
Use the point – slope formula to find the equation of the line.

Solution:
a) (95, 200), (105, 190)

$$m = \frac{200 - 190}{95 - 105} = \frac{10}{-10} = -1$$
$$y - y_1 = m(x - x_1)$$
$$y - 200 = -1(x - 95)$$
$$y - 200 = -1x + 95$$
$$y = -x + 295$$

The linear function is
$f(x) = -x + 295,\ 0 \le x \le 295$

b) Find the number of rooms occupied when the rate is \$120.
$f(120) = -1(120) + 295 = 175$
When the room rate is \$125, 175 rooms will be occupied.

33.

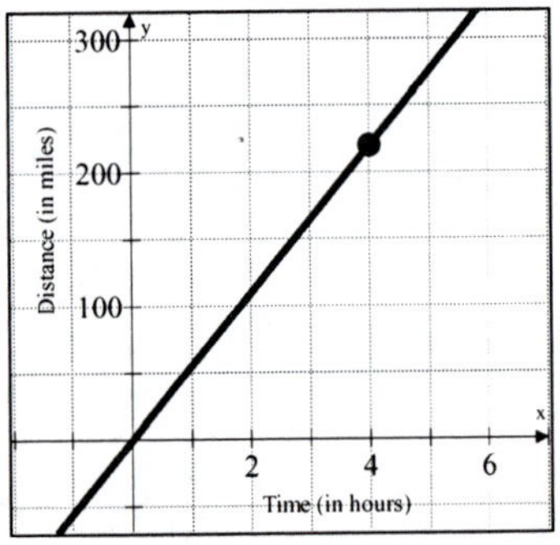

After 4 h the car will travel 220 mi.

34. (500, 12,000) and (200, 6000)

$$m = \frac{12{,}000 - 6000}{500 - 200} = \frac{6000}{300} = 20$$

The slope is 20. The manufacturing cost is \$20 per calculator.

35. a) The y-intercept is (0, 25,000).
The slope is 80.
$y = mx + b$
$y = 80x + 25{,}000$
The linear function is $f(x) = 80x + 25{,}000$.
b) Predict the cost of building a house with 2000 square feet.
$f(2000) = 80(2000) + 25{,}000$
$f(x) = 185{,}000$
The house will cost $185,000 to build.

Chapter 4 Test

1. $2x - 3y = 15$
$2(3) - 3y = 15$
$6 - 3y = 15$
$-3y = 9$
$y = -3$
(3, –3)

2. $y = -\frac{3}{2}x + 1$

x	y
–2	4
0	1
4	–5

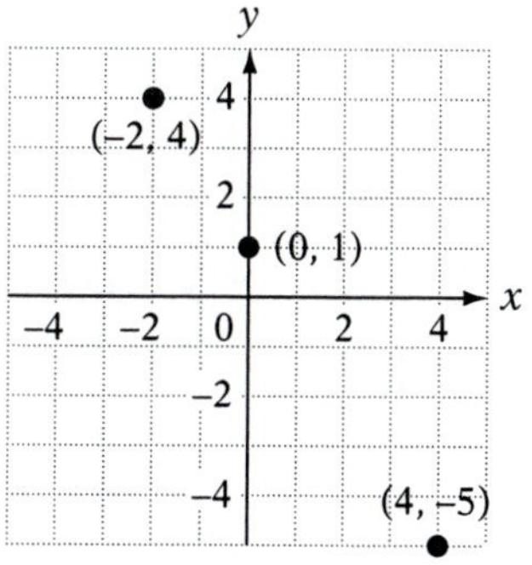

3. $y = \frac{2}{3}x - 4$

x	y
0	–4
3	–2
6	0

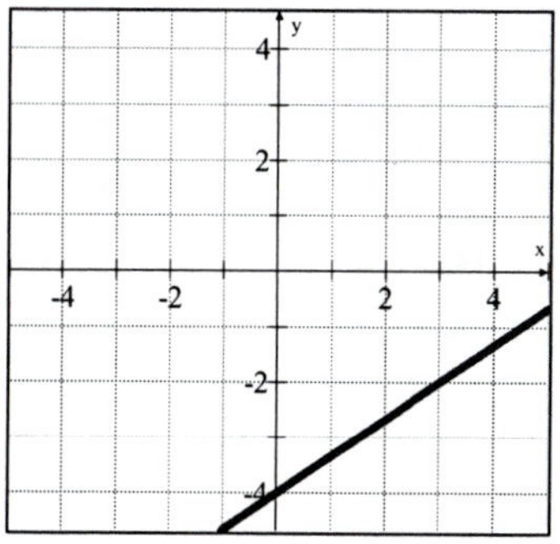

4. $2x + 3y = -3$

x	y
0	–1
3	–3
-3	1

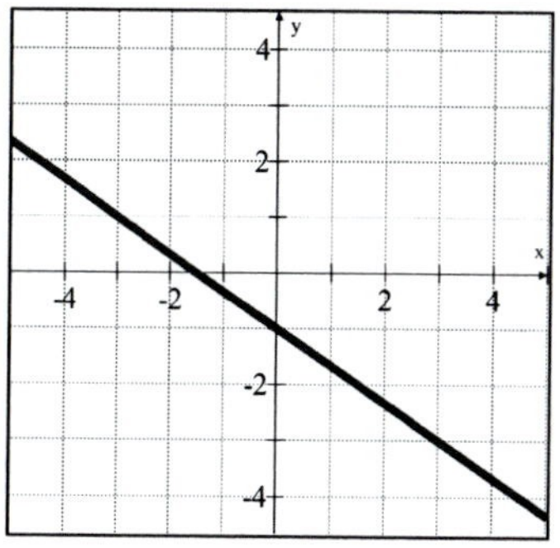

5. $f(x) = 4x - 12$
$0 = 4x - 12$
$12 = 4x$
$3 = x$

6. $f(t) = t^2 + t$
$f(2) = (2)^2 + 2$
$f(2) = 4 + 2$
$f(2) = 6$

7. (–2,3) and (4,2)
$$m = \frac{y_2 - y_1}{x_2 - x_1} = \frac{3-2}{-2-4} = \frac{1}{-6} = -\frac{1}{6}$$
The slope of the line is $-\frac{1}{6}$.

8. $P(x) = 3x^2 - 2x + 1$
$P(2) = 3(2)^2 - 2(2) + 1$
$P(2) = 9$

9. $2x - 3y = 6$
x-intercept:
$2x - 3(0) = 6$
$2x = 6$
$x = 3$
(3,0)
y-intercept:
$2(0) - 3y = 6$
$-3y = 6$
$y = -2$
(0, –2)

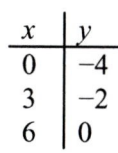

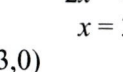

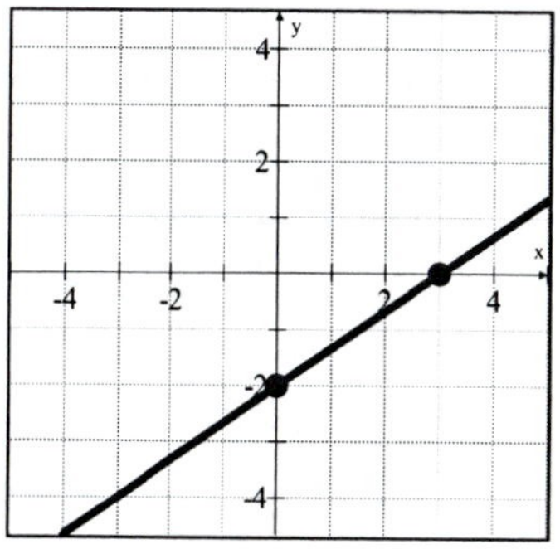

10. (−2, 3) and $m = -\frac{3}{2}$

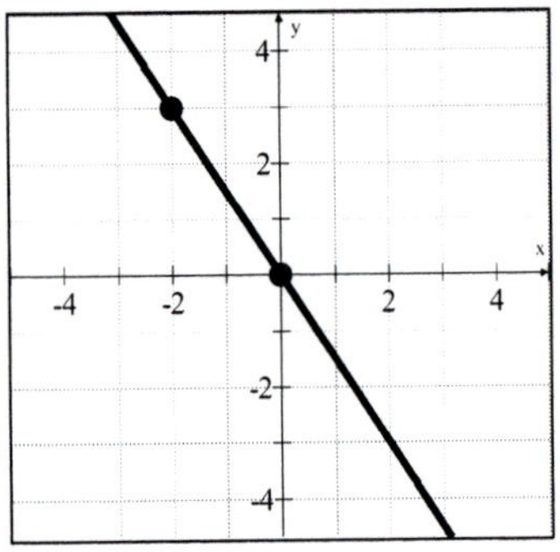

11. $m = \frac{2}{5}$ and (−5, 2)

$$y - 2 = \frac{2}{5}[x - (-5)]$$
$$y - 2 = \frac{2}{5}(x + 5)$$
$$y - 2 = \frac{2}{5}x + 2$$
$$y = \frac{2}{5}x + 4$$

The equation of the line is $y = \frac{2}{5}x + 4$.

12. $f(x) = \frac{2x + 1}{x}$

The function is not defined for zero in the denominator.
$x = 0$ is excluded from the domain of $f(x)$.

13. (3,−4) and (−2,3)

$$m = \frac{3 - (-4)}{-2 - 3} = \frac{7}{-5} = -\frac{7}{5}$$
$$y - (-4) = -\frac{7}{5}(x - 3)$$
$$y + 4 = -\frac{7}{5}x + \frac{21}{5}$$
$$y = -\frac{7}{5}x + \frac{1}{5}$$

The equation of the line is $y = -\frac{7}{5}x + \frac{1}{5}$.

14. $6x - 4y = 12$

x-intercept:
$$6x - 4(0) = 12$$
$$6x = 12$$
$$x = 2$$
(2, 0)

y-intercept:
$$6(0) - 4y = 12$$
$$-4y = 12$$
$$y = -3$$
(0, −3)

15. Domain = {−4, −2, 0,3}
Range = {0, 2, 5}

16. A line parallel to $y = -\frac{3}{2}x - 6$ has a slope of $m = -\frac{3}{2}$.

$m = -\frac{3}{2}$ and (1, 2)

$$y - 2 = -\frac{3}{2}(x - 1)$$
$$y - 2 = -\frac{3}{2}x + \frac{3}{2}$$
$$y = -\frac{3}{2}x + \frac{7}{2}$$

The equation of the line is $y = -\frac{3}{2}x + \frac{7}{2}$.

17. $y = -\frac{1}{2}x - 3$

$m_1 = -\frac{1}{2}$

$m_1 \cdot m_2 = -1$

$-\frac{1}{2} \cdot m_2 = -1$

$m_2 = 2$

The slope of the perpendicular line is $m = 2$
.

$m = 2$ and $(-2, -3)$

$y - (-3) = 2[x - (-2)]$

$y + 3 = 2(x + 2)$

$y + 3 = 2x + 4$

$y = 2x + 1$

The equation of the line is $y = 2x + 1$.

18. $3x - 4y > 8$

$-4y > -3x + 8$

$y < \frac{3}{4}x - 2$

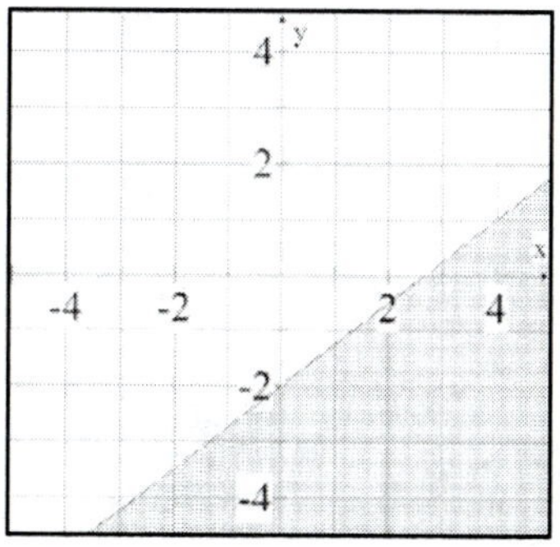

19.

x	y
0	128
1	96
2	64
3	32
4	0

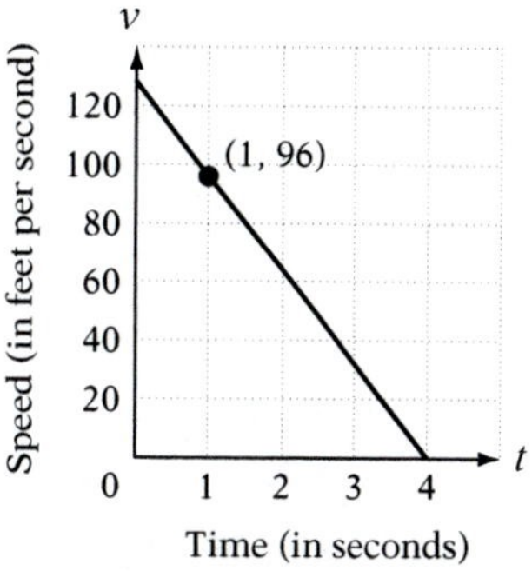

After 1 s, the ball is traveling 96 ft/s.

20. Strategy: Use two points on the graph to find the slope of the line.

Solution: (3, 120,000) and (12, 30,000)

$$m = \frac{120{,}000 - 30{,}000}{3 - 12} = \frac{90{,}000}{-9} = -10{,}000$$

The value of the house decreases by \$10,000 each year.

21. Strategy: Let x represent the tuition cost. Let y represent the number of students. Use the point – slope formula to find the equation of the line.

Solution:

a) $m = \frac{-6}{20} = -\frac{3}{10}$ and (250, 100)

$$y - y_1 = m(x - x_1)$$
$$y - 100 = -\frac{3}{10}(x - 250)$$
$$y - 100 = -\frac{3}{10}x + 75$$
$$y = -\frac{3}{10}x + 175$$

The linear function that will predict enrollment based on the cost of tuition is

$$f(x) = -\frac{3}{10}x + 175$$

b) Find the number of students who enroll when tuition is \$300.

$$f(300) = -\frac{3}{10}(300) + 175$$
$$f(300) = 85$$

When tuition is \$300, 85 students will enroll.

Cumulative Review Exercises

1. Commutative Property of Multiplication.

2. $3 - \frac{x}{2} = \frac{3}{4}$

$$4(3) - 4\left(\frac{x}{2}\right) = 4\left(\frac{3}{4}\right)$$
$$12 - 2x = 3$$
$$-2x = -9$$
$$x = \frac{9}{2}$$

The solution is $\frac{9}{2}$.

3. $3\sqrt{45} = 3\sqrt{3^2 \cdot 5} = 3\sqrt{3^2}\sqrt{5}$
$= 3^2\sqrt{5} = 9\sqrt{5}$

4. $\frac{1-3x}{2} + \frac{7x-2}{6} = \frac{4x+2}{9}$

$$18\left(\frac{1-3x}{2}\right) + 18\left(\frac{7x-2}{6}\right) = 18\left(\frac{4x+2}{9}\right)$$
$$9 - 27x + 21x - 6 = 8x + 4$$
$$-6x + 3 = 8x + 4$$
$$-1 = 14x$$
$$-\frac{1}{14} = x$$

The solution is $-\frac{1}{14}$.

5. $x - 3 < -4$ or $2x + 2 > 3$

$x < -1$ $2x > 1$

$x > \frac{1}{2}$

The solution is $\{x \mid x < -1\} \cup \left\{x \mid x > \frac{1}{2}\right\}$.

6. $8 - |2x - 1| = 4$

$$-|2x - 1| = -4$$
$$|2x - 1| = 4$$

$2x - 1 = 4$ $2x - 1 = -4$

$2x = 5$ and $2x = -3$

$x = \frac{5}{2}$ $x = -\frac{3}{2}$

The solutions are $-\frac{3}{2}$ and $\frac{5}{2}$.

7. $|3x - 5| < 5$

$$-5 < 3x - 5 < 5$$
$$-5 + 5 < 3x - 5 + 5 < 5 + 5$$
$$0 < 3x < 10$$
$$\frac{0}{3} < \frac{3x}{3} < \frac{10}{3}$$
$$0 < x < \frac{10}{3}$$
$$\left\{x \mid 0 < x < \frac{10}{3}\right\}$$

8. $4 - 2(4-5)^3 + 2 = 4 - 2(-1)^3 + 2$

$$= 4 - 2(-1) + 2$$
$$= 4 + 2 + 2 = 8$$

9. $(a-b)^2 \div (ab)$

$$\begin{aligned}(4-(-2))^2 \div (4(-2)) &= (4+2)^2 \div (-8)\\ &= (6)^2 \div (-8)\\ &= 36 \div (-8) = -4.5\end{aligned}$$

10. $\{x|\, x<-2\} \cup \{x|\, x>0\}$

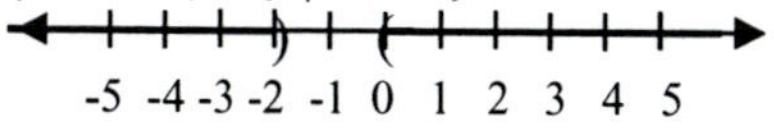

11.
$$\begin{aligned}3x-2[x-3(2-3x)] &= x-7\\ 3x-2[x-6+9x] &= x-7\\ 3x-2[10x-6] &= x-7\\ 3x-20x+12 &= x-7\\ -17x+12 &= x-7\\ -18x &= -19\\ x &= \frac{19}{18}\end{aligned}$$

The solution is $\frac{19}{18}$.

12. $6\frac{2}{3}\% = 6\frac{2}{3} \div 100 = \frac{20}{3}\cdot\frac{1}{100} = \frac{1}{15}$

13. $3x-1<4$ and $x-2>2$

$3x<5$ $\quad x>4$

$x<\frac{5}{3}$

$\{x|\ x<\frac{5}{3}\} \cap \{x|\ x>4\} = \emptyset$

14.
$$\begin{aligned}P(x) &= x^2+5\\ P(-3) &= (-3)^2+5\\ P(-3) &= 9+5\\ P(-3) &= 14\end{aligned}$$

15.
$$\begin{aligned}y &= -\frac{5}{4}x+3\\ y &= -\frac{5}{4}(-8)+3\\ y &= 10+3\\ y &= 13\end{aligned}$$

$(-8, 13)$

16. $(-1,3)$ and $(3,-4)$

$$m = \frac{y_2-y_1}{x_2-x_1} = \frac{-4-3}{3-(-1)} = \frac{-7}{4}$$

17. $m=\frac{3}{2}$ and $(-1, 5)$

$$\begin{aligned}y-y_1 &= m(x-x_1)\\ y-5 &= \frac{3}{2}(x+1)\\ y-5 &= \frac{3}{2}x+\frac{3}{2}\\ y &= \frac{3}{2}x+\frac{13}{2}\end{aligned}$$

The equation of the line is $y=\frac{3}{2}x+\frac{13}{2}$.

18. $m = \frac{y_2-y_1}{x_2-x_1} = \frac{3+2}{0-4} = -\frac{5}{4}$

$m=-\frac{5}{4}$ and $(0, 3)$

$$\begin{aligned}y-y_1 &= m(x-x_1)\\ y-3 &= -\frac{5}{4}(x-0)\\ y-3 &= -\frac{5}{4}x\\ y &= -\frac{5}{4}x+3\end{aligned}$$

The equation of the line is $y=-\frac{5}{4}x+3$.

19. $m = -\dfrac{3}{2}$ and (2, 4)

$y - y_1 =,(x - x_1)$

$y - 4 = -\dfrac{3}{2}(x - 2)$

$y - 4 = -\dfrac{3}{2}x + 3$

$y = -\dfrac{3}{2}x + 7$

The equation of the line is $y = -\dfrac{3}{2}x + 7.$

20. $3x - 2y = 5$

$-2y = -3x + 5$

$y = \dfrac{3}{2}x - \dfrac{5}{2}$

Find the slope that is perpendicular to $\dfrac{3}{2}.$

$\dfrac{3}{2} \cdot m_2 = -1$

$m_2 = -\dfrac{2}{3}$

$m = -\dfrac{2}{3}$ and (4, 0)

$y - y_1 = m(x - x_1)$

$y - 0 = -\dfrac{2}{3}(x - 4)$

$y = -\dfrac{2}{3}x + \dfrac{8}{3}$

The equation of the line is $y = -\dfrac{2}{3}x + \dfrac{8}{3}.$

21. $f(x) = -2x + 6$

$0 = -2x + 6$

$-6 = -2x$

$3 = x$

22. $3x - 5y = 15$

x-intercept:	*y*-intercept:
$3x - 5(0) = 15$	$3(0) - 5y = 15$
$3x = 15$	$-5y = 15$
$x = 5$	$y = -3$
(5, 0)	(0, –3)

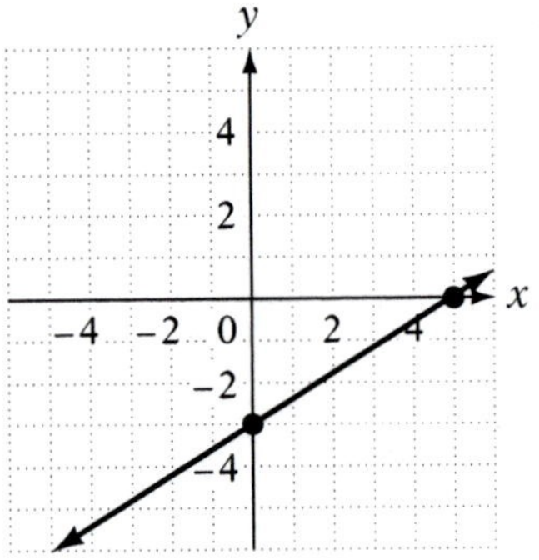

23. $m = -\dfrac{3}{2}$ and (–3, 1)

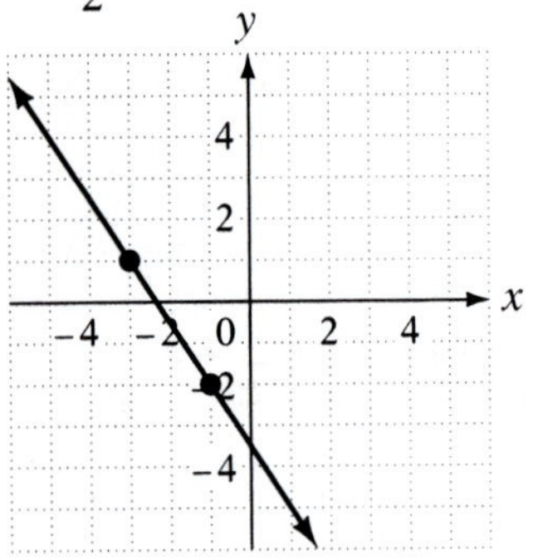

24. $3x - 2y \ge 6$

$-2y \ge -3x + 6$

$y \le \dfrac{3}{2}x - 3$

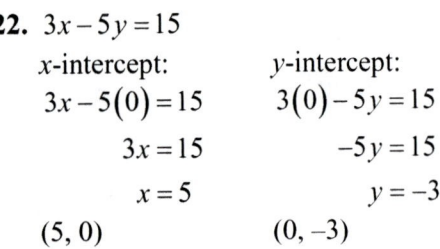

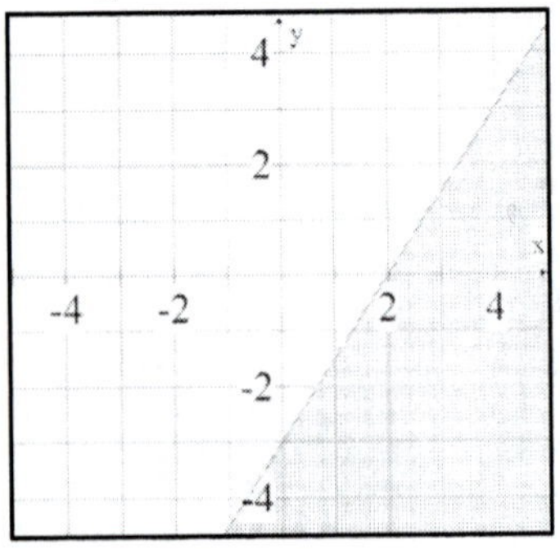

25. Strategy: Let r represent the speed of 1st plane.
The speed of the 2nd plane is $2r$.

	Rate	Time	Distance
1st plane	r	3	$3r$
2nd plane	$2r$	3	$3(2r)$

The total distance traveled by the two planes is 1800 mi.

Solution: $3r + 3(2r) = 1800$

$$3r + 6r = 1800$$
$$9r = 1800$$
$$r = 200$$

$2r = 2(200) = 400$

The first plane is traveling at 200 mph and the second plane is traveling at 400 mph.

26. Strategy: Let x represent the pounds of coffee costing \$9.00.
Pounds of coffee costing \$6.00 is 60 - x

	Amount	Cost	Value
\$9 coffee	x	9	$9x$
\$6 coffee	$60 - x$	6	$6(60 - x)$
Mixture	60	8	$8(60)$

The sum of the values before mixing is equal to the quantity after mixing.

Solution:

$$9x + 6(60 - x) = 8(60)$$
$$9x + 360 - 6x = 480$$
$$3x + 360 = 480$$
$$3x = 120$$
$$x = 40$$
$$60 - x = 60 - 40 = 20$$

The mixture contains 40 lb of \$9.00 coffee and 20 lb of \$6.00 coffee.

27. a. Strategy: Use two points on the graph to find the slope of the line. Write the equation of the line.

Solution: (0, 30,000) and (6, 0)

$$m = \frac{30{,}000 - 0}{0 - 6} = \frac{30{,}000}{-6} = -5000$$

$m = -5000$ and (0, 30,000)

$y = -5000x + 30{,}000$

b. The value of the truck decreases by \$5000 per year.

Chapter 5: Systems of Linear Equations and Inequalities

Prep Test

1. $10\left(\frac{3}{5}x+\frac{1}{2}y\right)=10\left(\frac{3}{5}x\right)+10\left(\frac{1}{2}y\right)=6x+5y$

2. $3x+2y-z$
$$\begin{aligned}&3(-1)+2(4)-(-2)\\&=-3+8+2\\&=7\end{aligned}$$

3. $$\begin{aligned}3x-2z&=4\\3x-2(-2)&=4\\3x+4&=4\\3x&=0\\x&=0\end{aligned}$$

4. $$\begin{aligned}3x+4(-2x-5)&=-5\\3x-8x-20&=-5\\-5x-20&=-5\\-5x&=15\\x&=-3\end{aligned}$$
The solution is −3.

5. $$\begin{aligned}0.45x+0.06(-x+4000)&=630\\0.45x-0.06x+240&=630\\0.39x+240&=630\\0.39x&=390\\x&=1000\end{aligned}$$
The solution is 1000.

6. $$\begin{aligned}3x-2y&=6\\-2y&=-3x+6\end{aligned}$$

$y=\frac{3}{2}x-3$

x	y
0	−3
2	0
4	3

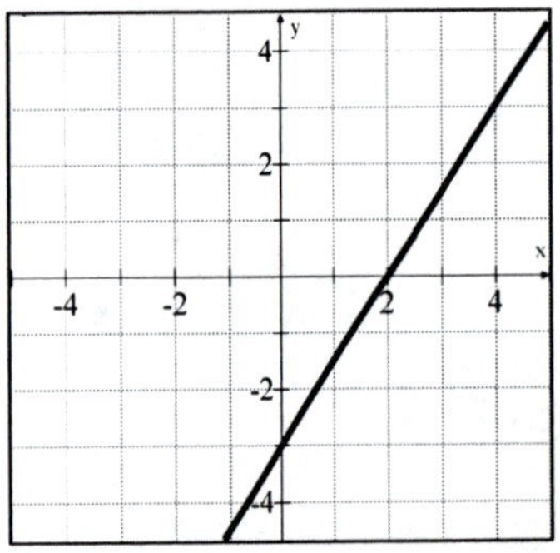

7. $y>-\frac{3}{5}x+1$

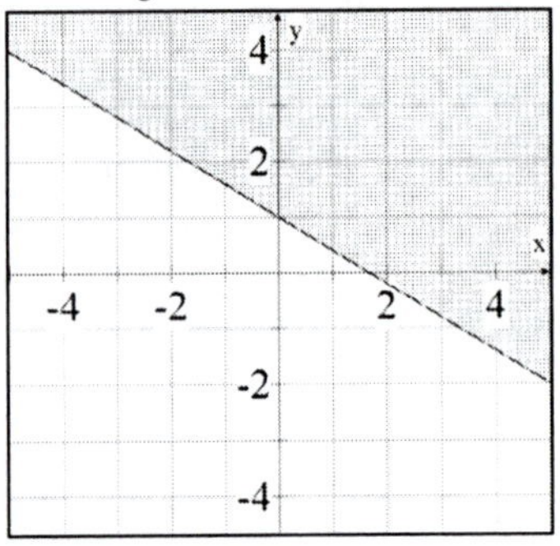

Section 5.1

Concept Check

1. $$\begin{aligned}x+y&=3\\2+1&=3\\3&=3\\2x-3y&=1\\2(2)-3(1)&=1\\4-3&=1\\1&=1\end{aligned}$$
Yes, (2,1) is a solution of the system of equations.

3. $$\begin{aligned}3x-y&=4\\3(1)-(-1)&=4\\3+1&=4\\4&=4\\7x+2y&=-5\\7(1)+2(-1)&=-5\\7-2&=-5\\5&\neq-5\end{aligned}$$
No, (1,−1) is not a solution of the system of equations.

5. The two graphs represent the same line. The system is a dependent system of equations.

7. (4, −1)

Objective A Exercises

9. $x + y = 2$
$x - y = 4$

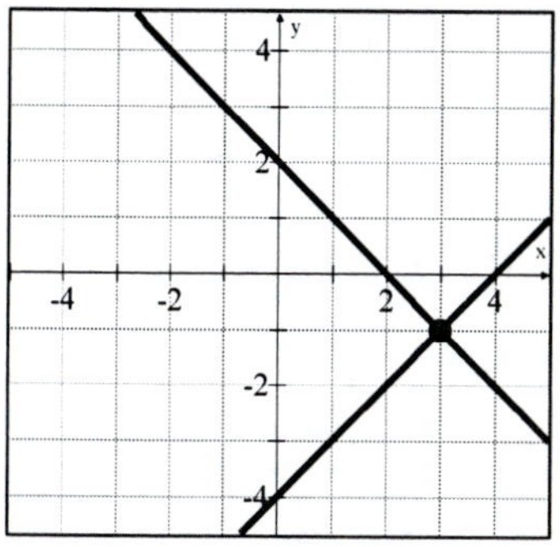

The solution is (3, −1).

11. $x - y = -2$
$x + 2y = 10$

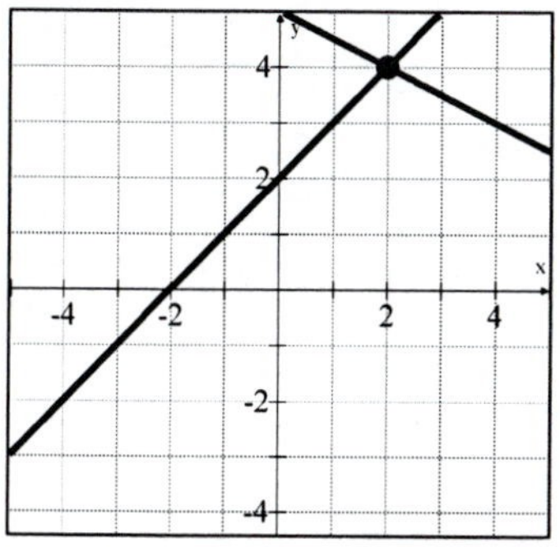

The solution is (2, 4).

13. $3x - 2y = 6$
$y = 3$

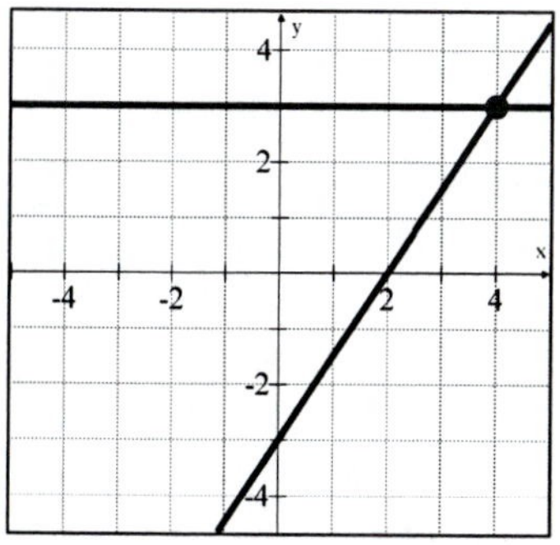

The solution is (4, 3).

15. $2x + 4y = 4$
$-3x - 6y = -6$

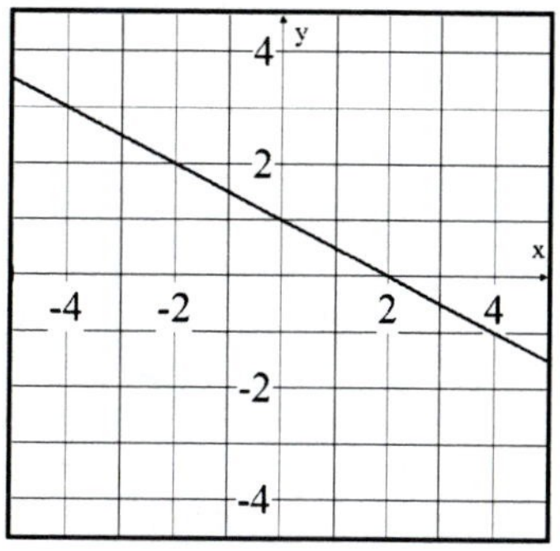

The two equations represent the same line. The system of equations is dependent. The solutions are the ordered pairs

$\left(x, -\frac{1}{2}x + 1\right)$.

17. $x - y = 6$
$x + y = 2$

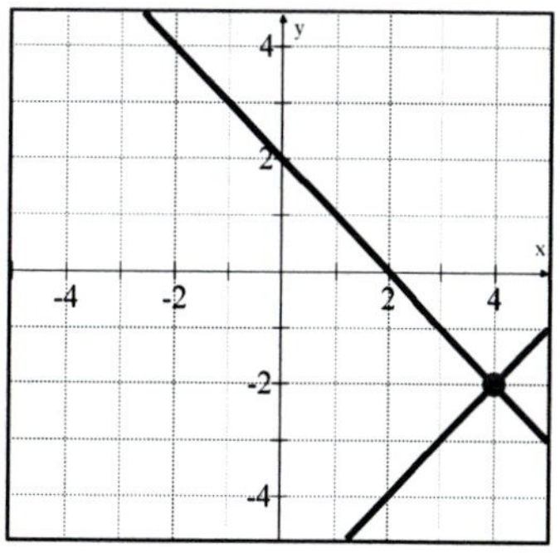

The solution is (4, −2).

19. $y = x - 5$
$2x + y = 4$

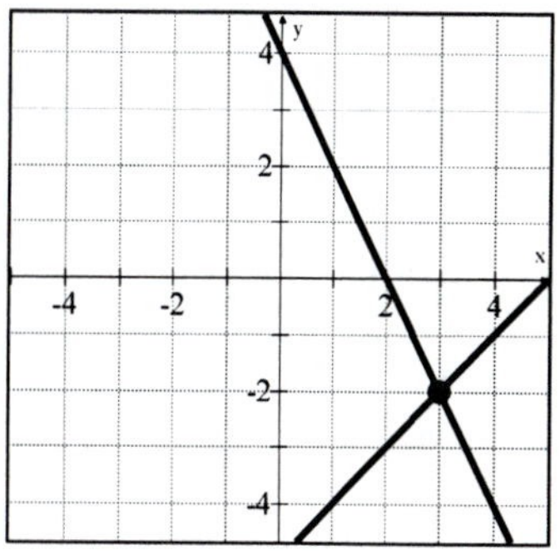

The solution is (3, −2).

21. $y = \frac{1}{2}x - 2$

$x - 2y = 8$

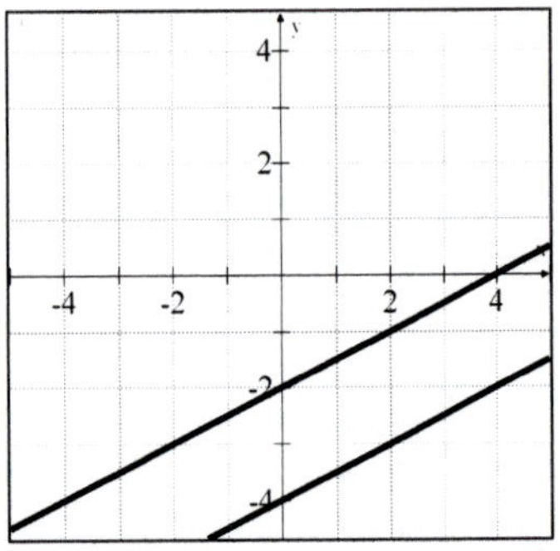

The lines are parallel and do not intersect so there is no solution.

23. $2x - 5y = 10$

$y = \frac{2}{5}x - 2$

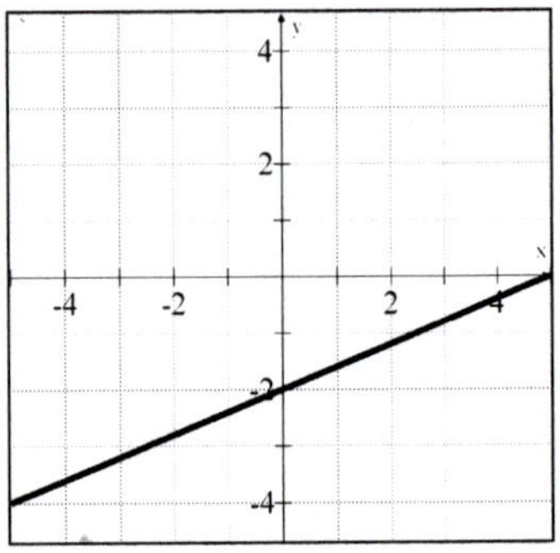

The two equations represent the same line. The system of equations is dependent. The solutions are the ordered pairs $\left(x, \frac{2}{5}x - 2\right)$.

Objective B Exercises

25. (1) $x = 2y - 3$

(2) $3x + y = 5$

Substitute $2y - 3$ for x in equation (2).

$$3x + y = 5$$
$$3(2y - 3) + y = 5$$
$$6y - 9 + y = 5$$
$$7y - 9 = 5$$
$$7y = 14$$
$$y = 2$$

Substitute 2 for y in equation (1).

$x = 2y - 3$

$x = 2(2) - 3$

$x = 4 - 3$

$x = 1$

The solution is (1, 2).

27. (1) $4x - 3y = 2$

(2) $y = 2x + 1$

Substitute $2x + 1$ for y in equation (1).

$$4x - 3y = 2$$
$$4x - 3(2x + 1) = 2$$
$$4x - 6x - 3 = 2$$
$$-2x - 3 = 2$$
$$-2x = 5$$
$$x = -\frac{5}{2}$$

Substitute $-\frac{5}{2}$ for x in equation (2).

$y = 2x + 1$

$y = 2\left(-\frac{5}{2}\right) + 1$

$y = -5 + 1$

$y = -4$

The solution is $\left(-\frac{5}{2}, -4\right)$.

29. (1) $3x - 2y = -11$

(2) $x = 2y - 9$

Substitute $2y - 9$ for x in equation (1).

$$3x - 2y = -11$$
$$3(2y - 9) - 2y = -11$$
$$6y - 27 - 2y = -11$$
$$4y - 27 = -11$$
$$4y = 16$$
$$y = 4$$

Substitute 4 for y in equation (2).

$x = 2y - 9$

$x = 2(4) - 9$

$x = 8 - 9$

$x = -1$

The solution is $(-1, 4)$.

31. (1) $3x + 2y = 4$
(2) $y = 1 - 2x$
Substitute $1 - 2x$ for y in equation (1).
$$3x + 2y = 4$$
$$3x + 2(1 - 2x) = 4$$
$$3x + 2 - 4x = 4$$
$$-x + 2 = 4$$
$$-x = 2$$
$$x = -2$$
Substitute -2 for x in equation (2).
$y = 1 - 2x$
$y = 1 - 2(-2)$
$y = 1 + 4$
$y = 5$
The solution is $(-2, 5)$.

33. (1) $5x + 2y = 15$
(2) $x = 6 - y$
Substitute $6 - y$ for x in equation (1).
$$5x + 2y = 15$$
$$5(6 - y) + 2y = 15$$
$$30 - 5y + 2y = 15$$
$$30 - 3y = 15$$
$$-3y = -15$$
$$y = 5$$
Substitute 5 for y in equation (2).
$x = 6 - y$
$x = 6 - 5$
$x = 1$
The solution is $(1, 5)$.

35. (1) $3x - 4y = 6$
(2) $x = 3y + 2$
Substitute $3y + 2$ for x in equation (1).
$$3x - 4y = 6$$
$$3(3y + 2) - 4y = 6$$
$$9y + 6 - 4y = 6$$
$$5y + 6 = 6$$
$$5y = 0$$
$$y = 0$$
Substitute 0 for y in equation (2).
$x = 3y + 2$
$x = 2(0) + 2$
$x = 0 + 2$
$x = 2$
The solution is $(2, 0)$.

37. (1) $3x + 7y = -5$
(2) $y = 6x - 5$
Substitute $6x - 5$ for y in equation (1).
$$3x + 7y = -5$$
$$3x + 7(6x - 5) = -5$$
$$3x + 42x - 35 = -5$$
$$45x - 35 = -5$$
$$45x = 30$$
$$x = \frac{2}{3}$$
Substitute $\frac{2}{3}$ for x in equation (2).
$y = 6x - 5$
$y = 6 \cdot \frac{2}{3} - 5$
$y = 4 - 5$
$y = -1$
The solution is $(\frac{2}{3}, -1)$.

39. (1) $3x - y = 10$
(2) $6x - 2y = 5$
Solve equation (1) for y.
$3x - y = 10$
$-y = -3x + 10$
$y = 3x - 10$
Substitute $3x - 10$ for y in equation (2).
$$6x - 2y = 5$$
$$6x - 2(3x - 10) = 5$$
$$6x - 6x + 20 = 5$$
$$20 = 5$$
No solution. This is not a true equation. The lines are parallel and the system is inconsistent.

41. (1) $3x + 4y = 14$
(2) $2x + y = 1$
Solve equation (2) for y.
$2x + y = 1$
$y = -2x + 1$
Substitute $-2x + 1$ for y in equation (1).
$$3x + 4y = 14$$
$$3x + 4(-2x + 1) = 14$$
$$3x - 8x + 4 = 14$$
$$-5x + 4 = 14$$
$$-5x = 10$$
$$x = -2$$

Substitute −2 for x in equation (2).
$2x + y = 1$
$2(-2) + y = 1$
$-4 + y = 1$
$y = 5$
The solution is (−2, 5).

43. (1) $3x + 5y = 0$
(2) $x - 4y = 0$
Solve equation (2) for x.
$x - 4y = 0$
$x = 4y$
Substitute $4y$ for x in equation (1).
$3x + 5y = 0$
$3(4y) + 5y = 0$
$12y + 5y = 0$
$17y = 0$
$y = 0$
Substitute 0 for y in equation (2).
$x - 4y = 0$
$x - 4(0) = 0$
$x - 0 = 0$
$x = 0$
The solution is (0, 0).

45. (1) $2x - 4y = 16$
(2) $-x + 2y = -8$
Solve equation (2) for x.
$-x + 2y = -8$
$x = 2y + 8$
Substitute $2y + 8$ for x in equation (1).
$2x - 4y = 16$
$2(2y + 8) - 4y = 16$
$4y + 16 - 4y = 16$
$16 = 16$
This is a true equation. The equations are dependent. The solutions are the ordered pairs $\left(x, \frac{1}{2}x - 4\right)$.

47. (1) $y = 3x + 2$
(2) $y = 2x + 3$
Substitute $2x + 3$ for y in equation (1).
$y = 3x + 2$
$2x + 3 = 3x + 2$
$3 = x + 2$
$x = 1$
Substitute 1 for x in equation (2).
$y = 2x + 3$
$y = 2(1) + 3$
$y = 5$
The solution is (1, 5).

49. (1) $y = 3x + 1$
(2) $y = 6x - 1$
Substitute $6x - 1$ for y in equation (1).
$y = 3x + 1$
$6x - 1 = 3x + 1$
$3x - 1 = 1$
$3x = 2$
$x = \frac{2}{3}$
Substitute $\frac{2}{3}$ for x in equation (2).
$y = 6x - 1$
$y = 6 \cdot \frac{2}{3} - 1$
$y = 3$
The solution is $\left(\frac{2}{3}, 3\right)$.

51. The value of $\frac{a}{b}$ is $\frac{2}{3}$.

Objective C Exercises

53. The interest rates on the two accounts are 5.5% and 7.2%.

55. **Strategy**: Let x represent the amount invested at 4.2%.
$2800 is invested at 3.5%.

	Principal	Rate	Interest
Amount at 3.5%	2800	0.035	0.035(2800)
Amount at 4.2%	x	0.042	$0.042x$

The sum of the interest earned is $329.

Solution: $0.035(2800) + 0.042x = 329$
$98 + 0.042x = 329$
$0.042x = 231$
$x = 5500$
$5500 is invested at 4.2%.

57. Strategy: Let x represent the amount invested at 6.5%.
Let y represent the total amount invested at 5%.
$6000 is invested at 4%.

	Principal	Rate	Interest
Amount at 4%	6000	0.04	0.04(6000)
Amount at 6.5%	x	0.065	$0.065x$
Total invested	y	0.05	$0.05y$

The total amount invested is y.
$y = 6000 + x$
The total interest earned is equal to 5% of the total investment.
$0.04(6000) + 0.065x = 0.05y$

Solution: (1) $y = 6000 + x$
(2) $0.04(6000) + 0.065x = 0.05y$
Substitute $6000 + x$ for y in equation (2).
$0.04(6000) + 0.065x = 0.05(6000 + x)$
$240 + 0.065x = 300 + 0.05x$
$240 + 0.015x = 300$
$0.015x = 60$
$x = 4000$
$4000 must be invested at 6.5% .

59. Strategy: Let x represent the amount invested at 3.5%.
Let y represent the amount invested at 4.5%.

	Principal	Rate	Interest
Amount at 3.5%	x	0.035	$0.035x$
Amount at 4.5%	y	0.045	$0.045y$

The total amount invested is $42,000.
$x + y = 42{,}000$
The interest earned from the 3.5% investment is equal to the interest earned from the 4.5% investment.
$0.035x = 0.045y$
Solution: (1) $x + y = 42{,}000$
(2) $0.035x = 0.045y$
Solve equation (1) for y and substitute for y in equation (2).
$y = 42{,}000 - x$
$0.035x = 0.045(42{,}000 - x)$
$0.035x = 1890 - 0.045x$
$0.080x = 1890$
$x = 23{,}625$

$y = 42{,}000 - x = 42{,}000 - 23{,}625 = 18{,}375$

$23,625 is invested at 3.5% and $18,375 is invested at 4.5%.

61. Strategy: Let x represent the amount invested at 4.5%.
Let y represent the amount invested at 8%.

	Principal	Rate	Interest
Amount at 4.5%	x	0.045	$0.045x$
Amount at 8%	y	0.08	$0.08y$

The total amount invested is $16,000.
$x + y = 16{,}000$
The total interest earned is $1070.
$0.045x + 0.08y = 1070$

Solution: (1) $x + y = 16{,}000$
(2) $0.045x + 0.08y = 1070$
Solve equation (1) for y and substitute for y in equation (2).
$y = 16{,}000 - x$
$0.045x + 0.08(16{,}000 - x) = 1070$
$0.045x + 1280 - 0.08x = 1070$
$-0.035x + 1280 = 1070$
$-0.035x = -210$
$x = 6000$
$6000 is invested at 4.5% .

Critical Thinking

63. The system of equations will be independent for k equal to any real number except 3.

Projects or Group Activities

65. Student answers will vary. Possible systems of equations for each case are:
a) $2x + y = -1$
$x + y = 2$
b) $-x + y = 10$
$-2x + 2y = 20$
c) $2x + y = 0$
$2x + y = 10$

Section 5.2

Concept Check

1. Student answers may vary. Possible answers are 6 and −5.

Objective A Exercises

3. (1) $x - y = 5$
(2) $x + y = 7$
Eliminate y. Add the two equations.
$2x = 12$
$x = 6$
Replace x with 6 in equation (1).
$x - y = 5$
$6 - y = 5$
$-y = -1$
$y = 1$
The solution is (6,1).

5. (1) $3x + y = 4$
(2) $x + y = 2$
Eliminate y.
$3x + y = 4$
$-1(x + y) = -1(2)$

$3x + y = 4$
$-x - y = -2$
Add the equations.
$2x = 2$
$x = 1$
Replace x with 1 in equation (2).
$x + y = 2$
$1 + y = 2$
$y = 1$
The solution is (1,1).

7. (1) $3x + y = 7$
(2) $x + 2y = 4$
Eliminate y.
$-2(3x + y) = -2(7)$
$x + 2y = 4$

$-6x - 2y = -14$
$x + 2y = 4$
Add the equations.
$-5x = -10$
$x = 2$
Replace x with 2 in equation (2).
$x + 2y = 4$
$2 + 2y = 4$
$2y = 2$
$y = 1$
The solution is (2,1).

9. (1) $2x + 3y = -1$
(2) $x + 5y = 3$
Eliminate x.
$2x + 3y = -1$
$-2(x + 5y) = -2(3)$

$2x + 3y = -1$
$-2x - 10y = -6$
Add the equations.
$-7y = -7$
$y = 1$
Replace y with 1 in equation (2).
$x + 5y = 3$
$x + 5(1) = 3$
$x + 5 = 3$
$x = -2$
The solution is (−2,1).

11. (1) $3x - y = 4$
(2) $6x - 2y = 8$
Eliminate y.
$-2(3x - y) = -2(4)$
$-6x + 2y = -8$

$6x - 2y = 8$
$6x - 2y = 8$
Add the equations.
$0 = 0$
This is a true equation. The equations are dependent. The solutions are the ordered pairs $(x, 3x - 4)$.

13. (1) $2x + 5y = 9$
(2) $4x - 7y = -16$
Eliminate x.
$-2(2x + 5y) = -2(9)$
$4x - 7y = -16$

$-4x - 10y = -18$
$4x - 7y = -16$
Add the equations.
$-17y = -34$
$y = 2$
Replace y with 2 in equation (1).
$2x + 5y = 9$
$2x + 5(2) = 9$

$2x + 10 = 9$
$2x = -1$
$x = -\frac{1}{2}$
The solution is $\left(-\frac{1}{2}, 2\right)$.

15. (1) $4x - 6y = 5$
(2) $2x - 3y = 7$
Eliminate y.
$4x - 6y = 5$
$-2(2x - 3y) = -2(7)$

$4x - 6y = 5$
$-4x + 6y = -14$
Add the equations.
$0 = -9$
This is not a true equation. The system of equations is inconsistent and therefore has no solution.

17. (1) $3x - 5y = 7$
(2) $x - 2y = 3$
Eliminate x.
$3x - 5y = 7$
$-3(x - 2y) = -3(3)$

$3x - 5y = 7$
$-3x + 6y = -9$
Add the equations.
$y = -2$
Replace y with -2 in equation (2).
$x - 2y = 3$
$x - 2(-2) = 3$
$x + 4 = 3$
$x = -1$
The solution is $(-1, -2)$.

19. (1) $x + 3y = 7$
(2) $-2x + 3y = 22$
Eliminate x.
$2(x + 3y) = 2(7)$
$-2x + 3y = 22$

$2x + 6y = 14$
$-2x + 3y = 22$
Add the equations.
$9y = 36$
$y = 4$
Replace y with 4 in equation (1).
$x + 3y = 7$
$x + 3(4) = 7$
$x + 12 = 7$
$x = -5$
The solution is $(-5, 4)$.

21. (1) $3x + 2y = 16$
(2) $2x - 3y = -11$
Eliminate x.
$-2(3x + 2y) = -2(16)$
$3(2x - 3y) = 3(-11)$
$-6x - 4y = -32$
$6x - 9y = -33$
Add the equations.
$-13y = -65$
$y = 5$
Replace y with 5 in equation (1).
$3x + 2y = 16$
$3x + 2(5) = 16$
$3x + 10 = 16$
$3x = 6$
$x = 2$
The solution is $(2, 5)$.

23. (1) $4x + 4y = 5$
(2) $2x - 8y = -5$
Eliminate x.
$4x + 4y = 5$
$-2(2x - 8y) = -2(-5)$

$4x + 4y = 5$
$-4x + 16y = 10$
Add the equations.
$20y = 15$
$y = \frac{3}{4}$
Replace y with $\frac{3}{4}$ in equation (1).
$4x + 4y = 5$
$4x + 4\left(\frac{3}{4}\right) = 5$
$4x + 3 = 5$
$4x = 2$
$x = \frac{1}{2}$
The solution is $\left(\frac{1}{2}, \frac{3}{4}\right)$.

25. (1) $5x + 4y = 0$
(2) $3x + 7y = 0$
Eliminate x.
$-3(5x + 4y) = -3(0)$
$5(3x + 7y) = 5(0)$
$-15x - 12y = 0$
$15x + 35y = 0$
Add the equations.
$23y = 0$
$y = 0$
Replace y with 0 in equation (1).
$5x + 4y = 0$
$5x + 4(0) = 0$
$5x + 0 = 0$
$5x = 0$
$x = 0$
The solution is (0,0).

27. (1) $5x + 2y = 1$
(2) $2x + 3y = 7$
Eliminate x.
$-2(5x + 2y) = -2(1)$
$5(2x + 3y) = 5(7)$

$-10x - 4y = -2$
$10x + 15y = 35$
Add the equations.
$11y = 33$
$y = 3$
Replace y with 3 in equation (1).
$5x + 2y = 1$
$5x + 2(3) = 1$
$5x + 6 = 1$
$5x = -5$
$x = -1$
The solution is $(-1,3)$.

29. (1) $3x - 6y = 6$
(2) $9x - 3y = 8$
Eliminate y.
$3x - 6y = 6$
$-2(9x - 3y) = -2(8)$

$3x - 6y = 6$
$-18x + 6y = -16$
Add the equations.
$-15x = -10$
$x = \frac{2}{3}$

Replace x with $\frac{2}{3}$ in equation (1).
$3x - 6y = 6$
$3\left(\frac{2}{3}\right) - 6y = 6$
$2 - 6y = 6$
$-6y = 4$
$y = -\frac{2}{3}$
The solution is $\left(\frac{2}{3}, -\frac{2}{3}\right)$.

31. (1) $\frac{3}{4}x + \frac{1}{3}y = -\frac{1}{2}$
(2) $\frac{1}{2}x - \frac{5}{6}y = -\frac{7}{2}$
Clear the fractions.
$12\left(\frac{3}{4}x + \frac{1}{3}y\right) = 12\left(-\frac{1}{2}\right)$
$6\left(\frac{1}{2}x - \frac{5}{6}y\right) = 6\left(-\frac{7}{2}\right)$

$9x + 4y = -6$
$3x - 5y = -21$
Eliminate x.
$9x + 4y = -6$
$-3(3x - 5y) = -3(-21)$

$9x + 4y = -6$
$-9x + 15y = 63$
Add the equations.
$19y = 57$
$y = 3$
Replace y with 3 in equation (1).
$\frac{3}{4}x + \frac{1}{3}y = -\frac{1}{2}$
$\frac{3}{4}x + \frac{1}{3}(3) = -\frac{1}{2}$
$\frac{3}{4}x + 1 = -\frac{1}{2}$
$\frac{3}{4}x = -\frac{3}{2}$
$x = -2$
The solution is $(-2,3)$.

33. (1) $\frac{5x}{6}+\frac{y}{3}=\frac{4}{3}$

(2) $\frac{2x}{3}-\frac{y}{2}=\frac{11}{6}$

Clear the fractions.

$6\left(\frac{5x}{6}+\frac{y}{3}\right)=6\left(\frac{4}{3}\right)$

$6\left(\frac{2x}{3}-\frac{y}{2}\right)=6\left(\frac{11}{6}\right)$

$5x+2y=8$
$4x-3y=11$

Eliminate y.

$3(5x+2y)=3(8)$
$2(4x-3y)=2(11)$

$15x+6y=24$
$8x-6y=22$

Add the equations.

$23x=46$
$x=2$

Replace x with 2 in equation (1).

$\frac{5x}{6}+\frac{y}{3}=\frac{4}{3}$

$\frac{5(2)}{6}+\frac{y}{3}=\frac{4}{3}$

$\frac{5}{3}+\frac{y}{3}=\frac{4}{3}$

$\frac{y}{3}=-\frac{1}{3}$

$y=-1$

The solution is $(2,-1)$.

35. (1) $\frac{2x}{5}-\frac{y}{2}=\frac{13}{2}$

(2) $\frac{3x}{4}-\frac{y}{5}=\frac{17}{2}$

Clear the fractions.

$10\left(\frac{2x}{5}-\frac{y}{2}\right)=10\left(\frac{13}{2}\right)$

$20\left(\frac{3x}{4}-\frac{y}{5}\right)=20\left(\frac{17}{2}\right)$

$4x-5y=65$
$15x-4y=170$

Eliminate y.

$4(4x-5y)=4(65)$
$-5(15x-4y)=-5(170)$

$16x-20y=260$
$-75x+20y=-850$

Add the equations.

$-59x=-590$
$x=10$

Replace x with 10 in equation (1).

$\frac{2x}{5}-\frac{y}{2}=\frac{13}{2}$

$\frac{2(10)}{5}-\frac{y}{2}=\frac{13}{2}$

$4-\frac{y}{2}=\frac{13}{2}$

$-\frac{y}{2}=\frac{5}{2}$

$y=-5$

The solution is $(10,-5)$.

37. (1) $\frac{3x}{2}-\frac{y}{4}=-\frac{11}{12}$

(2) $\frac{x}{3}-y=-\frac{5}{6}$

Clear the fractions.

$12\left(\frac{3x}{2}-\frac{y}{4}\right)=12\left(-\frac{11}{12}\right)$

$6\left(\frac{x}{3}-y\right)=6\left(-\frac{5}{6}\right)$

$18x-3y=-11$
$2x-6y=-5$

Eliminate y.

$-2(18x-3y)=-2(-11)$
$2x-6y=-5$

$-36x+6y=22$
$2x-6y=-5$

Add the equations.

$-34x=17$

$x=-\frac{1}{2}$

Replace x with $-\frac{1}{2}$ in equation (1).

$\frac{3x}{2}-\frac{y}{4}=-\frac{11}{12}$

$\frac{3}{2}\left(-\frac{1}{2}\right)-\frac{y}{4}=-\frac{11}{12}$

$-\frac{3}{4}-\frac{y}{4}=-\frac{11}{12}$

$-\frac{y}{4}=-\frac{1}{6}$

$y=\frac{2}{3}$

The solution is $(-\frac{1}{2},\frac{2}{3})$.

39. (1) $4x-5y=3y+4$
(2) $2x+3y=2x+1$
Write the equations in the form $Ax+By=C$.
Solve the system of equations.
(1) $4x-8y=4$
(2) $3y=1$

Solve equation (2) for y.
$3y=1$
$y=\frac{1}{3}$

Replace y with $\frac{1}{3}$ in equation (1).
$4x-5y=3y+4$

$4x-5\cdot\frac{1}{3}=3\cdot\frac{1}{3}+4$

$4x-\frac{5}{3}=1+4$

$4x-\frac{5}{3}=5$

$4x=\frac{20}{3}$

$x=\frac{5}{3}$

The solution is $\left(\frac{5}{3},\frac{1}{3}\right)$.

41. (1) $2x+5y=5x+1$
(2) $3x-2y=3y+3$
Write the equations in the form $Ax+By=C$.
Solve the system of equations.
(1) $-3x+5y=1$
(2) $3x-5y=3$

Add the equations.
$0=4$
This is not a true equation. The system of equations is inconsistent and therefore has no solution.

43. (1) $5x+2y=2x+1$
(2) $2x-3y=3x+2$
Write the equations in the form $Ax+By=C$.
Solve the system of equations.
(1) $3x+2y=1$
(2) $-x-3y=2$

Eliminate x.
$3x+2y=1$
$3(-x-3y)=3(2)$

$3x+2y=1$
$-3x-9y=6$
Add the equations.
$-7y=7$
$y=-1$
Replace y with -1 in equation (1).
$5x+2y=2x+1$
$5x+2(-1)=2x+1$
$5x-2=2x+1$
$3x=3$
$x=1$
The solution is $(1,-1)$.

Objective B Exercises

45. (1) $x+2y-z=1$
(2) $2x-y+z=6$
(3) $x+3y-z=2$
Eliminate z. Add equations (1) and (2).
$x+2y-z=1$
$2x-y+z=6$

(4) $3x+y=7$

Add equations (2) and (3).
$2x-y+z=6$
$x+3y-z=2$

(5) $3x+2y=8$

Use equations (4) and (5) solve for x and y.
$3x+y=7$
$3x+2y=8$

Eliminate x.
$-1(3x+y)=-1(7)$
$3x+2y=8$

$-3x-y=-7$
$3x+2y=8$
$y=1$
Replace y with 1 in equation (4).
$3x+y=7$
$3x+1=7$
$3x=6$
$x=2$
Replace x with 2 and y with 1 in equation (1).
$x+2y-z=1$
$2+2(1)-z=1$
$2+2-z=1$
$4-z=1$
$-z=-3$
$z=3$
The solution is (2, 1, 3).

47. (1) $2x-y+2z=7$
(2) $x+y+z=2$
(3) $3x-y+z=6$
Eliminate y. Add equations (1) and (2).
$2x-y+2z=7$
$x+y+z=2$

(4) $3x+3z=9$

Add equations (2) and (3).
$x+y+z=2$
$3x-y+z=6$

(5) $4x+2z=8$

Use equations (4) and (5) solve for x and z.
$3x+3z=9$
$4x+2z=8$
Eliminate z.
$-2(3x+3z)=-2(9)$
$3(4x+2z)=3(8)$

$-6x-6z=-18$
$12x+6z=24$
$6x=6$
$x=1$
Replace x with 1 in equation (4).
$3x+3z=9$
$3(1)+3z=9$
$3+3z=9$
$3z=6$
$z=2$
Replace x with 1 and z with 2 in equation (1).
$2x-y+2z=7$
$2(1)-y+2(2)=7$
$2-y+4=7$
$6-y=7$
$-y=1$
$y=-1$
The solution is (1, −1, 2).

49. (1) $3x+y=5$
(2) $3y-z=2$
(3) $x+z=5$
Eliminate z. Add equations (2) and (3).
$3y-z=2$
$x+z=5$

(4) $x+3y=7$

Use equations (1) and (4). Solve for x and y.
$3x+y=5$
$x+3y=7$

Eliminate y.
$-3(3x+y)=-3(5)$
$x+3y=7$

$-9x-3y=-15$
$x+3y=7$
$-8x=-8$
$x=1$
Replace x with 1 in equation (1).
$3x+y=5$
$3(1)+y=5$
$3+y=5$
$y=2$
Replace y with 2 in equation (2).
$3y-z=2$
$3(2)-z=2$
$6-z=2$
$-z=-4$
$z=4$
The solution is (1, 2, 4).

51. (1) $x-y+z=1$
(2) $2x+3y-z=3$
(3) $-x+2y-4z=4$
Eliminate z. Add equations (1) and (2).

$x - y + z = 1$
$2x + 3y - z = 3$

(4) $3x + 2y = 4$

Multiply equation (1) by 4 and add to equation (3).
$4(x - y + z) = 4(1)$
$-x + 2y - 4z = 4$

$4x - 4y + 4z = 4$
$-x + 2y - 4z = 4$

(5) $3x - 2y = 8$
Use equations (4) and (5) solve for x and y.
Add equation (4) and (5) to eliminate x.
$3x + 2y = 4$
$3x - 2y = 8$
$6x = 12$
$x = 2$
Replace x with 2 in equation (4).
$3x + 2y = 4$
$3(2) + 2y = 4$
$6 + 2y = 4$
$2y = -2$
$y = -1$
Replace x with 2 and y with -1 in equation (1).
$x - y + z = 1$
$2 - (-1) + z = 1$
$2 + 1 + z = 1$
$3 + z = 1$
$z = -2$
The solution is $(2, -1, -2)$.

53. (1) $2x + 3z = 5$
(2) $3y + 2z = 3$
(3) $3x + 4y = -10$
Eliminate z. Use equations (1) and (2).
$2x + 3z = 5$
$3y + 2z = 3$

$-2(2x + 3z) = -2(5)$
$3(3y + 2z) = 3(3)$

$-4x - 6z = -10$
$9y + 6z = 9$

(4) $-4x + 9y = -1$

Use equations (3) and (4), solve for x and y.
$3x + 4y = -10$
$-4x + 9y = -1$
Eliminate x.
$4(3x + 4y) = 4(-10)$
$3(-4x + 9y) = 3(-1)$

$12x + 16y = -40$
$-12x + 27y = -3$
$43y = -43$
$y = -1$
Replace y with -1 in equation (2).
$3y + 2z = 3$
$3(-1) + 2z = 3$
$-3 + 2z = 3$
$2z = 6$
$z = 3$
Replace z with 3 in equation (1).
$2x + 3z = 5$
$2x + 3(3) = 5$
$2x + 9 = 5$
$2x = -4$
$x = -2$
The solution is $(-2, -1, 3)$.

55. (1) $2x + 4y - 2z = 3$
(2) $x + 3y + 4z = 1$
(3) $x + 2y - z = 4$
Eliminate x. Use equations (1) and (2).
$2x + 4y - 2z = 3$
$x + 3y + 4z = 1$

$2x + 4y - 2z = 3$
$-2(x + 3y + 4z) = -2(1)$

$2x + 4y - 2z = 3$
$-2x - 6y - 8z = -2$

(4) $-2y - 10z = 1$

Use equations (2) and (3).
$x + 3y + 4z = 1$
$x + 2y - z = 4$

$x + 3y + 4z = 1$
$-1(x + 2y - z) = -1(4)$

$x + 3y + 4z = 1$
$-x - 2y + z = -4$

(5) $y + 5z = -3$

Use equations (4) and (5) solve for y and z.
$-2y - 10z = 1$
$y + 5z = -3$
Eliminate y.
$-2y - 10z = 1$
$2(y + 5z) = 2(-3)$

$-2y - 10z = 1$
$2y + 10z = -6$
$0 = -6$
This is not a true equation. The system of equations is inconsistent and therefore has no solution.

57. (1) $2x + y - z = 5$
(2) $x + 3y + z = 14$
(3) $3x - y + 2z = 1$
Eliminate z. Add equations (1) and (2).
$2x + y - z = 5$
$x + 3y + z = 14$

(4) $3x + 4y = 19$

Use equations (2) and (3).
$x + 3y + z = 14$
$3x - y + 2z = 1$

$-2(x + 3y + z) = -2(14)$
$3x - y + 2z = 1$

$-2x - 6y - 2z = -28$
$3x - y + 2z = 1$

(5) $x - 7y = -27$

Use equations (4) and (5) solve for x and y.
$3x + 4y = 19$
$x - 7y = -27$
Eliminate x.
$3x + 4y = 19$
$-3(x - 7y) = -3(-27)$

$3x + 4y = 19$
$-3x + 21y = 81$
$25y = 100$
$y = 4$
Replace y with 4 in equation (4).
$3x + 4y = 19$
$3x + 4(4) = 19$
$3x + 16 = 19$
$3x = 3$
$x = 1$
Replace x with 1 and y with 4 in equation (1).
$2x + y - z = 5$
$2(1) + 4 - z = 5$
$2 + 4 - z = 5$
$6 - z = 5$
$-z = -1$
$z = 1$
The solution is (1, 4, 1).

59. (1) $3x + y - 2z = 2$
(2) $x + 2y + 3z = 13$
(3) $2x - 2y + 5z = 6$
Eliminate x. Add equations (1) and (2).
$3x + y - 2z = 2$
$x + 2y + 3z = 13$

$3x + y - 2z = 2$
$-3(x + 2y + 3z) = -3(13)$

$3x + y - 2z = 2$
$-3x - 6y - 9z = -39$

(4) $-5y - 11z = -37$

Use equations (2) and (3).
$x + 2y + 3z = 13$
$2x - 2y + 5z = 6$

$-2(x + 2y + 3z) = -2(13)$
$2x - 2y + 5z = 6$

$-2x - 4y - 6z = -26$
$2x - 2y + 5z = 6$

(5) $-6y - z = -20$

Use equations (4) and (5) solve for y and z.
$-5y - 11z = -37$
$-6y - z = -20$
Eliminate z.
$-5y - 11z = -37$
$-11(-6y - z) = -11(-20)$

$-5y - 11z = -37$
$66y + 11z = 220$
$61y = 183$
$y = 3$
Replace y with 3 in equation (4).
$-5y - 11z = -37$

$-5(3) - 11z = -37$
$-15 - 11z = -37$
$-11z = -22$
$z = 2$
Replace y with 3 and z with 2 in equation (1).
$3x + y - 2z = 2$
$3x + 3 - 2(2) = 2$
$3x + 3 - 4 = 2$
$3x - 1 = 2$
$3x = 3$
$x = 1$
The solution is (1, 3, 2.)

61. (1) $2x - y + z = 6$
(2) $3x + 2y + z = 4$
(3) $x - 2y + 3z = 12$
Eliminate y. Use equations (1) and (2).
$2x - y + z = 6$
$3x + 2y + z = 4$

$2(2x - y + z) = 2(6)$
$3x + 2y + z = 4$

$4x - 2y + 2z = 12$
$3x + 2y + z = 4$

(4) $7x + 3z = 16$

Add equations (2) and (3).
$3x + 2y + z = 4$
$x - 2y + 3z = 12$

(5) $4x + 4z = 16$

Use equations (4) and (5) solve for x and z.
$7x + 3z = 16$
$4x + 4z = 16$
Eliminate z.
$4(7x + 3z) = 4(16)$
$-3(4x + 4z) = -3(16)$

$28x + 12z = 64$
$-12x - 12z = -48$
$16x = 16$
$x = 1$
Replace x with 1 in equation (4).
$7x + 3z = 16$
$7(1) + 3z = 16$
$7 + 3z = 16$
$3z = 9$
$z = 3$
Replace x with 1 and z with 3 in equation (1).
$2x - y + z = 6$
$2(1) - y + 3 = 6$
$-y + 5 = 6$
$-y = 1$
$y = -1$
The solution is $(1, -1, 3)$.

63. (1) $3x - 2y + 3z = -4$
(2) $2x + y - 3z = 2$
(3) $3x + 4y + 5z = 8$
Eliminate y. Use equations (1) and (2).
$3x - 2y + 3z = -4$
$2x + y - 3z = 2$

$3x - 2y + 3z = -4$
$2(2x + y - 3z) = 2(2)$

$3x - 2y + 3z = -4$
$4x + 2y - 6z = 4$

(4) $7x - 3z = 0$

Use equations (2) and (3).
$2x + y - 3z = 2$
$3x + 4y + 5z = 8$

$-4(2x + y - 3z) = -4(2)$
$3x + 4y + 5z = 8$

$-8x - 4y + 12z = -8$
$3x + 4y + 5z = 8$

(5) $-5x + 17z = 0$

Use equations (4) and (5) solve for x and z.
$7x - 3z = 0$
$-5x + 17z = 0$
Eliminate x.
$5(7x - 3z) = 5(0)$
$7(-5x + 17z) = 7(0)$

$35x - 15z = 0$
$-35x + 119z = 0$
$104z = 0$
$z = 0$

Replace z with 0 in equation (4).
$7x - 3z = 0$

$7x - 3(0) = 0$
$7x = 0$
$x = 0$
Replace x with 0 and z with 0 in equation (1).
$3x - 2y + 3z = -4$
$3(0) - 2y + 3(0) = -4$
$0 - 2y + 0 = -4$
$-2y = -4$
$y = 2$
The solution is (0, 2, 0).

65. (1) $3x - y + 2z = 2$
(2) $4x + 2y - 7z = 0$
(3) $2x + 3y - 5z = 7$
Eliminate y. Use equations (1) and (2).
$3x - y + 2z = 2$
$4x + 2y - 7z = 0$

$2(3x - y + 2z) = 2(2)$
$4x + 2y - 7z = 0$

$6x - 2y + 4z = 4$
$4x + 2y - 7z = 0$

(4) $10x - 3z = 4$

Use equations (1) and (3).
$3x - y + 2z = 2$
$2x + 3y - 5z = 7$

$3(3x - y + 2z) = 3(2)$
$2x + 3y - 5z = 7$

$9x - 3y + 6z = 6$
$2x + 3y - 5z = 7$

(5) $11x + z = 13$

Use equations (4) and (5) solve for x and z.
$10x - 3z = 4$
$11x + z = 13$
Eliminate z.
$10x - 3z = 4$
$3(11x + z) = 3(13)$

$10x - 3z = 4$
$33x + 3z = 39$
$43x = 43$
$x = 1$

Replace x with 1 in equation (4).
$10x - 3z = 4$
$10(1) - 3z = 4$
$-3z = -6$
$z = 2$
Replace x with 1 and z with 2 in equation (1).
$3x - y + 2z = 2$
$3(1) - y + 2(2) = 2$
$3 - y + 4 = 2$
$7 - y = 2$
$-y = -5$
$y = 5$
The solution is (1, 5, 2).

67. (1) $2x - 3y + 7z = 0$
(2) $x + 4y - 4z = -2$
(3) $3x + 2y + 5z = 1$
Eliminate x. Use equations (1) and (2).
$2x - 3y + 7z = 0$
$x + 4y - 4z = -2$

$2x - 3y + 7z = 0$
$-2(x + 4y - 4z) = -2(-2)$

$2x - 3y + 7z = 0$
$-2x - 8y + 8z = 4$

(4) $-11y + 15z = 4$

Use equations (2) and (3).
$x + 4y - 4z = -2$
$3x + 2y + 5z = 1$

$-3(x + 4y - 4z) = -3(-2)$
$3x + 2y + 5z = 1$

$-3x - 12y + 12z = 6$
$3x + 2y + 5z = 1$

(5) $-10y + 17z = 7$

Use equations (4) and (5) solve for y and z.
$-11y + 15z = 4$
$-10y + 17z = 7$
Eliminate y.
$10(-11y + 15z) = 10(4)$
$-11(-10y + 17z) = -11(7)$

$-110y + 150z = 40$
$110y - 187z = -77$

$-37z = -37$
$z = 1$

Replace z with 1 in equation (4).
$-11y + 15z = 4$
$-11y + 15(1) = 4$
$-11y = -11$
$y = 1$
Replace y with 1 and z with 1 in equation (1).
$2x - 3y + 7z = 0$
$2x - 3(1) + 7(1) = 0$
$2x - 3 + 7 = 0$
$2x + 4 = 0$
$2x = -4$
$x = -2$
The solution is $(-2, 1, 1)$.

69. a) (iii) no points
b) (ii) more than one point
c) (i) exactly one point

Critical Thinking

71. (1) $\frac{1}{x} - \frac{2}{y} = 3$

(2) $\frac{2}{x} + \frac{3}{y} = -1$

Clear the fractions.

$xy\left(\frac{1}{x} - \frac{2}{y}\right) = xy(3)$

$xy\left(\frac{2}{x} + \frac{3}{y}\right) = xy(-1)$

$y - 2x = 3xy$
$2y + 3x = -xy$
Eliminate y.
$-2(y - 2x) = -2(3xy)$
$2y + 3x = -xy$

$-2y + 4x = -6xy$
$2y + 3x = -xy$
$7x = -7xy$
$y = -1$
Replace y with -1 in equation (1).

$\frac{1}{x} - \frac{2}{y} = 3$

$\frac{1}{x} - \frac{2}{-1} = 3$

$\frac{1}{x} + 2 = 3$

$\frac{1}{x} = 1$

$x = 1$
The solution is $(1,-1)$.

73. (1) $\frac{3}{x} + \frac{2}{y} = 1$

(2) $\frac{2}{x} + \frac{4}{y} = -2$

Clear the fractions.

$xy\left(\frac{3}{x} + \frac{2}{y}\right) = xy(1)$

$xy\left(\frac{2}{x} + \frac{4}{y}\right) = xy(-2)$

$3y + 2x = xy$
$2y + 4x = -2xy$
Eliminate x.
$-2(3y + 2x) = -2(xy)$
$2y + 4x = -2xy$

$-6y - 4x = -2xy$
$2y + 4x = -2xy$
$-4y = -4xy$
$x = 1$
Replace x with 1 in equation (1).

$\frac{3}{x} + \frac{2}{y} = 1$

$\frac{3}{1} + \frac{2}{y} = 1$

$\frac{2}{y} = -2$

$y = -1$
The solution is $(1,-1)$.

75. P(3, –2, 4)

$$\begin{aligned}&Ax + 3y + 2z = 8\\&A(3) + 3(-2) + 2(4) = 8\\&3A + 2 = 8\\&3A = 6\\&A = 2\end{aligned}$$

$$\begin{aligned}&2x + By - 3z = -12\\&2(3) + B(-2) - 3(4) = -12\\&-2B - 6 = -12\\&-2B = -6\\&B = 3\end{aligned}$$

$$\begin{aligned}&3x - 2y + Cz = 1\\&3(3) - 2(-2) + C(4) = 1\\&13 + 4C = 1\\&4C = -12\\&C = -3\end{aligned}$$

Projects or Group Activities

77. a) The graph would be a plane which is parallel to the *yz* axis.
b) The graph would be a plane which is parallel to the *xz* axis.
c) The graph would be a plane which is parallel to the *xy* axis.
d) The graph would be a plane which is perpendicular to the *xy* plane.

Check Your Progress: Chapter 5

1.

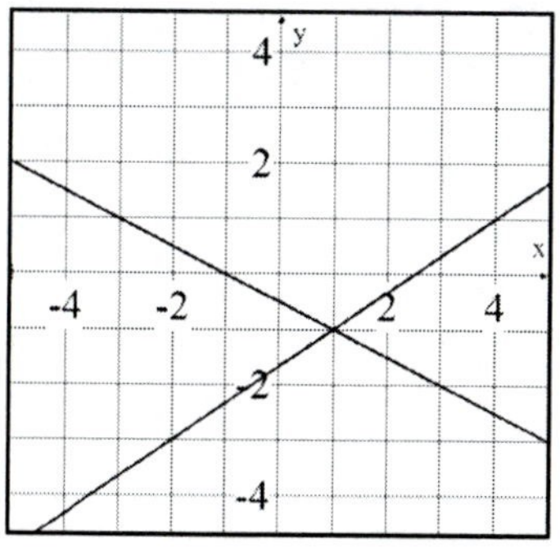

The solution is (1, –1).

2.

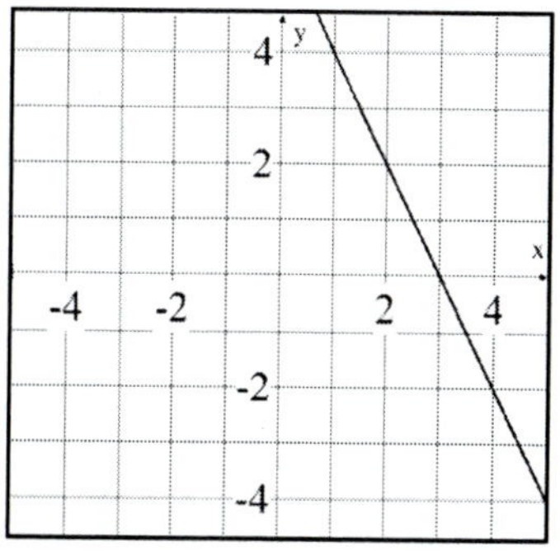

The solution is $(x, -2x + 6)$.

3. (1) $4x - 2y = 16$
(2) $3x - y = 11$
Solve equation (2) for *y*.

$$\begin{aligned}3x - y &= 11\\ y &= 3x - 11\end{aligned}$$

Substitute in equation (1).

$$\begin{aligned}4x - 2y &= 16\\ 4x - 2(3x - 11) &= 16\\ 4x - 6x + 22 &= 16\\ -2x &= -6\\ x &= 3\end{aligned}$$

Substitute 3 for *x* in equation (2).

$$\begin{aligned}&3x - y = 11\\&3(3) - y = 11\\&9 - y = 11\\&y = -2\end{aligned}$$

The solution is (3, –2).

4. (1) $9x + 12y = 11$
(2) $6x + 8y = 9$
Solve equation (2) for *x*.

$$\begin{aligned}&6x + 8y = 9\\&6x = 9 - 8y\\&x = \frac{3}{2} - \frac{4}{3}y\end{aligned}$$

Substitute in equation (1).

$$\begin{aligned}9x + 12y &= 11\\ 9\left(\frac{3}{2} - \frac{4}{3}y\right) + 12y &= 11\\ \frac{27}{2} - 12y + 12y &= 11\\ \frac{27}{2} &\neq 11\end{aligned}$$

There is no solution.

5. (1) $3x + 5y = 9$
(2) $2x - y = 7$
Solve equation (2) for y
$2x - y = 7$
$y = 2x - 7$
Substitute in equation (1)
$3x + 5y = 9$
$3x + 5(2x - 7) = 9$
$3x + 10x - 35 = 9$
$13x - 35 = 9$
$13x = 44$
$x = \frac{44}{13}$
Substitute $\frac{44}{13}$ for x in equation (2).
$2x - y = 7$
$2\left(\frac{44}{13}\right) - y = 7$
$\frac{88}{13} - y = 7$
$88 - 13y = 91$
$-13y = 3$
$y = -\frac{3}{13}$
The solution is $\left(\frac{44}{13}, -\frac{3}{13}\right)$.

6. (1) $3x + 5y = 14$
(2) $-2x + 3y = 16$
Eliminate x.
$2(3x + 5y) = 2(14)$
(3) $6x + 10y = 28$

$3(-2x + 3y) = 3(16)$
(4) $-6x + 9y = 48$

Add equations (3) and (4).
$6x + 10y = 28$
$-6x + 9y = 48$
$19y = 76$
$y = 4$

Substitute the value for y in equation (1)
$3x + 5y = 14$
$3x + 5(4) = 14$
$3x = -6$
$x = -2$
The solution is $(-2, 4)$.

7. (1) $2x + 5y = 10$
(2) $4x + 10y = 20$
Eliminate x.
$-2(2x + 5y) = -2(10)$
$-4x - 10y = -20$

Add the equations.
$-4x - 10y = -20$
$4x + 10y = 20$
$0 = 0$
This is a true equation. The equations are dependent. The solutions are the ordered pairs $\left(x, -\frac{2}{5}x + 2\right)$.

8. (1) $x + 3y - 2z = -7$
(2) $2x + y + z = 6$
(3) $-3x - y + 3z = 4$
Eliminate z. Use equations (1) and (2).
$x + 3y - 2z = -7$
$2x + y + z = 6$

$2(2x + y + z) = 2(6)$
$4x + 2y + 2z = 12$

$x + 3y - 2z = -7$
$4x + 2y + 2z = 12$
(4) $5x + 5y = 5$

Use equations (2) and (3).
$2x + y + z = 6$
$-3x - y + 3z = 4$

$-3(2x + y + z) = -3(6)$
$-6x - 3y - 3z = -18$

$-6x - 3y - 3z = -18$
$-3x - y + 3z = 4$
(5) $-9x - 4y = -14$

Use equations (4) and (5) to solve for x and y.
$5x + 5y = 5$
$-9x - 4y = -14$
Eliminate x.
$5x + 5y = 5$
$9(5x + 5y) = 9(5)$

$45x + 45y = 45$

$-9x - 4y = -13$
$5(-9x - 4y) = 5(-14)$
$-45x - 20y = -70$
Add the equations
$45x + 45y = 45$
$-45x - 20y = -70$
$25y = -25$
$y = -1$
Replace y with -1 in equation (4).
$5x + 5y = 5$
$5x + 5(-1) = 5$
$5x = 10$
$x = 2$
Replace x with 2 and y with -1 in equation (1).
$x + 3y - 2z = -7$
$2 + 3(-1) - 2z = -7$
$-1 - 2z = -7$
$z = 3$
The solution is $(2, -1, 3)$.

9. (1) $4x + 5y - z = 22$
(2) $3x - 6y + 2z = -28$
(3) $x + 2y - 2z = 12$
Eliminate z. Use equations (1) and (2).
$4x + 5y - z = 22$
$3x - 6y + 2z = -28$

$2(4x + 5y - z) = 2(22)$
$8x + 10y - 2z = 44$

$8x + 10y - 2z = 44$
$3x - 6y + 2z = -28$

(4) $11x + 4y = 16$

Use equations (2) and (3).
$3x - 6y + 2z = -28$
$x + 2y - 2z = 12$

(5) $4x - 4y = -16$

Use equations (4) and (5) to solve for x and y.
$11x + 4y = 16$
$4x - 4y = -16$
Eliminate y.
$15x = 0$
$x = 0$

Replace x with 0 in equation (5).
$4x - 4y = -16$
$4(0) - 4y = -16$
$-4y = -16$
$y = 4$
Replace x with 0 and y with 4 in equation (1).
$4x + 5y - z = 22$
$4(0) + 5(4) - z = 22$
$20 - z = 22$
$z = -2$
The solution is $(0, 4, -2)$.

10. (1) $x - 3y + 2z = -2$
(2) $2x + y - z = 4$
(3) $x - y - 5z = 17$
Eliminate z. Use equations (1) and (2).
$x - 3y + 2z = -2$
$2x + y - z = 4$

$x - 3y + 2z = -2$
$2(2x + y - z) = 2(4)$

$x - 3y + 2z = -2$
$4x + 2y - 2z = 8$
(4) $5x - y = 6$

Use equations (2) and (3).
$2x + y - z = 4$
$x - y - 5z = 17$

$-5(2x + y - z) = -5(4)$
$x - y - 5z = 17$

$-10x - 5y + 5z = -20$
$x - y - 5z = 17$
(5) $-9x - 6y = -3$

Use equations (4) and (5) to solve for x and y.
$5x - y = 6$
$-9x - 6y = -3$
Eliminate y.
$-6(5x - y) = -6(6)$
$-9x - 6y = -3$

$-30x + 6y = -36$
$-9x - 6y = -3$
$-39x = -39$
$x = 1$

Replace x with 1 in equation (4).
$5x - y = 6$
$5(1) - y = 6$
$y = -1$
Replace x with 1 and y with -1 in equation (1).
$x - 3y + 2z = -2$
$1 - 3(-1) + 2z = -2$
$4 + 2z = -2$
$z = -3$
The solution is $(1, -1, -3)$.

11. Strategy: Let x represent the amount invested at 6%.
Let y represent the amount invested at 4.5%.

	Principal	Rate	Interest
Amount at 6%	x	0.06	$0.06x$
Amount at 4.5%	y	0.045	$0.045y$

The total amount invested is $20,000.
$x + y = 20{,}000$
The sum of the interest earned is $1080.
$0.06x + 0.045y = 1080$

Solution: (1) $x + y = 20{,}000$
(2) $0.06x + 0.045y = 1080$
Solve equation (1) for y and substitute for y in equation (2).
$y = 20000 - x$
$0.06x + 0.045(20000 - x) = 1080$
$0.06x + 900 - 0.045x = 1080$
$0.015x + 900 = 1080$
$0.015x = 180$
$x = 12000$
$y = 20000 - x = 20000 - 12000 = 8000$
$12,000 is invested at 6% and $8000 is invested at 4.5%.

Section 5.3

Concept Check

1. 450 mph

3. $50x + 100y$

Objective A Exercises

5. n is less than m.

7. Strategy: Let x represent the rate of the motorboat in calm water.
The rate of the current is y.

	Rate	Time	Distance
with current	$x + y$	2	$2(x + y)$
against current	$x - y$	3	$3(x - y)$

The distance traveled with the current is 36 miles. The distance traveled against the current is 36 miles.
$2(x + y) = 36$
$3(x - y) = 36$

Solution: $2(x + y) = 36$
$3(x - y) = 36$
$\frac{1}{2} \cdot 2(x + y) = \frac{1}{2} \cdot 36$
$\frac{1}{3} \cdot 3(x - y) = \frac{1}{3} \cdot 36$
$x + y = 18$
$x - y = 12$
$2x = 30$
$x = 15$

$x + y = 18$
$15 + y = 18$
$y = 3$

The rate of the motorboat in calm water is 15 mph. The rate of the current is 3 mph.

9. Strategy: Let p represent the rate of the plane in calm air.
The rate of the wind is w.

	Rate	Time	Distance
with wind	$p + w$	4	$4(p + w)$
against wind	$p - w$	4	$4(p - w)$

The distance traveled with the wind is 2200 miles. The distance traveled against the wind is 1820 miles.
$4(p + w) = 2200$
$4(p - w) = 1820$

Solution: $4(p + w) = 2200$
$4(p - w) = 1820$

$$\frac{1}{4}\cdot 4(p+w)=\frac{1}{4}\cdot 2200$$
$$\frac{1}{4}\cdot 4(p-w)=\frac{1}{4}\cdot 1820$$
$p+w=550$
$p-w=455$
$2p=1005$
$p=502.5$

$p+w=550$
$502.5+w=550$
$w=47.5$

The rate of the plane in calm air is 502.5 mph. The rate of the wind is 47.5 mph.

11. Strategy: Let x represent the rate of the team in calm water.
The rate of the current is y.

	Rate	Time	Distance
with current	$x+y$	2	$2(x+y)$
against current	$x-y$	2	$2(x-y)$

The distance traveled with the current is 20 km. The distance traveled against the current is 12 km.
$2(x+y)=20$
$2(x-y)=12$

Solution: $2(x+y)=20$
$2(x-y)=12$
$$\frac{1}{2}\cdot 2(x+y)=\frac{1}{2}\cdot 20$$
$$\frac{1}{2}\cdot 2(x-y)=\frac{1}{2}\cdot 12$$
$x+y=10$
$x-y=6$
$2x=16$
$x=8$

$x+y=10$
$8+y=10$
$y=2$
The rate of the team in calm water is 8 km/h.
The rate of the current is 2 km/h.

13. Strategy: Let x represent the rate of the plane in calm air.
The rate of the wind is y.

	Rate	Time	Distance
with wind	$x+y$	4	$4(x+y)$
against wind	$x-y$	5	$5(x-y)$

The distance traveled with the wind is 800 miles. The distance traveled against the wind is 800 miles.
$4(x+y)=800$
$5(x-y)=800$

Solution: $4(x+y)=800$
$5(x-y)=800$
$$\frac{1}{4}\cdot 4(x+y)=\frac{1}{4}\cdot 800$$
$$\frac{1}{5}\cdot 5(x-y)=\frac{1}{5}\cdot 800$$
$x+y=200$
$x-y=160$
$2x=360$
$x=180$

$x+y=200$
$180+y=200$
$y=20$

The rate of the plane in calm air is 180 mph.
The rate of the wind is 20 mph.

15. Strategy: Let x represent the rate of the plane in calm air.
The rate of the wind is y.

	Rate	Time	Distance
with wind	$x+y$	5	$5(x+y)$
against wind	$x-y$	6	$6(x-y)$

The distance traveled with the wind is 600 miles. The distance traveled against the wind is 600 miles.
$5(x+y)=600$
$6(x-y)=600$

Solution: $5(x+y)=600$
$6(x-y)=600$

$\frac{1}{5} \cdot 5(x+y) = \frac{1}{5} \cdot 600$
$\frac{1}{6} \cdot 6(x-y) = \frac{1}{6} \cdot 600$
$x + y = 120$
$x - y = 100$
$2x = 220$
$x = 110$

$x + y = 120$
$110 + y = 120$
$y = 10$

The rate of the plane in calm air is 110 mph. The rate of the wind is 10 mph.

Objective B Exercises

17. The cost per pound of dark roast coffee is greater than the cost per pound of light roast coffee.

19. Strategy: Let x represent the cost of redwood.
The cost of pine is y.
First purchase.

	Amount	Rate	Total Value
Redwood	60	x	$60x$
Pine	80	y	$80y$

Second purchase.

	Amount	Rate	Total Value
Redwood	100	x	$100x$
Pine	60	y	$60y$

The first purchase costs \$286. The second purchase costs \$396.

$60x + 80y = 286$
$100x + 60y = 396$

Solution: $60x + 80y = 286$
$100x + 60y = 396$

$3(60x + 80y) = 3(286)$
$-4(100x + 60y) = -4(396)$

$180x + 240y = 858$
$-400x - 240y = -1584$

$-220x = -726$
$x = 3.3$

$60x + 80y = 286$
$60(3.3) + 80y = 286$
$198 + 80y = 286$
$80y = 88$
$y = 1.1$
The cost of the pine is \$1.10/ft. The cost of the redwood is \$3.30/ft.

21. Strategy: Let x represent the cost of nylon carpet.
The cost of wool carpet is y.
First purchase.

	Amount	Rate	Total Cost
Nylon	16	x	$16x$
Wood	20	y	$20y$

Second purchase.

	Amount	Rate	Total Cost
Nylon	18	x	$18x$
Wool	25	y	$25y$

The first purchase costs \$1840. The second purchase costs \$2200.

$16x + 20y = 1840$
$18x + 25y = 2200$

Solution: $16x + 20y = 1840$
$18x + 25y = 2200$

$5(16x + 20y) = 5(1840)$
$-4(18x + 25y) = -4(2200)$
$80x + 100y = 9200$
$-72x - 100y = -8800$
$8x = 400$
$x = 50$

$16x + 20y = 1840$
$16(50) + 20y = 1840$
$800 + 20y = 1840$
$20y = 1040$
$y = 52$
The cost of the wool carpet is \$52/yd.

23. Strategy: Let m represent the number of mountain bikes to be manufactured.
The number of trail bikes to be manufactured is t.

Cost of materials.

Type	Number	Cost	Total Cost
Mountain	m	70	$70m$
Trail	t	50	$50t$

Cost of labor.

Type	Number	Cost	Total Cost
Mountain	m	80	$80m$
Trail	t	40	$40t$

The company has budgeted \$2500 for materials and \$2600 for labor.

$70m + 50t = 2500$
$80m + 40t = 2600$

Solution: $70m + 50t = 2500$
$80m + 40t = 2600$

$4(70m + 50t) = 4(2500)$
$-5(80m + 40t) = -5(2600)$

$280m + 200t = 10{,}000$
$-400m - 200t = -13{,}000$

$-120m = -3000$
$m = 25$

The company plans to manufacture 25 mountain bikes during the week.

25. Strategy: Let x represent the number of miles driven in the city.
The number of miles driven on the highway is $394 - x$.
Cost of hybrid driving.

	Number	Cost	Total Cost
City	x	0.09	$0.09x$
Highway	$394 - x$	0.08	$0.08(394 - x)$

The total amount spent on gasoline was \$34.74.

Solution:
$0.09x + 0.08(394 - x) = 34.74$
$0.09x + 31.52 - 0.08x = 34.74$
$0.01x = 3.22$
$x = 322$

$394 - x = 394 - 322 = 72$

The owner drives 322 miles in the city and 72 on the highway.

27. Strategy: Let x represent the amount of the first alloy.
The amount of the second alloy is y.
Gold.

	Amount	Percent	Quantity
1st alloy	x	0.10	$0.10x$
2nd alloy	y	0.30	$0.30y$

Lead.

	Amount	Percent	Quantity
1st alloy	x	0.15	$0.15x$
2nd alloy	y	0.40	$0.40y$

The resulting alloy contains 60 g of gold and 88 g of lead.
$0.10x + 0.30y = 60$
$0.15x + 0.40y = 88$

Solution: $0.10x + 0.30y = 60$
$0.15x + 0.40y = 88$
$3(0.10x + 0.30y) = 3(60)$
$-2(0.15x + 0.40y) = -2(88)$

$0.30x + 0.90y = 180$
$-0.30x - 0.80y = -176$
$0.10y = 4$
$y = 40$

$0.10x + 0.30y = 60$
$0.10x + 0.30(40) = 60$
$0.10x + 12 = 60$
$0.10x = 48$
$x = 480$

The chemist should use 480 g of the first alloy and 40 g of the second alloy.

29. Strategy: Let x represent the cost of the Model II computer.
The cost of the Model IV computer is y.
The cost of the Model IX computer is z.

First shipment

	Number	Unit Cost	Value
Model II	4	x	$4x$
Model IV	6	y	$6y$
Model IX	10	z	$10z$

Second shipment

	Number	Unit Cost	Value
Model II	8	x	$8x$
Model IV	3	y	$3y$
Model IX	5	z	$5z$

Third shipment

	Number	Unit Cost	Value
Model II	2	x	$2x$
Model IV	9	y	$9y$
Model IX	5	z	$5z$

The value of the first shipment was \$114,000. The value of the second shipment was \$72,000. The value of the third shipment was \$81,000.

Solution: (1) $4x + 6y + 10z = 114{,}000$
(2) $8x + 3y + 5z = 72{,}000$
(3) $2x + 9y + 5z = 81{,}000$

Multiply equation (2) by −2 and add to equation (1).
$4x + 6y + 10z = 114{,}000$
$-16x - 6y - 10z = -144{,}000$
$-12x = -30{,}000$
$x = 2500$

Multiply equation (3) by −1 and add to equation (2).
$8x + 3y + 5z = 72{,}000$
$-2x - 9y - 5z = -81{,}000$
$6x - 6y = -9000$

$6(2500) - 6y = -9000$
$-6y = -24{,}000$
$y = 4000$

The Model IV computer costs \$4000.

31. Strategy: Let x represent the amount deposited at 8%.
The amount deposited at 6% is y.
The amount deposited at 4% is z.

	Principal	Rate	Interest
8%	x	0.08	$0.08x$
6%	y	0.06	$0.06y$
4%	z	0.04	$0.04z$

The amount deposited in the 8% account is twice the amount deposited in the 6% account. The total amount invested is \$25,000. The total interest earned is \$1520.

Solution: (1) $x = 2y$
(2) $x + y + z = 25{,}000$
(3) $0.08x + 0.06y + 0.04z = 1520$

Substitute $2y$ for x in equation (2) and equation (3).
$2y + y + z = 25000$
(4) $3y + z = 25000$
$0.08(2y) + 0.06y + 0.04z = 1520$
(5) $0.22y + 0.04z = 1520$

Solve equation (4) for z and substitute into equation (5).
$z = 25{,}000 - 3y$
$0.22y + 0.04(25{,}000 - 3y) = 1520$
$0.22y + 1000 - 0.12y = 1520$
$0.10y = 520$
$y = 5200$

$z = 25{,}000 - 3(5200)$
$z = 9400$

$x = 2(5200)$
$x = 10{,}400$

The investor placed \$10,400 in the 8% account, \$5200 in the 6% account and \$9400 in the 4% account.

Critical Thinking

33. **Strategy**: Let n represent the measure of the smaller angle.
The measure of the larger angle is m.
First relationship $m + n = 180$
Second relationship $m = 3n + 40$

Solution: Solve for n by substitution:
$(3n + 40) + n = 180$
$3n + 40 + n = 180$
$4n = 140$
$n = 35$
$m + n = 180$
$m + 35 = 180$
$m = 145$
The angles have measures of $35°$ and $145°$.

Projects or Group Activities

35. $w_1 d_1 = w_2 d_2 + w_3 d_3$
$d_3 = 3d_2$

$5d_1 = d_2 + 3d_3$
$5d_1 = d_2 + 3(3d_2)$
$5d_1 = 10d_2$
$d_1 = 2d_2$

$d_1 + d_3 = 15$
$2d_2 + 3d_2 = 15$
$5d_2 = 15$
$d_2 = 3$

$d_3 = 3d_2 = 3(3) = 9$
$d_1 = 2d_2 = 2(3) = 9$

$d_1 = 6$ in, $d_2 = 3$ in, $d_3 = 9$ in

Section 5.4

Concept Check

1. $2x - y < 4$
$x - 3y \geq 6$

(i) (5, 1)
$2(5) - 1 < 4$
$9 < 3$ is not a true statement.
(5, 1) is not a solution of the system of inequalities.

Objective A Exercises

3. Solve each inequality for y.
$x - y \geq 3$
$-y \geq -x + 3$
$y \leq x - 3$

$x + y \leq 5$
$y \leq 5 - x$

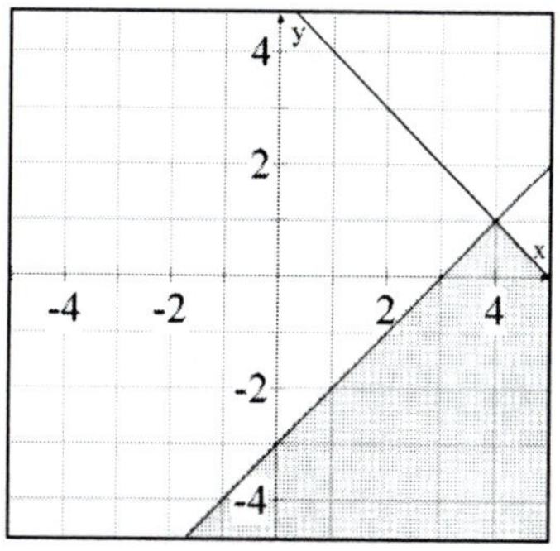

5. Solve each inequality for y.
$3x - y < 3$
$-y < -3x + 3$
$y > 3x - 3$

$2x + y \geq 2$
$y \geq 2 - 2x$

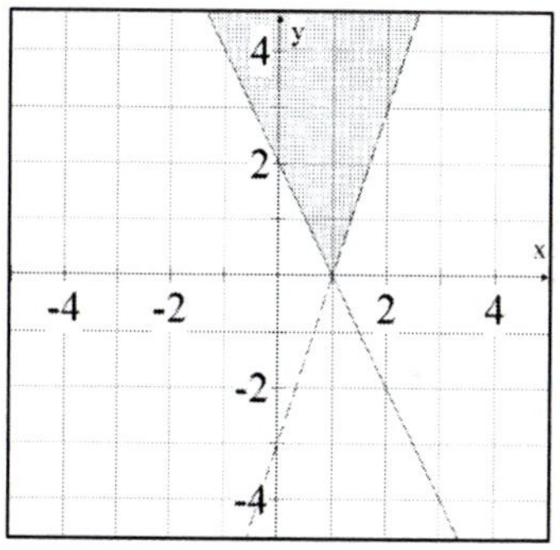

7. Solve each inequality for y.
$2x + y \geq -2$
$y \geq -2x - 2$

$6x + 3y \leq 6$
$3y \leq -6x + 6$
$y \leq -2x + 2$

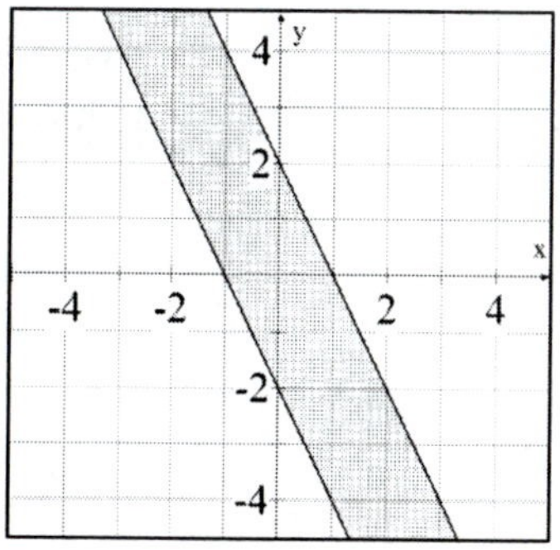

9. Solve the first inequality for y.
$3x - 2y < 6$
$-2y < -3x + 6$

$y > \frac{3}{2}x - 3$
$y \leq 3$

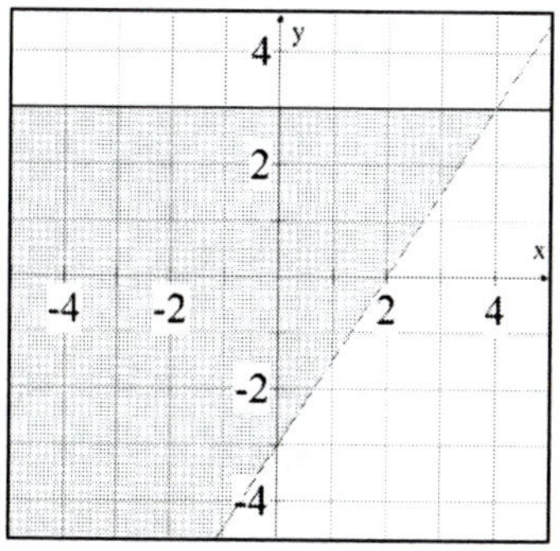

11. Solve each inequality for the variable.

$x+1\geq 0$
$x\geq -1$

$y-3\leq 0$
$y\leq 3$

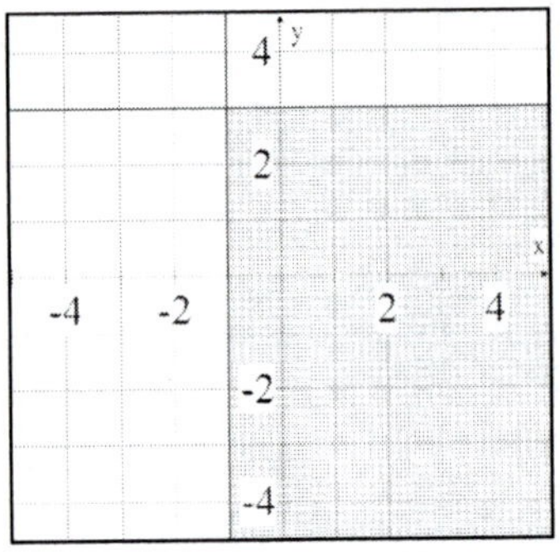

13. Solve each inequality for y.

$2x+y\geq 4$
$y\geq -2x+4$

$3x-2y<6$
$-2y<-3x+6$
$y>\frac{3}{2}x-3$

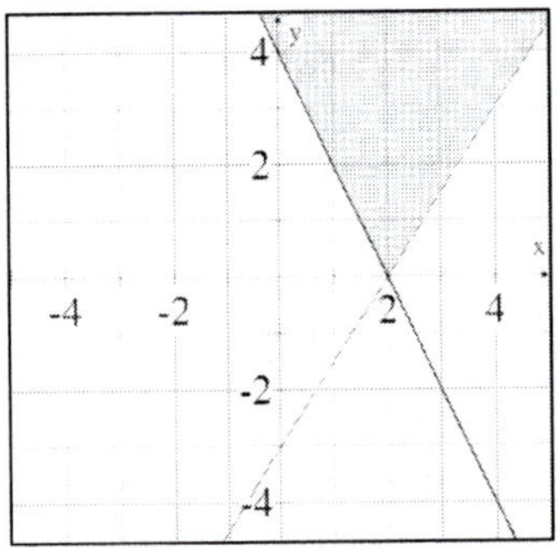

15. Solve each inequality for y.

$x-2y\leq 6$
$-2y\leq -x+6$
$y\geq \frac{1}{2}x-3$

$2x+3y\leq 6$
$3y\leq -2x+6$
$y\geq -\frac{2}{3}x+2$

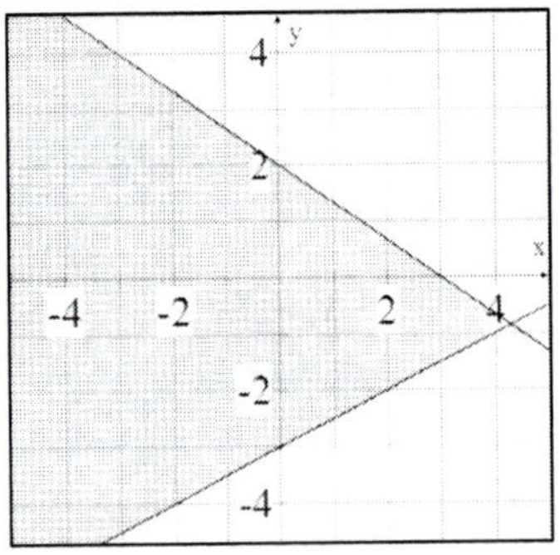

17. Solve each inequality for y.

$x-2y\leq 4$
$-2y\leq -x+4$
$y\geq \frac{1}{2}x-2$

$3x+2y\leq 8$
$2y\leq -3x+8$
$y\leq -\frac{3}{2}x+4$

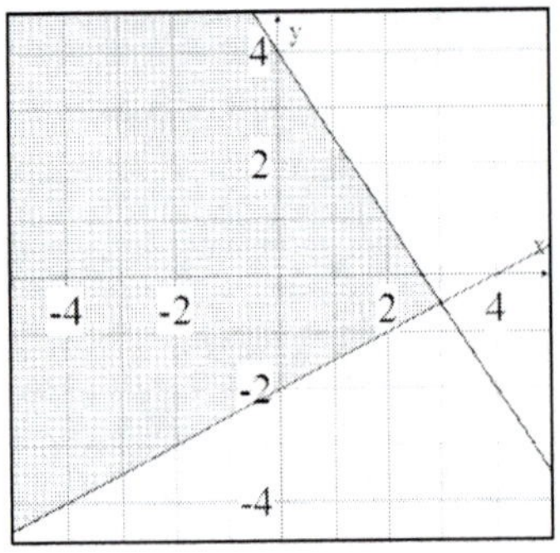

19. Region between the parallel lines $x+y=a$ and $x+y=b$.

Critical Thinking

21. Solve each inequality for y.

$2x+3y\leq 15$
$3y\leq -2x+15$
$y\leq -\frac{2}{3}x+5$

$3x-y\leq 6$
$-y\leq -3x+6$
$y\geq 3x-6$

$y\geq 0$

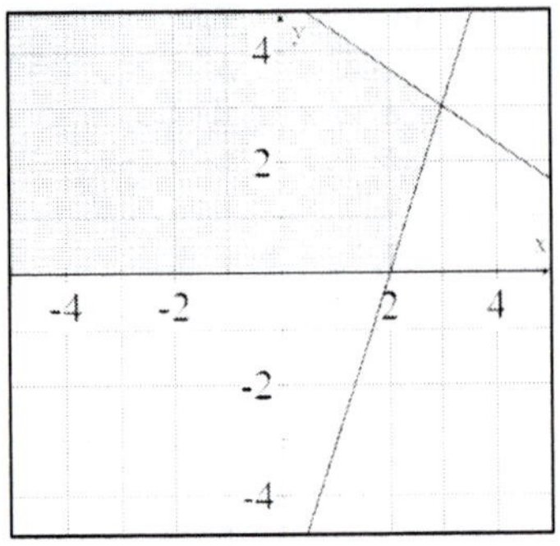

Projects or Group Activities

23. (ii) and (iii) are convex sets.

Chapter 5 Review Exercises

1. (1) $2x-6y=15$
(2) $x=4y+8$

Substitute $4y+8$ for x in equation (1).

$2(4y+8)-6y=15$
$8y+16-6y=15$
$2y=-1$
$y=-\frac{1}{2}$

Substitute $-\frac{1}{2}$ for y in equation (2).

$x=4\left(-\frac{1}{2}\right)+8$
$x=6$

The solution is $\left(6,-\frac{1}{2}\right)$.

2. (1) $3x + 2y = 2$
(2) $x + y = 3$

Multiply equation (2) by −2 and add to equation (1).

$3x + 2y = 2$
$-2x - 2y = -6$
$x = -4$
Replace x with −4 in equation (2).
$-4 + y = 3$
$y = 7$
The solution is (−4, 7).

3. $x + y = 3$
$3x - 2y = -6$

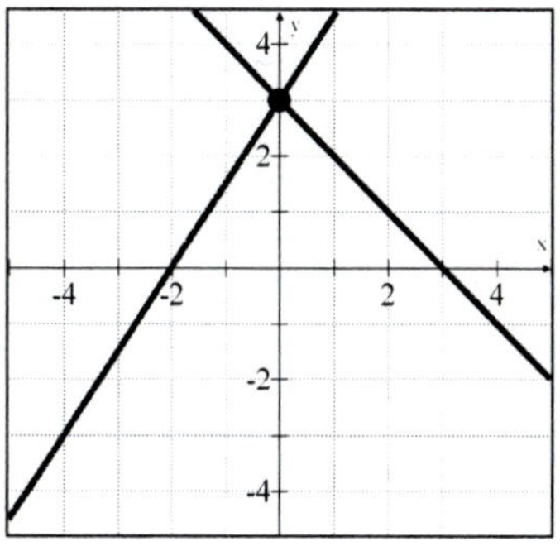

The solution is (0,3).

4. $2x - y = 4$
$y = 2x - 4$

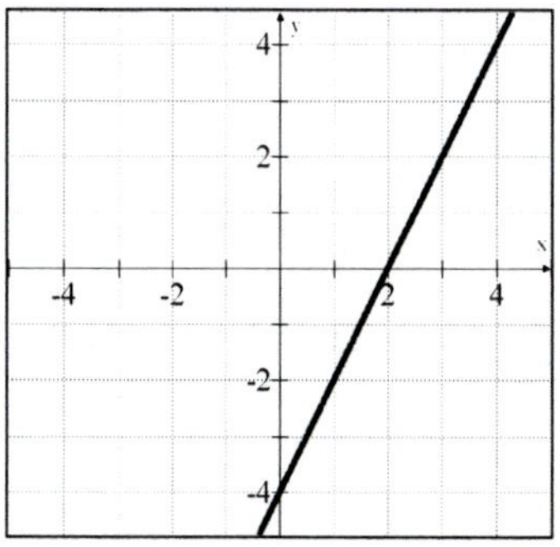

The two equations represent the same line. The system of equations is dependent. The solutions are the ordered pair $(x, 2x - 4)$.

5. (1) $3x + 12y = 18$
(2) $x + 4y = 6$

Solve equation (2) for x.
$x + 4y = 6$
$x = -4y + 6$
Substitute $-4y + 6$ for x in equation (1).
$3(-4y + 6) + 12y = 18$
$-12y + 18 + 12y = 18$
$18 = 18$
This is a true equation. The equations are dependent. The solutions are the ordered pairs $\left(x, -\frac{1}{4}x + \frac{3}{2}\right)$.

6. (1) $5x - 15y = 30$
(2) $x - 3y = 6$

Multiply equation (2) by −5 and add to equation (1).
$5x - 15y = 30$
$-5x + 15y = -30$
$0 = 0$
This is a true equation. The equations are dependent. The solutions are the ordered pairs $\left(x, \frac{1}{3}x - 2\right)$.

7. (1) $3x - 4y - 2z = 17$
(2) $4x - 3y + 5z = 5$
(3) $5x - 5y + 3z = 14$

Eliminate z. Multiply equation (1) by 3 and equation (3) by 2. Then add the equations.
$3(3x - 4y - 2z) = 3(17)$
$2(5x - 5y + 3z) = 2(14)$

$9x - 12y - 6z = 51$
$10x - 10y + 6z = 28$

(4) $19x - 22y = 79$

Multiply equation (1) by 5 and equation (2) by 2. Then add the equations
$5(3x - 4y - 2z) = 5(17)$
$2(4x - 3y + 5z) = 2(5)$

$15x - 20y - 10z = 85$
$8x - 6y + 10z = 10$
(5) $23x - 26y = 95$

Multiply equation (4) by 23 and equation (5) by −19. Then add the equations

$23(19x-22y)=23(79)$
$-19(23x-26y)=-19(95)$

$437x-506y=1817$
$-437x+494y=-1805$
$-12y=12$
$y=-1$

Replace y with -1 in equation (4).
$19x-22(-1)=79$
$19x+22=79$
$19x=57$
$x=3$
Replace x with 3 and y with -1 in equation (1).
$3(3)-4(-1)-2z=17$
$9+4-2z=17$
$-2z=4$
$z=-2$
The solution is $(3, -1, -2)$.

8. (1) $3x+y=13$
(2) $2y+3z=5$
(3) $x+2z=11$

Eliminate y. Multiply equation (1) by -2 then add to equation (2).
$-2(3x+y)=-2(13)$
$2y+3z=5$

$-6x-2y=-26$
$2y+3z=5$

(4) $-6x+3z=-21$

Multiply equation (3) by 6 then add to equation (4).
$6(x+2z)=6(11)$
$-6x+3z=-21$

$6x+12z=66$
$-6x+3z=-21$
$15z=45$
$z=3$

Replace z with 3 in equation (3).
$x+2(3)=11$
$x=5$
Replace x with 5 in equation (1).
$3(5)+y=13$
$y=-2$

The solution is $(5, -2, 3)$.

9.

$6x+y=4$	
$6(1)+(-2)$	4
$6-2$	4
$4=4$	

$2x-5y=12$	
$2(1)-5(-2)$	12
$2+10$	12
$12=12$	

Yes $(1, -2)$ is a solution of the system of equations.

10. (1) $2x-4y=11$
(2) $y=3x-4$
Substitute $3x-4$ for y in equation (1).
$2x-4y=11$
$2x-4(3x-4)=11$
$2x-12x+16=11$
$-10x+16=11$
$-10x=-5$
$x=\frac{1}{2}$
Substitute into equation (2).
$y=3x-4$
$y=3\left(\frac{1}{2}\right)-4$
$y=-\frac{5}{2}$
The solution is $\left(\frac{1}{2},-\frac{5}{2}\right)$.

11. (1) $2x-y=7$
(2) $3x+2y=7$
Solve equation (1) for y.
$2x-y=7$
$-y=-2x+7$
$y=2x-7$
Substitute into equation (2).
$3x+2y=7$
$3x+2(2x-7)=7$
$3x+4x-14=7$
$7x-14=7$
$7x=21$
$x=3$
Substitute into equation (1).
$2x-y=7$

$2(3) - y = 7$
$6 - y = 7$
$-y = 1$
$y = -1$
The solution is $(3,-1)$.

12. (1) $3x - 4y = 1$
(2) $2x + 5y = 16$
Eliminate y.
$5(3x - 4y) = 5(1)$
$4(2x + 5y) = 4(16)$
$15x - 20y = 5$
$8x + 20y = 64$
Add the equations.
$23x = 69$
$x = 3$
Substitute the value of x in equation (2).

$2x + 5y = 16$
$2(3) + 5y = 16$
$6 + 5y = 16$
$5y = 10$
$y = 2$
The solution is (3, 2).

13. (1) $x + y + z = 0$
(2) $x + 2y + 3z = 5$
(3) $2x + y + 2z = 3$

Eliminate x. Multiply equation (1) by -1 and add to equation (2).
$-1(x + y + z) = -1(0)$
$x + 2y + 3z = 5$

$-x - y - z = 0$
$x + 2y + 3z = 5$

(4) $y + 2z = 5$
Multiply equation (2) by -2 and add to equation (3).
$-2(x + 2y + 3z) = -2(5)$
$2x + y + 2z = 3$

$-2x - 4y - 6z = -10$
$2x + y + 2z = 3$

(5) $-3y - 4z = -7$
Multiply equation (4) by 2 and add to equation (5).
$2(y + 2z) = 2(5)$
$-3y - 4z = -7$

$2y + 4z = 10$
$-3y - 4z = -7$
$-y = 3$
$y = -3$
Replace y with -3 in equation (4).
$-3 + 2z = 5$
$2z = 8$
$z = 4$
Replace y with -3 and z with 4 in equation (1).
$x + y + z = 0$
$x - 3 + 4 = 0$
$x + 1 = 0$
$x = -1$
The solution is $(-1, -3, 4)$.

14. (1) $x + 3y + z = 6$
(2) $2x + y - z = 12$
(3) $x + 2y - z = 13$

Eliminate z. Add equations (1) and (2).
$x + 3y + z = 6$
$2x + y - z = 12$

(4) $3x + 4y = 18$

Add equations (1) and (3).
$x + 3y + z = 6$
$x + 2y - z = 13$

(5) $2x + 5y = 19$
Multiply equation (4) by -2 and equation (5) by 3. Then add the equations.
$-2(3x + 4y) = -2(18)$
$3(2x + 5y) = 3(19)$

$-6x - 8y = -36$
$6x + 15y = 57$
$7y = 21$
$y = 3$
Replace y with 3 in equation (4).
$3x + 4(3) = 18$
$3x + 12 = 18$
$3x = 6$
$x = 2$
Replace x with 2 and y with 3 in equation (1).
$x + 3y + z = 6$
$2 + 3(3) + z = 6$

$2 + 9 + z = 6$
$11 + z = 6$
$z = -5$
The solution is (2, 3, −5).

15. Solve each inequality for y.
$x + 3y \leq 6$
$3y \leq -x + 6$
$y \leq -\frac{1}{3}x + 2$
$2x - y \geq 4$
$-y \geq -2x + 4$
$y \leq 2x - 4$

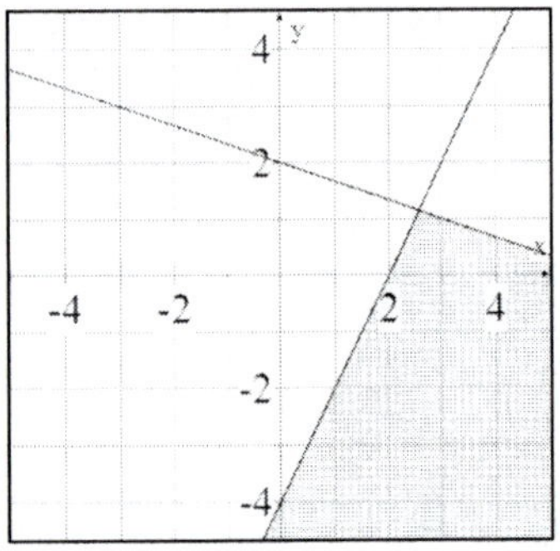

16. Solve each inequality for y.
$2x + 4y \geq 8$
$4y \geq -2x + 8$
$y \geq -\frac{1}{2}x + 2$
$x + y \leq 3$
$y \leq -x + 3$

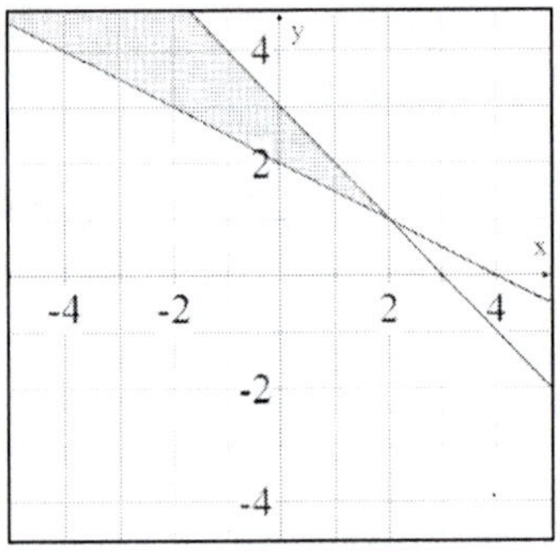

17. Strategy: Let x represent the rate of the cabin cruiser in calm water.
The rate of the current is y.

	Rate	Time	Distance
with current	$x + y$	3	$3(x + y)$
against current	$x - y$	5	$5(x - y)$

The distance traveled with the current is 60 mi. The distance traveled against the current is 60 mi.
$3(x + y) = 60$
$5(x - y) = 60$

Solution: $3(x + y) = 60$
$5(x - y) = 60$

$\frac{1}{3} \cdot 3(x + y) = \frac{1}{3} \cdot 60$
$\frac{1}{5} \cdot 5(x - y) = \frac{1}{5} \cdot 60$
$x + y = 20$
$x - y = 12$
$2x = 32$
$x = 16$

$x + y = 20$
$16 + y = 20$
$y = 4$
The rate of the boat in calm water is 16 mph.
The rate of the current is 4 mph.

18. Strategy: Let p represent the rate of the plane in calm air.
The rate of the wind is w.

	Rate	Time	Distance
with wind	$p + w$	3	$3(p + w)$
against wind	$p - w$	4	$4(p - w)$

The distance traveled with the wind is 600 mi. The distance traveled against the wind is 600 mi.
$3(p + w) = 600$
$4(p - w) = 600$

Solution: $3(p + w) = 600$
$4(p - w) = 600$
$\frac{1}{3} \cdot 3(p + w) = \frac{1}{3} \cdot 600$
$\frac{1}{4} \cdot 4(p - w) = \frac{1}{4} \cdot 600$
$p + w = 200$
$p - w = 150$
$2p = 350$
$p = 175$

$p + w = 200$
$175 + w = 200$
$w = 25$
The rate of the plane in calm air is 175 mph.
The rate of the wind is 25 mph.

19. Strategy: Let x represent the number of children's tickets sold.
The number of adult tickets sold is y.

Friday:

	Amount	Rate	Quantity
Children	x	5	$5x$
Adult	y	8	$8y$

Saturday:

	Amount	Rate	Quantity
Children	$3x$	5	$5(3x)$
Adult	$\frac{1}{2}y$	8	$8(\frac{1}{2})y$

The total receipts for Friday were \$2500.
The total receipts for Saturday were \$2500.
$5x + 8y = 2500$
$15x + 4y = 2500$

Solution: (1) $5x + 8y = 2500$
(2) $15x + 4y = 2500$
Multiply equation (2) by −2 then add to equation (1).
$5x + 8y = 2500$
$-2(15x + 4y) = -2(2500)$
$5x + 8y = 2500$
$-30x - 8y = -5000$
$-25x = -2500$
$x = 100$
On Friday, 100 children attended.

20. Strategy: Let x represent the amount invested at 3%.
The amount invested at 7% is y.

	Amount	Rate	Quantity
Amount at 3%	x	0.03	$0.03x$
Amount at 7%	y	0.07	$0.07y$

The total amount invested is \$20,000.
$x + y = 20{,}000$
The total annual interest earned is \$1200.
$0.03x + 0.07y = 1200$

Solution: (1) $x + y = 20{,}000$
(2) $0.03x + 0.07y = 1200$
Multiply equation (1) by −0.07 then add to equation (2).

$-0.07x - 0.07y = -1400$
$0.03x + 0.07y = 1200$
$-0.04x = -200$
$x = 5000$
Substitute 5000 for x in equation (1).
$x + y = 20{,}000$
$5000 + y = 20{,}000$
$y = 15{,}000$
The amount invested at 3% is \$5000.
The amount invested at 7% is \$15,000.

Chapter 5 Test

1. $2x - 3y = -6$
$2x - y = 2$

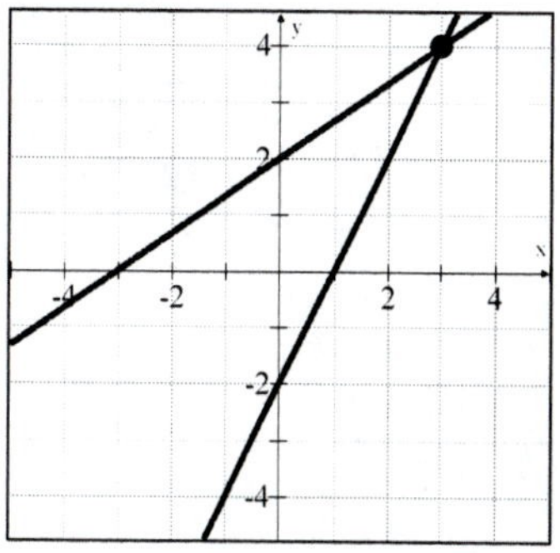

The solution is (3,4).

2. $x - 2y = -6$
$y = \frac{1}{2}x - 4$

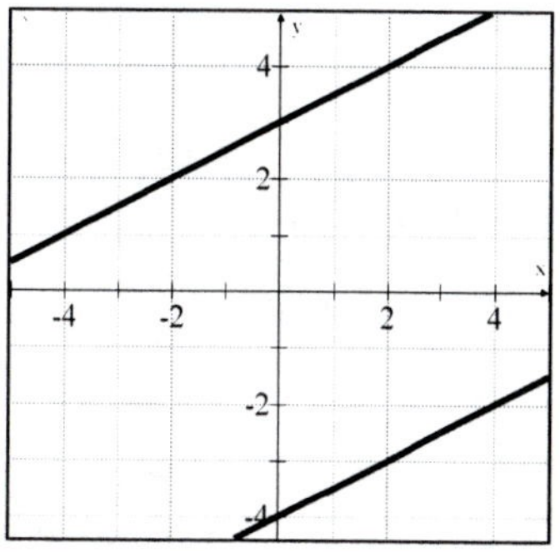

No solution.

3. Solve each inequality for y.

$2x - y < 3$
$-y < -2x + 3$
$y > 2x - 3$

$4x + 3y < 11$
$3y < -4x + 11$
$y < -\frac{4}{3}x + \frac{11}{3}$

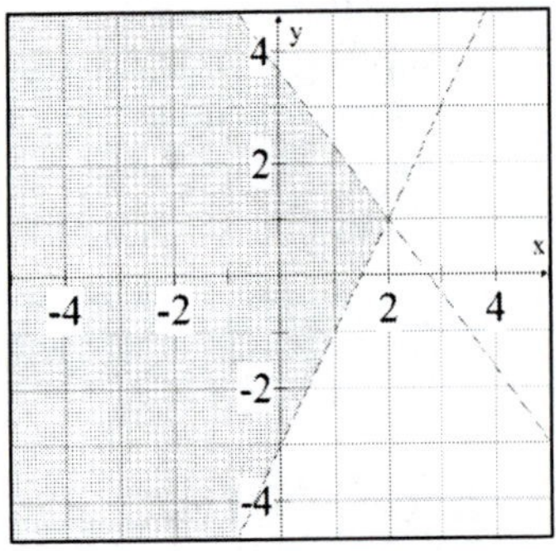

4. Solve each inequality for y.

$x + y > 2$
$y > -x + 2$

$2x - y < -1$
$-y < -2x - 1$
$y > 2x + 1$

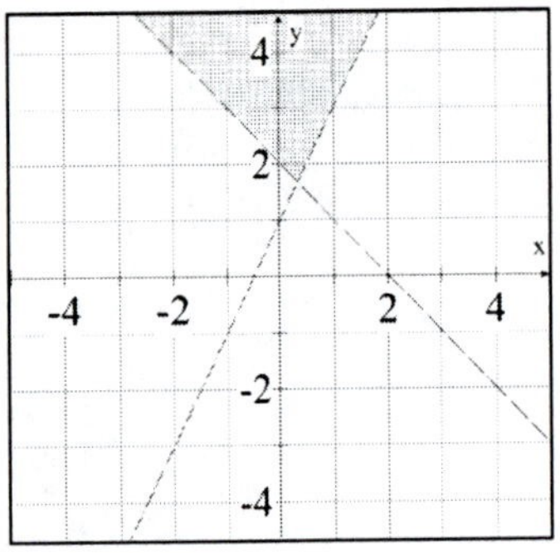

5. (1) $3x + 2y = 4$
(2) $x = 2y - 1$

Substitute $2y - 1$ for x in equation (1).
$3(2y - 1) + 2y = 4$
$6y - 3 + 2y = 4$
$8y = 7$
$y = \frac{7}{8}$
Substitute into equation (2).
$x = 2\left(\frac{7}{8}\right) - 1 = \frac{7}{4} - 1 = \frac{3}{4}$
The solution is $\left(\frac{3}{4}, \frac{7}{8}\right)$.

6. (1) $5x + 2y = -23$
(2) $2x + y = -10$

Solve equation (2) for y.
$y = -2x - 10$
Substitute $-2x - 10$ for y in equation (1).
$5x + 2(-2x - 10) = -23$
$5x - 4x - 20 = -23$
$x = -3$
Substitute in equation (2).
$2(-3) + y = -10$
$-6 + y = -10$
$y = -4$
The solution is $(-3, -4)$.

7. (1) $y = 3x - 7$
(2) $y = -2x + 3$

Substitute equation (2) into equation (1).
$-2x + 3 = 3x - 7$
$-5x + 3 = -7$
$-5x = -10$
$x = 2$
Substitute into equation (1).
$y = 3(2) - 7$
$y = -1$
The solution is $(2, -1)$.

8. (1) $3x + 4y = -2$
(2) $2x + 5y = 1$

Multiply equation (1) by 2 and equation (2) by -3 then add the new equations.
$2(3x + 4y) = 2(-2)$
$-3(2x + 5y) = -3(1)$

$6x + 8y = -4$
$-6x - 15y = -3$
$-7y = -7$
$y = 1$
Substitute into equation (1).
$3x + 4(1) = -2$
$3x = -6$
$x = -2$
The solution is $(-2, 1)$.

9. (1) $4x - 6y = 5$
(2) $6x - 9y = 4$

Multiply equation (1) by 3 and equation (2) by -2 then add the new equations.
$3(4x - 6y) = 3(5)$
$-2(6x - 9y) = -2(4)$

$12x - 18y = 15$
$-12x + 18y = -8$
$0 = 7$
This is not a true equation. The system of equations is inconsistent and therefore has no solution.

10. (1) $3x - y = 2x + y - 1$
(2) $5x + 2y = y + 6$
Write the equations in the form $Ax + By = C$
(3) $x - 2y = -1$
(4) $5x + y = 6$
$x - 2y = -1$
$10x + 2y = 12$
$11x = 11$
$x = 1$
Substitute in equation (4).
$5(1) + y = 6$
$y = 1$
The solution is (1,1).

11. (1) $2x + 4y - z = 3$
(2) $x + 2y + z = 5$
(3) $4x + 8y - 2z = 7$
Eliminate z. Add equations (1) and (2).
$2x + 4y - z = 3$
$x + 2y + z = 5$

(4) $3x + 6y = 8$
Multiply equation (2) by 2 and add to equation (3).
$2(x + 2y + z) = 2(5)$
$4x + 8y - 2z = 7$

$2x + 4y + 2z = 10$
$4x + 8y - 2z = 7$

(5) $6x + 12y = 17$
Multiply equation (4) by −2 then add to equation (5).
$-2(3x + 6y) = -2(8)$
$6x + 12y = 17$
$-6x - 12y = -16$
$6x + 12y = 17$
$0 = 1$
This is not a true equation. The system of equations is inconsistent and therefore has no solution.

12. (1) $x - y - z = 5$
(2) $2x + z = 2$
(3) $3y - 2z = 1$

Multiply equation (1) by 3 and then add to equation (3).
$3(x - y - z) = 3(5)$
$3x - 3y - 3z = 15$
$3y - 2z = 1$

Multiply equation (4) by 2 then add to equation (3).
$x - 2y = -1$
$2(5x + y) = 2(6)$

(4) $3x - 5z = 16$
Multiply equation (2) by 5 and add to equation (4).
$5(2x + z) = 5(2)$

$10x + 5z = 10$
$3x - 5z = 16$
$13x = 26$
$x = 2$

Substitute 2 in for x in equation (2).
$2(2) + z = 2$
$z = -2$
Substitute 2 in for x and −2 in for z in equation (1).
$2 - y - (-2) = 5$
$4 - y = 5$
$-y = 1$
$y = -1$

The solution is (2, −1, −2)

13. (1) $x - y = 3$
(2) $2x + y = -4$
Solve equation (2) for y.
$2x + y = -4$
$y = -2x - 4$

Substitute into equation (1).
$x - y = 3$
$x - (-2x - 4) = 3$
$x + 2x + 4 = 3$
$3x + 4 = 3$
$3x = -1$
$$x = -\frac{1}{3}$$

Substitute into equation (2).
$2x + y = -4$
$$2\left(-\frac{1}{3}\right) + y = -4$$

$-\frac{2}{3}+y=-4$

$y=-\frac{10}{3}$

The solution is $\left(-\frac{1}{3},-\frac{10}{3}\right)$.

14.

$5x+2y=6$	
$5(2)+2(-2)$	6
$10-4$	6

$6=6$

$3x+5y=-4$	
$3(2)+5(-2)$	-4
$6-10$	-4

$-4=-4$

Yes $(2, -2)$ is a solution of the system of equations.

15. (1) $x-y+z=2$
(2) $2x-y-z=1$
(3) $x+2y-3z=-4$
Eliminate x. Multiply equation (1) by −2 and add to equation (2).
$-2(x-y+z)=-2(2)$
$2x-y-z=1$

$-2x+2y-2z=-4$
$2x-y-z=1$

(4) $y-3z=-3$
Multiply equation (1) by −1 and add to equation (3).
$-1(x-y+z)=-1(2)$
$x+2y-3z=-4$

$-x+y-z=-2$
$x+2y-3z=-4$

(5) $3y-4z=-6$
Multiply equation (4) by −3 and add to equation (5).
$-3(y-3z)=-3(-3)$
$3y-4z=-6$

$-3y+9z=9$
$3y-4z=-6$
$5z=3$

$z=\frac{3}{5}$

Substitute into equation (4).
$y-3z=-3$

$y-3z=-3$

$y-3\left(\frac{3}{5}\right)=-3$

$y-\frac{9}{5}=-3$

$y=-\frac{6}{5}$

Substitute into equation (1).
$x-y+z=2$

$x-\left(-\frac{6}{5}\right)+\left(\frac{3}{5}\right)=2$

$x+\frac{9}{5}=2$

$x=\frac{1}{5}$

The solution is $\left(\frac{1}{5},-\frac{6}{5},\frac{3}{5}\right)$.

16. Strategy: Let x represent the rate of the plane in calm air.
The rate of the wind is y.

	Rate	Time	Distance
with wind	$x+y$	2	$2(x+y)$
against wind	$x-y$	2.8	$2.8(x-y)$

The distance traveled with the wind is 350 mi. The distance traveled against the wind is 350 mi.
$2(x+y)=350$
$2.8(x-y)=350$

Solution: $2(x+y)=350$
$2.8(x-y)=350$

$\frac{1}{2}\cdot 2(x+y)=\frac{1}{2}\cdot 350$

$\frac{1}{2.8}\cdot 2.8(x-y)=\frac{1}{2.8}\cdot 350$

$x+y=175$
$x-y=125$
$2x=300$
$x=150$

$x+y=175$
$150+y=175$
$y=25$

The rate of the plane in calm air is 150 mph.
The rate of the wind is 25 mph.

17. Strategy: Let x represent the cost per yard of cotton.
The cost per yard of wool is y.
First purchase:

	Amount	Rate	Total Value
Cotton	60	x	$60x$
Wool	90	y	$90y$

Second purchase:

	Amount	Rate	Total Value
Cotton	80	x	$80x$
Wool	20	y	$20y$

The total cost of the first purchase was \$1800. The total cost of the second purchase was \$1000.

$60x+90y=1800$
$80x+20y=1000$

Solution: $-4(60x+90y)=-4(1800)$
$3(80x+20y)=3(1000)$

$-240x-360y=-7200$
$240x+60y=3000$
$-300y=-4200$
$y=14$

$60x+90(14)=1800$
$60x+1260=1800$
$60x=540$
$x=9$

The cost of cotton is \$9.00/yd.
The cost of wool is \$14.00/yd.

18. Strategy: Let x represent the amount invested at 2.7%.
The amount invested at 5.1% is y.

	Amount	Rate	Quantity
Amount at 2.7%	x	0.027	$0.027x$
Amount at 5.1%	y	0.051	$0.051y$

The total amount invested is \$15,000.
$x+y=15{,}000$
The total annual interest earned is \$549.
$0.027x+0.051y=549$

Solution: (1) $x+y=15{,}000$
(2) $0.027x+0.051y=549$
Multiply equation (1) by -0.051 then add to equation (2).

$-0.051x-0.051y=-765$
$0.027x+0.051y=549$
$-0.024x=-216$
$x=9000$
Substitute 9000 for x in equation (1).
$x+y=15{,}000$
$9000+y=15{,}000$
$y=6000$
The amount invested at 2.7% is \$9000.
The amount invested at 5.1% is \$6000.

Cumulative Review Exercises

1. $\frac{3}{2}x - \frac{3}{8} + \frac{1}{4}x = \frac{7}{12}x - \frac{5}{6}$

$$24\left(\frac{3}{2}x - \frac{3}{8} + \frac{1}{4}x\right) = 24\left(\frac{7}{12}x - \frac{5}{6}\right)$$
$$36x - 9 + 6x = 14x - 20$$
$$42x - 9 = 14x - 20$$
$$28x - 9 = -20$$
$$28x = -11$$
$$x = -\frac{11}{28}$$

The solution is $-\frac{11}{28}$.

2. $(2, -1)\ (3, 4)$

$$m = \frac{y_2 - y_1}{x_2 - x_1} = \frac{4-(-1)}{3-2} = \frac{5}{1} = 5$$
$$y - y_1 = m(x - x_1)$$
$$y - 4 = 5(x - 3)$$
$$y - 4 = 5x - 15$$
$$y = 5x - 11$$

The equation of the line is $y = 5x - 11$.

3. $3[x - 2(5 - 2x) - 4x] + 6$
$= 3(x - 10 + 4x - 4x) + 6$
$= 3(x - 10) + 6$
$= 3x - 30 + 6$
$= 3x - 24$

4. $a + bc \div 2$
$4 + 8(-2) \div 2 = 4 - 16 \div 2 = 4 - 8 = -4$

5. $2x - 3 < 9$ or $5x - 1 < 4$
Solve each inequality.
$2x - 3 < 9 \qquad 5x - 1 < 4$
$2x < 12 \qquad 5x < 5$
$x < 6$ or $x < 1$
$(-\infty, 6)$

6. $|x - 2| - 4 < 2$
$|x - 2| < 6$
$-6 < x - 2 < 6$
$-6 + 2 < x - 2 + 2 < 6 + 2$
$-4 < x < 8$
$\{x \mid -4 < x < 8\}$

7. $|2x - 3| > 5$
Solve each inequality.
$2x - 3 < -5$ or $2x - 3 > 5$
$2x < -2$ or $2x > 8$
$x < -1$ or $x > 4$
$\{x \mid x < -1\} \cup \{x \mid x > 4\}$
$\{x \mid x < -1 \text{ or } x > 4\}$

8. $f(x) = 3x^3 - 2x^2 + 1$
$f(-3) = 3(-3)^3 - 2(-3)^2 + 1$
$f(-3) = 3(-27) - 2(9) + 1$
$f(-3) = -98$

9. $f(x) = 3x^2 - 2x$
Domain $= \{x \mid -\infty < x < \infty\}$

10. $F(x) = x^2 - 3$
$F(2) = (2)^2 - 3 = 1$

11. $f(x) = 3x - 4$
$f(2 + h) = 3(2 + h) - 4 = 6 + 3h - 4 = 2 + 3h$
$f(2) = 3(2) - 4 = 6 - 4 = 2$
$f(2 + h) - f(2) = 2 + 3h - 2 = 3h$

12. $\{x \mid x \le 2\} \cap \{x \mid x > -3\}$

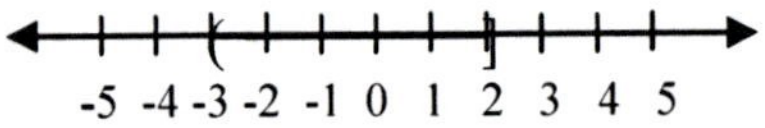

13. $(-2, 3),\ m = -\frac{2}{3}$

$$y - y_1 = m(x - x_1)$$
$$y - 3 = -\frac{2}{3}(x - (-2))$$
$$y - 3 = -\frac{2}{3}x - \frac{4}{3}$$
$$y = -\frac{2}{3}x - \frac{4}{3} + 3$$
$$y = -\frac{2}{3}x + \frac{5}{3}$$

14. The slope of the line $2x - 3y = 7$ is found by solving the equation for y.
$-3y = -2x + 7$
$$y = \frac{2}{3}x - \frac{7}{3}$$
The slope is $\frac{2}{3}$.

Use $(-1, 2)$ and the perpendicular slope $-\frac{3}{2}$

$y - y_1 = m(x - x_1)$

$y - 2 = -\frac{3}{2}(x - (-1))$

$y - 2 = -\frac{3}{2}x - \frac{3}{2}$

$y = -\frac{3}{2}x - \frac{3}{2} + 2$

$y = -\frac{3}{2}x + \frac{1}{2}$

The equation of the line is $y = -\frac{3}{2}x + \frac{1}{2}$.

15. $2x - 5y = 10$

$-5y = -2x + 10$

$y = \frac{2}{5}x - 2$

slope is $\frac{2}{5}$; y-intercept is -2

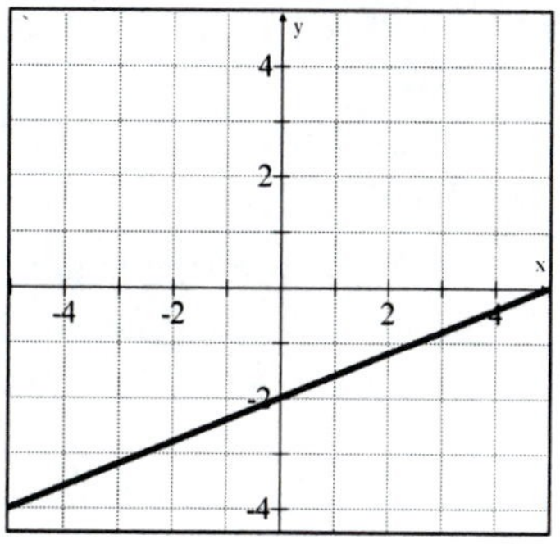

16. $3x - 4y \geq 8$

$-4y \geq -3x + 8$

$y \leq \frac{3}{4}x - 2$

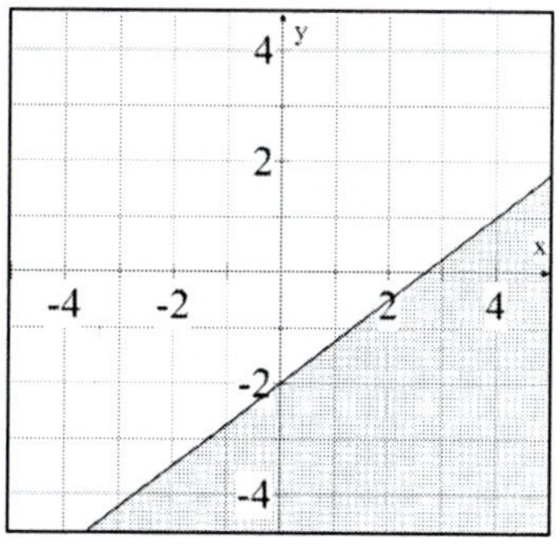

17. $5x - 2y = 10$

$3x + 2y = 6$

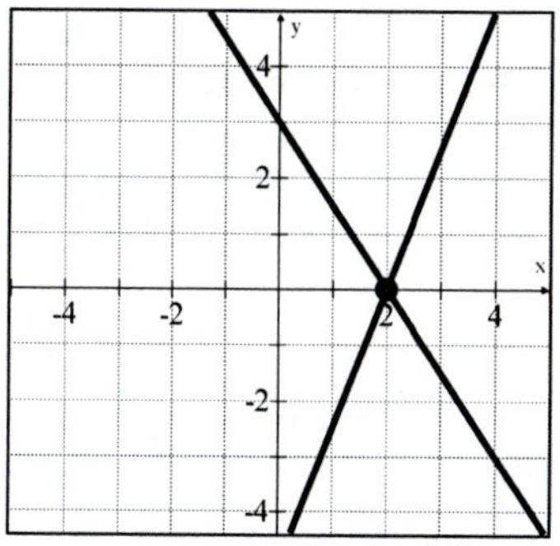

The solution is (2, 0).

18. Solve each inequality for y.

$3x - 2y \geq 4$

$-2y \geq -3x + 4$

$y \leq \frac{3}{2}x - 2$

$x + y < 3$

$y < -x + 3$

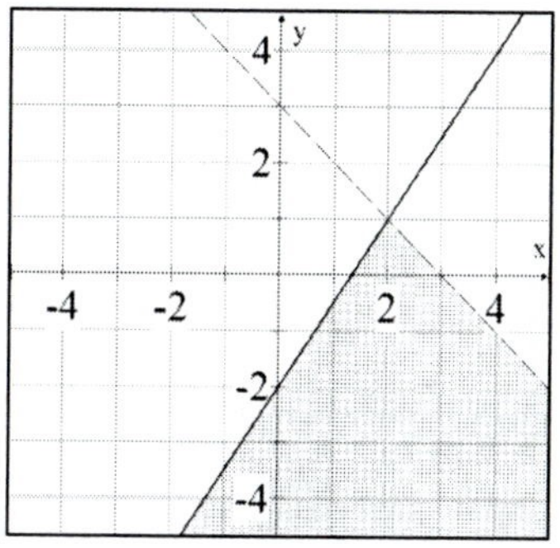

19. (1) $3x + 2z = 1$

(2) $2y - z = 1$

(3) $x + 2y = 1$

Multiply equation (2) by -1 and add to equation (3).

$-1(2y - z) = -1(1)$

$x + 2y = 1$

$-2y + z = -1$

$x + 2y = 1$

(4) $x + z = 0$

Multiply equation (4) by -2 and add to equation (1).

$-2(x + z) = -2(0)$

$3x + 2z = 1$

$-2x - 2z = 0$

$3x + 2z = 1$

$x = 1$

Substitute 1 for x in equation (3).

$1 + 2y = 1$

$2y = 0$

$y = 0$

Substitute 0 for y in equation (2).

$2(0) - z = 1$

$-z = 1$

$z = -1$

The solution is $(1, 0, -1)$.

20. (1) $2x - y + z = 2$

(2) $3x + y + 2z = 5$

(3) $3x - y + 4z = 1$

Eliminate y. Add equation (1) to equation (2).

$2x - y + z = 2$

$3x + y + 2z = 5$

(4) $5x + 3z = 7$

Add equation (2) to equation (3).
$3x + y + 2z = 5$
$3x - y + 4z = 1$

(5) $6x + 6z = 6$
Multiply equation (4) by −2 and add to equation (5).
$-2(5x + 3z) = -2(7)$
$6x + 6z = 6$

$-10x - 6z = -14$
$6x + 6z = 6$
$-4x = -8$
$x = 2$
Substitute into equation (4).
$5x + 3z = 7$
$5(2) + 3z = 7$
$10 + 3z = 7$
$3z = -3$
$z = -1$

Substitute into equation (1).
$2x - y + z = 2$
$2(2) - y + (-1) = 2$
$4 - y - 1 = 2$
$3 - y = 2$
$-y = -1$
$y = 1$
The solution is (2, 1, −1).

21. (1) $4x - 3y = 17$
(2) $3x - 2y = 12$
Multiply equation (1) by −2 and equation (2) by 3. Then add the equations together.
$-2(4x - 3y) = -2(17)$
$3(3x - 2y) = 3(12)$

$-8x + 6y = -34$
$9x - 6y = 36$
$x = 2$

Substitute 2 for x in equation (1) and solve for y.
$4x - 3y = 17$
$4(2) - 3y = 17$
$8 - 3y = 17$
$-3y = 9$
$y = -3$
The solution is (2, −3).

22. (1) $3x - 2y = 7$
(2) $y = 2x - 1$

Solve by substitution.
$3x - 2(2x - 1) = 7$
$3x - 4x + 2 = 7$
$-x + 2 = 7$
$-x = 5$
$x = -5$

Substitute −5 for x in equation (2).
$y = 2(-5) - 1$
$y = -10 - 1$
$y = -11$
The solution is (−5,−11).

23. Let x represent the amount of pure water.

	Amount	Percent	Quantity
Water	x	0	$0x$
4%	100	0.04	0.04(100)
2.5%	$100 + x$	0.025	$0.025(100 + x)$

The sum of the quantities before mixing is equal to the quantity after mixing.
$0x + 0.04(100) = 0.025(100 + x)$

Solution: $0x + 0.04(100) = 0.025(100 + x)$
$0 + 4 = 2.5 + 0.025x$
$1.5 = 0.025x$
$x = 60$
The amount of water that should be added is 60ml.

24. **Strategy:** Let x represent the rate of the plane in calm air.
The rate of the wind is y.

	Rate	Time	Distance
with wind	$x + y$	2	$2(x + y)$
against wind	$x - y$	3	$3(x - y)$

The distance traveled with the wind is 150 mi. The distance traveled against the wind is 150 mi.
$2(x + y) = 150$
$3(x - y) = 150$

Solution: $2(x + y) = 150$
$3(x - y) = 150$
$$\frac{1}{2} \cdot 2(x + y) = \frac{1}{2} \cdot 150$$

$\frac{1}{3} \cdot 3(x - y) = \frac{1}{3} \cdot 150$
$x + y = 75$
$x - y = 50$
$2x = 125$
$x = 62.5$

$x + y = 75$
$62.5 + y = 75$
$y = 12.5$
The rate of the wind is 12.5 mph.

25. Let x represent the cost per pound of hamburger.
The cost per pound of steak is y.
First purchase:

	Amount	Cost	Quantity
Hamburger	100	x	$100x$
Steak	50	y	$50y$

Second purchase:

	Amount	Cost	Quantity
Hamburger	150	x	$150x$
Steak	100	y	$100y$

The total cost of the first purchase is $541.
The total cost of the second purchase is $960.
$100x + 50y = 540$
$150x + 100y = 960$

Solution: $100x + 50y = 540$
$150x + 100y = 960$

$-2(100x + 50y) = -2(540)$
$150x + 100y = 960$

$-200x - 100y = -1080$
$150x + 100y = 960$
$-50x = -120$
$x = 2.4$

$100(2.4) + 50y = 540$
$240 + 50y = 540$
$50y = 300$
$y = 6$
The cost of steak is $6.00/lb.

26. **Strategy:** Let M represent the number of ohms, T the tolerance and r the value of the resistor. Find the tolerance and solve $|M - 12{,}000| \le T$ for M.

Solution: $T = 0.15 \cdot 12{,}000 = 1800$ ohms
$|M - 12{,}000| \le 1800$
$-1800 \le M - 12{,}000 \le 1800$
$-1800 + 12{,}000 \le M - 12{,}000 + 12{,}000$
$\le 1800 + 12{,}000$
$10{,}200 \le M \le 13{,}800$

The lower and upper limits of the resistor are 10,200 ohms and 13,800 ohms.

27. The slope of the line is
$\frac{5000 - 1000}{100 - 0} = \frac{4000}{100} = 40$.
The commission rate of the executive is $40 for every $1000 in sales.

Chapter 6: Polynomials

Prep Test

1. $-2-(-3) = -2+3 = 1$

2. $-3(6) = -18$

3. $-\dfrac{24}{-36} = \dfrac{2}{3}$

4. $3n^4 = 3(-2)^4 = 48$

5. $b \neq 0$

6. No

7. $3x^2 - 4x + 1 + 2x^2 - 5x - 7$
 $= 5x^2 - 9x - 6$

8. $-4y + 4y = 0$

9. $-3(2x-8) = -6x + 24$

10. $3xy - 4y - 2(5xy - 7y)$
 $= 3xy - 4y - 10xy + 14y$
 $= -7xy + 10y$

Section 6.1

Concept Check

1. (i), (ii) and (iv) are monomials.

3. No, the Rule for Simplifying Powers of Products does not apply because the expression inside the parentheses is a sum, not a product.

5. $-z^0 = -1;\ z^0 = 1$

Objective A Exercises

7. $(ab^3)(a^3b) = a^4b^4$

9. $(9xy^2)(-2x^2y^2) = -18x^3y^4$

11. $(x^2y^4)^4 = x^8y^{16}$

13. $(-3x^2y^3)^4 = (-3)^4x^8y^{12} = 81x^4y^{12}$

15. $(27a^5b^3)^2 = (27)^2a^{10}b^6 = 729a^{10}b^6$

17. $[(2a^4b^3)^3]^2 = (2a^4b^3)^6 = (2)^6a^{24}b^{18} = 64a^{24}b^{18}$

19. $(x^2y^2)(xy^3)^3 = (x^2y^2)(x^3y^9) = x^5y^{11}$

21. $(-5ab)(3a^3b^2)^2 = (-5ab)(3^2a^6b^4)$
 $= (-5ab)(9a^6b^4)$
 $= -45a^7b^5$

23. $(3x^5y)(-4x^3)^3 = (3x^5y)((-4)^3x^9)$
 $= (3x^5y)(-64x^9)$
 $= -192x^{14}y$

25. $(-6a^4b^2)(-7a^2c^5) = 42a^6b^2c^5$

27. $(-2ab^2)(-3a^4b^5)^3 = (-2ab^2)((-3)^3a^{12}b^{15})$
 $= (-2ab^2)(-27a^{12}b^{15})$
 $= 54a^{13}b^{17}$

29. $(-3ab^3)^3(-2^2a^2b)^2 = ((-3)^3a^3b^9)(2^4a^4b^2)$
 $= (-27a^3b^9)(16a^4b^2)$
 $= -432a^7b^{11}$

31. $(-2x^2y^3z)(3x^2yz^4) = -6x^4y^4z^5$

33. $(2xy)(-3x^2yz)(x^2y^3z^3) = -6x^5y^5z^4$

35. $(3b^5)(2ab^2)(-2ab^2c^2) = -12a^2b^9c^2$

37. The value of n is 33.

Objective B Exercises

39. $4^{-2} = \dfrac{1}{4^2} = \dfrac{1}{16}$

41. $\dfrac{1}{2^{-7}} = 2^7 = 128$

43. $3a^{-5} = \dfrac{3}{a^5}$

45. $\dfrac{1}{3a^{-7}} = \dfrac{a^7}{3}$

47. $\dfrac{a^3}{4b^{-2}} = \dfrac{a^3b^2}{4}$

49. $xy^{-4} = \dfrac{x}{y^4}$

51. $\dfrac{1}{2x^0} = \dfrac{1}{2}$

53. $\dfrac{-3^{-2}}{(2y)^0} = \dfrac{-1}{3^2} = -\dfrac{1}{9}$

55. $(x^3y^5)^{-2} = x^{-6}y^{-10} = \dfrac{1}{x^6y^{10}}$

57. $(-3a^{-4}b^{-5})(-5a^{-2}b^4) = 15a^{-6}b^{-1} = \dfrac{15}{a^6b}$

59. $(4y^{-3}z^{-4})(-3y^3z^{-3})^{-2}$

$= (4y^{-3}z^{-4})((-3)^{-2}y^{-6}z^6)$

$= (4)(-3)^{-2}y^{-9}z^2 = \dfrac{4z^2}{(-3)^2y^9}$

$= \dfrac{4z^2}{9y^9}$

61. $(4x^{-3}y^2)^{-3}(2xy^{-3})^4$

$= (4^{-3}x^9y^{-6})(2^4x^4y^{-12})$

$= (4^{-3})(2)^4x^{13}y^{-18} = \dfrac{2^4x^{13}}{(4)^3y^{18}}$

$= \dfrac{16x^{13}}{64y^{18}} = \dfrac{x^{13}}{4y^{18}}$

63. $\dfrac{9x^5}{12x^8} = \dfrac{3}{4x^3}$

65. $\dfrac{-6x^2y}{12x^4y} = -\dfrac{1}{2x^2}$

67. $\dfrac{y^{-2}}{y^6} = y^{-2-(6)} = y^{-8} = \dfrac{1}{y^8}$

69. $\dfrac{a^6b^{-4}}{a^{-2}b^5} = a^8b^{-9} = \dfrac{a^8}{b^9}$

71. $\dfrac{-3ab^2}{(9a^2b^4)^3} = \dfrac{-3ab^2}{9^3a^6b^{12}} = \dfrac{-3ab^2}{729a^6b^{12}}$

$= -\dfrac{1}{243a^5b^{10}}$

73. $\dfrac{(3a^2b)^3}{(-6ab^3)^2} = \dfrac{3^3a^6b^3}{(-6)^2a^2b^6} = \dfrac{27a^6b^3}{36a^2b^6} = \dfrac{3a^4}{4b^3}$

75. $\dfrac{(-8x^2y^2)^4}{(16x^3y^7)^2} = \dfrac{(-8)^4x^8y^8}{(16)^2x^6y^{14}} = \dfrac{4096x^8y^8}{256x^6y^{14}}$

$= \dfrac{16x^2}{y^6}$

77. $\dfrac{(3a^4b^{-2})^{-2}}{(2a^{-3}b)^3} = \dfrac{(3)^{-2}a^{-8}b^4}{(2)^3a^{-9}b^3} = \dfrac{ab}{(2)^3(3)^2} = \dfrac{ab}{72}$

79. $\left(\dfrac{9ab^{-2}}{8a^{-2}b}\right)^{-2}\left(\dfrac{3a^{-2}b}{2a^2b^{-2}}\right)^3$

$= \left(\dfrac{9^{-2}a^{-2}b^4}{8^{-2}a^4b^{-2}}\right)\left(\dfrac{3^3a^{-6}b^3}{2^3a^6b^{-6}}\right)$

$= \left(\dfrac{9^{-2}b^6}{8^{-2}a^6}\right)\left(\dfrac{3^3b^9}{2^3a^{12}}\right)$

$= \left(\dfrac{8^2b^6}{9^2a^6}\right)\left(\dfrac{3^3b^9}{2^3a^{12}}\right) = \dfrac{8b^{15}}{3a^{18}}$

81. The value of $p - q$ is 0.

Objective C Exercises

83. $4.67 \text{ x } 10^{-6}$

85. $1.7 \text{ x } 10^{-10}$

87. $2 \text{ x } 10^{11}$

89. 0.000000123

91. 8,200,000,000,000,000

93. 0.039

95. $(3 \text{ x } 10^{-12})(5 \text{ x } 10^{16})$
$= (3)(5) \text{ x } 10^{-12+16}$
$= 15 \text{ x } 10^{4}$
$= 150{,}000$

97. (0.0000065)(3,200,000,000,000)
$= (6.5 \text{ x } 10^{-6})(3.2 \text{ x } 10^{12})$
$= (6.5)(3.2) \text{ x } 10^{-6+12}$
$= 20.8 \text{ x } 10^{6}$
$= 20{,}800{,}000$

99. $\dfrac{9\times10^{-3}}{6\times10^{5}} = 1.5\times10^{-3-5}$
$= 1.5\times10^{-8} = 0.000000015$

101. $\dfrac{0.0089}{500{,}000{,}000} = \dfrac{8.9\times10^{-3}}{5\times10^{8}}$
$= 1.78\times10^{-3-8} = 1.78\times10^{-11}$
$= 0.0000000000178$

103. $\dfrac{(3.3\times10^{-11})(2.7\times10^{15})}{8.1\times10^{-3}}$
$= \dfrac{(3.3)(2.7)\times10^{-11+15-(-3)}}{8.1}$
$= 1.1\times10^{7} = 11{,}000{,}000$

105. $\dfrac{(0.00000004)(84{,}000)}{(0.0003)(1{,}400{,}000)}$
$= \dfrac{4\times10^{-8}\times8.4\times10^{4}}{3\times10^{-4}\times1.4\times10^{6}}$
$= \dfrac{4(8.4)\times10^{-8+4-(-4)-6}}{3(1.4)}$
$= 8\times10^{-6} = 0.000008$

107. Greater than zero

Objective D Exercises

109. **Strategy:** To find the number of years needed to cross the galaxy, divide the width of the galaxy by the product of the rate of the space ship and the number of hours in a year.
Solution:
$$\frac{5.6\times10^{19}}{2.5\times10^{4}\times8.76\times10^{3}} \approx 2.6\times10^{11}$$
It would takes a space ship 2.6×10^{11} years to cross the galaxy.

111. **Strategy:** To find the number of times larger the mass of the proton is, divide the mass of the proton by the mass of an electron.
Solution: $\dfrac{1.673\times10^{-27}}{9.109\times10^{-31}} \approx 1.837\times10^{3}$
The mass of a proton is approximately 1.837×10^{3} times larger than an electron.

113. **Strategy:** To find how fast the radio signal travels, divide the distance from Mars to Earth by the speed of the radio signal.
Solution: $\dfrac{154{,}000{,}000}{13.8} = 11159420.29$
$\approx 1.12 \times 10^{7}$ mi/day
The radio signal travel approximately 1.12×10^{7} mi/day.

115. **Strategy:** To find the number of times larger the mass of the sun is, divide the mass of the sun by the mass of the earth.
Solution: $\dfrac{2\times10^{30}}{5.9\times10^{24}} \approx 3.39\times10^{5}$
The sun is approximately 3.39×10^{5} times larger than the earth.

117. **Strategy:** To find the number of seeds produced, divide the number of seeds by the number of pine seedlings growing.
Solution:

$$\frac{2{,}000{,}000}{12{,}000}=\frac{2\times10^6}{1.2\times10^4}=1.\bar{6}\times10^2$$

$1.\bar{6}\times10^2$ seeds are produced.

Critical Thinking

119. a) $3^{x^2}=3^{2^2}=3^4=81$

b) $3^{x^2}=3^{3^2}=3^9=19{,}683$

c) $3^{x^2}=3^{0^2}=3^0=1$

d) $3^{x^2}=3^{(-2)^2}=3^4=81$

121. $x^{3n}x^{4n}=x^{7n}$

123. $\dfrac{x^n y^{5m}}{x^{3n}y^m}=\dfrac{y^{4m}}{x^{2n}}$

Projects or Group Activities

125. a)

$$1+[1+(1+2^{-1})^{-1}]^{-1}=1+\left(1+\left(1+\frac{1}{2}\right)^{-1}\right)^{-1}$$
$$=1+\left(1+\left(\frac{3}{2}\right)^{-1}\right)^{-1}$$
$$=1+\left(1+\frac{2}{3}\right)^{-1}$$
$$=1+\left(\frac{5}{3}\right)^{-1}$$
$$=1+\left(\frac{3}{5}\right)$$
$$=\frac{8}{5}$$

b)

$$2-[2-(2-2^{-1})^{-1}]^{-1}=2-\left(2-\left(2-\frac{1}{2}\right)^{-1}\right)^{-1}$$
$$=2-\left(2-\left(\frac{3}{2}\right)^{-1}\right)^{-1}$$
$$=2-\left(2-\frac{2}{3}\right)^{-1}$$
$$=2-\left(\frac{4}{3}\right)^{-1}$$
$$=2-\left(\frac{3}{4}\right)$$
$$=\frac{5}{4}$$

Section 6.2

Concept Check

1. a) binomial
b) binomial
c) monomial
d) trinomial
e) none of these
f) monomial

3. All real numbers

5. This is a polynomial.
a) –1
b) 8
c) 2

7. This is not a polynomial.

9. This is a polynomial.
a) –1
b) 0
c) 6

Objective A Exercises

11. $P(x) = 3x^2 - 2x - 8$

$P(3) = 3(3)^2 - 2(3) - 8$

$P(3) = 13$

13. $R(x) = 2x^3 - 3x^2 + 4x - 2$

$R(2) = 2(2)^3 - 3(2)^2 + 4(2) - 2$

$R(2) = 10$

15. $f(x) = x^4 - 2x^2 - 10$

$f(-1) = (-1)^4 - 2(-1)^2 - 10$

$f(-1) = -11$

17. $P(x) = x^2 - 3x - 3$

x	y
−2	7
−1	1
0	−3
1	−5
2	−5

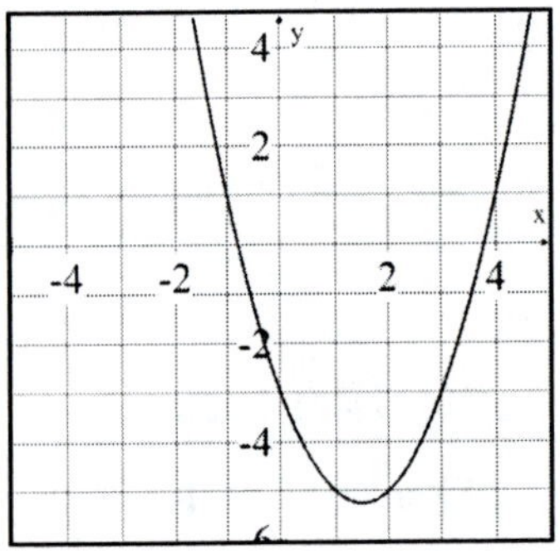

19. $P(x) = x^3 + 2$

x	y
−2	−6
−1	1
0	2
1	3
2	10

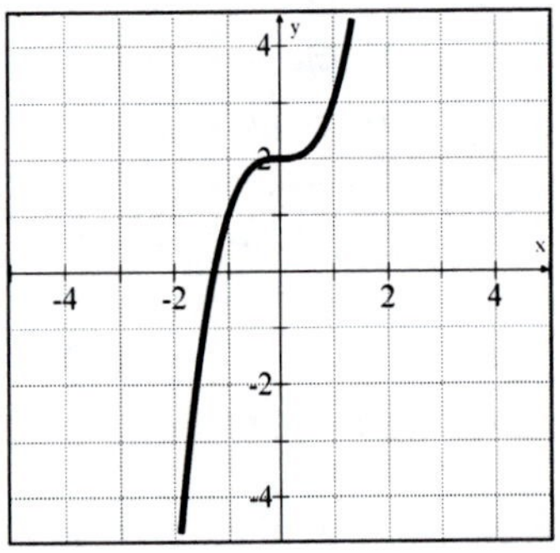

21. $f(x) = x^3 - 4x^2 - 4x + 16$

x	y
−2	0
−1	15
0	16
1	9
2	0

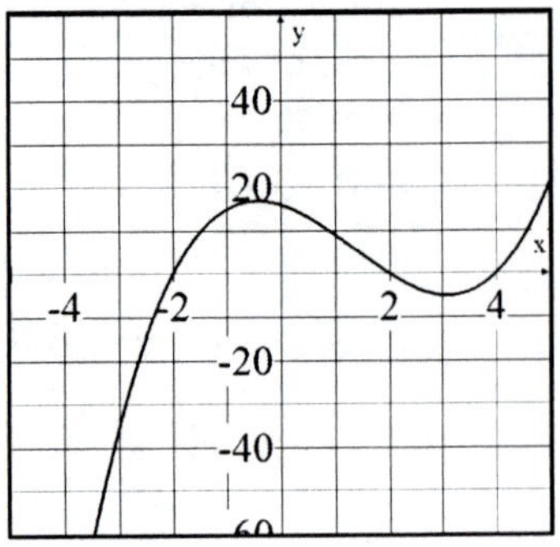

23. (a) $f(c) - g(c) > 0$

(b) $f(c) - g(c) < 0$

Objective B Exercises

25.
$$\begin{array}{r} 5x^2 + 2x - 7 \\ \underline{x^2 - 8x + 12} \\ 6x^2 - 6x + 5 \end{array}$$

27.
$$\begin{array}{r} x^2 - 3x + 8 \\ \underline{-2x^2 + 3x - 7} \\ -x^2 \qquad + 1 \end{array}$$

29. $(3y^2 - 7y) + (2y^2 - 8y + 2)$

$= (3y^2 + 2y^2) + (-7y - 8y) + 2$

$= 5y^2 - 15y + 2$

31. $(2a^2 - 3a - 7) - (-5a^2 - 2a - 9)$

$= (2a^2 + 5a^2) + (-3a + 2a) + (-7 + 9)$

$= 7a^2 - a + 2$

33. $P(x) + R(x) = (x^2 - 3xy + y^2) + (2x^2 - 3y^2)$

$= (x^2 + 2x^2) - 3xy + (y^2 - 3y^2)$

$= 3x^2 - 3xy - 2y^2$

35. $P(x) - R(x)$

$= (3x^2 + 2y^2) - (-5x^2 + 2xy - 3y^2)$

$= (3x^2 + 5x^2) - 2xy + (2y^2 + 3y^2)$

$= 8x^2 - 2xy + 5y^2$

Critical Thinking

37. $(2x^3 + 3x^2 + kx + 5) - (x^3 + x^2 - 5x - 2) = x^3 + 2x^2 + 3x + 7$
$(2x^3 - x^3) + (3x^2 - x^2) + (kx + 5x) + (5 + 2) = x^3 + 2x^2 + 3x + 7$
$x^3 + 2x^2 + (k + 5)x + 7 = x^3 + 2x^2 + 3x + 7$
$(k + 5)x = 3x$
$k + 5 = 3$
$k = -2$

39. $P(-1) = -3;\ P(x) = 4x^4 - 3x^2 + 6x + c$
$-3 = 4(-1)^4 - 3(-1)^2 + 6(-1) + c$
$-3 = 4 - 3 - 6 + c$
$-3 = -5 + c$
$2 = c$

41. If $P(x)$ is a third degree polynomial and $Q(x)$ is a fourth degree polynomial, then $P(x) + Q(x)$ is a fourth degree polynomial.
Example: $P(x) = 3x^3 - 5x - 8$
$Q(x) = 3x^4 - 2x + 1$
$P(x) + Q(x) = 3x^4 + 3x^3 - 7x - 7$

Projects and Group Activities

43. The graph of $k(x)$ is the graph of $f(x)$ moved 2 units down

Check Your Progress: Chapter 6

1. $(-12a^6b)(6a^4b^3) = -72a^{10}b^4$

2. $(-4x^6y^8)(-3x^{-1}y^{-8}) = 12x^5$

3. $(2x^3)^4(3x^2) = (16x^{12})(3x^2) = 48x^{14}$

4. $(2a^3b^{-4})^4(3a^{-3}b^2)^{-2} = \dfrac{16a^{12}b^{-16}}{9a^{-6}b^4} = \dfrac{16a^{18}}{9b^{20}}$

5. $\dfrac{x^4y}{xy^5} = \dfrac{x^3}{y^4}$

6. $\dfrac{2x^{-3}}{4x^{-5}} = \dfrac{x^2}{2}$

7. $\dfrac{3a^4b^2c^8}{6a^7b^{-2}c^8} = \dfrac{b^4}{2a^3}$

8. $\dfrac{(3x^3y^{-2})^{-2}}{(2x^{-1}y^3)^3} = \dfrac{3^{-2}x^{-6}y^4}{8x^{-3}y^9} = \dfrac{1}{72x^3y^5}$

9. 6.83×10^{-7}

10. 0.002607

11. 2,140,000

12. $P(-2) = 4(-2)^3 - 6(-2) + 1 = -19$

13. $3x^2 - 6x + 7$
$\underline{2x^2 + x - 9}$
$5x^2 - 5x - 2$

14. $(-5x^2 + 7x - 8) - (6x^2 + 7x - 7)$
$= -5x^2 + 7x - 8 - 6x^2 - 7x + 7$
$= -11x^2 - 1$

15.

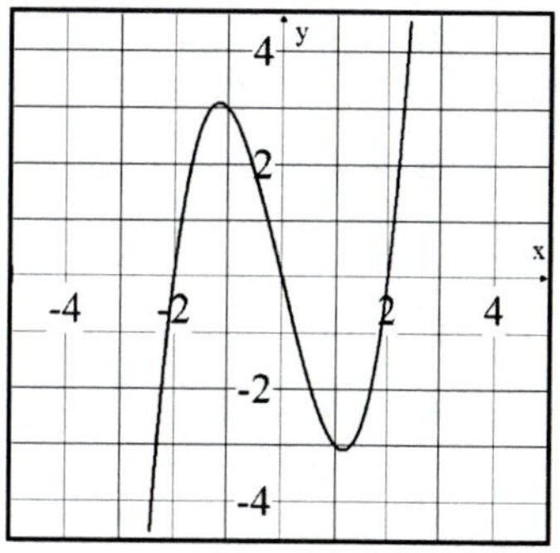

Section 6.3

Concept Check

1. Multiply $2(2x + 1)$. By the Order of Operation Agreement we do multiplication before addition.

3. FOIL is a method used to find the product of two binomials. It is based on the Distributive Property. The letters of FOIL stand for First, Outer, Inner, and Last.

Objective A Exercises

5. $2x(x - 3) = 2x^2 - 6x$

7. $3x^2(2x^2 - x) = 6x^4 - 3x^3$

9. $3xy(2x - 3y) = 6x^2y - 9xy^2$

11. $-3xy^2(4x - 5y) = -12x^2y^2 + 15xy^3$

13. $2b + 4b(2 - b) = 2b + 8b - 4b^2 = -4b^2 + 10b$

15. $-2a^2(3a^2 - 2a + 3) = -6a^4 + 4a^3 - 6a^2$

17. $(-3y^2 - 4y + 2)(y^2) = -3y^4 - 4y^3 + 2y^2$

19. $-5x^2(4 - 3x + 3x^2 + 4x^3)$
$= -20x^2 + 15x^3 - 15x^4 - 20x^5$

21. $-2x^2y(x^2 - 3xy + 2y^2) = -2x^4y + 6x^3y^2 - 4x^2y^3$

23. $5x^3 - 4x(2x^2 + 3x - 7)$
$= 5x^3 - 8x^3 - 12x^2 + 28x$
$= -3x^3 - 12x^2 + 28x$

25. $2y^2 - y[3 - 2(y - 4) - y]$
$= 2y^2 - y(3 - 2y + 8 - y)$
$= 2y^2 - y(11 - 3y)$
$= 2y^2 - 11y + 3y^2$
$= 5y^2 - 11y$

27. $2y - 3[y - 2y(y - 3) + 4y]$
$= 2y - 3(y - 2y^2 + 6y + 4y)$
$= 2y - 3(-2y^2 + 11y)$
$= 2y + 6y^2 - 33y$
$= 6y^2 - 31y$

29. $P(b) = 3b$ and $Q(b) = 3b^4 - 3b^2 + 8$
$P(b) \cdot Q(b) = 3b(3b^4 - 3b^2 + 8)$
$= 9b^5 - 9b^3 + 24b$

Objective B Exercises

31. $(x - 2)(x + 7) = x^2 + 7x - 2x - 14$
$= x^2 + 5x - 14$

33. $(2y - 3)(4y + 7) = 8y^2 + 14y - 12y - 21$
$= 8y^2 + 2y - 21$

35. $(a + 3c)(4a - 5c) = 4a^2 - 5ac + 12ac - 15c^2$
$= 4a^2 + 7ac - 15c^2$

37. $(5x - 7)(5x - 7) = 25x^2 - 35x - 35x + 49$
$= 25x^2 - 70x + 49$

39. $2(2x - 3y)(2x + 5y)$
$= 2(4x^2 + 10xy - 6xy - 15y^2)$
$= 2(4x^2 + 4xy - 15y^2)$
$= 8x^2 + 8xy - 30y^2$

41. $(xy + 4)(xy - 3) = x^2y^2 - 3xy + 4xy - 12$
$= x^2y^2 + xy - 12$

43. $(2x^2 - 5)(x^2 - 5) = 2x^4 - 10x^2 - 5x^2 + 25$
$= 2x^4 - 15x^2 + 25$

45. $(5x^2 - 5y)(2x^2 - y)$
$= 10x^4 - 5x^2y - 10x^2y + 5y^2$
$= 10x^4 - 15x^2y + 5y^2$

47. $(x + 5)(x^2 - 3x + 4)$
$= x(x^2 - 3x + 4) + 5(x^2 - 3x + 4)$
$= x^3 - 3x^2 + 4x + 5x^2 - 15x + 20$
$= x^3 + 2x^2 - 11x + 20$

49. $(2a-3b)(5a^2-6ab+4b^2)$
$=2a(5a^2-6ab+4b^2)-3b(5a^2-6ab+4b^2)$
$=10a^3-12a^2b+8ab^2-15a^2b+18ab^2-12b^3$
$=10a^3-27a^2b+26ab^2-12b^3$

51. $(2x^3+3x^2-2x+5)(2x-3)$
$=(2x^3+3x^2-2x+5)2x+(2x^3+3x^2-2x+5)(-3)$
$=4x^4+6x^3-4x^2+10x-6x^3-9x^2+6x-15$
$=4x^4-13x^2+16x-15$

53. $(2x-5)(2x^4-3x^3-2x+9)$
$=2x(2x^4-3x^3-2x+9)-5(2x^4-3x^3-2x+9)$
$=4x^5-6x^4-4x^2+18x-10x^4+15x^3+10x-45$
$=4x^5-16x^4+15x^3-4x^2+28x-45$

55. $(x^2+2x-3)(x^2-5x+7)$
$=x^2(x^2-5x+7)+2x(x^2-5x+7)-3(x^2-5x+7)$
$=x^4-5x^3+7x^2+2x^3-10x^2+14x-3x^2+15x-21$
$=x^4-3x^3-6x^2+29x-21$

57. $(a-2)(2a-3)(a+7)$
$=(2a^2-3a-4a+6)(a+7)$
$=(2a^2-7a+6)(a+7)$
$=(2a^2-7a+6)a+(2a^2-7a+6)7$
$=2a^3-7a^2+6a+14a^2-49a+42$
$=2a^3+7a^2-43a+42$

59. $(2x+3)(x-4)(3x+5)$
$=(2x^2-8x+3x-12)(3x+5)$
$=(2x^2-5x-12)(3x+5)$
$=(2x^2-5x-12)3x+(2x^2-5x-12)5$
$=6x^3-15x^2-36x+10x^2-25x-60$
$=6x^3-5x^2-61x-60$

61. $P(y)=2y^2-1$ and $Q(y)=y^3-5y^2-3$
$P(y)\cdot Q(y)=(2y^2-1)(y^3-5y^2-3)$
$=2y^2(y^3-5y^2-3)-1(y^3-5y^2-3)$
$=2y^5-10y^4-6y^2-y^3+5y^2+3$
$=2y^5-10y^4-y^3-y^2+3$

63. *mn*

Objective C Exercises

65. $(3x-2)(3x+2)=9x^2-4$

67. $(6-x)(6+x)=36-x^2$

69. $(2a-3b)(2a+3b)=4a^2-9b^2$

71. $(3ab+4)(3ab-4)=9a^2b^2-16$

73. $(x^2+1)(x^2-1)=x^4-1$

75. $(x-5)^2=x^2-10x+25$

77. $(3a+5b)^2=9a^2+30ab+25b^2$

79. $(x^2-3)^2=x^4-6x^2+9$

81. $(2x^2-3y^2)^2=4x^4-12x^2y^2+9y^4$

83. $(3mn-5)^2=9m^2n^2-30mn+25$

85. $y^2-(x-y)^2=y^2-(x^2-2xy+y^2)$
$=y^2-x^2+2xy-y^2$
$=-x^2+2xy$

87. $(x-y)^2-(x+y)^2$
$=x^2-2xy+y^2-(x^2+2xy+y^2)$
$=x^2-2xy+y^2-x^2-2xy-y^2$
$=-4xy$

89. False

Objective D Exercises

91. ft^2

93. Strategy: To find the area substitute the given values for L and W in the equation $A = L \cdot W$ and solve for A.

Solution: $A = L \cdot W$
$A = (3x - 2)(x + 4)$
$A = 3x^2 + 12x - 2x - 8$
$A = 3x^2 + 10x - 8$
The area is $(3x^2 + 10x - 8)$ ft^2.

95. Strategy: To find the area, add the area of the small rectangle to the area of the large rectangle.
Larger rectangle:
Length = $L_1 = x + 5$
Width = $W_1 = x - 2$
Smaller rectangle:
Length = $L_2 = 5$
Width = $W_2 = 2$

Solution: $A = L_1 \cdot W_1 + L_2 \cdot W_2$
$A = (x + 5)(x - 2) + (5)(2)$
$A = x^2 - 2x + 5x - 10 + 10$
$A = x^2 + 3x$
The area is $(x^2 + 3x)$ ft^2.

97. Strategy: To find the volume substitute the given value for s in the equation $V = s^3$ and solve for V.

Solution: $V = s^3$
$V = (x + 3)^3$
$V = (x + 3)(x + 3)(x + 3)$
$V = (x^2 + 6x + 9)(x + 3)$
$V = x^3 + 9x^2 + 27x + 27$
The volume is $(x^3 + 9x^2 + 27x + 27)$ cm^3.

99. Strategy: To find the volume subtract the volume of the small rectangular solid from the volume of the large rectangular solid.
Large rectangular solid:
Length = $L_1 = x + 2$
Width = $W_1 = 2x$
Height = $H_1 = x$
Small rectangular solid:
Length = $L_2 = x$
Width = $W_2 = 2x$
Height = $H_2 = 2$

Solution: $V = (L_1 \cdot W_1 \cdot H_1) - (L_2 \cdot W_2 \cdot H_2)$
$V = (x + 2)(2x)(x) - (x)(2x)(2)$
$V = (2x^2 + 4x)(x) - (2x^2)(2)$
$V = 2x^3 + 4x^2 - 4x^2$
$V = 2x^3$
The volume is $(2x^3)$ in^3.

101. Strategy: To find the area substitute the given value for r into the equation $A = \pi r^2$ and solve for A.

Solution: $A = \pi r^2$
$A = 3.14(5x + 4)^2$
$A = 3.14(25x^2 + 40x + 16)$
$A = 78.5x^2 + 125.6x + 50.24$
The area is $(78.5x^2 + 125.6x + 50.24)$ in^2.

Critical Thinking

103. a) $(3x - k)(2x + k) = 6x^2 + 5x + k^2$
$6x^2 + xk - k^2 = 6x^2 + 5x + k^2$
$xk = 5x$
$k = 5$

b) $(4x + k)^2 = 16x^2 + 8x + k^2$
$16x^2 + 8xk + k^2 = 16x^2 + 8x + k^2$
$8xk = 8x$
$k = 1$

105. $(2x - 3)(x + 7) = 2x^2 + 14x - 3x - 21$
$2x^2 + 11x - 21$

Projects of Group Activities

107. Answers will vary. One example is $(x+3)(2x^2-1)$

Section 6.4

Concept Check

1. degree 4

3. If the polynomial $P(x)$ is divided by $x-a$, the remainder is $P(a)$.

Objective A Exercises

5. $\dfrac{3x^2-6x}{3x}=\dfrac{3x^2}{3x}-\dfrac{6x}{3x}=x-2$

7. $\dfrac{5x^2-10x}{-5x}=\dfrac{5x^2}{-5x}-\dfrac{10x}{-5x}=-x+2$

9. $\dfrac{5x^2y^2+10xy}{5xy}=\dfrac{5x^2y^2}{5xy}+\dfrac{10xy}{5xy}=xy+2$

11. $\dfrac{x^3+3x^2-5x}{x}=\dfrac{x^3}{x}+\dfrac{3x^2}{x}-\dfrac{5x}{x}=x^2+3x-5$

13. $\dfrac{9b^5+12b^4+6b^3}{3b^2}=\dfrac{9b^5}{3b^2}+\dfrac{12b^4}{3b^2}+\dfrac{6b^3}{3b^2}$
$=3b^3+4b^2+2b$

15. $\dfrac{a^5b-6a^3b+ab}{ab}=\dfrac{a^5b}{ab}-\dfrac{6a^3b}{ab}+\dfrac{ab}{ab}$
$=a^4-6a^2+1$

17. $P(x)=6x^3+21x^2-15x$

Objective B Exercises

19.
$$\begin{array}{r}
x+8 \\
x-5\overline{)x^2+3x-40} \\
\underline{x^2-5x} \\
8x-40 \\
\underline{8x-40} \\
0
\end{array}$$

$(x^2+3x-40)\div(x-5)=x+8$

21.
$$\begin{array}{r}
x^2+3x+6 \\
x-3\overline{)x^3+0x^2-3x+2} \\
\underline{x^3-3x^2} \\
3x^2-3x \\
\underline{3x^2-9x} \\
6x+2 \\
\underline{6x-18} \\
20
\end{array}$$

$(x^3-3x+2)\div(x-3)=x^2+3x+6+\dfrac{20}{x-3}$

23.
$$\begin{array}{r}
3x+5 \\
2x+1\overline{)6x^2+13x+8} \\
\underline{6x^2+3x} \\
10x+8 \\
\underline{10x+5} \\
3
\end{array}$$

$(6x^2+13x+8)\div(2x+1)=3x+5+\dfrac{3}{2x+1}$

25.
$$\begin{array}{r} 5x+7 \\ 2x-1\overline{\big)10x^2+9x-5} \\ \underline{10x^2-5x} \\ 14x-5 \\ \underline{14x-7} \\ 2 \end{array}$$

$$(10x^2+9x-5)\div(2x-1)=5x+7+\frac{2}{2x-1}$$

27.
$$\begin{array}{r} 4x^2+6x+9 \\ 2x-3\overline{\big)8x^3+0x^2+0x-9} \\ \underline{8x^3-12x^2} \\ 12x^2+0x \\ \underline{12x^2-18x} \\ 18x-9 \\ \underline{18x-27} \\ 18 \end{array}$$

$$(8x^3-9)\div(2x-3)=4x^2+6x+9+\frac{18}{2x-3}$$

29.
$$\begin{array}{r} 3x^2+1 \\ 2x^2-5\overline{\big)6x^4+0x^3-13x^2+0x-4} \\ \underline{6x^4\qquad -15x^2} \\ 2x^2+0x-4 \\ \underline{2x^2\qquad -5} \\ 1 \end{array}$$

$$(6x^4-13x^2-4)\div(2x^2-5)$$
$$=3x^2+1+\frac{1}{2x^2-5}$$

31.
$$\begin{array}{r} x^2-3x-10 \\ 3x+1\overline{\big)3x^3-8x^2-33x-10} \\ \underline{3x^3+x^2} \\ -9x^2-33x \\ \underline{-9x^2-3x} \\ -30x-10 \\ \underline{-30x-10} \\ 0 \end{array}$$

$$\frac{3x^3-8x^2-33x-10}{3x+1}=x^2-3x-10$$

33.
$$\begin{array}{r} x^2-2x+1 \\ x-3\overline{\big)x^3-5x^2+7x-4} \\ \underline{x^3-3x^2} \\ -2x^2+7x \\ \underline{-2x^2+6x} \\ x-4 \\ \underline{x-3} \\ -1 \end{array}$$

$$\frac{x^3-5x^2+7x-4}{x-3}=x^2-2x+1-\frac{1}{x-3}$$

35.

$$
\begin{array}{r}
2x^3-3x^2+x-4 \\
x-5\overline{\smash{)}2x^4-13x^3+16x^2-9x+20} \\
\underline{2x^4-10x^3} \\
-3x^3+16x^2 \\
\underline{-3x^3+15x^2} \\
x^2-9x \\
\underline{x^2-5x} \\
-4x+20 \\
\underline{-4x+20} \\
0
\end{array}
$$

$$\frac{2x^4-13x^3+16x^2-9x+20}{x-5}=2x^3-3x^2+x-4$$

37.

$$
\begin{array}{r}
2x \\
x^2+2x-1\overline{\smash{)}2x^3+4x^2-x+2} \\
\underline{2x^3+4x^2-2x} \\
x+2
\end{array}
$$

$$\frac{2x^3+4x^2-x+2}{x^2+2x-1}=2x+\frac{x+2}{x^2+2x-1}$$

39.

$$
\begin{array}{r}
x^2+4x+6 \\
x^2-2x-1\overline{\smash{)}x^4+2x^3-3x^2-6x+2} \\
\underline{x^4-2x^3-x^2} \\
4x^3-2x^2-6x \\
\underline{4x^3-8x^2-4x} \\
6x^2-2x+2 \\
\underline{6x^2-12x-6} \\
10x+8
\end{array}
$$

$$\frac{x^4+2x^3-3x^2-6x+2}{x^2-2x-1}=x^2+4x+6+\frac{10x+8}{x^2-2x-1}$$

41. $\dfrac{P(x)}{Q(x)}=\dfrac{2x^3+x^2+8x+7}{2x+1}$

$$
\begin{array}{r}
x^2+4 \\
2x+1\overline{\smash{)}2x^3+x^2+8x+7} \\
\underline{2x^3+x^2} \\
8x+7 \\
\underline{8x+4} \\
3
\end{array}
$$

$$\frac{2x^3+x^2+8x+7}{2x+1}=x^2+4+\frac{3}{2x+1}$$

43. $6x^3+27x^2+18x-30$

Objective C Exercise

45. 3

47.

$$
\begin{array}{r|rrr}
-1 & 2 & -6 & -8 \\
 & & -2 & 8 \\
\hline
 & 2 & -8 & 0
\end{array}
$$

$$(2x^2-6x-8)\div(x+1)=2x-8$$

49.

$$
\begin{array}{r|rrr}
2 & 3 & -14 & 16 \\
 & & 6 & -16 \\
\hline
 & 3 & -8 & 0
\end{array}
$$

$$(3x^2-14x+16)\div(x-2)=3x-8$$

51.

$$
\begin{array}{r|rrr}
1 & 3 & 0 & -4 \\
 & & 3 & 3 \\
\hline
 & 3 & 3 & -1
\end{array}
$$

$$(3x^2-4)\div(x-1)=3x+3-\frac{1}{x-1}$$

53.

$$
\begin{array}{r|rrrr}
-1 & 2 & -1 & 6 & 9 \\
 & & -2 & 3 & -9 \\
\hline
 & 2 & -3 & 9 & 0
\end{array}
$$

$$(2x^3-x^2+6x+9)\div(x+1)=2x^2-3x+9$$

55.

2	4	0	−1	−18
		8	16	30
	4	8	15	12

$$(4x^3 - x - 18) \div (x-2) = 4x^2 + 8x + 15 + \frac{12}{x-2}$$

57.

−4	2	5	−5	20
		−8	12	−28
	2	−3	7	−8

$$(2x^3 + 5x^2 - 5x + 20) \div (x+4) = 2x^2 - 3x + 7 - \frac{8}{x+4}$$

59.

2	3	−4	8	−5	−5
		6	4	24	38
	3	2	12	19	33

$$(3x^4 - 4x^3 + 8x^2 - 5x - 5) \div (x-2)$$
$$= 3x^3 + 2x^2 + 12x + 19 + \frac{33}{x-2}$$

61.

−1	3	3	−1	3	2
		−3	0	1	−4
	3	0	−1	4	−2

$$(3x^4 + 3x^3 - x^2 + 3x + 2) \div (x+1)$$
$$= 3x^3 - x + 4 - \frac{2}{x+1}$$

63. $\frac{P(x)}{Q(x)} = \frac{3x^2 - 5x + 6}{x-2}$

2	3	−5	6
		6	2
	3	1	8

$$(3x^2 - 5x + 6) \div (x-2) = 3x + 1 + \frac{8}{x-2}$$

Objective D Exercises

65. $x - 3$

67.

3	2	−3	−1
		6	9
	2	3	8

$P(3) = 8$

69.

4	1	−2	3	−1
		4	8	44
	1	2	11	43

$R(4) = 43$

71.

−2	2	−4	3	−1
		−4	16	−38
	2	−8	19	−39

$P(-2) = -39$

73.

−3	2	−1	0	3
		−6	21	−63
	2	−7	21	−60

$Z(-3) = -60$

75.

2	1	3	−2	4	−9
		2	10	16	40
	1	5	8	20	31

$Q(2) = 31$

77.

$$\begin{array}{r|rrrrr} -3 & 2 & -1 & 0 & 2 & -5 \\ & & -6 & 21 & -63 & 183 \\ \hline & 2 & -7 & 21 & -61 & 178 \end{array}$$

$F(-3) = 178$

79.

$$\begin{array}{r|rrrr} 5 & 1 & 0 & 0 & -3 \\ & & 5 & 25 & 125 \\ \hline & 1 & 5 & 25 & 122 \end{array}$$

$P(5) = 122$

81.

$$\begin{array}{r|rrrrr} -3 & 4 & 0 & -3 & 0 & 5 \\ & & -12 & 36 & -99 & 297 \\ \hline & 4 & -12 & 33 & -99 & 302 \end{array}$$

$R(-3) = 302$

83.

$$\begin{array}{r|rrrrrr} 2 & 1 & 0 & -4 & -2 & 5 & -2 \\ & & 2 & 4 & 0 & -4 & 2 \\ \hline & 1 & 2 & 0 & -2 & 1 & 0 \end{array}$$

$Q(2) = 0$

Critical Thinking

85. Use the Remainder Theorem to find k so that $P(3) = 0$.
$(3)^3 - 3(3)^2 - 3 + k = 0$
$27 - 27 - 3 + k = 0$
$-3 + k = 0$
$k = 3$

87. Use the Remainder Theorem to find k so that $P(3) = 0$.
$(3)^2 + 3k - 6 = 0$
$9 + 3k - 6 = 0$
$3 + 3k = 0$
$3k = -3$
$k = -1$

89. Possible degrees of $p(x)$ and $q(x)$ are:
1, 5; 2, 4; 3, 3

Projects or Groups Activities

91. $P(-5) = (-5)^4 + (-5)^3 - 21(-5)^2 - (-5) + 20$
$= 0$
Yes, $x + 5$ is a factor.

Chapter 6 Review Exercises

1. $(12y^2 + 17y - 4) + (9y^2 - 13y + 3)$
$= 21y^2 + 4y - 1$

2.

$$\begin{array}{r} 5x+4 \\ 3x-2\overline{\smash{)}15x^2+2x-2} \\ \underline{15x^2-10x} \\ 12x-2 \\ \underline{12x-8} \\ 6 \end{array}$$

$$\frac{15x^2+2x-2}{3x-2} = 5x+4+\frac{6}{3x-2}$$

3. $(2x^{-1}y^2z^5)^4(-3x^3yz^{-3})^2$
$= (16x^{-4}y^8z^{20})(9x^6y^2z^{-6})$
$= 144x^2y^{10}z^{14}$

4. $(5y - 7)^2 = (5y - 7)(5y - 7)$
$= 25y^2 - 35y - 35y + 49$
$= 25y^2 - 70y + 49$

5. $\dfrac{a^{-1}b^3}{a^3b^{-3}} = \dfrac{b^6}{a^4}$

6.

$$\begin{array}{r|rrrr} 2 & 1 & -2 & 3 & -5 \\ & & 2 & 0 & 6 \\ \hline & 1 & 0 & 3 & 1 \end{array}$$

$P(2) = 1$

7. $(5x^2 - 8xy + 2y^2) - (x^2 - 3y^2)$
$= (5x^2 - x^2) - 8xy + (2y^2 + 3y^2)$
$= 4x^2 - 8xy + 5y^2$

8. $\dfrac{12b^7 + 36b^5 - 3b^3}{3b^3} = \dfrac{12b^7}{3b^3} + \dfrac{36b^5}{3b^3} - \dfrac{3b^3}{3b^3}$
$= 4b^4 + 12b^2 - 1$

9. $\left(\dfrac{3ab^4}{-6a^2b^4} = -\dfrac{1}{2a}\right.$

10. $(-2a^2b^4)(3ab^2) = -6a^3b^6$

11. $\dfrac{8x^{12}}{12x^9} = \dfrac{2x^3}{3}$

12.

$$\begin{array}{r|rrrr} -6 & 4 & 27 & 10 & 2 \\ & & -24 & -18 & 48 \\ \hline & 4 & 3 & -8 & 50 \end{array}$$

$$\frac{4x^3+27x^2+10x+2}{x+6} = 4x^2+3x-8+\frac{50}{x+6}$$

13. $P(x) = 2x^3 - x + 7$
$P(-2) = 2(-2)^3 - (-2) + 7$
$P(-2) = -16 + 2 + 7$
$P(-2) = -7$

14. $(13y^3 - 7y - 2) - (12y^2 - 2y - 1)$
$= (13y^3 - 7y - 2) + (-12y^2 + 2y + 1)$
$= 13y^3 - 12y^2 - 5y - 1$

15.

$$\begin{array}{r|rrrr} 7 & 1 & -2 & -33 & -7 \\ & & 7 & 35 & 14 \\ \hline & 1 & 5 & 2 & 7 \end{array}$$

$$\frac{b^3-2b^2-33b-7}{b-7} = b^2+5b+2+\frac{7}{b-7}$$

16. $4x^2y(3x^3y^2 + 2xy - 7y^3)$
$= 12x^5y^3 + 8x^3y^2 - 28x^2y^4$

17. $(2a - b)(x - 2y) = 2ax - 4ay - bx + 2by$

18. $(2b - 3)(4b + 5) = 8b^2 + 10b - 12b - 15$
$= 8b^2 - 2b - 15$

19. $5x^2 - 4x[x - 3(3x + 2) + x]$
$= 5x^2 - 4x(x - 9x - 6 + x)$
$= 5x^2 - 4x(-7x - 6)$
$= 5x^2 + 28x^2 + 24x$
$= 33x^2 + 24x$

20. $(xy^5z^3)(x^3y^3z) = x^4y^8z^4$

21. $(4x - 3y)^2 = 16x^2 - 24xy + 9y^2$

22.

$$\begin{array}{r|rrrrr} 4 & 1 & 0 & 0 & 0 & -4 \\ & & 4 & 16 & 64 & 256 \\ \hline & 1 & 4 & 16 & 64 & 252 \end{array}$$

$$\frac{x^4-4}{x-4} = x^3+4x^2+16x+64+\frac{252}{x-4}$$

23. $(3x^2 - 2x - 6) + (-x^2 - 3x + 4)$
$= 2x^2 - 5x - 2$

24. $(5x^2yz^4)(2xy^3z^{-1})\,(7x^{-2}y^{-2}z^3)$
$= (10x^3y^4z^3)(\,7x^{-2}y^{-2}z^3)$
$= 70xy^2z^6$

25. $\dfrac{3x^4yz^{-1}}{-12xy^3z^2} = -\dfrac{x^3}{4y^2z^3}$

26. 9.48×10^8

27. $\dfrac{3\times10^{-3}}{15\times10^2} = 0.2\times10^{-5} = 2\times10^{-6}$

28.

$$\begin{array}{r|rrrr} -3 & -2 & 2 & 0 & -4 \\ & & 6 & -24 & 72 \\ \hline & -2 & 8 & -24 & 68 \end{array}$$

$P(-3) = 68$

29. $\dfrac{16x^5-8x^3+20x}{4x} = \dfrac{16x^5}{4x} - \dfrac{8x^3}{4x} + \dfrac{20x}{4x}$
$= 4x^4 - 2x^2 + 5$

30.

$$\begin{array}{r} 2x-3 \\ 6x+1 \overline{) 12x^2-16x-7} \\ \underline{12x^2+2x} \quad \\ -18x-7 \\ \underline{-18x-3} \\ -4 \end{array}$$

$$\frac{12x^2-16x-7}{6x+1}=2x-3-\frac{4}{6x+1}$$

31. $a^{2n+3}(a^n-5a+2)=a^{3n+3}-5a^{2n+4}+2a^{2n+3}$

32. $(x+6)(x^3-3x^2-5x+1)$
$=x(x^3-3x^2-5x+1)+6(x^3-3x^2-5x+1)$
$=x^4-3x^3-5x^2+x+6x^3-18x^2-30x+6$
$=x^4+3x^3-23x^2-29x+6$

33. $-2x(4x^2+7x-9)=-8x^3-14x^2+18x$

34.

$$\begin{array}{r} 3y^2+4y-7 \\ \times \qquad\qquad \underline{2y+3} \\ 9y^2+12y-21 \\ \underline{6y^3+8y^2-14y \qquad\quad} \\ 6y^3+17y^2-2y-21 \end{array}$$

35. $(-2u^3v^4)^4=16u^{12}v^{16}$

36.

$$\begin{array}{r} 2x^3+7x^2+x \qquad \\ + \qquad \underline{2x^2-4x-12} \\ 2x^3+9x^2-3x-12 \end{array}$$

37. $(5x^2-2x-1)-(3x^2-5x+7)$
$=(5x^2-2x-1)+(-3x^2+5x-7)$
$=2x^2+3x-8$

38. $(a+7)(a-7)=a^2-7a+7a-49=a^2-49$

39. $(5a^7b^6)^2(4ab)=(25a^{14}b^{12})(4ab)=100a^{15}b^{13}$

40. $1.46\times10^7=14{,}600{,}000$

41. $(-2x^3)^2(-3x^4)^3=(4x^6)(-27x^{12})=-108x^{18}$

42.

$$\begin{array}{r} 2y-9 \\ 3y-4 \overline{) 6y^2-35y+36} \\ \underline{6y^2-8y} \qquad \\ -27y+36 \\ \underline{-27y+36} \\ 0 \end{array}$$

$$\frac{6y^2-35y+36}{3y-4}=2y-9$$

43. $-4^{-2}=-\frac{1}{4^2}=-\frac{1}{16}$

44. $(5a-7)(2a+9)=10a^2+45a-14a-63$
$=10a^2+31a-63$

45.

$$\begin{array}{r|rrr} -3 & -1 & -1 & 7 \\ & & 3 & -6 \\ \hline & -1 & 2 & 1 \end{array}$$

$$\frac{7-x-x^2}{x+3}=-x+2+\frac{1}{x+3}$$

46. $0.000000127=1.27\times10^{-7}$

47. $\frac{16y^2-32y}{-4y}=\frac{16y^2}{-4y}+\frac{32y}{4y}=-4y+8$

48. $\frac{(2a^4b^{-3}c^2)^3}{(2a^3b^2c^{-1})^4}=\frac{8a^{12}b^{-9}c^6}{16a^{12}b^8c^{-4}}=\frac{c^{10}}{2b^{17}}$

49. $(x-4)(3x+2)(2x-3)$
$=(x-4)(6x^2-5x-6)$
$=x(6x^2-5x-6)-4(6x^2-5x-6)$
$=6x^3-5x^2-6x-24x^2+20x+24$
$=6x^3-29x^2+14x+24$

50. $(-3x^{-2}y^{-3})^{-2}=\frac{x^4y^6}{9}$

51. $(2a^{12}b^3)(-9b^2c^6)(3ac)=-54a^{13}b^5c^7$

52. $(5a + 2b)(5a - 2b) = 25a^2 - 4b^2$

53. 0.00254

54. $2ab^3(4a^2 - 2ab - 3b^2) = 8a^3b^3 - 4a^2b^4 + 6ab^5$

55. $y = x^2 + 1$

x	x
−2	5
−1	2
0	1
1	2
2	5

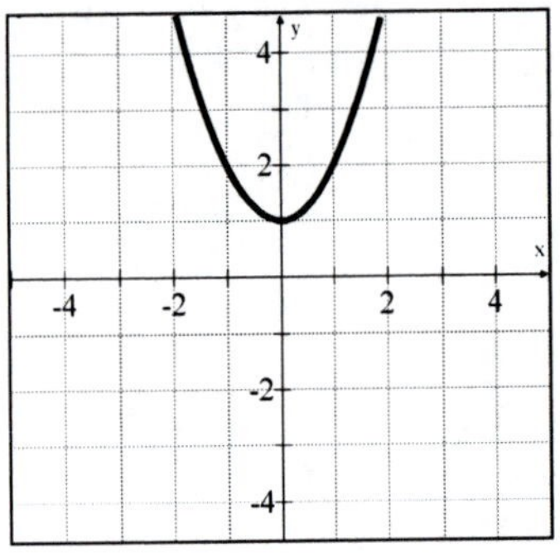

56. a) 3
b) 8
c) 5

57. Strategy: To find the mass of the moon multiply the mass of the sun by 3.7×10^{-8}.

Solution:
$(2.19\times10^{27})(3.7\times10^{-8}) = 8.103\times10^{19}$
The mass of the moon is 8.103×10^{19} tons

58. Strategy: Let $3x - 2$ represent the side of a square. Find the area of the square.

Solution: $S^2 = A$
$(3x - 2)^2 = A$
$9x^2 - 6x - 6x + 4 = A$
$9x^2 - 12x + 4 = A$
The area is $(9x^2 - 12x + 4)$ in^2.

59. Strategy: To find how far Earth is from the Great Galaxy of Andromeda, use the equation $d = rt$ where $r = 6.7\times10^8$ mph and $t = 2.2\times10^6$ years.
$2.2\times10^6 \times 24 \times 365 = 1.9272\times10^{10}$ hours.

Solution: $d=rt$
$d = (6.7\times10^8)(1.9272\times10^{10})$
$= 1.291224\times10^{19}$
The distance from the Earth to the Great Galaxy of Andromeda is 1.291224×10^{19} miles.

60. Strategy: To find the area substitute the given values for L and W in the equation $A = LW$ and solve for A.

Solution: $A = LW$
$A = (5x + 3)(2x - 7) = 10x^2 - 29x - 21$
The area is $(10x^2 - 29x - 21)$ cm^2.

Chapter 6 Test

1. $2x(2x^2 - 3x) = 4x^3 - 6x^2$

2.

−2	−1	0	4	−8
		2	−4	0
	−1	2	0	−8

$P(-2) = -8$

3. $\dfrac{12x^2}{-3x^8} = -\dfrac{4}{x^6}$

4. $(-2xy^2)(3x^2y^4) = -6x^3y^6$

5.
$$\begin{array}{r} x-1 \\ x+1\overline{)\,x^2+0+1} \\ \underline{x^2+x} \\ -x+1 \\ \underline{-x-1} \\ 2 \end{array}$$

$\dfrac{x^2+1}{x+1} = x - 1 + \dfrac{2}{x+1}$

6. $(x - 3)(x^2 - 4x + 5)$
$= x^3 - 4x^2 + 5x - 3x^2 + 12x - 15$
$= x^3 - 7x^2 + 17x - 15$

7. $(-2a^2b)^3 = -8a^6b^3$

8. $\dfrac{(3x^{-2}y^3)^3}{3x^4y^{-1}} = \dfrac{27x^{-6}y^9}{3x^4y^{-1}} = \dfrac{9y^{10}}{x^{10}}$

9. $(a-2b)(a+5b) = a^2 + 5ab - 2ab - 10b^2$
$= a^2 + 3ab - 10b^2$

10. $P(x) = 3x^2 - 8x + 1$
$P(2) = 3(2)^2 - 8(2) + 1$
$P(2) = 12 - 16 + 1$
$P(2) = -3$

11.
$$\begin{array}{r} x+7 \\ x-1\overline{)x^2+6x-7} \\ \underline{x^2-1x} \\ 7x-7 \\ \underline{7x-7} \end{array}$$

$\dfrac{x^2+6x-7}{x-1} = x+7$

12. $-3y^2(-2y^2 + 3y - 6) = 6y^4 - 9y^3 + 18y^2$

13. $(-2x^3 + x^2 - 7)(2x - 3)$
$= -4x^4 + 6x^3 + 2x^3 - 3x^2 - 14x + 21$
$= -4x^4 + 8x^3 - 3x^2 - 14x + 21$

14. $(4y-3)(4y+3) = 16y^2 + 12y - 12y - 9$
$= 16y^2 - 9$

15. $\dfrac{18x^5 + 9x^4 - 6x^3}{3x^2} = 6x^3 + 3x^2 - 2x$

16. $\dfrac{2a^{-1}b}{2^{-2}a^{-2}b^{-3}} = 2^3ab^4$

17. $\dfrac{(2a^{-4}b^2)^3}{4a^{-2}b^{-1}} = \dfrac{8a^{-12}b^6}{4a^{-2}b^{-1}} = \dfrac{2b^7}{a^{10}}$

18. $(3a^2 - 2a - 7) - (5a^3 + 2a - 10)$
$= -5a^3 + 3a^2 - 4a + 3$

19. $(2x-5)^2 = (2x-5)(2x-5)$
$= 4x^2 - 10x - 10x + 25$
$= 4x^2 - 20x + 25$

20.
$$\begin{array}{r} x^2-5x+10 \\ x+3\overline{)x^3-2x^2-5x+7} \\ \underline{x^3+3x^2} \\ -5x^2-5x \\ \underline{-5x^2-15x} \\ 10x+7 \\ \underline{10x+30} \\ -23 \end{array}$$

$\dfrac{x^3-2x^2-5x+7}{x+3} = x^2-5x+10-\dfrac{23}{x+3}$

21. $(2x-7y)(5x-4y)$
$= 10x^2 - 8xy - 35xy + 28y^2$
$= 10x^2 - 43xy + 28y^2$

22. $(3x^3 - 2x^2 - 4) + (8x^2 - 8x + 7)$
$= 3x^3 + 6x^2 - 8x + 3$

23. $0.00000000302 = 3.02 \times 10^{-9}$

24. $10\ weeks \cdot \dfrac{7\ day}{1\ week} \cdot \dfrac{24\ h}{1\ day} \cdot \dfrac{60\text{ min}}{1\ h} \cdot \dfrac{60\ s}{1\text{ min}}$
$= 6.048 \times 10^6$ s
There are 6.048×10^6 s in 10 weeks.

25. $r = (x-5)$
$A = \pi r^2 = \pi(x-5)^2 = \pi(x-5)(x-5)$
$= \pi(x^2 - 5x - 5x + 25)$
$= \pi(x^2 - 10x + 25)$
The area of the circle is $\pi(x^2 - 10x + 25)$ m^2.

Cumulative Review Exercises

1. $8 - 2[-3-(-1)]^2 \div 4$
$= 8 - 2(-3+1)^2 \div 4$
$= 8 - 2(-2)^2 \div 4$
$= 8 - 2(4) \div 4$
$= 8 - 2$
$= 6$

2. $\dfrac{2a-b}{b-c}$

$\dfrac{2(4)-(-2)}{(-2)-6}=\dfrac{8+2}{-8}=\dfrac{10}{-8}=-\dfrac{5}{4}$

3. Inverse Property of Addition

4. $2x-4[x-2(3-2x)+4]$
$=2x-4(x-6+4x+4)$
$=2x-4(5x-2)$
$=2x-20x+8$
$=-18x+8$

5. $\dfrac{2}{3}-y=\dfrac{5}{6}$

$\dfrac{2}{3}-y-\dfrac{2}{3}=\dfrac{5}{6}-\dfrac{2}{3}$

$-y=\dfrac{1}{6}$

$(-1)(-y)=-\dfrac{1}{6}(-1)$

$y=-\dfrac{1}{6}$

The solution is $-\dfrac{1}{6}$.

6. $8x-3-x=-6+3x-8$
$7x-3=3x-14$
$4x-3=-14$
$4x=-11$
$x=-\dfrac{11}{4}$

The solution is $-\dfrac{11}{4}$.

7.

3	1	0	0	−3
		3	9	27
	1	3	9	24

$\dfrac{x^3-3}{x-3}=x^2+3x+9+\dfrac{24}{x-3}$

8. $3-|2-3x|=-2$
$-|2-3x|=-5$
$|2-3x|=5$
$2-3x=5 \quad 2-3x=-5$
$-3x=3 \quad -3x=-7$
$x=-1 \quad x=\dfrac{7}{3}$

The solutions are -1 and $\dfrac{7}{3}$.

9. $P(x)=3x^2-2x+2$
$P(-2)=3(-2)^2-2(-2)+2$
$P(-2)=3(4)+4+2$
$P(-2)=18$

10. Domain $\{x|\, x\neq -2\}$

11. $F(x)=3x-4$
$3x-4=0$
$x=\dfrac{4}{3}$

The zero of the function is $\dfrac{4}{3}$.

12. $m=\dfrac{y_2-y_1}{x_2-x_1}=\dfrac{2-3}{4-(-2)}=-\dfrac{1}{6}$

13. Use the point-slope formula
$y-y_1=m(x-x_1)$

$y-2=-\dfrac{3}{2}[x-(-1)]$

$y-2=-\dfrac{3}{2}x-\dfrac{3}{2}$

$y=-\dfrac{3}{2}x+\dfrac{1}{2}$

14. Solve the equation $3x+2y=4$ for y to find the slope of this line.
$3x+2y=4$
$2y=-3x+4$
$y=-\dfrac{3}{2}x+2$
$m=-\dfrac{3}{2}$

The perpendicular line will have a slope that is the negative reciprocal of $-\frac{3}{2}$.

$m = \frac{2}{3}$ and $(-2, 4)$

$y - y_1 = m(x - x_1)$

$y - 4 = \frac{2}{3}[x - (-2)]$

$y - 4 = \frac{2}{3}x + \frac{4}{3}$

$y = \frac{2}{3}x + \frac{16}{3}$

The equation of the perpendicular line is $y = \frac{2}{3}x + \frac{16}{3}$.

15. $2x - 3y = 4$

$x + y = -3$

$x = -y - 3$

$2(-y - 3) - 3y = 4$

$-2y - 6 - 3y = 4$

$-5y = 10$

$y = -2$

$x + (-2) = -3$

$x - 2 = -3$

$x = -1$

The solution is $(-1, -2)$.

16. (1) $x - y + z = 0$

(2) $2x + y - 3z = -7$

(3) $-x + 2y + 2z = 5$

Add equations (1) and (3) to eliminate x.

$x - y + z = 0$

$-x + 2y + 2z = 5$

(4) $y + 3z = 5$

Add -2 times equation (1) and equation (2) to eliminate x.

$-2x + 2y - 2z = 0$

$2x + y - 3z = -7$

(5) $3y - 5z = -7$

Add -3 times equation (4) to equation (5) to eliminate y.

$-3y - 9z = -15$

$3y - 5z = -7$

$-14z = -22$

$z = \frac{11}{7}$

Substitute $\frac{11}{7}$ for z in equation (4).

$y + 3(\frac{11}{7}) = 5$

$y = \frac{2}{7}$

Substitute in values for y and z.

$x - y + z = 0$

$x - \frac{2}{7} + \frac{11}{7} = 0$

$x = -\frac{9}{7}$

The solution is $\left(-\frac{9}{7}, \frac{2}{7}, \frac{11}{7}\right)$.

17. $3x - 4y = 12$

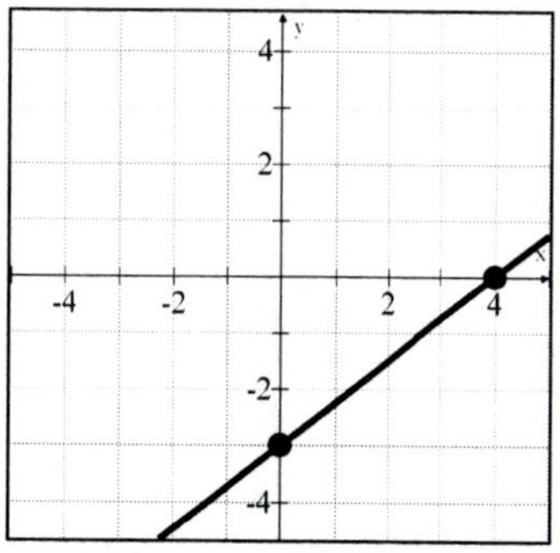

18. $-3x + 2y < 6$

$2y < 3x + 6$

$y < \frac{3}{2}x + 3$

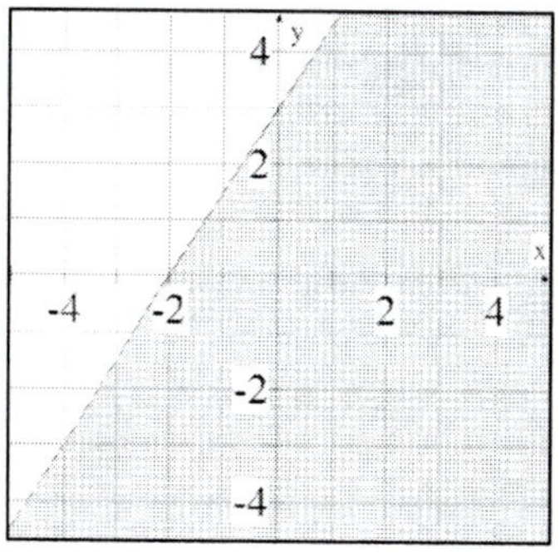

19. $x - 2y = 3$
$-2y = -x + 3$
$y = \frac{1}{2}x - \frac{3}{2}$
$-2x + y = -3$
$y = 2x - 3$

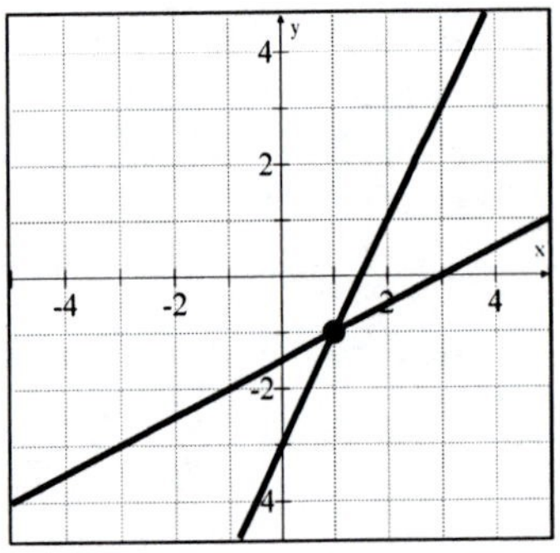

The solution is $(1, -1)$.

20. Solve each inequality for y.
$2x + y < 2$
$y < -2x + 2$

$-6x + 3y \geq 6$
$3y \geq 6x + 6$
$y \geq 2x + 2$

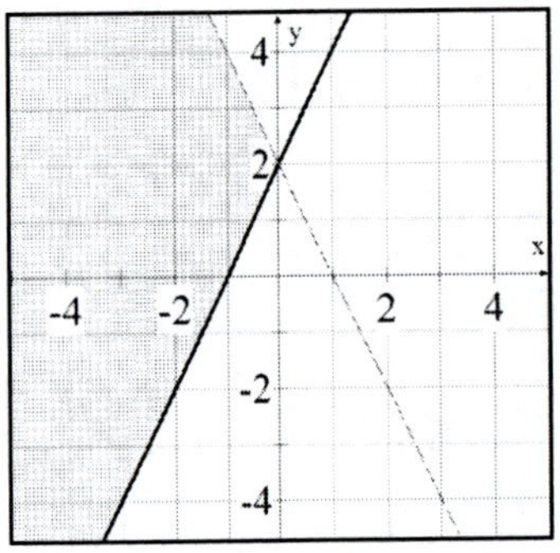

21. $(4a^{-2}b^3)(2a^3b^{-1})^{-2} = 4a^{-2}b^3(2^{-2}a^{-6}b^2)$
$= 4(2^{-2})a^{-8}b^5$
$= \frac{b^5}{a^8}$

22. $\frac{(5x^3y^{-3}z)^{-2}}{y^4z^{-2}} = \frac{5^{-2}x^{-6}y^6z^{-2}}{y^4z^{-2}} = \frac{y^2}{25x^6}$

23. $3-(3-3^{-1})^{-1} = 3-\left(3-\frac{1}{3}\right)^{-1}$
$= 3-\left(\frac{8}{3}\right)^{-1} = 3-\frac{3}{8}$
$= \frac{21}{8}$

24. $(2x+3)(2x^2-3x+1)$
$= 2x(2x^2-3x+1)+3(2x^2-3x+1)$
$= 4x^3-6x^2+2x+6x^2-9x+3$
$= 4x^3-7x+3$

25. Strategy Let x represent the smaller integer.
The larger integer is $24 - x$.
The difference between four times the smaller and nine is 3 less than twice the larger.
$4x - 9 = 2(24 - x) - 3$

Solution: $4x - 9 = 2(24 - x) - 3$
$4x - 9 = 48 - 2x - 3$
$4x - 9 = 45 - 2x$
$6x - 9 = 45$
$6x = 54$
$x = 9$
$24 - x = 24 - 9 = 15$
The integers are 9 and 15.

26. Strategy: Let x represent the amount invested at 4%.

	Principal	Rate	Interest
Amount at 4%	x	0.04	$0.04x$
Amount at 4.5%	$12{,}000 - x$	0.045	$0.045(12{,}000 - x)$

The total amount of interest earned is $530.

Solution: $0.045(12{,}000 - x) + 0.04x = 530$
$540 - 0.045x + 0.04x = 530$
$-0.005x = -10$
$x = 2000$
The amount invested at 4% is $2000.

27. Strategy: Let x represent the speed of the slower cyclist.
The speed of the faster cyclist is $1.5x$.

	Rate	Time	Distance
Faster cyclist	$1.5x$	2	$2(1.5x)$
Slower cyclist	x	2	$2x$

The sum of the distances is 25 mi.

Solution: $2x + 2(1.5x) = 25$

$$2x + 3x = 25$$
$$5x = 25$$
$$x = 5$$

$1.5x = 1.5(5) = 7.5$

The faster cyclist travels at 7.5 mph and the slower cyclist travels at 5 mph.

28. Strategy: Let x represent the number of ounces of pure gold.

	Amount	Cost	Value
Pure gold	x	360	$360x$
Alloy	80	120	$80(120)$
Mixture	$x + 80$	200	$200(x + 80)$

The sum of values before mixing is equal to the value after mixing.

Solution: $360x + 80(120) = 200(x + 80)$

$$360x + 9600 = 200x + 16{,}000$$
$$160x + 9600 = 16{,}000$$
$$160x = 6400$$

$x = 40$

40oz of pure gold must be mixed with the alloy.

29. $m = \dfrac{y_2 - y_1}{x_2 - x_1} = \dfrac{300 - 100}{6 - 2} = \dfrac{200}{4} = 50$

The average speed is 50 mph.

30. Strategy: To find the time, use the equation $d = rt$, where r is the speed of the space vehicle and d is the distance from Earth to the moon.

Solution: $d = rt$

$$2.4 \times 10^5 = (2 \times 10^4)t$$
$$\frac{2.4 \times 10^5}{2 \times 10^4} = t$$
$$1.2 \times 10^1 = t$$

The vehicle will reach the moon in 12 h.

31. Strategy: Let x represent the amount invested at 4%.

	Principal	Rate	Interest
Amount at 4%	x	0.04	$0.04x$
Amount at 4.5%	$12{,}000 - x$	0.045	$0.045(12{,}000 - x)$

The total amount of interest earned is \$530.

Solution:
$$\begin{aligned} 0.045(12{,}000 - x) + 0.04x &= 530 \\ 540 - 0.045x + 0.04x &= 530 \\ -0.005x &= -10 \\ x &= 2000 \end{aligned}$$
The amount invested at 4% is \$2000.

32. $m = \dfrac{y_2 - y_1}{x_2 - x_1} = \dfrac{300 - 100}{6 - 2} = \dfrac{200}{4} = 50$

The average speed is 50 mph.

33. Strategy: Let x represent the length. The width is $0.40x$.
Use the formula for perimeter of a rectangle.

Solution:
$$\begin{aligned} 2L + 2W &= P \\ 2x + 2(0.40x) &= 42 \\ 2x + 0.80x &= 42 \\ 2.8x &= 42 \\ x &= 15 \end{aligned}$$
$0.40(x) = 0.40(15) = 6$

The length is 15 m and the width is 6 m.

34. $A = s^2$

$A = (2x + 3)^2$

$A = (2x + 3)(2x + 3)$

$A = 4x^2 + 6x + 6x + 9$

$A = 4x^2 + 12x + 9$

The area is $(4x^2 + 12x + 9)$ m^2.

Chapter 7 Factoring

Prep Test

1. $30 = 2 \cdot 3 \cdot 5$

2. $-3(4y-5) = -12y+15$

3. $-(a-b) = -a+b$

4. $2(a-b)-5(a-b) = 2a-2b-5a+5b$
$= -3a+3b$

5. $4x = 0$
$\frac{4x}{4} = \frac{0}{4}$
$x = 0$
The solution is 0.

6. $2x+1=0$
$2x+1-1=0-1$
$2x=-1$
$\frac{2x}{2}=\frac{-1}{2}$
$x=-\frac{1}{2}$
The solution is $-\frac{1}{2}$.

7. $(x+4)(x-6) = x^2-6x+4x-24$
$= x^2-2x-24$

8. $(2x-5)(3x+2) = 6x^2+4x-15x-10$
$= 6x^2-11x-10$

9. $\frac{x^5}{x^2} = x^{5-2} = x^3$

10. $\frac{6x^4y^3}{2xy^2} = 3x^{4-1}y^{3-2} = 3x^3y$

Section 7.1

Concept Check4

1. 4

3. **a.** x

   b. $2x - 1$

5. $(2x^3-x^2)+(6x-3)$

Objective A Exercises

7. $5a+5 = 5(a)+5(1) = 5(a+1)$

9. $16-8a^2 = 8(2)+8(-a^2) = 8(2-a^2)$

11. $8x+12 = 4(2x)+4(3) = 4(2x+3)$

13. $7x^2-3x = x(7x)+x(-3) = x(7x-3)$

15. $3a^2+5a^5 = a^2(3)+a^2(5a^3) = a^2(3+5a^3)$

17. $2x^4-4x = 2x(x^3)+2x(-2) = 2x(x^3-2)$

19. $10x^4-12x^2 = 2x^2(5x^2)+2x^2(-6) = 2x^2(5x^2-6)$

21. $8a^8-4a^5 = 4a^5(2a^3)+4a^5(-1) = 4a^5(2a^3-1)$

23. $x^2y^2-xy = xy(xy)+xy(-1) = xy(xy-1)$

25. $3x^2y^4-6xy = 3xy(xy^3)+3xy(-2) = 3xy(xy^3-2)$

27. $3x^3+6x^2+9x = 3x(x^2)+3x(2x)+3x(3)$
$= 3x(x^2+2x+3)$

29. $2x^4-4x^3+6x^2 = 2x^2(x^2)+2x^2(-2x)+2x^2(3)$
$= 2x^2(x^2-2x+3)$

31. $2x^3+6x^2-14x = 2x(x^2)+2x(3x)+2x(-7)$
$= 2x(x^2+3x-7)$

33. $2y^5-3y^4+7y^3 = y^3(2y^2)+y^3(-3y)+y^3(7)$
$= y^3(2y^2-3y+7)$

35. $x^3y-3x^2y^2+7xy^3 = xy(x^2)+xy(-3xy)+xy(7y^2)$
$= xy(x^2-3xy+7y^2)$

37. $5y^3+10y^2-25y = 5y(y^2)+5y(2y)+5y(-5)$
$= 5y(y^2+2y-5)$

39. $3a^2b^2-9ab^2+15b^2$
$= 3b^2(a^2)+3b^2(-3a)+3b^2(5)$
$= 3b^2(a^2-3a+5)$

41. The GCF of $x^a + x^b + x^c$ is x^c, where c is the smallest exponent.

Objective B Exercises

43. $x(b+4)+3(b+4)=(b+4)(x+3)$

45. $a(y-x)-b(y-x)=(y-x)(a-b)$

47. $x(x-2)+y(2-x)=x(x-2)-y(x-2)=(x-2)(x-y)$

49. $$\begin{aligned}&8c(2m-3n)+(3n-2m)\\&=8c(2m-3n)-1(2m-3n)\\&=(2m-3n)(8c-1)\end{aligned}$$

51. $$\begin{aligned}x^2+2x+2xy+4y&=(x^2+2x)+(2xy+4y)\\&=x(x+2)+2y(x+2)\\&=(x+2)(x+2y)\end{aligned}$$

53. $$\begin{aligned}p^2-2p-3rp+6r&=(p^2-2p)+(-3rp+6r)\\&=p(p-2)-3r(p-2)\\&=(p-2)(p-3r)\end{aligned}$$

55. $$\begin{aligned}ab+6b-4a-24&=(ab+6b)+(-4a-24)\\&=b(a+6)-4(a+6)\\&=(a+6)(b-4)\end{aligned}$$

57. $$\begin{aligned}2z^2-z+2yz-y&=(2z^2-z)+(2yz-y)\\&=z(2z-1)+y(2z-1)\\&=(2z-1)(z+y)\end{aligned}$$

59. $$\begin{aligned}2x^2-5x-6xy+15y&=(2x^2-5x)+(-6xy+15y)\\&=x(2x-5)-3y(2x-5)\\&=(2x-5)(x-3y)\end{aligned}$$

61. $$\begin{aligned}3y^2-6y-ay+2a&=(3y^2-6y)+(-ay+2a)\\&=3y(y-2)-a(y-2)\\&=(y-2)(3y-a)\end{aligned}$$

63. $$\begin{aligned}3xy-y^2-y+3x&=(3xy-y^2)+(-y+3x)\\&=y(3x-y)+1(-y+3x)\\&=y(3x-y)+1(3x-y)\\&=(3x-y)(y+1)\end{aligned}$$

65. $$\begin{aligned}3st+t^2-2t-6s&=(3st+t^2)+(-2t-6s)\\&=t(3s+t)-2(t+3s)\\&=t(3s+t)-2(3s+t)\\&=(3s+t)(t-2)\end{aligned}$$

Critical Thinking

67. -1

69. $b-3a$

Section 7.2

Concept Check

1. -8

3. -2 and 6

5. Different

Objective A Exercises

7.

Factors	*Sum*
+1, +2	+3

$x^2+3x+2=(x+1)(x+2)$

9.

Factors	*Sum*
−1, +2	+1
+1, −2	−1

$x^2-x-2=(x+1)(x-2)$

11.

Factors	*Sum*
−1, +12	+11
+1, −12	−11
−2, +6	+4
+2, −6	−4
−3, +4	+1
+3, −4	−1

$a^2+a-12=(a+4)(a-3)$

13.

Factors	*Sum*
−1, −2	−3

$a^2-3a+2=(a-1)(a-2)$

15.

Factors	*Sum*
−1, +2	+1
+1, −2	−1

$a^2+a-2=(a+2)(a-1)$

17.

Factors	*Sum*
−1, −9	−10
−3, −3	−6

$b^2 - 6b + 9 = (b-3)(b-3)$

19.

Factors	*Sum*
−1, +8	+7
+1, −8	−7
−2, +4	+2
+2, −4	−2

$b^2 + 7b - 8 = (b+8)(b-1)$

21.

Factors	*Sum*
−1, +55	+54
+1, −55	−54
−5, +11	+6
+5, −11	−6

$y^2 + 6y - 55 = (y+11)(y-5)$

23.

Factors	*Sum*
−1, −6	−7
−2, −3	−5

$y^2 - 5y + 6 = (y-2)(y-3)$

25.

Factors	*Sum*
−1, −45	−46
−3, −15	−18
−5, −9	−14

$z^2 - 14z + 45 = (z-5)(z-9)$

27.

Factors	*Sum*
−1, +160	+159
+1, −160	−159
−2, +80	+78
+2, −80	−78
−4, +40	+36
+4, −40	−36
−5, +32	+27
+5, −32	−27
−8, +20	+12
+8, −20	−12
−10, +16	+6
+10, −16	−6

$z^2 - 12z - 160 = (z+8)(z-20)$

29.

Factors	*Sum*
+1, +27	+28
+3, +9	+12

$p^2 + 12p + 27 = (p+3)(p+9)$

31.

Factors	*Sum*
+1, +100	+101
+2, +50	+52
+4, +25	+29
+5, +20	+25
+10, +10	+20

$x^2 + 20x + 100 = (x+10)(x+10)$

33.

Factors	*Sum*
−1, +20	+19
+1, −20	−19
−2, +10	+8
+2, −10	−8
−4, +5	+1
+4, −5	−1

$b^2 - b - 20 = (b+4)(b-5)$

35.

Factors	*Sum*
−1, +51	+50
+1, −51	−50
−3, +17	+14
+3, −17	−14

$y^2 - 14y - 51 = (y+3)(y-17)$

37.

Factors	*Sum*
−1, +21	+20
+1, −21	−20
−3, +7	+4
+3, −7	−4

$p^2 - 4p - 21 = (p+3)(p-7)$

39.

Factors	*Sum*
−1, −32	−33
−2, −16	−18
−4, −8	−12

$y^2 - 8y + 32$ is nonfactorable over the integers.

41.

Factors	*Sum*
−1, −75	−76
−3, −25	−28
−5, −15	−20

$x^2 - 20x + 75 = (x-5)(x-15)$

43.

Factors	*Sum*
+1, +63	+64
+3, +21	+24

$p^2 + 24p + 63 = (p+3)(p+21)$

45.

Factors	*Sum*
+1, +38	+39
+2, +19	+21

$x^2 + 21x + 38 = (x+2)(x+19)$

47.

Factors	*Sum*
+1, −3	−2
−1, +3	+2

x^2+5x-3 is nonfactorable over the integers.

49.

Factors	*Sum*
−1, +44	+43
+1, −44	−43
−2, +22	+20
+2, −22	−20
−4, +11	+7
+4, −11	−7

$a^2-7a-44=(a+4)(a-11)$

51.

Factors	*Sum*
−1, −54	−55
−2, −27	−29
−3, −18	−21
−6, −9	−15

$a^2-21a+54=(a-3)(a-18)$

53.

Factors	*Sum*
−1, +147	+146
+1, −147	−146
−3, +49	+46
+3, −49	−46
−7, +21	+14
+7, −21	−14

$z^2+14z-147=(z+21)(z-7)$

55.

Factors	*Sum*
−1, +180	+179
+1, −180	−179
−2, +90	+88
+2, −90	−88
−3, +60	+57
+3, −60	−57
−4, +45	+41
+4, −45	−41
−5, +36	+31
+5, −36	−31
−6, +30	+24
+6, −30	−24
−9, +20	+11
+9, −20	−11
−10, +18	+8
+10, −18	−8
−12, +15	+3
+12, −15	−3

$c^2-3c-180=(c+12)(c-15)$

57.

Factors	*Sum*
+1, +135	+136
+3, +45	+48
+5, +27	+32
+9, +15	+24

$p^2+24p+135=(p+9)(p+15)$

59.

Factors	*Sum*
+1, +18	+19
+2, +9	+11
+3, +6	+9

$c^2+11c+18=(c+2)(c+9)$

61.

Factors	*Sum*
−1, +75	+74
+1, −75	−74
−3, +25	+22
+3, −25	−22
−5, +15	+10
+5, −15	−10

$x^2+10x-75=(x+15)(x-5)$

63.

Factors	*Sum*
−1, +100	+99
+1, −100	−99
−2, +50	+48
+2, −50	−48
−4, +25	+21
+4, −25	−21
−5, +20	+15
+5, −20	−15
−10, +10	0

$x^2+21x-100=(x+25)(x-4)$

65.

Factors	*Sum*
−1, −72	−73
−2, −36	−38
−3, −24	−27
−4, −18	−22
−6, −12	−18
−8, −9	−17

$b^2-22b+72=(b-4)(b-18)$

67.

Factors	*Sum*
$-1, +135$	$+134$
$+1, -135$	-134
$-3, +45$	$+42$
$+3, -45$	-42
$-5, +27$	$+22$
$+5, -27$	-22
$-9, +15$	$+6$
$+9, -15$	-6

$a^2 + 42a - 135 = (a + 45)(a - 3)$

69.

Factors	*Sum*
$-1, -126$	-127
$-2, -63$	-65
$-3, -42$	-45
$-6, -21$	-27
$-7, -18$	-25

$b^2 - 25b + 126 = (b-7)(b-18)$

71.

Factors	*Sum*
$+1, +144$	$+145$
$+2, +72$	$+74$
$+3, +48$	$+51$
$+4, +36$	$+40$
$+6, +24$	$+30$
$+8, +18$	$+26$
$+9, +16$	$+25$
$+12, +12$	$+24$

$z^2 + 24z + 144 = (z + 12)(z + 12)$

73.

Factors	*Sum*
$-1, -100$	-101
$-2, -50$	-52
$-4, -25$	-29
$-5, -20$	-25
$-10, -10$	-20

$x^2 - 29x + 100 = (x - 4)(x - 25)$

75.

Factors	*Sum*
$-1, +112$	$+111$
$+1, -112$	-111
$-2, +56$	$+54$
$+2, -56$	-54
$-4, +28$	$+24$
$+4, -28$	-24
$-7, +16$	$+9$
$+7, -16$	-9
$-8, +14$	$+6$
$+8, -14$	-6

$x^2 + 9x - 112 = (x + 16)(x - 7)$

77. Positive. The sum of two positive numbers is positive.

Objective B Exercises

79. The GCF is 3.

$3x^2 + 15x + 18 = 3(x^2 + 5x + 6)$

Factor the trinomial

$x^2 + 5x + 6$.

Factors	*Sum*
$+1, +6$	$+7$
$+2, +3$	$+5$

$3x^2 + 15x + 18 = 3(x + 2)(x + 3)$

81. The GCF is -1.

$-x^2 - 4x + 12 = -(x^2 + 4x - 12)$

Factor the trinomial $x^2 + 4x - 12$.

Factors	*Sum*
$-1, +12$	$+11$
$+1, -12$	-11
$-2, +6$	$+4$
$+2, -6$	-4

$12 - 4x - x^2 = -(x-2)(x+6)$

83. The GCF is a.

$ab^2 + 7ab - 8a = a(b^2 + 7b - 8)$

Factor the trinomial $b^2 + 7b - 8$.

Factors	*Sum*
$-1, +8$	$+7$
$+1, -8$	-7
$-2, +4$	$+2$
$+2, -4$	-2

$ab^2 + 7ab - 8a = a(b + 8)(b - 1)$

85. The GCF is x.

$xy^2+8xy+15x=x(y^2+8y+15)$

Factor the trinomial $y^2+8y+15$.

Factors	*Sum*
+1, +15	+16
+3, +5	+8

$xy^2+8xy+15x=x(y+3)(y+5)$

87. The GCF is $-2a$.

$-2a^3-6a^2-4a=-2a(a^2+3a+2)$

Factor the trinomial a^2+3a+2.

Factors	*Sum*
+1, +2	+3

$-2a^3-6a^2-4a=-2a(a+1)(a+2)$

89. The GCF is $4y$.

$4y^3+12y^2-72y=4y(y^2+3y-18)$

Factor the trinomial $y^2+3y-18$.

Factors	*Sum*
−1, +18	+17
+1, −18	−17
−2, +9	+7
+2, −9	−7
−3, +6	+3
+3, −6	−3

$4y^3+12y^2-72y=4y(y+6)(y-3)$

91. The GCF is $2x$.

$2x^3-2x^2+4x=2x(x^2-x+2)$

Factor the trinomial x^2-x+2.

Factors	*Sum*
−1, −2	−3

x^2-x+2 is nonfactorable over the integers.

$2x^3-2x^2+4x=2x(x^2-x+2)$

93. The GCF is 6.

$6z^2+12z-90=6(z^2+2z-15)$

Factor the trinomial $z^2+2z-15$.

Factors	*Sum*
−1, +15	+14
+1, −15	−14
−3, +5	+2
+3, −5	−2

$6z^2+12z-90=6(z+5)(z-3)$

95. The GCF is $3a$.

$3a^3-9a^2-54a=3a(a^2-3a-18)$

Factor the trinomial $a^2-3a-18$.

Factors	*Sum*
−1, +18	+17
+1, −18	−17
−2, +9	+7
+2, −9	−7
−3, +6	+3
+3, −6	−3

$3a^3-9a^2-54a=3a(a+3)(a-6)$

97. There is no common factor.

Factor the trinomial $x^2+4xy-21y^2$.

Factors	*Sum*
−1, +21	+20
+1, −21	−20
−3, +7	+4
+3, −7	−4

$x^2+4xy-21y^2=(x+7y)(x-3y)$

99. There is no common factor.

Factor the trinomial $a^2-15ab+50b^2$.

Factors	*Sum*
−1, −50	−51
−2, −25	−27
−5, −10	−15

$a^2-15ab+50b^2=(a-5b)(a-10b)$

101. There is no common factor.

Factor the trinomial $s^2 + 2st - 48t^2$.

Factors	Sum
−1, +48	+47
+1, −48	−47
−2, +24	+22
+2, −24	−22
−3, +16	+13
+3, −16	−13
−4, +12	+8
+4, −12	−8
−6, +8	+2
+6, −8	−2

$s^2 + 2st - 48t^2 = (s + 8t)(s - 6t)$

103. There is no common factor.

Factor the trinomial $y^2 + 85yz + 36z^2$.

Factors	Sum
+1, +36	+37
+2, +18	+20
+3, +12	+15
+4, +9	+13
+6, +6	+12

$y^2 + 85yz + 36z^2$ is nonfactorable over the integers.

105. The GCF is z^2.

$z^4 + 2z^3 - 80z^2 = z^2(z^2 + 2z - 80)$

Factor the trinomial $z^2 + 2z - 80$.

Factors	Sum
−1, +80	+79
+1, −80	−79
−2, +40	+38
+2, −40	−38
−4, +20	+16
+4, −20	−16
−5, +16	+11
+5, −16	−11
−8, +10	+2
+8, −10	−2

$z^4 + 2z^3 - 80z^2 = z^2(z + 10)(z - 8)$

107. The GCF is b^2.

$b^4 - 3b^3 - 10b^2 = b^2(b^2 - 3b - 10)$

Factor the trinomial $b^2 - 3b - 10$.

Factors	Sum
−1, +10	+9
+1, −10	−9
−2, +5	+3
+2, −5	−3

$b^4 - 3b^3 - 10b^2 = b^2(b + 2)(b - 5)$

109. The GCF is $3y^2$.

$3y^4 + 54y^3 + 135y^2 = 3y^2(y^2 + 18y + 45)$

Factor the trinomial $y^2 + 18y + 45$.

Factors	Sum
+1, +45	+46
+3, +15	+18
+5, +9	+14

$3y^4 + 54y^3 + 135y^2 = 3y^2(y + 3)(y + 15)$

111. The GCF is $-x^2$.

$-x^4 + 11x^3 + 12x^2 = -x^2(x^2 - 11x - 12)$

Factor the trinomial $x^2 - 11x - 12$.

Factors	Sum
−1, +12	+11
+1, −12	−11
−2, +6	+4
+2, −6	−4
−3, +4	+1
+3, −4	−1

$-x^4 + 11x^3 + 12x^2 = -x^2(x + 1)(x - 12)$

113. The GCF is $3y$.

$3x^2y - 6xy - 45y = 3y(x^2 - 2x - 15)$

Factor the trinomial $x^2 - 2x - 15$.

Factors	Sum
−1, +15	+14
+1, −15	−14
−3, +5	+2
+3, −5	−2

$3x^2y - 6xy - 45y = 3y(x + 3)(x - 5)$

115. The GCF is $-3x$.

$-3x^3+36x^2-81x=-3x\left(x^2-12x+27\right)$

Factor the trinomial $x^2-12x+27$.

Factors	*Sum*
−1, −27	−28
−3, −9	−12

$-3x^3+36x^2-81x=-3x(x-3)(x-9)$

117. There is no common factor.

Factor the trinomial $x^2-8xy+15y^2$.

Factors	*Sum*
−1, −15	−16
−3, −5	−8

$x^2-8xy+15y^2=(x-3y)(x-5y)$

119. There is no common factor.

Factor the trinomial $a^2-13ab+42b^2$.

Factors	*Sum*
−1, −42	−43
−2, −21	−23
−3, −14	−17
−6, −7	−13

$a^2-13ab+42b^2=(a-6b)(a-7b)$

121. There is no common factor.

Factor the trinomial $y^2+8yz+7z^2$.

Factors	*Sum*
+1, +7	+8

$y^2+8yz+7z^2=(y+z)(y+7z)$

123. The GCF is $3y$.

$3x^2y+60xy-63y=3y\left(x^2+20x-21\right)$

Factor the trinomial $x^2+20x-21$.

Factors	*Sum*
−1, +21	+20
+1, −21	−20
−3, +7	+4
+3, −7	−4

$3x^2y+60xy-63y=3y(x+21)(x-1)$

125. The GCF is $3x$.

$3x^3+3x^2-36x=3x\left(x^2+x-12\right)$

Factor the trinomial x^2+x-12.

Factors	*Sum*
−1, +12	+11
+1, −12	−11
−2, +6	+4
+2, −6	−4
−3, +4	+1
+3, −4	−1

$3x^3+3x^2-36x=3x(x+4)(x-3)$

127. The GCF is 2.

$2t^2-24ts+70s^2=2\left(t^2-12ts+35s^2\right)$

Factor the trinomial $t^2-12ts+35s^2$.

Factors	*Sum*
−1, −35	−36
−5, −7	−12

$2t^2-24ts+70s^2=2(t-5s)(t-7s)$

129. The GCF is 3.

$3a^2-24ab-99b^2=3\left(a^2-8ab-33b^2\right)$

Factor the trinomial $a^2-8ab-33b^2$.

Factors	*Sum*
−1, +33	32
+1, −33	−32
+3, −11	−8
−3, +11	8

$3a^2-24ab-99b^2=3(a+3b)(a-11b)$

131. The GCF is $5x$.

$5x^3+30x^2y+40xy^2=5x\left(x^2+6xy+8y^2\right)$

Factor the trinomial $x^2+6xy+8y^2$.

Factors	*Sum*
+1, +8	+9
+2, +4	+6

$5x^3+30x^2y+40xy^2=5x(x+2y)(x+4y)$

133. **a.** The GCF is 2.

$2x^2 - 2xy - 4y^2 = 2(x^2 - xy - 2y^2)$

Factor the trinomial $x^2 - xy - 2y^2$.

Factors	*Sum*
+1, −2	−1
+2, −1	1

$2x^2 - 2xy - 4y^2 = 2(x + y)(x - 2y)$

Yes, $x + y$ is a factor.

b. The GCF is $2y$.

$2x^2y - 4xy - 4y = 2y(x^2 - 2x - 2)$

Factor the trinomial $x^2 - 2x - 2$.

Factors	*Sum*
+1, −2	−1
+2, −1	1

$x^2 - 2x - 2$ is nonfactorable over the integers.

$2x^2y - 4xy - 4y = 2y(x^2 - 2x - 2)$

No, $x + y$ is not a factor.

Critical Thinking

135. The GCF is $-2x$.

$-2x^3 - 6x^2 - 4x = -2x(x^2 + 3x + 2)$

Factor the trinomial $(x^2 + 3x + 2)$.

Factors	*Sum*
+1, +2	3
−1, −2	−3

$x^2 + 3x + 2 = (x + 1)(x + 2)$

$$-2x^3 - 6x^2 - 4x = -2x(x + 1)(x + 2)$$
$$a(x + 1)(x + 2) = -2x(x + 1)(x + 2)$$
$$a = -2x$$

137. The GCF is y. Then write in standard form.

$x^2y - 54y - 3xy = y(x^2 - 54 - 3x) = y(x^2 - 3x - 54)$

$x^2 - 3x - 54$

Factors	*Sum*
+1, −54	−53
−1, +54	+53
+2, −27	−25
−2, +27	+25
+3, −18	−15
−3, +18	+15
+6, −9	−3
−6, +9	+3

$x^2 - 3x - 54 = (x + 6)(x - 9)$

$x^2y - 54y - 3xy = y(x + 6)(x - 9)$

139. The GCF is $3p$. Then write in standard form.

$$12p^2 - 96p + 3p^3 = 3p(4p - 32 + 3p^2)$$
$$= 3p(3p^2 + 4p - 32)$$

$p^2 + 4p - 32$

Factors	*Sum*
+1, −32	−31
−1, +32	+31
+2, −16	−14
−2, +16	+14
+4, −8	−4
−4, +8	+8

$p^2 + 4p - 32 = (p - 4)(p + 8)$

$12p^2 - 96p + 3p^3 = 3p(p - 4)(p + 8)$

Projects or Group Activities

141. $x^2 + kx + 18$. The factors of 18 must sum to k.

Factors	*Sum*
−1, −18	−19
+1, +18	+19
−2, −9	−11
+2, +9	+11
−3, −6	−9
+3, +6	+9

k can be −19, 19, −11, 11, −9, or 9.

143. $x^2 - kx + 14$

Find all possible factors of 14 and their sums.

Integers	*Sum*
+1, +14	+15
−1, −14	−15
+2, +7	+9
−2, −7	−9

k can be 15, −15, 9 or −9.

145. $z^2 + 7z + k,\ k > 0$

Find two positive integers that sum to 7. Their product is k.

Integers	*Product*
+1, +6	6
+2, +5	10
+3, +4	12

k can be 6, 10, or 12.

147. $c^2 - 7c + k,\ k > 0$

Find two negative integers that sum to −7. Their product is k.

Integers	*Product*
−1, −6	6
−2, −5	10
−3, −4	12

k can be 6, 10, or 12.

149. $y^2 + 5y + k,\ k > 0$

Find two positive integers that sum to 5. Their product is k.

Integers	*Product*
+2, +3	6
+1, +4	4

k can be 4 or 6.

Section 7.3

Concept Check

1. $2x + 5$

3. $4x - 3$

5. 4; −5

7. $-2x - 6x$

Objective A Exercises

9.

Positive	Positive
Factors of 2: 1, 2	Factors of 1: +1, +1
Trial Factors	*Middle Term*
$(1x + 1)(2x + 1)$	$x + 2x = 3x$

$2x^2 + 3x + 1 = (x + 1)(2x + 1)$

11.

Positive	Positive
Factors of 2: 1, 2	Factors of 3: +1, +3
Trial Factors	*Middle Term*
$(1y + 1)(2y + 3)$	$3y + 2y = 5y$
$(1y + 3)(2y + 1)$	$y + 6y = 7y$

$2y^2 + 7y + 3 = (y + 3)(2y + 1)$

13.

Positive	Negative
Factors of 2: 1, 2	Factors of 1: −1, −1
Trial Factors	*Middle Term*
$(1a - 1)(2a - 1)$	$-a - 2a = -3a$

$2a^2 - 3a + 1 = (a - 1)(2a - 1)$

15.

Positive	Negative
Factors of 2: 1, 2	Factors of 5: −1, −5
Trial Factors	*Middle Term*
$(1b - 1)(2b - 5)$	$-5b - 2b = -7b$
$(1b - 5)(2b - 1)$	$-b - 10b = -11b$

$2b^2 - 11b + 5 = (b - 5)(2b - 1)$

17.

Positive	
Factors of 2: 1, 2	Factors of −1: −1, +1
Trial Factors	*Middle Term*
$(1x - 1)(2x + 1)$	$x - 2x = -x$
$(1x + 1)(2x - 1)$	$-x + 2x = x$

$2x^2 + x - 1 = (x + 1)(2x - 1)$

19.

Positive	
Factors of 2: 1, 2	Factors of −3: −1, +3
	+1, −3
Trial Factors	*Middle Term*
$(1x - 1)(2x + 3)$	$3x - 2x = x$
$(1x + 3)(2x - 1)$	$-x + 6x = 5x$
$(1x + 1)(2x - 3)$	$-3x + 2x = -x$
$(1x - 3)(2x + 1)$	$x - 6x = -5x$

$2x^2 - 5x - 3 = (x - 3)(2x + 1)$

21. Positive

Factors of 2: 1, 2 Factors of −10: −1, +10

+1, −10

−2, +5

+2, −5

Trial Factors	*Middle Term*
$(1t-1)(2t+10)$	Common factor
$(1t+10)(2t-1)$	$-t+20t=19t$
$(1t+1)(2t-10)$	Common factor
$(1t-10)(2t+1)$	$t-20t=-19t$
$(1t-2)(2t+5)$	$5t-4t=t$
$(1t+5)(2t-2)$	Common factor
$(1t+2)(2t-5)$	$-5t+4t=-t$
$(1t-5)(2t+2)$	Common factor

$2t^2-t-10=(t+2)(2t-5)$

23. Positive Negative

Factors of 3: 1, 3 Factors of 5: −1, −5

Trial Factors	*Middle Term*
$(1p-1)(3p-5)$	$-5p-3p=-8p$
$(1p-5)(3p-1)$	$-p-15p=-16p$

$3p^2-16p+5=(p-5)(3p-1)$

25. Positive Negative

Factors of 12: 1, 12 Factors of 1: −1, −1

2, 6

3, 4

Trial Factors	*Middle Term*
$(1y-1)(12y-1)$	$-y-12y=-13y$
$(2y-1)(6y-1)$	$-2y-6y=-8y$
$(3y-1)(4y-1)$	$-3y-4y=-7y$

$12y^2-7y+1=(3y-1)(4y-1)$

27. Positive Negative

Factors of 6: 1, 6 Factors of 3: −1, −3

2, 3

Trial Factors	*Middle Term*
$(1z-1)(6z-3)$	Common factor
$(1z-3)(6z-1)$	$-z-18z=-19z$
$(2z-1)(3z-3)$	Common factor
$(2z-3)(3z-1)$	$-2z-9z=-11z$

$6z^2-7z+3$ is nonfactorable over the integers.

29. Positive Negative

Factors of 6: 1, 6 Factors of 4: −1, −4

2, 3 −2, −2

Trial Factors	*Middle Term*
$(1t-1)(6t-4)$	Common factor
$(1t-4)(6t-1)$	$-t-24t=-25t$
$(1t-2)(6t-2)$	Common factor
$(2t-1)(3t-4)$	$-8t-3t=-11t$
$(2t-4)(3t-1)$	Common factor
$(2t-2)(3t-2)$	Common factor

$6t^2-11t+4=(2t-1)(3t-4)$

31. Positive Positive

Factors of 8: 1, 8 Factors of 4: +1, +4

2, 4 +2, +2

Trial Factors	*Middle Term*
$(1x+1)(8x+4)$	Common factor
$(1x+4)(8x+1)$	$x+32x=33x$
$(1x+2)(8x+2)$	Common factor
$(2x+1)(4x+4)$	Common factor
$(2x+4)(4x+1)$	Common factor
$(2x+2)(4x+2)$	Common factor

$8x^2+33x+4=(x+4)(8x+1)$

33. Positive

Factors of 5: 1, 5 Factors of −7: −1, +7

+1, −7

Trial Factors	*Middle Term*
$(1x-1)(5x+7)$	$7x-5x=2x$
$(1x+7)(5x-1)$	$-x+35x=34x$
$(1x+1)(5x-7)$	$-7x+5x=-2x$
$(1x-7)(5x+1)$	$x-35x=-34x$

$5x^2-62x-7$ is nonfactorable over the integers.

35. Positive Positive

Factors of 12: 1, 12 Factors of 5: +1, +5

2, 6

3, 4

Trial Factors	*Middle Term*
$(1y+1)(12y+5)$	$5y+12y=17y$
$(1y+5)(12y+1)$	$y+60y=61y$
$(2y+1)(6y+5)$	$10y+6y=16y$
$(2y+5)(6y+1)$	$2y+30y=32y$
$(3y+1)(4y+5)$	$15y+4y=19y$
$(3y+5)(4y+1)$	$3y+20y=23y$

$12y^2+19y+5=(3y+1)(4y+5)$

37. Positive

Factors of 2: 1, 2 Factors of −14: −1, +14; +1, −14; −2, +7; +2, −7

Trial Factors	*Middle Term*
$(1z-1)(2z+14)$	Common factor
$(1z+14)(2z-1)$	$-z+28z=27z$
$(1z+1)(2z-14)$	Common factor
$(1z-14)(2z+1)$	$z-28z=-27z$
$(1z-2)(2z+7)$	$7z-4z=3z$
$(1z+7)(2z-2)$	Common factor
$(1z+2)(2z-7)$	$-7z+4z=-3z$
$(1z-7)(2z+2)$	Common factor

$2z^2-27z-14=(z-14)(2z+1)$

39. Positive

Factors of 3: 1, 3 Factors of −16: −1, +16; +1, −16; −2, +8; +2, −8; −4, +4

Trial Factors	*Middle Term*
$(1p-1)(3p+16)$	$16p-3p=13p$
$(1p+16)(3p-1)$	$-p+48p=47p$
$(1p+1)(3p-16)$	$-16p+3p=-13p$
$(1p-16)(3p+1)$	$p-48p=-47p$
$(1p-2)(3p+8)$	$8p-6p=2p$
$(1p+8)(3p-2)$	$-2p+24p=22p$
$(1p+2)(3p-8)$	$-8p+6p=-2p$
$(1p-8)(3p+2)$	$2p-24p=-22p$
$(1p-4)(3p+4)$	$4p-12p=-8p$
$(1p+4)(3p-4)$	$-4p+12p=8p$

$3p^2+22p-16=(p+8)(3p-2)$

41. The GCF is 2.

$4x^2+6x+2=2(2x^2+3x+1)$

Factor the trinomial.

Positive Positive

Factors of 2: 1, 2 Factors of 1: +1, +1

Trial Factors	*Middle Term*
$(1x+1)(2x+1)$	$x+2x=3x$

$4x^2+6x+2=2(x+1)(2x+1)$

43. The GCF is 5.

$15y^2-50y+35=5(3y^2-10y+7)$

Factor the trinomial.

Positive Negative

Factors of 3: 1, 3 Factors of 7: −1, −7

Trial Factors	*Middle Term*
$(1y-1)(3y-7)$	$-7y-3y=-10y$
$(1y-7)(3y-1)$	$-y-21y=-22y$

$15y^2-50y+35=5(y-1)(3y-7)$

45. The GCF is x.

$2x^3-11x^2+5x=x(2x^2-11x+5)$

Factor the trinomial $2x^2-11x+5$.

Positive Negative

Factors of 2: 1, 2 Factors of 5: −1, −5

Trial Factors	*Middle Term*
$(1x-1)(2x-5)$	$-5x-2x=-7x$
$(1x-5)(2x-1)$	$-x-10x=-11x$

$2x^3-11x^2+5x=x(x-5)(2x-1)$

47. The GCF is b.

$3a^2b-16ab+16b=b(3a^2-16a+16)$

Factor the trinomial $3a^2-16a+16$.

Positive Negative

Factors of 3: 1, 3 Factors of 16: −1, −16; −2, −8; −4, −4

Trial Factors	*Middle Term*
$(1a-1)(3a-16)$	$-16a-3a=-19a$
$(1a-16)(3a-1)$	$-a-48a=-49a$
$(1a-2)(3a-8)$	$-8a-6a=-14a$
$(1a-8)(3a-2)$	$-2a-24a=-26a$
$(1a-4)(3a-4)$	$-4a-12a=-16a$

$3a^2b-16ab+16b=b(a-4)(3a-4)$

49. There is no common factor.

Factor the trinomial.

Positive		Positive	
Factors of 3:	1, 3	Factors of 10:	+1, +10
			+2, +5

Trial Factors	*Middle Term*
$(1z+1)(3z+10)$	$10z+3z=13z$
$(1z+10)(3z+1)$	$z+30z=31z$
$(1z+2)(3z+5)$	$5z+6z=11z$
$(1z+5)(3z+2)$	$2z+15z=17z$

$3z^2+95z+10$ is nonfactorable over the integers.

51. The GCF is $-3x$.

$36x-3x^2-3x^3=-3x(x^2+x-12)$

Factor the trinomial.

Postive			
Factors of 1:	1, 1	Factors of -12:	$-1, +12$
			$+1, -12$
			$+2, -6$
			$-2, +6$
			$+3, -4$
			$-3, +4$

Trial Factors	*Middle Term*
$(1x-1)(1x+12)$	$12x-1x=11x$
$(1x+1)(1x-12)$	$-12x+1x=-11x$
$(1x+2)(1x-6)$	$-6x+2x=-4x$
$(1x-2)(1x+6)$	$6x-2x=4x$
$(1x+3)(1x-4)$	$-4x+3x=-x$
$(1x-3)(1x+4)$	$4x-3x=x$

$36x-3x^2-3x^3=-3x(x-3)(x+4)$

53. The GCF is 4.

$80y^2-36y+4=4(20y^2-9y+1)$

Factor the trinomial.

Positive		Negative	
Factors of 20:	1, 20	Factors of 1:	$-1, -1$
	2, 10		
	4, 5		

Trial Factors	*Middle Term*
$(1y-1)(20y-1)$	$-y-20y=-21y$
$(2y-1)(10y-1)$	$-2y-10y=-12y$
$(4y-1)(5y-1)$	$-4y-5y=-9y$

$80y^2-36y+4=4(4y-1)(5y-1)$

55. The GCF is z.

$8z^3+14z^2+3z=z(8z^2+14z+3)$

Factor the trinomial.

Positive		Positive	
Factors of 8:	1, 8	Factors of 3:	+1, +3
	2, 4		

Trial Factors	*Middle Term*
$(1z+1)(8z+3)$	$3z+8z=11z$
$(1z+3)(8z+1)$	$z+24z=25z$
$(2z+1)(4z+3)$	$6z+4z=10z$
$(2z+3)(4z+1)$	$2z+12z=14z$

$8z^3+14z^2+3z=z(2z+3)(4z+1)$

57. The GCF is y.

$6x^2y-11xy-10y=y(6x^2-11x-10)$

Factor the trinomial.

Positive			
Factors of 6:	1, 6	Factors of -10:	$-1, +10$
	2, 3		$+1, -10$
			$-2, +5$
			$+2, -5$

Trial Factors	*Middle Term*
$(1x-1)(6x+10)$	Common factor
$(1x+10)(6x-1)$	$-x+60x=59x$
$(1x+1)(6x-10)$	Common factor
$(1x-10)(6x+1)$	$x-60x=-59x$
$(1x-2)(6x+5)$	$5x-12x=-7x$
$(1x+5)(6x-2)$	Common factor
$(1x+2)(6x-5)$	$-5x+12x=7x$
$(1x-5)(6x+2)$	Common factor
$(2x-1)(3x+10)$	$20x-3x=17x$
$(2x+10)(3x-1)$	Common factor
$(2x+1)(3x-10)$	$-20x+3x=-17x$
$(2x-10)(3x+1)$	Common factor
$(2x-2)(3x+5)$	Common factor
$(2x+5)(3x-2)$	$-4x+15x=11x$
$(2x+2)(3x-5)$	Common factor
$(2x-5)(3x+2)$	$4x-15x=-11x$

$6x^2y-11xy-10y=y(2x-5)(3x+2)$

59. The GCF is 5. $10t^2 - 5t - 50 = 5(2t^2 - t - 10)$

Factor the trinomial.

Positive Factors of 2: 1, 2 Factors of −10: −1, +10; +1, −10; −2, +5; +2, −5

Trial Factors	Middle Term
$(1t - 1)(2t + 10)$	Common factor
$(1t + 10)(2t - 1)$	$-t + 20t = 19t$
$(1t + 1)(2t - 10)$	Common factor
$(1t - 10)(2t + 1)$	$t - 20t = -19t$
$(1t - 2)(2t + 5)$	$5t - 4t = t$
$(1t + 5)(2t - 2)$	Common factor
$(1t + 2)(2t - 5)$	$-5t + 4t = -t$
$(1t - 5)(2t + 2)$	Common factor

$10t^2 - 5t - 50 = 5(t + 2)(2t - 5)$

61. The GCF is p. $3p^3 - 16p^2 + 5p = p(3p^2 - 16p + 5)$

Factor the trinomial.

Positive Factors of 3: 1, 3 Negative Factors of 5: −1, −5

Trial Factors	Middle Term
$(1p - 1)(3p - 5)$	$-5p - 3p = -8p$
$(1p - 5)(3p - 1)$	$-p - 15p = -16p$

$3p^3 - 16p^2 + 5p = p(p - 5)(3p - 1)$

63. The GCF is 2. $26z^2 + 98z - 24 = 2(13z^2 + 49z - 12)$

Factor the trinomial.

Positive Factors of 13: 1, 13 Factors of −12: −1, +12; +1, −12; −2, +6; +2, −6; −3, +4; +3, −4

Trial Factors	Middle Term
$(1z - 1)(13z + 12)$	$2z - 13z = -z$
$(1z + 12)(13z - 1)$	$-z + 156z = 155z$
$(1z + 1)(13z - 12)$	$-12z + 13z = z$
$(1z - 12)(13z + 1)$	$z - 156z = -155z$
$(1z - 2)(13z + 6)$	$6z - 26z = -20z$
$(1z + 6)(13z - 2)$	$-2z + 78z = 76z$
$(1z + 2)(13z - 6)$	$-6z + 26z = 20z$
$(1z - 6)(13z + 2)$	$2z - 78z = -76z$
$(1z - 3)(13z + 4)$	$4z - 39z = -35z$
$(1z + 4)(13z - 3)$	$-3z + 52z = 49z$
$(1z + 3)(13z - 4)$	$-4z + 39z = 35z$
$(1z - 4)(13z + 3)$	$3z - 52z = -49z$

$26z^2 + 98z - 24 = 2(z + 4)(13z - 3)$

65. The GCF is $2y$.

$10y^3 - 44y^2 + 16y = 2y(5y^2 - 22y + 8)$

Factor the trinomial.

Positive Factors of 5: 1, 5 Negative Factors of 8: −1, −8; −2, −4

Trial Factors	Middle Term
$(1y - 1)(5y - 8)$	$-8y - 5y = -13y$
$(1y - 8)(5y - 1)$	$-y - 40y = -41y$
$(1y - 2)(5y - 4)$	$-4y - 10y = -14y$
$(1y - 4)(5y - 2)$	$-2y - 20y = -22y$

$10y^3 - 44y^2 + 16y = 2y(y - 4)(5y - 2)$

67. The GCF is yz.

$4yz^3 + 5yz^2 - 6yz = yz(4z^2 + 5z - 6)$

Factor the trinomial.

Positive			
Factors of 4:	1, 4	Factors of −6:	−1, +6
	2, 2		+1, −6
			−2, +3
			+2, −3

Trial Factors	*Middle Term*
$(1z-1)(4z+6)$	Common factor
$(1z+6)(4z-1)$	$-z+24z=23z$
$(1z+1)(4z-6)$	Common factor
$(1z-6)(4z+1)$	$z-24z=-23z$
$(1z-2)(4z+3)$	$3z-8z=-5z$
$(1z+3)(4z-2)$	Common factor
$(1z+2)(4z-3)$	$-3z+8z=5z$
$(1z-3)(4z+2)$	Common factor
$(2z-1)(2z+6)$	Common factor
$(2z+1)(2z-6)$	Common factor
$(2z-2)(2z+3)$	Common factor
$(2z+2)(2z-3)$	Common factor

$4yz^3 + 5yz^2 - 6yz = yz(z+2)(4z-3)$

69. The GCF is $3a$.

$42a^3 + 45a^2 - 27a = 3a(14a^2 + 15a - 9)$

Factor the trinomial.

Positive			
Factors of 14:	1, 14	Factors of −9:	−1, +9
	2, 7		+1, −9
			−3, +3

Trial Factors	*Middle Term*
$(1a-1)(14a+9)$	$9a-14a=-5a$
$(1a+9)(14a-1)$	$-a+126a=125a$
$(1a+1)(14a-9)$	$-9a+14a=5a$
$(1a-9)(14a+1)$	$a-126a=-125a$
$(1a-3)(14a+3)$	$3a-42a=-39a$
$(1a+3)(14a-3)$	$-3a+42a=39a$
$(2a-1)(7a+9)$	$18a-7a=11a$
$(2a+9)(7a-1)$	$-2a+63a=61a$
$(2a+1)(7a-9)$	$-18a+7a=-11a$
$(2a-9)(7a+1)$	$2a-63a=-61a$
$(2a-3)(7a+3)$	$6a-21a=-15a$
$(2a+3)(7a-3)$	$-6a+21a=15a$

$42a^3 + 45a^2 - 27a = 3a(2a+3)(7a-3)$

71. The GCF is y.

$9x^2y - 30xy^2 + 25y^3 = y(9x^2 - 30xy + 25y^2)$

Factor the trinomial.

Positive		Negative	
Factors of 9:	1, 9	Factors of 25:	−1, −25
	3, 3		−5, −5

Trial Factors	*Middle Term*
$(1x-1y)(9x-25y)$	$-25xy-9xy=-34xy$
$(1x-25y)(9x-1y)$	$-xy-225xy=-226xy$
$(1x-5y)(9x-5y)$	$-5xy-45xy=-50xy$
$(3x-1y)(3x-25y)$	$-75xy-3xy=-78xy$
$(3x-5y)(3x-5y)$	$-15xy-15xy=-30xy$

$9x^2y - 30xy^2 + 25y^3 = y(3x-5y)(3x-5y)$

73. The GCF is xy.

$9x^3y - 24x^2y^2 + 16xy^3 = xy(9x^2 - 24xy + 16y^2)$

Factor the trinomial.

Positive		Negative	
Factors of 9:	1, 9	Factors of 16:	−1, −16
	3, 3		−2, −8
			−4, −4

Trial Factors	*Middle Term*
$(1x-1y)(9x-16y)$	$-16xy-9xy=-25xy$
$(1x-16y)(9x-1y)$	$-xy-144xy=-145xy$
$(1x-2y)(9x-8y)$	$-8xy-18xy=-26xy$
$(1x-8y)(9x-2y)$	$-2xy-72xy=-74xy$
$(1x-4y)(9x-4y)$	$-4xy-36xy=-40xy$
$(3x-1y)(3x-16y)$	$-48xy-3xy=-51xy$
$(3x-2y)(3x-8y)$	$-24xy-6xy=-30xy$
$(3x-4y)(3x-4y)$	$-12xy-12xy=-24xy$

$9x^3y - 24x^2y^2 + 16xy^3 = xy(3x-4y)(3x-4y)$

75. p must be odd. If p were even, $(nx+p)$ would have a common factor of 2 since n is even.

Objective B Exercises

77. $6x^2 - 17x + 12 \qquad 6 \cdot 12 = 72$

Factors of 72 whose sum is −17: −9 and −8

$$\begin{aligned} 6x^2 - 17x + 12 &= 6x^2 - 9x - 8x + 12 \\ &= (6x^2 - 9x) + (-8x + 12) \\ &= 3x(2x-3) - 4(2x-3) \\ &= (2x-3)(3x-4) \end{aligned}$$

79. $5b^2+33b-14 \qquad 5(-14)=-70$

Factors of -70 whose sum is 33: 35 and –2

$$\begin{aligned}5b^2+33b-14&=5b^2+35b-2b-14\\&=\left(5b^2+35b\right)+(-2b-14)\\&=5b(b+7)-2(b+7)\\&=(b+7)(5b-2)\end{aligned}$$

81. $6a^2+7a-24 \qquad 6(-24)=-144$

Factors of -144 whose sum is 7: 16 and –9

$$\begin{aligned}6a^2+7a-24&=6a^2+16a-9a-24\\&=\left(6a^2+16a\right)+(-9a-24)\\&=2a(3a+8)-3(3a+8)\\&=(3a+8)(2a-3)\end{aligned}$$

83. $4z^2+11z+6 \qquad 4\cdot 6=24$

Factors of 24 whose sum is 11: 8 and 3

$$\begin{aligned}4z^2+11z+6&=4z^2+8z+3z+6\\&=\left(4z^2+8z\right)+(3z+6)\\&=4z(z+2)+3(z+2)\\&=(z+2)(4z+3)\end{aligned}$$

85. $22p^2+51p-10 \qquad 22(-10)=-220$

Factors of -220 whose sum is 51: 55 and –4

$$\begin{aligned}22p^2+51p-10&=22p^2+55p-4p-10\\&=\left(22p^2+55p\right)+(-4p-10)\\&=11p(2p+5)-2(2p+5)\\&=(2p+5)(11p-2)\end{aligned}$$

87. $8y^2+17y+9 \qquad 8\cdot 9=72$

Factors of 72 whose sum is 17: 9 and 8

$$\begin{aligned}8y^2+17y+9&=8y^2+8y+9y+9\\&=\left(8y^2+8y\right)+(9y+9)\\&=8y(y+1)+9(y+1)\\&=(y+1)(8y+9)\end{aligned}$$

89. $18t^2-9t-5 \qquad 18(-5)=-90$

Factors of -90 whose sum is –9: –15 and 6

$$\begin{aligned}18t^2-9t-5&=18t^2-15t+6t-5\\&=\left(18t^2-15t\right)+(6t-5)\\&=3t(6t-5)+1(6t-5)\\&=(6t-5)(3t+1)\end{aligned}$$

91. $6b^2+71b-12 \qquad 6(-12)=-72$

Factors of -72 whose sum is 71: 72 and –1

$$\begin{aligned}6b^2+71b-12&=6b^2+72b-b-12\\&=\left(6b^2+72b\right)+(-b-12)\\&=6b(b+12)-1(b+12)\\&=(b+12)(6b-1)\end{aligned}$$

93. $9x^2+12x+4 \qquad 9\cdot 4=36$

Factors of 36 whose sum is 12: 6 and 6

$$\begin{aligned}9x^2+12x+4&=9x^2+6x+6x+4\\&=\left(9x^2+6x\right)+(6x+4)\\&=3x(3x+2)+2(3x+2)\\&=(3x+2)(3x+2)\end{aligned}$$

95. $6b^2-13b+6 \qquad 6\cdot 6=36$

Factors of 36 whose sum is –13: –9 and –4

$$\begin{aligned}6b^2-13b+6&=6b^2-9b-4b+6\\&=\left(6b^2-9b\right)+(-4b+6)\\&=3b(2b-3)-2(2b-3)\\&=(2b-3)(3b-2)\end{aligned}$$

97. $33b^2+34b-35 \qquad 33(-35)=-1155$

Factors of -1155 whose sum is 34: 55 and –21

$$\begin{aligned}33b^2+34b-35&=33b^2+55b-21b-35\\&=\left(33b^2+55b\right)+(-21b-35)\\&=11b(3b+5)-7(3b+5)\\&=(3b+5)(11b-7)\end{aligned}$$

99. $18y^2-39y+20 \qquad 18\cdot 20=360$

Factors of 360 whose sum is –39: –24 and –15

$$\begin{aligned}18y^2-39y+20&=18y^2-24y-15y+20\\&=\left(18y^2-24y\right)+(-15y+20)\\&=6y(3y-4)-5(3y-4)\\&=(3y-4)(6y-5)\end{aligned}$$

101. $15a^2+26a-21 \quad 15(-21)=-315$

Factors of -315 whose sum is 26: 35 and –9

$$\begin{aligned}15a^2+26a-21&=15a^2+35a-9a-21\\&=(15a^2+35a)+(-9a-21)\\&=5a(3a+7)-3(3a+7)\\&=(3a+7)(5a-3)\end{aligned}$$

103. $8y^2-26y+15 \quad 8\cdot 15=120$

Factors of 120 whose sum is –26: –20 and –6

$$\begin{aligned}8y^2-26y+15&=8y^2-20y-6y+15\\&=(8y^2-20y)+(-6y+15)\\&=4y(2y-5)-3(2y-5)\\&=(2y-5)(4y-3)\end{aligned}$$

105. $8z^2+2z-15 \quad 8(-15)=-120$

Factors of -120 whose sum is 2: 12 and -10

$$\begin{aligned}8z^2+2z-15&=8z^2+12z-10z-15\\&=(8z^2+12z)+(-10z-15)\\&=4z(2z+3)-5(2z+3)\\&=(2z+3)(4z-5)\end{aligned}$$

107. $15x^2-82x+24 \quad 15\cdot 24=360$

Factors of 360 whose sum is –82: none

$15x^2-82x+24$ is nonfactorable over the integers.

109. $10z^2-29z+10 \quad 10\cdot 10=100$

Factors of 100 whose sum is –29: –25 and –4

$$\begin{aligned}10z^2-29z+10&=10z^2-25z-4z+10\\&=(10z^2-25z)+(-4z+10)\\&=5z(2z-5)-2(2z-5)\\&=(2z-5)(5z-2)\end{aligned}$$

111. $36z^2+72z+35 \quad 36\cdot 35=1260$

Factors of 1260 whose sum is 72: 30 and 42

$$\begin{aligned}36z^2+72z+35&=36z^2+30z+42z+35\\&=(36z^2+30z)+(42z+35)\\&=6z(6z+5)+7(6z+5)\\&=(6z+5)(6z+7)\end{aligned}$$

113. $3x^2+xy-2y^2 \quad 3(-2)=-6$

Factors of -6 whose sum is 1: 3 and -2

$$\begin{aligned}3x^2+xy-2y^2&=3x^2+3xy-2xy-2y^2\\&=(3x^2+3xy)+(-2xy-2y^2)\\&=3x(x+y)-2y(x+y)\\&=(x+y)(3x-2y)\end{aligned}$$

115. $3a^2+5ab-2b^2 \quad 3(-2)=-6$

Factors of -6 whose sum is 5: 6 and -1

$$\begin{aligned}3a^2+5ab-2b^2&=3a^2+6ab-ab-2b^2\\&=(3a^2+6ab)+(-ab-2b^2)\\&=3a(a+2b)-b(a+2b)\\&=(a+2b)(3a-b)\end{aligned}$$

117. $4y^2-11yz+6z^2 \quad 4\cdot 6=24$

Factors of 24 whose sum is –11: –8 and -3

$$\begin{aligned}4y^2-11yz+6z^2&=4y^2-8yz-3yz+6z^2\\&=(4y^2-8yz)+(-3yz+6z^2)\\&=4y(y-2z)-3z(y-2z)\\&=(y-2z)(4y-3z)\end{aligned}$$

119. $28+3z-z^2 \quad 28(-1)=-28$

Factors of -28 whose sum is : 7 and –4

$$\begin{aligned}28+3z-z^2&=-z^2+3z+28=-z^2-4z+7z+28\\&=(-z^2-4z)+(7z+28)\\&=-z(z+4)+7(z+4)\\&=-(z-7)(z+4)\end{aligned}$$

121. $8-7x-x^2 \quad 8(-1)=-8$

Factors of -8 whose sum is –7: –8 and 1

$$\begin{aligned}8-7x-x^2&=-x^2-7x+8=-x^2+x-8x+8\\&=(-x^2+x)+(-8x+8)\\&=-x(x-1)-8(x-1)\\&=-(x-1)(x+8)\end{aligned}$$

123. $9x^2+33x-60$

Common factor 3: $3(3x^2+11x-20)$

$3(-20)=-60$

Factors of -60 whose sum is 11: 15 and -4

$$\begin{aligned}3(3x^2+11x-20)&=3(3x^2+15x-4x-20)\\&=3\left[(3x^2+15x)+(-4x-20)\right]\\&=3\left[3x(x+5)-4(x+5)\right]\\&=3(x+5)(3x-4)\end{aligned}$$

125. $24x^2-52x+24$

Common factor 4: $4(6x^2-13x+6)$

$6\cdot 6=36$

Factors of 36 whose sum is -13: -9 and -4

$$\begin{aligned}4(6x^2-13x+6)&=4(6x^2-9x-4x+6)\\&=4\left[(6x^2-9x)+(-4x+6)\right]\\&=4\left[3x(2x-3)-2(2x-3)\right]\\&=4(2x-3)(3x-2)\end{aligned}$$

127. $35a^4+9a^3-2a^2$

Common factor a^2: $a^2(35a^2+9a-2)$

$35(-2)=-70$

Factors of -70 whose sum is 9: 14 and -5

$$\begin{aligned}a^2(35a^2+9a-2)&=a^2(35a^2+14a-5a-2)\\&=a^2\left[(35a^2+14a)+(-5a-2)\right]\\&=a^2\left[7a(5a+2)-1(5a+2)\right]\\&=a^2(5a+2)(7a-1)\end{aligned}$$

129. $15b^2-115b+70$

Common factor 5: $5(3b^2-23b+14)$

$3\cdot 14=42$

Factors of 42 whose sum is -23: -21 and -2

$$\begin{aligned}5(3b^2-23b+14)&=5(3b^2-21b-2b+14)\\&=5\left[(3b^2-21b)+(-2b+14)\right]\\&=5\left[3b(b-7)-2(b-7)\right]\\&=5(b-7)(3b-2)\end{aligned}$$

131. $3x^2-26xy+35y^2$ $\quad 3\cdot 35=105$

Factors of 105 whose sum is -26: -21 and -5

$$\begin{aligned}3x^2-25xy+36y^2&=3x^2-21xy-5xy+35y^2\\&=(3x^2-21xy)+(-5xy+35y^2)\\&=3x(x-7y)-5y(x-7y)\\&=(x-7y)(3x-5y)\end{aligned}$$

133. $216y^2-3y-3$

Common factor 3: $3(72y^2-y-1)$

$72(-1)=-72$

Factors of -72 whose sum is -1: -9 and 8

$$\begin{aligned}3(72y^2-y-1)&=3(72y^2-9y+8y-1)\\&=3\left[(72y^2-9y)+(8y-1)\right]\\&=3\left[9y(8y-1)+1(8y-1)\right]\\&=3(8y-1)(9y+1)\end{aligned}$$

135. If c is negative, there must be 1 positive and 1 negative sign.

137. If c is negative, there must be 1 positive and 1 negative sign.

Critical Thinking

139. $(x+1)^2-(x+1)-6$ $\quad$ Let $a=x+1$

$$\begin{aligned}&=a^2-a-6\\&=(a-3)(a+2)\\&=(x+1-3)(x+1+2)\\&=(x-2)(x+3)\end{aligned}$$

141. $(y+3)^2-5(y+3)+6$ $\quad$ Let $a=y+3$

$$\begin{aligned}&=a^2-5a+6\\&=(a-2)(a-3)\\&=[y+3-2][y+3-3]\\&=(y+1)y \text{ or } y(y+1)\end{aligned}$$

143. $3(a+2)^2-(a+2)-4$ $\quad$ Let $x=a+2$

$$\begin{aligned}&=3x^2-x-4\\&=(3x-4)(x+1)\\&=\left[3(a+2)-4\right]\left[a+2+1\right]\\&=(3a+6-4)(a+3)\\&=(3a+2)(a+3)\end{aligned}$$

145. The GCF is $2y$.

$$6y+8y^3-26y^2=2y\left(3+4y^2-13y\right)$$
$$=2y\left(4y^2-13y+3\right)$$
$$=2y(4y-1)(y-3)$$

147. The GCF is ab.

$$a^3b-24ab-2a^2b=ab\left(a^2-24-2a\right)$$
$$=ab\left(a^2-2a-24\right)$$
$$=ab(a-6)(a+4)$$

Projects or Group Activities

149. $2x^2+kx+3$

$2\times 3=6$

Factors	*Sum*
+1, +6	+7
−1, −6	−7
+2, +3	+5
−2, −3	−5

k can be 7, –7, 5, or –5.

151. $3x^2+kx+2$

$2\times 3=6$

Factors	*Sum*
+1, +6	+7
−1, −6	−7
+2, +3	+5
−2, −3	−5

k can be 7, –7, 5, or –5.

153. $2x^2+kx+5$

$2\times 5=10$

Factors	*Sum*
+1, +10	11
−1, −10	−11
+2, +5	7
−2, −5	−7

k can be 11, –11, 7, or –7.

155. $3x^2+x-2=(x+1)(3x-2)$

Length and width must both be positive.

$$3x-2>0$$
$$3x-2+2>0+2$$
$$3x>2$$
$$\frac{3x}{3}>\frac{2}{3}$$
$$x>\frac{2}{3}$$

The length must be greater than the width.

$3x-2$ if the length if $3x-2>x+1$.

$$3x-2>x+1$$
$$3x-x-2>x-x+1$$
$$2x-2>1$$
$$2x-2+2>1+2$$
$$2x>3$$
$$\frac{2x}{2}>\frac{3}{2}$$
$$x>\frac{3}{2}$$

$x+1$ is the length if $x+1>3x-2$.

$3x-2$ is the length if $\frac{2}{3}<x<\frac{3}{2}$ and $x+1$ is the length if $x>\frac{3}{2}$.

Check Your Progress: Chapter 7

1. The GCF is 5. $20b+5=5(4b+1)$

2. The GCF is $(b+7)$

$$2x(7+b)-y(b+7)=2x(b+7)-y(b+7)$$
$$=(b+7)(2x-y)$$

3. $x^2+20x+100$

Factors	*Sum*
+1, +100	+101
+2, +50	+52
+4, +25	+29
+5, +20	+25
+10, +10	+20

$x^2+20x+100=(x+10)(x+10)$

4. The GCF is y.

$x^2y-2xy-24y=y\left(x^2-2x-24\right)$

Factors	*Sum*
+1, −24	−23
−1, +24	+23
+2, −12	−10
−2, +12	+10
+3, −8	−5
−3, +8	+5
+4, −6	−2
−4, +6	+2

$x^2-2x-24=(x+4)(x-6)$

$x^2y-2xy-24y=y(x+4)(x-6)$

5. Factor −1.

$35+2x-x^2=-1\left(x^2-2x-35\right)$

Factors	*Sum*
+1, −35	−34
−1, +35	+34
+5, −7	−2
−5, +7	+2

$x^2-2x-35=(x+5)(x-7)$

$35+2x-x^2=-1(x+5)(x-7)$

6. x^2-8x-2

Factors	*Sum*
+1, −2	−1
−1, +2	+1

x^2-8x-2 is nonfactorable over the integers.

7. $21x^2+6xy-49x-14y$

$=\left(21x^2+6xy\right)+(-49x-14y)$

$=3x(7x+2y)-7(7x+2y)$

$=(7x+2y)(3x-7)$

8. The GCF is 3a. $6ab+9a=3a(2b+3)$

9. Negative

Factors of 5: 1, 5 Factors of 8: −1, −8

−2, −4

Trial Factors	*Middle Term*
$(y-1)(5y-8)$	$-8y-5y=-13y$
$(y-8)(5y-1)$	$-y-40y=-41y$
$(y-2)(5y-4)$	$-4y-10y=-14y$
$(y-4)(5y-2)$	$-2y-20y=-22y$

$5y^2-22y-8=(y-4)(5y-2)$

10. $12x^2+31x+9$ $\quad 12(9)=108$

Factors of 108 whose sum is 31: 27 and 4

$12x^2+27x+4x+9$

$=\left(12x^2+27x\right)+(4x+9)$

$=3x(4x+9)+(4x+9)$

$=(4x+9)(3x+1)$

$12x^2+31x+9=(4x+9)(3x+1)$

11. The GCF is x. $9x-5x^2=x(9-5x)$

12. $2x^2+x+2xy+y$

$=\left(2x^2+x\right)+(2xy+y)$

$=x(2x+1)+y(2x+1)$

$=(2x+1)(x+y)$

13. $8a^2-2ab-3b^2$ $\quad 8(-3)=-24$

Factors of −24 whose sum is −2: 4 and −6

$8a^2+4ab-6ab-3b^2$

$=\left(8a^2+4ab\right)+\left(-6ab-3b^2\right)$

$=4a(2a+b)-3b(2a+b)$

$=(2a+b)(4a-3b)$

14. $b^2+9b+20$

Factors	*Sum*
+1, +20	+21
+4, +5	+9

$b^2+9b+20=(b+4)(b+5)$

15. The GCF is $2a$.

$2a^3+24a^2+54a=2a\left(a^2+12a+27\right)$

$a^2+12a+27$

Factors	*Sum*
+1, +27	+28
+3, +9	+12

$a^2+12a+27=(a+3)(a+9)$

$2a^2+24a^2+54a=2a(a+3)(a+9)$

16. Factors of 11: 1, 11 Factors of -5: 1, -5
-1, 5

Trial Factors	*Middle Term*
$(a+1)(11a-5)$	$-5a+11a=-6a$
$(a-5)(11a+1)$	$a-55a=-54a$
$(a-1)(11a+5)$	$5a-11a=-6a$
$(a+5)(11a-1)$	$-a+55a=+54a$

$11a^2-54a-5=(a-5)(11a+1)$

17. The GCF is 4. $360y^2+4y-4=4(90y^2+y-1)$

$90y^2+y-1$ $\qquad 90(-1)=-90$

The factors of –90 whose sum is +1: +10 and –9

$90y^2+10y-9y-1$
$=(90y^2+10y)+(-9y-1)$
$=10y(9y+1)+-1(9y+1)$
$=(9y+1)(10y-1)$
$90y^2+y-1=(9y+1)(10y-1)$
$360y^2+4y-4=4(9y+1)(10y-1)$

18. The GCF is y.
$14y^3+5y^2+11y=y(14y^2+5y+11)$
$14y^2+5y+11$
The factors of 14: +1, +14 The factors of 11: +1, +11
+2, +7

Trial Factors	Middle Term
$(x+1)(14x+11)$	$11x+14x=25x$
$(x+11)(14x+1)$	$154x+x=155x$
$(2x+1)(7x+11)$	$22x+7x=29x$
$(2x+11)(7x+1)$	$14x+77x=91x$

$14y^2+5y+11$ is nonfactorable over the integers.
$14y^3+5y^2+11y=y(14y^2+5y+11)$

19. Negative
$x^2-7x+10$

Factors	*Sum*
$-1, -10$	-11
$-2, -5$	-7

$x^2-7x+10=(x-2)(x-5)$

20. Positive
$x^2+8xy+9y^2$

Factors	*Sum*
$+1, +9$	$+10$
$+3, +3$	$+6$

$x^2+8xy+9y^2$ is nonfactorable over the integers.

21. Positive
$b^2+13b+40$

Factors	*Sum*
$+1, +40$	$+41$
$+2, +20$	$+22$
$+4, +10$	$+14$
$+5, +8$	$+13$

$b^2+13b+40=(b+5)(b+8)$

22. $2x^2-5x-6xy+15y$
$=2x^2-6xy-5x+15y$
$=(2x^2-6xy)+(-5x+15y)$
$=2x(x-3y)-5(x-3y)$
$=(x-3y)(2x-5)$

23. The GCF is xy.
$x^2y-xy^3+x^3y=xy(x-y^2+x^2)$

24. Positive
Factors of 3: +1, +3 Factors of 16: +1, +16
+2, +8
+4, +4

Trail Factors	Middle Term
$(b+1)(3b+16)$	$16b+3b=19b$
$(b+16)(3b+1)$	$b+48b=49b$
$(b+2)(3b+8)$	$8b+6b=14b$
$(b+8)(3b+2)$	$2b+24b=26b$
$(b+4)(3b+4)$	$4b+12b=16b$

$3b^2+16b+16=(b+4)(3b+4)$

25. $x^2 - 11x - 42$

Factors	*Sum*
+1, − 42	−41
−1, + 42	+41
+2, − 21	−19
−2, + 21	+19
+3, − 14	−11
−3, + 14	+11
+6, − 7	−1
−6, + 7	+1

$x^2 - 11x - 42 = (x+3)(x-14)$

Section 7.4

Concept Check

1. **a.** No **b.** Yes **c.** No **d.** Yes

3. **a.** Yes **b.** No **c.** Yes **d.** No

5. **a.** $2x^3$ **b.** $3y^5$ **c.** $4a^2b^6$ **d.** $5c^4d$

7. **a.** Yes **b.** No **c.** Yes **d.** Yes

Objective A Exercises

9. $x^2 - 16 = x^2 - 4^2 = (x+4)(x-4)$

11. $4x^2 - 1 = (2x)^2 - 1^2 = (2x+1)(2x-1)$

13. $16x^2 - 121 = (4x)^2 - 11^2 = (4x+11)(4x-11)$

15. $1 - 9a^2 = 1^2 - (3a)^2 = (1+3a)(1-3a)$

17. $x^2y^2 - 100 = (xy)^2 - 10^2 = (xy+10)(xy-10)$

19. Not factorable

21. $25 - a^2b^2 = 5^2 - (ab)^2 = (5+ab)(5-ab)$

23. $x^2 - 12x + 36 = (x-6)^2$

25. $b^2 - 2b + 1 = (b-1)^2$

27. $16x^2 - 40x + 25 = (4x-5)^2$

29. Not factorable

31. Not factorable

33. $x^2 + 6xy + 9y^2 = (x+3y)^2$

35. $25a^2 - 40ab + 16b^2 = (5a-4b)^2$

37. $(x-4)^2 - 9 = [(x-4)-3][(x-4)+3]$
$= (x-7)(x-1)$

39. $(x-y)^2 - (a+b)^2$
$= [(x-y)-(a+b)][(x-y)+(a+b)]$
$= (x-y-a-b)(x-y+a+b)$

Objective B Exercises

41. $x^3 - 27 = x^3 - 3^3$
$= (x-3)(x^2+3x+9)$

43. $8x^3 - 1 = (2x)^3 - 1^3$
$= (2x-1)(4x^2+2x+1)$

45. $x^3 - y^3 = (x-y)(x^2+xy+y^2)$

47. $m^3 + n^3 = (m+n)(m^2-mn+n^2)$

49. $64x^3 + 1 = (4x)^3 + 1^3$
$= (4x+1)(16x^2-4x+1)$

51. $27x^3 - 8y^3 = (3x)^3 - (2y)^3$
$= (3x-2y)(9x^2+6xy+4y^2)$

53. $x^3y^3 + 64 = (xy)^3 + (4)^3$
$= (xy+4)(x^2y^2-4xy+16)$

55. Not factorable

57. Not factorable

59. $125 - c^3 = 5^3 - c^3$
$= (5-c)(25+5c+c^2)$

Objective C Exercises

61. No. It is a fourth degree polynomial.

63. Let $u = xy$
$x^2y^2 - 8xy - 33 = u^2 - 8u - 33$
$= (u+3)(u-11)$
$= (xy+3)(xy-11)$

65. Let $u = xy$

$$x^2y^2 - 17xy + 60 = u^2 - 17u + 60$$
$$= (u - 12)(u - 5)$$
$$= (xy - 12)(xy - 5)$$

67. Let $u = x^2$

$$x^4 - 9x^2 + 18 = u^2 - 9u + 18$$
$$= (u - 3)(u - 6)$$
$$= (x^2 - 3)(x^2 - 6)$$

69. Let $u = b^2$

$$b^4 - 13b^2 - 90 = u^2 - 13u - 90$$
$$= (u + 5)(u - 18)$$
$$= (b^2 + 5)(b^2 - 18)$$

71. Let $u = x^2y^2$

$$x^4y^4 - 8x^2y^2 + 12 = u^2 - 8u + 12$$
$$= (u - 2)(u - 6)$$
$$= (x^2y^2 - 2)(x^2y^2 - 6)$$

73. Let $u = \sqrt{x}$

$$x + 3\sqrt{x} + 2 = u^2 + 3u + 2$$
$$= (u + 2)(u + 1)$$
$$= (\sqrt{x} + 2)(\sqrt{x} + 1)$$

75. Let $u = xy$

$$3x^2y^2 - 14xy + 15 = 3u^2 - 14u + 15$$
$$= (3u - 5)(u - 3)$$
$$= (3xy - 5)(xy - 3)$$

77. Let $u = ab$

$$6a^2b^2 - 23ab + 21 = 6u^2 - 23u + 21$$
$$= (2u - 3)(3u - 7)$$
$$= (2ab - 3)(3ab - 7)$$

79. Let $u = x^2$

$$2x^4 - 13x^2 - 15 = 2u^2 - 13u - 15$$
$$= (2u - 15)(u + 1)$$
$$= (2x^2 - 15)(x^2 + 1)$$

81. Let $u = x^3$

$$x^6 - x^3 - 6 = u^2 - u - 6$$
$$= (u - 3)(u + 2)$$
$$= (x^3 - 3)(x^3 + 2)$$

83. Let $u = xy^2$

$$4x^2y^4 - 12xy^2 + 9 = 4u^2 - 12u + 9$$
$$= (2u - 3)^2$$
$$= (2xy^2 - 3)^2$$

Objective D Exercises

85. $12x^2 - 36x + 27 = 3(4x^2 - 12x + 9)$
$= 3(2x - 3)^2$

87. $27a^4 - a = a(27a^3 - 1)$
$= a(3a - 1)(9a^2 + 3a + 1)$

89. $20x^2 - 5 = 5(4x^2 - 1)$
$= 5(2x + 1)(2x - 1)$

91. $y^5 + 6y^4 - 55y^3 = y^3(y^2 + 6y - 55)$
$= y^3(y + 11)(y - 5)$

93. $16x^4 - 81 = (4x^2 + 9)(4x^2 - 9)$
$= (4x^2 + 9)(2x + 3)(2x - 3)$

95. $16a - 2a^4 = 2a(8 - a^3)$
$= 2a(2 - a)(4 + 2a + a^2)$

97. $a^3b^6 - b^3 = b^3(a^3b^3 - 1)$
$= b^3(ab - 1)(a^2b^2 + ab + 1)$

99. $8x^4 - 40x^3 + 50x^2 = 2x^2(4x^2 - 20x + 25)$
$= 2x^2(2x - 5)^2$

101. $x^4 - y^4 = (x^2 + y^2)(x^2 - y^2)$
$= (x^2 + y^2)(x + y)(x - y)$

103. $x^6 + y^6 = (x^2 + y^2)(x^4 - x^2y^2 + y^4)$

105. Not factorable

107. $16a^4 - 2a = 2a(8a^3 - 1)$
$= 2a(2a - 1)(4a^2 + 2a + 1)$

109. $a^4b^2 - 8a^3b^3 - 48a^2b^4$
$= a^2b^2(a^2 - 8ab - 48b^2)$
$= a^2b^2(a + 4b)(a - 12b)$

111. $x^3 - 2x^2 - 4x + 8 = x^2(x-2) - 4(x-2)$
$= (x-2)(x^2-4)$
$= (x-2)(x+2)(x-2)$
$= (x-2)^2(x+2)$

113. $2x^3 + x^2 - 32x - 16$
$= x^2(2x+1) - 16(2x+1)$
$= (2x+1)(x^2-16)$
$= (2x+1)(x+4)(x-4)$

115. $4x^4 - x^2 - 4x^2y^2 + y^2$
$= x^2(4x^2-1) - y^2(4x^2-1)$
$= (4x^2-1)(x^2-y^2)$
$= (2x+1)(2x-1)(x+y)(x-y)$

117. The coefficient is 8.

Critical Thinking

119. $9x^2 - kx + 1$

$9 = 3^2;\ 1 = 1^2$

$k = 2(3)(1) = 6$ or -6

121. $a^3 + (a+b)^3$
$= [a + (a+b)][a^2 - a(a+b) + (a+b)^2]$
$= (2a+b)(a^2 - a^2 - ab + a^2 + 2ab + b^2)$
$= (2a+b)(a^2 + ab + b^2)$

Projects or Group Activities

123. If 3 is a zero of the polynomial, $x-3$ is a factor of the polynomial. Divide the polynomial by $x-3$ to obtain the factorization.

$$
\begin{array}{r}
x^2 + 2x + 3 \\
x-3\overline{)\,x^3 - x^2 - 3x - 9} \\
\underline{x^3 - 3x^2} \\
2x^2 - 3x \\
\underline{2x^2 - 6x} \\
3x - 9 \\
\underline{3x - 9}
\end{array}
$$

$(x-3)(x^2+2x+3)$

Section 7.5

Concept Check

1. **a.** Yes **b.** No **c.** Yes

3. **a.** Yes **b.** Yes **c.** No **d.** Yes

e. No **f.** Yes

Objective A Exercises

5. $(y+3)(y+2) = 0$
$y+3=0 \quad y+2=0$
$y=-3 \quad y=-2$

The solutions are –2 and –3.

7. $(z-7)(z-3)=0$
$z-7=0 \quad z-3=0$
$z=7 \quad z=3$

The solutions are 3 and 7.

9. $x(x-5)=0$
$x=0 \quad x-5=0$
$x=5$

The solutions are 0 and 5.

11. $a(a-9)=0$
$a=0 \quad a-9=0$
$a=9$

The solutions are 0 and 9.

13. $y(2y+3)=0$
$y=0 \quad 2y+3=0$
$2y=-3$
$y=-\frac{3}{2}$

The solutions are 0 and $-\frac{3}{2}$.

15. $2a(3a-2)=0$
$2a=0 \quad 3a-2=0$
$a=0 \quad 3a=2$
$a=\frac{2}{3}$

The solutions are 0 and $\frac{2}{3}$.

17. $(b+2)(b-5)=0$
$b+2=0 \quad b-5=0$
$b=-2 \quad b=5$

The solutions are –2 and 5.

19. $x^2-81=0$

$(x+9)(x-9)=0$

$x+9=0 \quad x-9=0$

$x=-9 \quad x=9$

The solutions are −9 and 9.

21. $4x^2-49=0$

$(2x+7)(2x-7)=0$

$2x+7=0 \quad 2x-7=0$

$2x=-7 \quad 2x=7$

$x=-\frac{7}{2} \quad x=\frac{7}{2}$

The solutions are $-\frac{7}{2}$ and $\frac{7}{2}$.

23. $9x^2-1=0$

$(3x+1)(3x-1)=0$

$3x+1=0 \quad 3x-1=0$

$3x=-1 \quad 3x=1$

$x=-\frac{1}{3} \quad x=\frac{1}{3}$

The solutions are $-\frac{1}{3}$ and $\frac{1}{3}$.

25. $x^2+6x+8=0$

$(x+2)(x+4)=0$

$x+2=0 \quad x+4=0$

$x=-2 \quad x=-4$

The solutions are −2 and −4.

27. $z^2+5z-14=0$

$(z+7)(z-2)=0$

$z+7=0 \quad z-2=0$

$z=-7 \quad z=2$

The solutions are −7 and 2.

29. $2a^2-9a-5=0$

$(2a+1)(a-5)=0$

$2a+1=0 \quad a-5=0$

$2a=-1 \quad a=5$

$a=-\frac{1}{2}$

The solutions are $-\frac{1}{2}$ and 5.

31. $6z^2+5z+1=0$

$(3z+1)(2z+1)=0$

$3z+1=0 \quad 2z+1=0$

$3z=-1 \quad 2z=-1$

$z=-\frac{1}{3} \quad z=-\frac{1}{2}$

The solutions are $-\frac{1}{3}$ and $-\frac{1}{2}$.

33. $x^2-3x=0$

$x(x-3)=0$

$x=0 \quad x-3=0$

$x=3$

The solutions are 0 and 3.

35. $x^2-7x=0$

$x(x-7)=0$

$x=0 \quad x-7=0$

$x=7$

The solutions are 0 and 7.

37. $a^2+5a=-4$

$a^2+5a+4=0$

$(a+1)(a+4)=0$

$a+1=0 \quad a+4=0$

$a=-1 \quad a=-4$

The solutions are −1 and −4.

39. $y^2-5y=-6$

$y^2-5y+6=0$

$(y-2)(y-3)=0$

$y-2=0 \quad y-3=0$

$y=2 \quad y=3$

The solutions are 2 and 3.

41. $2t^2+7t=4$

$2t^2+7t-4=0$

$(2t-1)(t+4)=0$

$2t-1=0 \quad t+4=0$

$2t=1 \quad t=-4$

$t=\frac{1}{2}$

The solutions are $\frac{1}{2}$ and −4.

43. $3t^2 - 13t = -4$
$3t^2 - 13t + 4 = 0$
$(3t - 1)(t - 4) = 0$
$3t - 1 = 0 \quad t - 4 = 0$
$3t = 1 \qquad t = 4$
$t = \frac{1}{3}$

The solutions are $\frac{1}{3}$ and 4.

45. $x(x - 12) = -27$
$x^2 - 12x = -27$
$x^2 - 12x + 27 = 0$
$(x - 3)(x - 9) = 0$
$x - 3 = 0 \quad x - 9 = 0$
$x = 3 \qquad x = 9$

The solutions are 3 and 9.

47. $y(y - 7) = 18$
$y^2 - 7y = 18$
$y^2 - 7y - 18 = 0$
$(y + 2)(y - 9) = 0$
$y + 2 = 0 \quad y - 9 = 0$
$y = -2 \qquad y = 9$

The solutions are −2 and 9.

49. $p(p + 3) = -2$
$p^2 + 3p = -2$
$p^2 + 3p + 2 = 0$
$(p + 1)(p + 2) = 0$
$p + 1 = 0 \quad p + 2 = 0$
$p = -1 \qquad p = -2$

The solutions are −1 and −2.

51. $y(y + 4) = 45$
$y^2 + 4y = -45$
$y^2 + 4y - 45 = 0$
$(y + 9)(y - 5) = 0$
$y + 9 = 0 \quad y - 5 = 0$
$y = -9 \qquad y = 5$

The solutions are −9 and 5.

53. $x(x + 3) = 28$
$x^2 + 3x = 28$
$x^2 + 3x - 28 = 0$
$(x + 7)(x - 4) = 0$
$x + 7 = 0 \quad x - 4 = 0$
$x = -7 \qquad x = 4$

The solutions are −7 and 4.

55. $(x + 8)(x - 3) = -30$
$x^2 + 5x - 24 = -30$
$x^2 + 5x + 6 = 0$
$(x + 2)(x + 3) = 0$
$x + 2 = 0 \quad x + 3 = 0$
$x = -2 \qquad x = -3$

The solutions are −2 and −3.

57. $(z - 5)(z + 4) = 52$
$z^2 - z - 20 = 52$
$z^2 - z - 72 = 0$
$(z + 8)(z - 9) = 0$
$z + 8 = 0 \quad z - 9 = 0$
$z = -8 \qquad z = 9$

The solutions are −8 and 9.

59. $(z - 6)(z + 1) = -10$
$z^2 - 5z - 6 = -10$
$z^2 - 5z + 4 = 0$
$(z - 1)(z - 4) = 0$
$z - 1 = 0 \quad z - 4 = 0$
$z = 1 \qquad z = 4$

The solutions are 1 and 4.

61. $(a - 4)(a + 7) = -18$
$a^2 + 3a - 28 = -18$
$a^2 + 3a - 10 = 0$
$(a + 5)(a - 2) = 0$
$a + 5 = 0 \quad a - 2 = 0$
$a = -5 \qquad a = 2$

The solutions are −5 and 2.

63. To have one positive solution and one negative solution, the factors must have one negative sign and one positive sign, so c must be less than zero

Objective B Exercises

65. **Strategy** The positive number: x
The square of the positive number is six more than five times the positive number.

Solution

$$x^2 = 5x + 6$$
$$x^2 - 5x - 6 = 0$$
$$(x - 6)(x + 1) = 0$$
$$x - 6 = 0 \quad x + 1 = 0$$
$$x = 6 \quad x = -1$$

Because -1 is not a positive number, it is not a solution. The number is 6.

67. **Strategy** One of the numbers: x
The other number: $6 - x$
The sum of the squares of the numbers is twenty.

Solution

$$x^2 + (-x + 6)^2 = 20$$
$$2x^2 - 12x + 36 = 20$$
$$2x^2 - 12x + 16 = 0$$
$$(2x - 4)(x - 4) = 0$$
$$2x - 4 = 0 \quad x - 4 = 0$$
$$x = 2 \quad x = 4$$

The numbers are 2 and 4.

69. **Strategy** First positive integer: x
Next positive integer: $x + 1$
The sum of the squares of two consecutive positive integers is 113.

Solution

$x^2 + (x + 1)^2 = 113$ is (ii).

71. **Strategy** The first positive integer: x
The next positive integer: $x + 1$
The sum of the squares of two consecutive positive integers is forty-one.

Solution

$$x^2 + (x + 1)^2 = 41$$
$$2x^2 + 2x + 1 = 41$$
$$2x^2 + 2x - 40 = 0$$
$$(2x + 10)(x - 4) = 0$$
$$2x + 10 = 0 \quad x - 4 = 0$$
$$x = -5 \quad x = 4$$

Because -5 is not a positive number, it is not a solution. The numbers are 4 and 5.

73. **Strategy** The first positive integer: x
The next positive integer: $x + 1$
The product of two consecutive positive integers is two hundred forty.

Solution

$$x(x + 1) = 240$$
$$x^2 + x - 240 = 0$$
$$(x - 15)(x + 16) = 240$$
$$x - 15 = 0 \quad x + 16 = 0$$
$$x = 15 \quad x = -16$$

Because -16 is not a positive number, it is not a solution. The numbers are 15 and 16.

75. **Strategy** Height of the triangle: x
Base of the triangle: $3x$
The area of the rectangle is 54 ft2.
The equation for the area of a triangle is $A = \frac{1}{2}bh$. Substitute in the equation and solve for x.

Solution

$$A = \frac{1}{2}bh$$
$$54 = \frac{1}{2}3x \times x$$
$$54 = \frac{3}{2}x^2$$
$$\frac{2}{3} \cdot 54 = \frac{2}{3} \cdot \frac{3}{2}x^2$$
$$36 = x^2$$
$$0 = x^2 - 36$$
$$0 = (x-6)(x+6)$$
$$x - 6 = 0 \quad x + 6 = 0$$
$$x = 6 \quad x = -6$$

Because -6 is not positive, it can not be a solution.

$3 \cdot 6 = 18$

The height is 6 ft.
The base is 18 ft.

77. **Strategy** Width of the rectangle: x
Length of the rectangle: $2x + 2$
The area of the rectangle is144 ft^2.
The equation for the area of a rectangle is $A = LW$. Substitute in the equation and solve for x.

Solution

$$A = LW$$
$$144 = (2x+2)(x)$$
$$144 = 2x^2 + 2x$$
$$0 = 2x^2 + 2x - 144$$
$$0 = 2(x^2 + x - 72)$$
$$0 = 2(x+9)(x-8)$$
$$x + 9 = 0 \quad x - 8 = 0$$
$$x = -9 \quad x = 8$$

Because -9 is not positive, it can not be a solution.

$2x + 2 = 2(8) + 2 = 18$

The length is 8 ft.
The width is 18 ft.

79. **Strategy** Side of the original square: x
Side of the larger square: $x + 4$
The area of the larger square is 64 cm^2.
The equation for the area of a square is $A = s^2$. Substitute in the equation and solve for x.

Solution

$$A = s^2$$
$$64 = (x+4)^2$$
$$64 = x^2 + 8x + 16$$
$$0 = x^2 + 8x - 48$$
$$0 = (x+12)(x-4)$$
$$x + 12 = 0 \quad x - 4 = 0$$
$$x = -12 \quad x = 4$$

Because the length of a side of a square cannot be a negative number, -12 is not a solution.
The length of a side of the original square is 4 m.

81. **Strategy** Radius of the original circle: x
Radius of the larger circle: $x + 3$
The equation for the area of a circle is $A = \pi r^2$. The area of the new circle is 100 in^2 more than the area of the original circle.

Solution

$$\pi x^2 + 100 = \pi(x+3)^2$$
$$\pi x^2 + 100 = \pi(x^2 + 6x + 9)$$
$$\pi x^2 + 100 = \pi x^2 + 6\pi x + 9\pi$$
$$\pi x^2 - \pi x^2 + 100 = \pi x^2 + 6\pi x + 9\pi$$
$$100 - 9\pi = 6\pi x + 9\pi - 9\pi$$
$$100 - 9\pi = 6\pi x$$
$$\frac{100 - 9\pi}{6\pi} = \frac{6\pi x}{6\pi}$$
$$3.81 \approx x$$

The radius of the original circle was 3.81 in.

83. **Strategy** Length of the border : x
Width of type area: $6 - 2x$
Length of type area: $9 - 2x$
The area of the type area is 28 in^2.
The equation for the area of a square is $A = lw$. Substitute in the equation and solve for x.

Solution

$$28 = (6 - 2x)(9 - 2x)$$
$$28 = 54 - 30x + 4x^2$$
$$0 = 4x^2 - 30x + 26$$
$$0 = 2(2x^2 - 15x + 13)$$
$$0 = 2(2x - 13)(x - 1)$$
$$2x - 13 = 0 \qquad x - 1 = 0$$
$$x = \frac{13}{2} \qquad x = 1$$

Since $6 - 2x = 6 - 2\left(\frac{13}{2}\right) = 6 - 13 = -7$, $\frac{13}{2}$ cannot be the border.
Width: $6 - 2(1) = 4$ in
Length: $9 - 2(1) = 7$ in

85. **Strategy** Width of the rectangle: x
Length of the rectangle: $x + 3$
The area of the rectangle is 304 ft^2.
The equation for the area of a rectangle is $A = lw$. Substitute in the equation and solve for x.

Solution

$$A = LW$$
$$304 = (x+3)(x)$$
$$304 = x^2 + 3x$$
$$0 = x^2 + 3x - 304$$
$$0 = (x+19)(x-16)$$
$$x + 19 = 0 \qquad x - 16 = 0$$
$$x = -19 \qquad x = 16$$

Because -19 is not positive, it can not be a solution.
The width is 16 ft.

87. **Strategy** Known: $d = 320$; $v = 16$

Unknown: t

Solution

$$d = vt + 16t^2$$
$$320 = 16t + 16t^2$$
$$0 = 16t^2 + 16t - 320$$
$$0 = 16(t^2 + t - 20)$$
$$0 = 16(t - 4)(t + 5)$$
$$t + 5 = 0 \qquad t - 4 = 0$$
$$t = -5 \qquad t = 4$$

Time cannot be negative, so -5 cannot be a solution. In 4 s, the object will hit the ground.

89. **Strategy** Known: $S = 120$
Unknown: n

Solution

$$S = \frac{n^2 + n}{2}$$
$$120 = \frac{n^2 + n}{2}$$
$$240 = n^2 + n$$
$$n^2 + n - 240 = 0$$
$$(n+16)(n-15) = 0$$

$n + 16 = 0 \qquad n - 15 = 0$

$n = -16 \qquad n = 15$

There cannot be a negative number of numbers, so −16 is not a solution.
Fifteen consecutive numbers beginning with 1 and giving a sum of 120.

91. **Strategy** Known: N = 45
Unknown: t

Solution

$$N = \frac{t^2 - t}{2}$$
$$45 = \frac{t^2 - t}{2}$$
$$90 = t^2 - t$$
$$t^2 - t - 90 = 0$$
$$(t+9)(t-10) = 0$$

$t + 9 = 0 \qquad t - 10 = 0$

$t = -9 \qquad t = 10$

There cannot be a negative number of teams, so −9 is not a solution.
There are 10 teams in the league.

93. **Strategy** Known: $h = 0$; $v = 48$
Unknown: t

Solution

$$h = vt - 16t^2$$
$$0 = 48t - 16t^2$$
$$0 = -16t(t-3)$$

$t = 0 \qquad t - 3 = 0$

$t = 3$

The ball will hit home plate in 3 s.

Critical Thinking

95.
$$2y(y+4) = 3(y+4)$$
$$2y^2 + 8y = 3y + 12$$
$$2y^2 + 5y - 12 = 0$$
$$(2y-3)(y+4) = 0$$

$2y - 3 = 0 \qquad y + 4 = 0$

$2y = 3 \qquad y = -4$

$y = \frac{3}{2}$

The solutions are $\frac{3}{2}$ and −4.

97.
$$(b+5)^2 = 16$$
$$b^2 + 10b + 25 = 16$$
$$b^2 + 10b + 9 = 0$$
$$(b+1)(b+9) = 0$$

$b + 1 = 0 \qquad b + 9 = 0$

$b = -1 \qquad b = -9$

The solutions are −1 and −9.

99.
$$p^3 = 7p^2$$
$$p^3 - 7p^2 = 0$$
$$p^2(p-7) = 0$$

$p^2 = 0 \qquad p - 7 = 0$

$p = 0 \qquad p = 7$

The solutions are 0 and 7.

101.
$$(x+3)(2x-1) = (3-x)(5-3x)$$
$$2x^2 + 5x - 3 = 15 - 14x + 3x^2$$
$$-x^2 + 19x - 18 = 0$$
$$x^2 - 19x + 18 = 0$$
$$(x-1)(x-18) = 0$$

$x - 1 = 0 \qquad x - 18 = 0$

$x = 1 \qquad x = 18$

The solutions are 1 or 18.

103.
$$n(n+3) = 4$$
$$n^2 + 3n - 4 = 0$$
$$(n+4)(n-1) = 0$$

$n + 4 = 0 \qquad n - 1 = 0$

$n = -4 \qquad n = 1$

$2n^3 = 2(-4)^3 \qquad 2n^3 = 2(1)^3$

$= 2(-64) \qquad = 2(1)$

$= -128 \qquad = 2$

$2n^3$ equals −128 or 2.

105. The error of accidentally dividing by zero takes place in the solution

$$x^2 = x$$
$$\frac{x^2}{x} = \frac{x}{x}$$
$$x = 1$$

Since one of the solutions of $x^2 = x$ is $x = 0$, an error occurs when the quotients in the equation $\frac{x^2}{x} = \frac{x}{x}$ are introduced. Here is the correct solution.

$$x^2 = x$$
$$x^2 - x = 0$$
$$x(x-1) = 0$$
$$x = 0 \quad x - 1 = 0$$
$$x = 1$$

The solutions are 0 and 1.

Projects and Group Activities

107. **Strategy** Width: x
Length: $x + 10$
After 2 in is cut off both ends.
New width: $x - 4$
New length: $x + 10 - 4 = x + 6$
Height: 2

Solution
$V = lwh$.

$$2(x+6)(x-4) = 192$$
$$2(x^2+2x-24) = 192$$
$$2x^2+4x-48 = 192$$
$$2x^2+4x-240 = 0$$
$$2(x^2+2x-120) = 0$$
$$2(x-10)(x+12) = 0$$
$$x - 10 = 0 \qquad x + 12 = 0$$
$$x = 10 \qquad x = -12$$

The width cannot be –12.
Width: 10 cm
Length: 10 + 10 = 20 cm

Chapter 7 Review Exercises

1. $b^2 - 13b + 30 = b - 10b - 3b + 30$
$= (b - 10b) + (-3b + 30)$
$= b(b-10) - 3(b-10)$
$= (b-10)(b-3)$

2. $4x(x-3) - 5(3-x) = 4x(x-3) + 5(x-3)$
$= (x-3)(4x+5)$

3. $2x^2 - 5x + 6$

Factors of 2	*Negative Factors of* 6
1 and 2	−1 and −6
	−2 and −3

Trial Factors	*Middle Term*
$(x-1)(2x-6)$	Common factor
$(x-2)(2x-3)$	$-3x-4x=-7x$
$(2x-1)(x-6)$	$-12x-x=-13x$
$(2x-2)(x-3)$	Common factor

$2x^2 - 5x + 6$ is nonfactorable over the integers.

4. $21x^4y^4 + 23x^2y^2 + 6$
Let $u = x^2y^2$
$21u^2 + 23u + 6$
$(3u+2)(7u+3)$
$(3x^2y^2+2)(7x^2y^2+3)$

5. $14y^9 - 49y^6 + 7y^3$
The GCF is $7y^3$: $7y^3(2y^6 - 7y^3 + 1)$

6. $y^2 + 5y - 36$
Factors of –36 whose sum is 5: 9 and –4
$y^2 + 9y - 4y - 36 = y(y+9) - 4(y+9)$
$= (y+9)(y-4)$

7. $6x^2 - 29x + 28$

Factors of 6	*Negative Factors of* 28
1 and 6	−1 and −28
2 and 3	−2 and −14
	−4 and −7

Trial Factors	*Middle Term*
$(x-1)(6x-28)$	Common factor
$(x-4)(6x-7)$	$-7x-24x=-31x$
$(2x-1)(3x-28)$	$-56x-6x=-62x$
$(2x-4)(3x-7)$	Common factor
$(2x-7)(3x-4)$	$-8x-21x=-29x$

$6x^2 - 29x + 28 = (2x-7)(3x-4)$

8. $12a^2b + 3ab^2$
The GCF is $3ab$: $3ab(4a+b)$

9. $a^6 - 100 = (a^3)^2 - 10^2 = (a^3+10)(a^3-10)$

10. $n^4 - 2n^3 - 3n^2$

The GCF is n^2: $n^2(n^2 - 2n - 3)$

Factors of –3 whose sum is –2: –3 and 1

$$n^2(n^2 - 3n + n - 3) = n^2[n(n-3) + 1(n-3)]$$
$$= n^2(n-3)(n+1)$$

11. $12y^2 + 16y - 3$

Factors of 12	*Factors of* –3
1 and 12	1 and –3
2 and 6	–1 and 3
3 and 4	

Trial Factors	*Middle Term*
$(3y+1)(4y-3)$	$-9y + 4y = -5y$
$(3y-1)(4y+3)$	$9y - 4y = 5y$
$(2y+1)(6y-3)$	$-6y + 6y = 0$
$(6y-1)(2y+3)$	$18y - 2y = 16y$

$$12y^2 + 16y - 3 = (6y-1)(2y+3)$$

12. $12b^3 - 58b^2 + 56b$

The GCF is $2b$: $2b(6b^2 - 29b + 28)$

$6 \times 28 = 168$

Factors of 168 whose sum is –29: –21 and –8

$$2b(6b^2 - 21b - 8b + 28)$$
$$2b[3b(2b-7) - 4(2b-7)]$$
$$2b(2b-7)(3b-4)$$

$$12b^3 - 58b^2 + 56b = 2b(2b-7)(3b-4)$$

13.
$$9y^4 - 25z^2 = (3y^2)^2 - (5z)^2$$
$$= (3y^2 + 5z)(3y^2 - 5z)$$

14. $c^2 + 8c + 12$

Factors of 12 whose sum is 8: 6 and 2

$$c^2 + 6c + 2c + 12 = (c^2 + 6c) + (2c + 12)$$
$$= c(c+6) + 2(c+6)$$
$$= (c+6)(c+2)$$

15. $18a^2 - 3a - 10$

$18(-10) = -180$

Factors of –180 whose sum is –3: –15 and 12

$$18a^2 - 15a + 12a - 10 = 3a(6a-5) + 2(6a-5)$$
$$= (6a-5)(3a+2)$$
$$18a^2 - 3a - 10 = (6a-5)(3a+2)$$

16.
$$4x^2 + 27x = 7$$
$$4x^2 + 27x - 7 = 0$$
$$(4x-1)(x+7) = 0$$

$4x - 1 = 0 \quad x + 7 = 0$

$4x = 1 \quad x = -7$

$x = \frac{1}{4}$

The solutions are $\frac{1}{4}$ and –7.

17. $4x^3 - 20x^2 - 24x$

The GCF is $4x$: $4x(x^2 - 5x - 6)$

Factors of –6 whose sum is –5: –6 and 1

$$4x(x^2 - 5x - 6) = 4x(x^2 - 6x + x - 6)$$
$$= 4x[(x^2 - 6x) + (x - 6)]$$
$$= 4x[x(x-6) + 1(x-6)]$$
$$= 4x(x-6)(x+1)$$

18.
$$64a^3 - 27b^3 = (4a)^3 - (3b)^3$$
$$= (4a - 3b)(16a^2 + 12ab + 9b^2)$$

19. $2a^2 - 19a - 60$

$2(-60) = -120$

Factors of –120 whose sum is –19: –24 and 5

$$2a^2 - 24a + 5a - 60$$
$$= (2a^2 - 24a) + (5a - 60)$$
$$= 2a(a-12) + 5(a-12)$$
$$= (a-12)(2a+5)$$
$$2a^2 - 19a - 60 = (a-12)(2a+5)$$

20.
$$(x+1)(x-5) = 16$$
$$x^2 - 4x - 5 = 16$$
$$x^2 - 4x - 21 = 0$$
$$(x-7)(x+3) = 0$$

$x - 7 = 0 \quad x + 3 = 0$

$x = 7 \quad x = -3$

The solutions are 7 and –3.

21. $21ax - 35bx - 10by + 6ay$

$$= (21ax - 35bx) + (-10by + 6ay)$$
$$= 7x(3a - 5b) + 2y(-5b + 3a)$$
$$= 7x(3a - 5b) + 2y(3a - 5b)$$
$$= (3a - 5b)(7x + 2y)$$

22. $36x^8 - 36x^4 + 5$

Let $u = x^4$

$36u^2 - 36u + 5$

$(6u-5)(6u-1)$

$(6x^4-5)(6x^4-1)$

23. $10x^2 + 25x + 4xy + 10y$

$= (10x^2 + 25x) + (4xy + 10y)$

$= 5x(2x+5) + 2y(2x+5)$

$= (2x+5)(5x+2y)$

24. $5x^2 - 5x - 30$

The GCF is 5: $5(x^2 - x - 6)$

$1 \times (-6) = -6$ Factors of –6 whose sum is –1: –

3 and 2

$5(x^2 - 3x + 2x - 6)$

$= 5\left[(x^2 - 3x) + (2x - 6)\right]$

$= 5[x(x-3) + 2(x-3)] = 5(x-3)(x+2)$

$5x^2 - 5x - 30 = 5(x-3)(x+2)$

25. $3x^2 + 36x + 108$

The GCF is 3: $3(x^2 + 12x + 36)$

$\sqrt{x^2} = x$ $2(x \cdot 6) = 12x$

$\sqrt{36} = 6$ The trinomial is a perfect square.

$3x^2 + 36x + 108 = 3(x+6)^2$

26. $3x^2 - 17x + 10$

$3 \cdot 10 = 30$

Factors of 30 whose sum is –17: –15 and –2

$3x^2 - 15x - 2x + 10$

$= (3x^2 - 15x) + (-2x + 10)$

$= 3x(x-5) - 2(x-5) = (x-5)(3x-2)$

$3x^2 - 17x + 10 = (x-5)(3x-2)$

27. **Strategy** • Width: x

Length: $2x - 20$

• Use the equation for the area of a rectangle:

$A = LW$

Solution

$LW = A$

$x(2x - 20) = 6000$

$2x^2 - 20x - 6000 = 0$

$2(x^2 - 10x - 3000) = 0$

$2(x - 60)(x + 50) = 0$

$x - 60 = 0$ $x + 50 = 0$

$x = 60$ $x = -50$

Because the width of a rectangle cannot be a negative number, –50 is not a solution.

$2x - 20 = 120 - 20 = 100$

The width is 60 yd and the length is 100 yd.

28. **Strategy** Known: $S = 400$

Unknown: d

Solution

$S = d^2$

$400 = d^2$

$0 = d^2 - 400$

$0 = (d+20)(d-20)$

$d + 20 = 0$ $d - 20 = 0$

$d = -20$ $d = 20$

Since the distance cannot be negative, –20 is not a solution.

The distance is 20 ft.

29. **Strategy** •Width of the picture frame: x

New width: $12+2x$

New length: $15+2x$

•Use the equation for the area of a rectangle.

Solution

$$LW = A$$
$$(12+2x)(15+2x) = 270$$
$$180+24x+30x+4x^2 = 270$$
$$4x^2+54x-90 = 0$$
$$2(2x^2+27x-45) = 0$$
$$2(2x-3)(x+15) = 0$$

$2x-3=0 \quad x+15=0$

$2x = 3 \quad x = -15$

$x = 1.5$

Because the width of a picture frame cannot be a negative number, –15 is not a solution. The width of the frame is 1.5 in.

30. **Strategy** • First consecutive positive integer: n

Second consecutive positive integer: $n+1$

• The sum of the squares is 41.

Solution

$$n^2+(n+1)^2 = 41$$
$$n^2+n^2+2n+1 = 41$$
$$2n^2+2n-40 = 0$$
$$2(n^2+n-20) = 0$$
$$2(n+5)(n-4) = 0$$

$x+5=0 \quad x-4=0$

$x=-5 \quad x=4$

Since –5 is not positive, –5 is not a solution. The integers are 4 and 5.

Chapter 7 Test

1. $ab+6a-3b-18$

$=(ab+6a)+(-3b-18)$

$=a(b+6)-3(b+6)=(b+6)(a-3)$

2. $2y^4-14y^3-16y^2$

$=2y^2(y^2-7y-8)$

$=2y^2[y^2-8y+y-8]$

$=2y^2[(y^2-8y)+(y-8)]$

$=2y^2[y(y-8)+(y-8)]$

$=2y^2(y+1)(y-8)$

3. $8x^2+20x-48$

$8(-48)=-384$

Factors of –384 whose sum is 20: 32 and –12

$8x^2+20x-48$

$=8x^2-12x+32x-48$

$=(8x^2-12x)+(32x-48)$

$=4x(2x-3)+16(2x-3)$

$=(4x+16)(2x-3)=4(x+4)(2x-3)$

4. There is no common factor.

Factor the trinomial.

Positive

Factors of 6	*Positive Factors of* 8
1 and 6	1 and 8
2 and 3	2 and 4

Trial Factors	*Middle Term*
$(x+1)(6x+8)$	Common factor
$(x+8)(6x+1)$	$x+48x=49x$
$(2x+1)(3x+8)$	$16x+3x=19x$
$(2x+8)(3x+1)$	Common factor

$6x^2+19x+8=(2x+1)(3x+8)$

5.

Factors	*Sum*
–1, –48	–49
–2, –24	–26
–3, –16	–19
–4, –12	–16
–6, –8	–14

$a^2-19a+48=(a-3)(a-16)$

6. $6x^3-8x^2+10x=2x(3x^2-4x+5)$

7.

Factors	*Sum*
–1, +15	+14
+1, –15	–14
–3, +5	+2
+3, –5	–2

$x^2+2x-15=(x+5)(x-3)$

8. $4x^2 - 1 = 0$

$(2x - 1)(2x + 1) = 0$

$2x - 1 = 0 \quad 2x + 1 = 0$

$x = \frac{1}{2} \quad x = -\frac{1}{2}$

The solutions are $-\frac{1}{2}$ and $\frac{1}{2}$.

9. $5x^2 - 45x - 15 = 5(x^2 - 9x - 3)$

10. $p^2 + 12p + 36 = (p + 6)^2$

11. $x(x - 8) = -15$

$x^2 - 8x = -15$

$x^2 - 8x + 15 = 0$

$(x - 3)(x - 5) = 0$

$x - 3 = 0 \quad x - 5 = 0$

$x = 3 \quad x = 5$

The solutions are 3 and 5.

12. $3x^2 + 12xy + 12y^2 = 3(x^2 + 4xy + 4y^2)$

$3(x + 2y)(x + 2y) = 3(x + 2y)^2$

13. $b^2 - 16 = (b + 4)(b - 4)$

14. $6x^2y^2 + 9xy^2 + 3y^2$

Common Factor $3y^2$:

$3y^2(2x^2 + 3x + 1)$

$2 \cdot 2 = 2$

Factors of 2 whose sum is 3: 2, 1

$$3y^2(2x^2 + 3x + 1) = 3y^2(2x^2 + 2x + x + 1)$$
$$= 3y^2\left[(2x^2 + 2x) + (x + 1)\right]$$
$$= 3y^2\left[2x(x + 1) + (x + 1)\right]$$
$$= 3y^2(2x + 1)(x + 1)$$

15. $27x^3 - 8 = (3x)^2 - (2)^2$

$= (3x - 2)(9x^2 + 6x + 4)$

16. $6a^4 - 13a^2 - 5$

Let $u = a^2$

$6u^2 - 13u - 5$

$(2u - 5)(3u + 1)$

$(2a^2 - 5)(3a^2 + 1)$

17. $x(p + 1) - (p + 1) = (p + 1)(x - 1)$

18. $3a^2 - 75 = 3(a^2 - 25)$

$= 3(a + 5)(a - 5)$

19. There is no common factor.

Factor the trinomial.

Factors of 2	*Factors of* -5
1 and 2	-1 and 5
	1 and -5

Trial Factors	*Middle Term*
$(x - 1)(2x + 5)$	$5x - 2x = 3x$
$(x + 1)(2x - 5)$	$-5x + 2x = -3x$

$2x^2 + 4x - 5$ is nonfactorable over the integers.

20.

Factors	*Sum*
$-1, +36$	$+35$
$+1, -36$	-35
$-2, +18$	$+16$
$+2, -18$	-16
$-3, +12$	$+9$
$+3, -12$	-9
$-4, +9$	$+5$
$+4, -9$	-5
$-6, +6$	0

$x^2 - 9x - 36 = (x + 3)(x - 12)$

21. $4a^2 - 12ab + 9b^2 = (2a - 3b)^2$

22. $4x^2 - 49y^2 = (2x + 7y)(2x - 7y)$

23. $(2a - 3)(a + 7) = 0$

$2a - 3 = 0 \quad a + 7 = 0$

$a = \frac{3}{2} \quad a = -7$

The solutions are $\frac{3}{2}$ and -7.

24. **Strategy** The first number: x

The second number: $10 - x$

The sum of the squares of the two numbers is fifty-eight.

Solution

$x^2 + (-x + 10)^2 = 58$

$2x^2 - 20x + 100 = 58$

$2x^2 - 20x + 42 = 0$

$(2x - 6)(x - 7) = 0$

$2x - 6 = 0 \quad x - 7 = 0$

$x = 3 \quad x = 7$

The numbers are 3 and 7.

25. **Strategy** The width of the rectangle: w

The length of a rectangle is 3 cm more than twice its width.

Length of the rectangle: $2w + 3$

The area of the rectangle is 90 cm^2.

Solution

$$(2w+3)w = 90$$
$$2w^2 + 3w = 90$$
$$2w^2 + 3w - 90 = 0$$
$$(2w+15)(w-6) = 0$$
$$2w+15=0 \qquad w-6=0$$
$$w=-7.5 \qquad w=6$$

Because –7.5 is not positive, it cannot be a solution. Therefore the width is 6 cm and the length is 15 cm.

Cumulative Review Exercises

1. $$-2-(-3)-5-(-11) = -2+3-5+11 = 1-5+11 = -4+11 = 7$$

2. $$(3-7)^2 \div (-2) - 3(-4) = (-4)^2 \div (-2) - 3(-4) = 16 \div (-2) - (-12) = -8+12 = 4$$

3. $$-2a^2 \div (2b) - c$$
$$-2(-4)^2 \div [2(2)] - (-1)$$
$$= -2(16) \div 4 + 1$$
$$= -32 \div 4 + 1 = -8 + 1 = -7$$

4. $$-\frac{3}{4}(-20x^2) = \frac{3 \cdot 20x^2}{4} = 15x^2$$

5. $$-2[4x-2(3-2x)-8x] = -2[4x-6+4x-8x] = -2[-6] = 12$$

6. $$-\frac{5}{7}x = -\frac{10}{21}$$
$$-\frac{7}{5}\left(-\frac{5}{7}x\right) = -\frac{7}{5}\left(-\frac{10}{21}\right)$$
$$x = \frac{2}{3}$$

The solution is $\frac{2}{3}$.

7. $$3x - 2 = 12 - 5x$$
$$3x + 5x = 12 + 2$$
$$8x = 14$$
$$x = \frac{14}{8}$$
$$x = \frac{7}{4}$$

The solution is $\frac{7}{4}$.

8. $$-2+4[3x-2(4-x)-3] = 4x+2$$
$$-2+4[3x-8+2x-3] = 4x+2$$
$$-2+4[5x-11] = 4x+2$$
$$-2+20x-44 = 4x+2$$
$$20x-46 = 4x+2$$
$$16x = 48$$
$$x = 3$$

The solution is 3.

9. $$P \cdot B = A$$
$$120\% \cdot B = 54$$
$$1.2B = 54$$
$$B = \frac{54}{1.2}$$
$$B = 45$$

The number is 45.

10. $$f(x) = -x^2 + 3x - 1$$
$$f(2) = -(2)^2 + 3(2) - 1$$
$$f(2) = -4 + 6 - 1$$
$$f(2) = 1$$

11. $$y = \frac{1}{4}x + 3$$

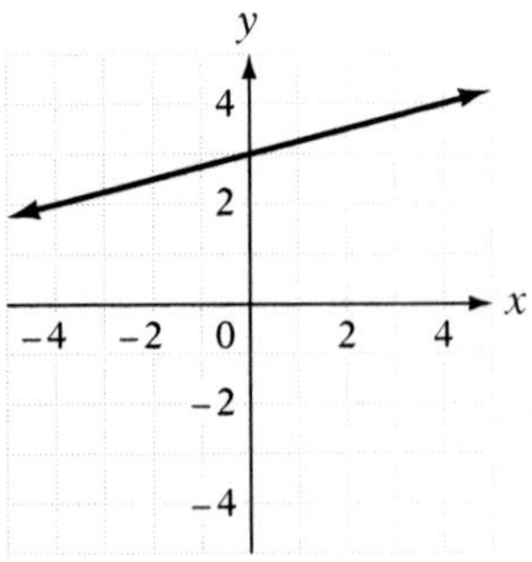

12. $5x+3y=15$

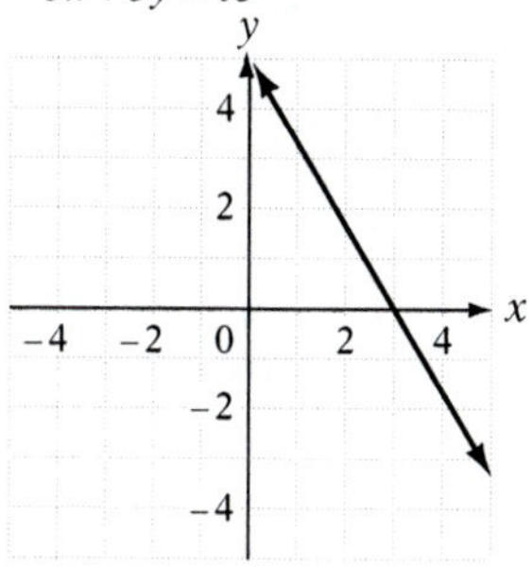

13.
$$y-4=\frac{2}{3}(x-(-3))$$
$$y-4=\frac{2}{3}(x+3)$$
$$y-4=\frac{2}{3}x+2$$
$$y=\frac{2}{3}x+6$$

14.
$$8x-y=2$$
$$y=5x+1$$

$$8x-(5x+1)=2$$
$$8x-5x-1=2$$
$$3x=3$$
$$x=1$$

$$y=5x+1$$
$$y=5(1)+1$$
$$y=6$$

The solution is (1, 6).

15.
$$5x+2y=-9 \qquad -12[5x+2y=-9]$$
$$12x-7y=2 \qquad 5[12x-7y=2]$$

$$-60x-24y=108$$
$$60x-35y=10$$
$$-59y=118$$
$$y=-2$$

$$5x+2y=-9$$
$$5x+2(-2)=-9$$
$$5x-4=-9$$
$$5x=-5$$
$$x=-1$$

The solution is (−1, −2)

16. $\left(-3a^3b^2\right)^2=9a^6b^4$

17.
$$\begin{array}{r} x^2-5x+4 \\ x+2 \\ \hline 2x^2-10x+8 \\ x^3-5x^2+4x \quad \\ \hline x^3-3x^2-6x+8 \end{array}$$

18.
$$\begin{array}{r} 4x+8 \\ 2x-3\overline{)8x^2+4x-3} \\ \underline{8x^2-12x} \quad \\ 16x-3 \\ \underline{16x-24} \\ 21 \end{array}$$

$$\left(8x^2+4x-3\right)\div(2x-3)=4x+8+\frac{21}{2x-3}$$

19. $\left(x^{-4}y^3\right)^2=x^{-8}y^6=\dfrac{y^6}{x^8}$

20.
$$3a-3b-ax+bx=(3a-3b)+(-ax+bx)$$
$$=3(a-b)-x(a-b)$$
$$=(a-b)(3-x)$$

21. $15xy^2-20xy^4$

The GCF is $5xy^2$: $5xy^2(3-4y^2)$

22. $x^2-5xy-14y^2$

Factors of −14 whose sum is −5: −7 and 2

$$x^2-7xy+2xy-14y^2=\left(x^2-7xy\right)+\left(2xy-14y^2\right)$$
$$=x(x-7y)+2y(x-7y)$$
$$=(x-7y)(x+2y)$$

23.
$$6x^2+60=39x$$
$$6x^2-39x+60=0$$
$$3\left(2x^2-13x+20\right)=0$$
$$3(2x-5)(x-4)=0$$
$$2x-5=0 \qquad x-4=0$$
$$2x=5 \qquad x=4$$
$$x=\frac{5}{2}$$

The solutions are $\frac{5}{2}$ and 4.

24.
$$18a^3+57a^2+30a$$
$$=3a\left(6a^2+19a+10\right)$$
$$=3a(3a+2)(2a+5)$$

25. $36a^2 - 49b^2 = (6a - 7b)(6a + 7b)$

26. $3x^2 + 19x - 14 = 0$
$(3x - 2)(x + 7) = 0$
$3x - 2 = 0 \quad x + 7 = 0$
$3x = 2 \quad x = -7$
$x = \frac{2}{3}$

The solutions are $\frac{2}{3}$ and –7.

27. **Strategy** • Shorter piece: x
Longer piece: $10 - x$
• Four times the length of the shorter piece is 2 ft less than three times the length of the longer piece.

Solution

$4x = 3(10 - x) - 2$
$4x = 30 - 3x - 2$
$4x = 28 - 3x$
$7x = 28$
$x = 4$
$10 - x = 10 - 4 = 6$

The shorter piece is 4 ft long and the longer piece is 6 ft long.

28. **Strategy** • To find the measure of $\angle a$, use the fact that $\angle a$ and the 72° angle are the alternate angles of parallel lines.
• To find the measure of b, use the fact that $\angle a$ and $\angle b$ are supplementary angles.

Solution

$m\angle a = 72^\circ$
$m\angle a + m\angle b = 180^\circ$
$72^\circ + m\angle b = 180^\circ$
$m\angle b = 108^\circ$
$m\angle a = 72^\circ$
$m\angle b = 108^\circ$

29. **Strategy** • Time driving to resort: x

	Rate	Time	Distance
To resort	42	x	$42x$
From resort	56	$7 - x$	$56(7 - x)$

• The distances are the same.

Solution

$42x = 56(7 - x)$
$42x = 392 - 56x$
$98x = 392$
$x = 4$
$42x = 42(4) = 168$

The distance to the resort is 168 mi.

30. **Strategy** • First integer: n
Middle integer: $n + 2$
Third integer: $n + 4$
• Five times the middle integer is twelve more than twice the sum of the first and third.

Solution

$5(n + 2) = 2(n + n + 4) + 12$
$5n + 10 = 2(2n + 4) + 12$
$5n + 10 = 4n + 8 + 12$
$5n + 10 = 4n + 20$
$n = 10$
$n + 2 = 12$
$n + 4 = 14$

The integers are 10, 12, and 14.

31. **Strategy** • Height: x

Base: $3x$

• Use the equation for the area of a triangle.

Solution

$$\frac{1}{2}bh = A$$

$$\frac{1}{2}(3x)(x) = 24$$

$$\frac{3}{2}x^2 = 24$$

$$\frac{2}{3}\left(\frac{3}{2}x^2\right) = 24 \cdot \frac{2}{3}$$

$$x^2 = 16$$

$$x = 4$$

$$3x = 3(4) = 12$$

The base of the triangle is 12 in.

Chapter 8 Rational Expressions

Prep Test

1. $10 = 2 \cdot 5$

$25 = 5 \cdot 5$

$\text{LCM} = 2 \cdot 5 \cdot 5 = 50$

2. $-\frac{3}{8} \cdot \frac{4}{9} = -\frac{\cancel{3} \cdot \cancel{2} \cdot \cancel{2}}{\cancel{2} \cdot \cancel{2} \cdot 2 \cdot \cancel{3} \cdot 3} = -\frac{1}{6}$

3. $-\frac{4}{5} \div \frac{8}{15} = -\frac{4}{5} \cdot \frac{15}{8} = -\frac{\cancel{2} \cdot \cancel{2} \cdot 3 \cdot \cancel{5}}{\cancel{5} \cdot \cancel{2} \cdot \cancel{2} \cdot 2} = -\frac{3}{2}$

4. $-\frac{5}{6} + \frac{7}{8} = -\frac{20}{24} + \frac{21}{24} = \frac{1}{24}$

5. $-\frac{3}{8} - \left(\frac{7}{12}\right) = -\frac{9}{24} + \frac{14}{24} = \frac{5}{24}$

6. $\frac{2x-3}{x^2-x+1}$

$\frac{2(2)-3}{(2)^2-2+1} = \frac{4-3}{4-2+1} = \frac{1}{3}$

7. $4(2x+1) = 3(x-2)$

$8x + 4 = 3x - 6$

$5x = -10$

$x = -2$

8. $10\left(\frac{t}{2} + \frac{t}{5}\right) = 10(1)$

$5t + 2t = 10$

$7t = 10$

$t = \frac{10}{7}$

9. Strategy • rate of second plane: r

• rate of first plane: $r - 20$

	Rate	Time	Distance
1st Plane	$r - 20$	2	$2(r - 20)$
2nd Plane	r	2	$2r$

•The sum of the two distances is 480 mi.

Solution

$2(r-20) + 2r = 480$

$2r - 40 + 2r = 480$

$4r = 520$

$r = 130$

$r - 20 = 130 - 20 = 110$

The first plane is flying 110 mph, the second is flying 130 mph.

Section 8.1

Concept Check

1. A rational expression is a fraction whose numerator and denominator are polynomials. For example: $\frac{x^2-2}{3x}$.

3. If x were equal to 4, then the denominator $x - 4$ would equal to zero and division by zero is undefined.

Objective A Exercises

5. $\frac{9x^3}{12x^4} = \frac{9}{12x^{4-3}} = \frac{3}{4x}$

7. $\frac{(x+3)^2}{(x+3)^3} = \frac{1}{(x+3)^{3-2}} = \frac{1}{x+3}$

9. $\frac{3n-4}{4-3n} = \frac{-\cancel{(4-3n)}}{\cancel{4-3n}} = -1$

11. $\frac{6y(y+2)}{9y^2(y+2)} = \frac{6\cancel{(y+2)}}{9y^{2-1}\cancel{(y+2)}} = \frac{2}{3y}$

13. $\frac{6x(x-5)}{8x^2(5-x)} = \frac{6\cancel{(x-5)}}{8x^{2-1}(-1)\cancel{(x-5)}} = -\frac{3}{4x}$

15. $\frac{a^2+4a}{ab+4b} = \frac{a\cancel{(a+4)}}{b\cancel{(a+4)}} = \frac{a}{b}$

17. $\frac{4-6x}{3x^2-2x} = \frac{-1(2)\cancel{(3x-2)}}{x\cancel{(3x-2)}} = -\frac{2}{x}$

19. $\frac{y^2-3y+2}{y^2-4y+3} = \frac{\cancel{(y-1)}(y-2)}{\cancel{(y-1)}(y-3)} = \frac{y-2}{y-3}$

21. $\frac{x^2+3x-10}{x^2+2x-8} = \frac{(x+5)\cancel{(x-2)}}{(x+4)\cancel{(x-2)}} = \frac{x+5}{x+4}$

23. $\frac{x^2+x-12}{x^2-6x+9} = \frac{(x+4)\cancel{(x-3)}}{(x-3)\cancel{(x-3)}} = \frac{x+4}{x-3}$

25. $\frac{x^2-3x-10}{25-x^2} = \frac{(x+2)\cancel{(x-5)}}{-1(x+5)\cancel{(x-5)}} = -\frac{x+2}{x+5}$

27. $\frac{2x^3+2x^2-4x}{x^3+2x^2-3x}$

$= \frac{2x(x^2+x-2)}{x(x^2+2x-3)} = \frac{2x(x+2)\cancel{(x-1)}}{x(x+3)\cancel{(x-1)}} = \frac{2(x+2)}{x+3}$

29. $\frac{6x^2-7x+2}{6x^2+5x-6} = \frac{\cancel{(3x-2)}(2x-1)}{\cancel{(3x-2)}(2x+3)} = \frac{2x-1}{2x+3}$

31. $\frac{x^2+3x-28}{24-2x-x^2} = \frac{(x+7)\cancel{(x-4)}}{-1(x+6)\cancel{(x-4)}} = -\frac{x+7}{x+6}$

Objective B Exercises

33. $\frac{8x^2}{9y^3}\cdot\frac{3y^2}{4x^3} = \frac{8\cdot 3}{9y^{3-2}4x^{3-2}} = \frac{2}{3xy}$

35. $\frac{12x^3y^4}{7a^2b^3}\cdot\frac{14a^3b^4}{9x^2y^2} = \frac{12x^{3-2}y^{4-2}\cdot 14a^{3-2}b^{4-3}}{7\cdot 9} = \frac{8abxy^2}{3}$

37. $\frac{3x-6}{5x-20}\cdot\frac{10x-40}{27x-54} = \frac{3(x-2)}{5(x-4)}\cdot\frac{10(x-4)}{27(x-2)}$

$= \frac{3\cancel{(x-2)}10\cancel{(x-4)}}{5\cancel{(x-4)}\cdot 27\cancel{(x-2)}} = \frac{2}{9}$

39. $\frac{3x^2+2x}{2xy-3y}\cdot\frac{2xy^3-3y^3}{3x^3+2x^2}$

$= \frac{x(3x+2)}{y(2x-3)}\cdot\frac{y^3(2x-3)}{x^2(3x+2)}$

$= \frac{\cancel{(3x+2)}}{\cancel{(2x-3)}}\cdot\frac{y^{3-2}\cancel{(2x-3)}}{x^{2-1}\cancel{(3x+2)}} = \frac{y^2}{x}$

41. $\frac{x^2+5x+4}{x^3y^2}\cdot\frac{x^2y^3}{x^2+2x+1}$

$= \frac{(x+4)\cancel{(x+1)}}{x^{3-2}}\cdot\frac{y^{3-2}}{\cancel{(x+1)}(x+1)} = \frac{y(x+4)}{x(x+1)}$

43. $\frac{x^4y^2}{x^2+3x-28}\cdot\frac{x^2-49}{xy^4}$

$= \frac{x^{4-1}}{\cancel{(x+7)}(x-4)}\cdot\frac{(x-7)\cancel{(x+7)}}{y^{4-2}} = \frac{x^3(x-7)}{y^2(x-4)}$

45. $\frac{2x^2-5x}{2xy+y}\cdot\frac{2xy^2+y^2}{5x^2-2x^3}$

$= \frac{x(2x-5)}{y(2x+1)}\cdot\frac{y^2(2x+1)}{-x^2(2x-5)}$

$= \frac{\cancel{(2x-5)}}{\cancel{(2x+1)}}\cdot\frac{y^{2-1}\cancel{(2x+1)}}{-1x^{2-1}\cancel{(2x-5)}} = -\frac{y}{x}$

47. $\frac{x^2-2x-24}{x^2-5x-6}\cdot\frac{x^2+5x+6}{x^2+6x+8}$

$= \frac{\cancel{(x+4)}\cancel{(x-6)}}{(x+1)\cancel{(x-6)}}\cdot\frac{\cancel{(x+2)}(x+3)}{\cancel{(x+2)}\cancel{(x+4)}} = \frac{x+3}{x+1}$

49. $\frac{x^2+2x-35}{x^2+4x-21}\cdot\frac{x^2+3x-18}{x^2+9x+18}$

$= \frac{\cancel{(x+7)}(x-5)}{\cancel{(x+7)}\cancel{(x-3)}}\cdot\frac{\cancel{(x+6)}\cancel{(x-3)}}{(x+3)\cancel{(x+6)}} = \frac{x-5}{x+3}$

51. $\dfrac{x^2-3x-4}{x^2+6x+5}\cdot\dfrac{x^2+5x+6}{8+2x-x^2}$

$=\dfrac{\overset{1}{\cancel{(x-4)}}\,\overset{1}{\cancel{(x+1)}}}{(x+5)\,\cancel{(x+1)}}\cdot\dfrac{\overset{1}{\cancel{(x+2)}}\,(x+3)}{\underset{-1}{\cancel{(4-x)}}\,\underset{1}{\cancel{(2+x)}}}=-\dfrac{x+3}{x+5}$

53. $\dfrac{16+6x-x^2}{x^2-10x-24}\cdot\dfrac{x^2-6x-27}{x^2-17x+72}$

$=\dfrac{\overset{-1}{\cancel{(8-x)}}\,\overset{1}{\cancel{(2+x)}}}{\underset{1}{\cancel{(x+2)}}\,(x-12)}\cdot\dfrac{(x+3)\,\overset{1}{\cancel{(x-9)}}}{\underset{1}{\cancel{(x-8)}}\,\underset{1}{\cancel{(x-9)}}}=-\dfrac{x+3}{x-12}$

55. $\dfrac{2x^2+5x+2}{2x^2+7x+3}\cdot\dfrac{x^2-7x-30}{x^2-6x-40}$

$=\dfrac{\overset{1}{\cancel{(2x+1)}}\,(x+2)}{\underset{1}{\cancel{(2x+1)}}\,\cancel{(x+3)}}\cdot\dfrac{\overset{1}{\cancel{(x+3)}}\,\overset{1}{\cancel{(x-10)}}}{(x+4)\,\underset{1}{\cancel{(x-10)}}}=\dfrac{x+2}{x+4}$

57. Since $a > d$, the x will be in the numerator, and since $c > b$, the y will also be in the numerator, the denominator will be 1.

59. Since $a < d$, the x will be in the denominator, and since $c = b$, there will be no y variable.   The numerator will be 1.

Objective C Exercises

61. $\dfrac{9x^3y^4}{16a^4b^2}\div\dfrac{45x^4y^2}{14a^7b}$

$=\dfrac{9x^3y^4}{16a^4b^2}\cdot\dfrac{14a^7b}{45x^4y^2}=\dfrac{9y^{4-2}\cdot14a^{7-4}}{16b^{2-1}\cdot45x^{4-3}}=\dfrac{7a^3y^2}{40bx}$

63. $\dfrac{28x+14}{45x-30}\div\dfrac{14x+7}{30x-20}$

$=\dfrac{14(2x+1)}{15(3x-2)}\cdot\dfrac{10(3x-2)}{7(2x+1)}$

$=\dfrac{14\,\overset{1}{\cancel{(2x+1)}}\cdot10\,\overset{1}{\cancel{(3x-2)}}}{15\,\underset{1}{\cancel{(3x-2)}}\cdot7\,\underset{1}{\cancel{(2x+1)}}}=\dfrac{4}{3}$

65. $\dfrac{5a^2y+3a^2}{2x^3+5x^2}\div\dfrac{10ay+6a}{6x^3+15x^2}$

$=\dfrac{a^2(5y+3)}{x^2(2x+5)}\cdot\dfrac{3x^2(2x+5)}{2a(5y+3)}$

$=\dfrac{a^{2-1}\,\overset{1}{\cancel{(5y+3)}}}{x^2\,\cancel{(2x+5)}}\cdot\dfrac{3x^2\,\overset{1}{\cancel{(2x+5)}}}{2\,\underset{1}{\cancel{(5y+3)}}}=\dfrac{3a}{2}$

67. $\dfrac{x^3y^2}{x^2-3x-10}\div\dfrac{xy^4}{x^2-x-20}$

$=\dfrac{x^3y^2}{(x-5)(x+2)}\cdot\dfrac{(x-5)(x+4)}{xy^4}$

$=\dfrac{x^{3-1}}{\cancel{(x-5)}\,(x+2)}\cdot\dfrac{\overset{1}{\cancel{(x-5)}}\,(x+4)}{y^{4-2}}=\dfrac{x^2(x+4)}{y^2(x+2)}$

69. $\dfrac{x^2y^5}{x^2-11x+30}\div\dfrac{xy^6}{x^2-7x+10}$

$=\dfrac{x^{2-1}}{\cancel{(x-5)}\,(x-6)}\cdot\dfrac{(x-2)\,\overset{1}{\cancel{(x-5)}}}{y^{6-5}}=\dfrac{x(x-2)}{y(x-6)}$

71. $\dfrac{3x^2y-9xy}{a^2b}\div\dfrac{3x^2-x^3}{ab^2}$

$=\dfrac{3xy(x-3)}{a^2b}\cdot\dfrac{ab^2}{x^2(3-x)}$

$=\dfrac{3xy\,\overset{1}{\cancel{(x-3)}}}{a^{2-1}}\cdot\dfrac{b^{2-1}}{x^{2-1}\,\underset{-1}{\cancel{(3-x)}}}=-\dfrac{3by}{ax}$

73. $\dfrac{x^2+3x-40}{x^2+2x-35}\div\dfrac{x^2+2x-48}{x^2+3x-18}$

$=\dfrac{\overset{1}{\cancel{(x+8)}}\,\overset{1}{\cancel{(x-5)}}}{(x+7)\,\cancel{(x-5)}}\cdot\dfrac{(x+6)(x-3)}{\underset{1}{\cancel{(x+8)}}\,(x-6)}=\dfrac{(x+6)(x-3)}{(x+7)(x-6)}$

75. $\dfrac{y^2-y-56}{y^2+8y+7}\div\dfrac{y^2-13y+40}{y^2-4y-5}$

$=\dfrac{\overset{1}{\cancel{(y+7)}}\,\overset{1}{\cancel{(y-8)}}}{\underset{1}{\cancel{(y+1)}}\,\underset{1}{\cancel{(y+7)}}}\cdot\dfrac{\overset{1}{\cancel{(y+1)}}\,\overset{1}{\cancel{(y-5)}}}{\underset{1}{\cancel{(y-5)}}\,\underset{1}{\cancel{(y-8)}}}=1$

77. $\dfrac{x^2-x-2}{x^2-7x+10} \div \dfrac{x^2-3x-4}{40-3x-x^2}$

$= \dfrac{\overset{1}{\cancel{(x+1)}}\overset{1}{\cancel{(x-2)}}}{\underset{1}{\cancel{(x-2)}}\underset{1}{\cancel{(x-5)}}} \cdot \dfrac{(8+x)\overset{-1}{\cancel{(5-x)}}}{\underset{1}{\cancel{(x+1)}}(x-4)} = -\dfrac{x+8}{x-4}$

79. $\dfrac{6n^2+13n+6}{4n^2-9} \div \dfrac{6n^2+n-2}{4n^2-1}$

$= \dfrac{\overset{1}{\cancel{(2n+3)}}\overset{1}{\cancel{(3n+2)}}}{\underset{1}{\cancel{(2n+3)}}(2n-3)} \cdot \dfrac{(2n+1)\overset{1}{\cancel{(2n-1)}}}{\underset{1}{\cancel{(3n+2)}}\underset{1}{\cancel{(2n-1)}}} = \dfrac{2n+1}{2n-3}$

81. $\dfrac{x+1}{x+6} \div \dfrac{x-1}{x-4} = \dfrac{x+1}{x+6} \cdot \dfrac{x-4}{x-1}$

$= \dfrac{(x+1)(x-4)}{(x+6)(x-1)} = \dfrac{x^2-3x-4}{x^2+5x-6}$ Yes.

83. $\dfrac{x-1}{x+1} \div \dfrac{x-4}{x+6} = \dfrac{x-1}{x+1} \cdot \dfrac{x+6}{x-4}$

$= \dfrac{(x-1)(x+6)}{(x+1)(x-4)} = \dfrac{x^2+5x-6}{x^2-3x-4}$ No.

Critical Thinking

85. Strategy To find where the expression is undefined, find where the denominator x^2-4x-5 is equal to zero.

Solution

$x^2-4x-5=0$

$(x-5)(x+1)=0$

$x-5=0 \qquad x+1=0$

$x=5 \qquad x=-1$

When x is 5 or -1, the expression is undefined.

87. $\dfrac{\text{shaded area}}{\text{total area}} = \dfrac{\pi(2x)^2}{\pi(5x)^2} = \dfrac{4x^2\pi}{25x^2\pi} = \dfrac{4}{25}$

89. $\dfrac{2x^2+7x-4}{3x^2-8x-3} = \dfrac{(2x-1)(x+4)}{(3x+1)(x-3)} = \dfrac{2x-1}{3x+1} \cdot \dfrac{x+4}{x-3}$

or $\dfrac{2x-1}{x-3} \cdot \dfrac{x+4}{3x+1}$

Projects or Group Activities

91. Yes. For example: $x = 3.000000001$

$\dfrac{1}{y-3}$

$\dfrac{1}{3.000000001-3} = \dfrac{1}{0.000000001}$

$= 1{,}000{,}000{,}000 > 10{,}000{,}000$

Section 8.2

Concept Check

1. **a.** $2x+6=2(x+3)$

$7x+21=7(x+3)$

$\text{LCM}: 2\cdot 7(x+3)=14(x+3)$

b. $x+4$

$x-6$

LCM: $(x+4)(x-6)$

c. $x^2-4=(x-2)(x+2)$

$x^2+3x-10=(x-2)(x+5)$

LCM: $(x-2)(x+2)(x+5)$

Objective A Exercises

3. a) six
b) four

5. The LCM is $12x^2y^4$.

$\dfrac{3}{4x^2y} = \dfrac{3}{4x^2y} \cdot \dfrac{3y^3}{3y^3} = \dfrac{9y^3}{12x^2y^4}$

$\dfrac{17}{12xy^4} = \dfrac{17}{12xy^4} \cdot \dfrac{x}{x} = \dfrac{17x}{12x^2y^4}$

7. The LCM is $6x^2(x-2)$.

$\dfrac{x-2}{3x(x-2)} = \dfrac{x-2}{3x(x-2)} \cdot \dfrac{2x}{2x} = \dfrac{2x^2-4x}{6x^2(x-2)}$

$\dfrac{3}{6x^2} = \dfrac{3}{6x^2} \cdot \dfrac{x-2}{x-2} = \dfrac{3x-6}{6x^2(x-2)}$

9. The LCM is $2x(x-5)$.

$$\frac{3x-1}{2x(x-5)}$$

$$-3x = \frac{-3x}{1} = \frac{-3x}{1}\cdot\frac{2x(x-5)}{2x(x-5)} = -\frac{6x^3-30x^2}{2x(x-5)}$$

11. The LCM is $(2x-3)(2x+3)$.

$$\frac{3x}{2x-3} = \frac{3x}{2x-3}\cdot\frac{2x+3}{2x+3} = \frac{6x^2+9x}{(2x-3)(2x+3)}$$

$$\frac{5x}{2x+3} = \frac{5x}{2x+3}\cdot\frac{2x-3}{2x-3} = \frac{10x^2-15x}{(2x-3)(2x+3)}$$

13. The LCM is $(x-3)(x+3)$.

$$\frac{2x}{x^2-9} = \frac{2x}{(x-3)(x+3)}$$

$$\frac{x+1}{x-3} = \frac{x+1}{x-3}\cdot\frac{x+3}{x+3} = \frac{x^2+4x+3}{(x-3)(x+3)}$$

15. $3x^2-12y^2 = 3(x+2y)(x-2y)$

$6x-12y = 6(x-2y)$

The LCM is $6(x+2y)(x-2y)$.

$$\frac{3}{3x^2-12y^2} = \frac{3}{3(x+2y)(x-2y)}\cdot\frac{2}{2}$$

$$= \frac{6}{6(x+2y)(x-2y)}$$

$$\frac{5}{6x-12y} = \frac{5}{6(x-2y)}\cdot\frac{x+2y}{x+2y}$$

$$= \frac{5x+10y}{6(x-2y)(x+2y)}$$

17. $x^2-1 = (x+1)(x-1)$

$x^2-2x+1 = (x-1)(x-1)$

The LCM is $(x+1)(x-1)(x-1) = (x+1)(x-1)^2$.

$$\frac{3x}{x^2-1} = \frac{3x}{(x+1)(x-1)}\cdot\frac{x-1}{x-1}$$

$$= \frac{3x^2-3x}{(x+1)(x-1)^2}$$

$$\frac{5x}{x^2-2x+1} = \frac{5x}{(x-1)(x-1)}\cdot\frac{x+1}{x+1}$$

$$= \frac{5x^2+5x}{(x+1)(x-1)^2}$$

19. $8-x^3 = -(x^3-8) = -(x-2)(x^2+2x+4)$

The LCM is $-(x-2)(x^2+2x+4)$.

$$\frac{x-3}{8-x^3} = -\frac{x-3}{(x-2)(x^2+2x+4)}$$

$$\frac{2}{4+2x+x^2} = \frac{2}{x^2+2x+4}\cdot\frac{x-2}{x-2}$$

$$= \frac{2x-4}{(x-2)(x^2+2x+4)}$$

21. $x^2+2x-3 = (x+3)(x-1)$

$x^2+6x+9 = (x+3)(x+3)$

The LCM is

$(x-1)(x+3)(x+3) = (x-1)(x+3)^2$.

$$\frac{2x}{x^2+2x-3} = \frac{2x}{(x+3)(x-1)}\cdot\frac{x+3}{x+3}$$

$$= \frac{2x^2+6x}{(x-1)(x+3)^2}$$

$$\frac{-x}{x^2+6x+9} = \frac{-x}{(x+3)(x+3)}\cdot\frac{x-1}{x-1}$$

$$= -\frac{x^2-x}{(x-1)(x+3)^3}$$

23. $4x^2 - 16x + 15 = (2x-3)(2x-5)$

$6x^2 - 19x + 10 = (2x-5)(3x-2)$

The LCM is $(2x-3)(2x-5)(3x-2)$.

$$\frac{-4x}{4x^2-16x+15} = \frac{-4x}{(2x-3)(2x-5)} \cdot \frac{3x-2}{3x-2}$$
$$= -\frac{12x^2-8x}{(2x-3)(2x-5)(3x-2)}$$
$$\frac{3x}{6x^2-19x+10} = \frac{3x}{(2x-5)(3x-2)} \cdot \frac{2x-3}{2x-3}$$
$$= \frac{6x^2-9x}{(2x-3)(2x-5)(3x-2)}$$

25. $6x^2 - 17x + 12 = (3x-4)(2x-3)$

$4 - 3x = -(3x-4)$

The LCM is $(3x-4)(2x-3)$.

$$\frac{5}{6x^2-17x+12} = \frac{5}{(3x-4)(2x-3)}$$
$$\frac{2x}{4-3x} = -\frac{2x}{3x-4} \cdot \frac{2x-3}{2x-3}$$
$$= -\frac{4x^2-6x}{(3x-4)(2x-3)}$$
$$\frac{x+1}{2x-3} = \frac{x+1}{2x-3} \cdot \frac{3x-4}{3x-4} = \frac{3x^2-x-4}{(3x-4)(2x-3)}$$

27. $15 - 2x - x^2 = -(x^2 + 2x - 15)$

$= -(x+5)(x-3)$

The LCM is $(x+5)(x-3)$.

$$\frac{2x}{x-3} = \frac{2x}{x-3} \cdot \frac{x+5}{x+5} = \frac{2x^2+10x}{(x-3)(x+5)}$$
$$\frac{-2}{x+5} = \frac{-2}{x+5} \cdot \frac{x-3}{x-3} = -\frac{2x-6}{(x-3)(x+5)}$$
$$\frac{x-1}{20-x-x^2} = -\frac{x-1}{(x-3)(x+5)}$$

Objective B Exercises

29. True

31. The LCM is $4x^2$.

$$-\frac{3}{4x^2} + \frac{8}{4x^2} - \frac{3}{4x^2} = \frac{-3+8-3}{4x^2}$$
$$= \frac{2}{4x^2} = \frac{1}{2x^2}$$

33. The LCM is $3x^2 + x - 10$.

$$\frac{3x}{3x^2+x-10} - \frac{5}{3x^2+x-10} = \frac{3x-5}{3x^2+x-10}$$
$$= \frac{3x-5}{(3x-5)(x+2)} = \frac{1}{x+2}$$

35. The LCM is $30a^2b^2$.

$$\frac{2}{5ab} - \frac{3}{10a^2b} + \frac{4}{15ab^2}$$
$$= \frac{2}{5ab} \cdot \frac{6ab}{6ab} - \frac{3}{10a^2b} \cdot \frac{3b}{3b} + \frac{4}{15ab^2} \cdot \frac{2a}{2a}$$
$$= \frac{12ab - 9b + 8a}{30a^2b^2}$$

37. The LCM is $40ab$.

$$\frac{3}{4ab} - \frac{2}{5a} + \frac{3}{10b} - \frac{5}{8ab}$$
$$= \frac{3}{4ab} \cdot \frac{10}{10} - \frac{2}{5a} \cdot \frac{8b}{8b} + \frac{3}{10b} \cdot \frac{4a}{4a} - \frac{5}{8ab} \cdot \frac{5}{5}$$
$$= \frac{30 - 16b + 12a - 25}{30ab}$$
$$= \frac{5 - 16b + 12a}{40ab}$$

39. The LCM is $12x$.

$$\frac{3x-4}{6x} - \frac{2x-5}{4x} = \frac{3x-4}{6x} \cdot \frac{2}{2} - \frac{2x-5}{4x} \cdot \frac{3}{3}$$
$$= \frac{2(3x-4) - 3(2x-5)}{12x} = \frac{6x - 8 - 6x + 15}{12x}$$
$$= \frac{7}{12x}$$

41. The LCM is $10x^2y^2$.

$$\frac{2y-4}{5xy^2}+\frac{3-2x}{10x^2y}=\frac{2y-4}{5xy^2}\cdot\frac{2x}{2x}+\frac{3-2x}{10x^2y}\cdot\frac{y}{y}$$
$$=\frac{2x(2y-4)+y(3-2x)}{10x^2y^2}=\frac{4xy-8x+3y-2xy}{10x^2y^2}$$
$$=\frac{2xy-8x+3y}{10x^2y^2}$$

43. The LCM is $(a-2)(a+1)$.

$$\frac{3a}{a-2}-\frac{5a}{a+1}=\frac{3a}{a-2}\cdot\frac{a+1}{a+1}-\frac{5a}{a+1}\cdot\frac{a-2}{a-2}$$
$$=\frac{3a(a+1)-5a(a-2)}{(a-2)(a+1)}=\frac{3a^2+3a-5a^2+10a}{(a-2)(a+1)}$$
$$=\frac{-2a^2+13a}{(a-2)(a+1)}=-\frac{a(2a-13)}{(a-2)(a+1)}$$

45. The LCM is $(2x-5)(5x-2)$.

$$\frac{x}{2x-5}-\frac{2}{5x-2}=\frac{x}{2x-5}\cdot\frac{5x-2}{5x-2}-\frac{2}{5x-2}\cdot\frac{2x-5}{2x-5}$$
$$=\frac{x(5x-2)-2(2x-5)}{(2x-5)(5x-2)}=\frac{5x^2-2x-4x+10}{(2x-5)(5x-2)}$$
$$=\frac{5x^2-6x+10}{(2x-5)(5x-2)}$$

47. The LCM is $b(a-b)$.

$$\frac{1}{a-b}+\frac{1}{b}=\frac{1}{a-b}\cdot\frac{b}{b}+\frac{1}{b}\cdot\frac{a-b}{a-b}$$
$$=\frac{b+a-b}{b(a-b)}=\frac{a}{b(a-b)}$$

49. The LCM is $a(a-3)$.

$$\frac{6a}{a-3}-5+\frac{3}{a}=\frac{6a}{a-3}\cdot\frac{a}{a}-\frac{5}{1}\cdot\frac{a(a-3)}{a(a-3)}+\frac{3}{a}\cdot\frac{a-3}{a-3}$$
$$=\frac{6a^2-5a^2+15a+3a-9}{a(a-3)}$$
$$=\frac{a^2+18a-9}{a(a-3)}$$

51. The LCM is $x(6x-5)$.

$$\frac{5}{x}-\frac{5x}{5-6x}+2$$
$$=\frac{5}{x}\cdot\frac{6x-5}{6x-5}-\frac{-5x}{6x-5}\cdot\frac{x}{x}+\frac{2}{1}\cdot\frac{x(6x-5)}{x(6x-5)}$$
$$=\frac{5(6x-5)+5x(x)+2(x)(6x-5)}{x(6x-5)}$$
$$=\frac{30x-25+5x^2+12x^2-10x}{x(6x-5)}$$
$$=\frac{17x^2+20x-25}{x(6x-5)}$$

53. $x^2-6x+9=(x-3)(x-3)$

$x^2-9=(x+3)(x-3)$

The LCM is

$(x+3)(x-3)(x-3)=(x+3)(x-3)^2$.

$$\frac{1}{x^2-6x+9}-\frac{1}{x^2-9}$$
$$=\frac{1}{(x-3)(x-3)}\cdot\frac{x+3}{x+3}-\frac{1}{(x+3)(x-3)}\cdot\frac{x-3}{x-3}$$
$$=\frac{x+3-x+3}{(x+3)(x-3)^2}$$
$$=\frac{6}{(x+3)(x-3)^2}$$

55. $x^2+4x+4=(x+2)(x+2)$

The LCM is $(x+2)(x+2)=(x+2)^2$.

$$\frac{1}{x+2}-\frac{3x}{x^2+4x+4}$$
$$=\frac{1}{(x+2)}\cdot\frac{x+2}{x+2}-\frac{3x}{(x+2)(x+2)}$$
$$=\frac{x+2-3x}{(x+2)(x+2)}=\frac{-2x+2}{(x+2)^2}$$
$$=-\frac{2(x-1)}{(x+2)^2}$$

57. $x^2+2x-8=(x+4)(x-2)$

The LCM is $(x+4)(x-2)$.

$$\frac{-3x^2+8x+2}{x^2+2x-8}-\frac{2x-5}{x+4}$$
$$=\frac{-3x^2+8x+2}{(x+4)(x-2)}-\frac{2x-5}{x+4}\cdot\frac{x-2}{x-2}$$
$$=\frac{-3x^2+8x+2-2x^2+9x-10}{(x+4)(x-2)}$$
$$=\frac{-5x^2+17x-8}{(x+4)(x-2)}$$
$$=-\frac{5x^2-17x+8}{(x+4)(x-2)}$$

59. $4x^2-36=4(x^2-9)=4(x-3)(x+3)$

The LCM is $4(x-3)(x+3)$.

$$\frac{x^2+4}{4x^2-36}-\frac{13}{x+3}$$
$$=\frac{x^2+4}{4(x-3)(x+3)}-\frac{13}{x+3}\cdot\frac{4(x-3)}{4(x-3)}$$
$$=\frac{x^2+4-13(4)(x-3)}{4(x-3)(x+3)}=\frac{x^2+4-52x+156}{4(x-3)(x+3)}$$
$$=\frac{x^2-52x+160}{4(x-3)(x+3)}$$

61. $4x^2+9x+2=(4x+1)(x+2)$

The LCM is $(4x+1)(x+2)$.

$$\frac{3x-4}{4x+1}+\frac{3x+6}{4x^2+9x+2}$$
$$=\frac{3x-4}{4x+1}\cdot\frac{x+2}{x+2}+\frac{3x+6}{(4x+1)(x+2)}$$
$$=\frac{3x^2+2x-8+3x+6}{(4x+1)(x+2)}=\frac{3x^2+5x-2}{(4x+1)(x+2)}$$
$$=\frac{(3x-1)(x+2)}{(4x+1)(x+2)}$$
$$=\frac{3x-1}{4x+1}$$

63. $x^2+x-12=(x+4)(x-3)$

$x^2+7x+12=(x+4)(x+3)$

The LCM is $(x+4)(x-3)(x+3)$.

$$\frac{x+1}{x^2+x-12}-\frac{x-3}{x^2+7x+12}$$
$$=\frac{x+1}{(x+4)(x-3)}\cdot\frac{x+3}{x+3}-\frac{x-3}{(x+4)(x+3)}\cdot\frac{x-3}{x-3}$$
$$=\frac{x^2+4x+3-x^2+6x-9}{(x+4)(x-3)(x+3)}$$
$$=\frac{10x-6}{(x+4)(x-3)(x+3)}=\frac{2(5x-3)}{(x+4)(x-3)(x+3)}$$

65. $x^2-2x-15=(x-5)(x+3)$

$5-x=-(x-5)$

The LCM is $(x-5)(x+3)$.

$$\frac{2x^2-2x}{x^2-2x-15}-\frac{2}{x+3}+\frac{x}{5-x}$$
$$=\frac{2x^2-2x}{(x-5)(x+3)}-\frac{2}{x+3}\cdot\frac{x-5}{x-5}+\frac{-(x)}{x-5}\cdot\frac{x+3}{x+3}$$
$$=\frac{2x^2-2x-2x+10-x^2-3x}{(x-5)(x+3)}$$
$$=\frac{x^2-7x+10}{(x-5)(x+3)}=\frac{(x-5)(x-2)}{(x-5)(x+3)}$$
$$=\frac{x-2}{x+3}$$

67. $3x^2-11x-20=(3x+4)(x-5)$

The LCM is $(3x+4)(x-5)$.

$$\frac{x}{3x+4}+\frac{3x+2}{x-5}-\frac{7x^2+24x+28}{3x^2-11x-20}$$
$$=\frac{x}{3x+4}\cdot\frac{x-5}{x-5}+\frac{3x+2}{x-5}\cdot\frac{3x+4}{3x+4}-\frac{7x^2+24x+28}{(3x+4)(x-5)}$$
$$=\frac{x^2-5x+9x^2+18x+8-7x^2-24x-28}{(3x+4)(x-5)}$$
$$=\frac{3x^2-11x-20}{(2x-5)(x-2)}=\frac{(3x+4)(x-5)}{(3x+4)(x-5)}$$
$$=1$$

69. $8x^2-10x+3=(4x-3)(2x-1)$

$1-2x=-(2x-1)$

The LCM is $(4x-3)(2x-1)$.

$$\frac{x+1}{1-2x}-\frac{x+3}{4x-3}+\frac{10x^2+7x-9}{8x^2-10x+3}$$
$$=\frac{-(x+1)}{2x-1}\cdot\frac{4x-3}{4x-3}-\frac{x+3}{4x-3}\cdot\frac{2x-1}{2x-1}+\frac{10x^2+7x-9}{(4x-3)(2x-1)}$$
$$=\frac{-4x^2-x+3-2x^2-5x+3+10x^2+7x-9}{(4x-3)(2x-1)}$$
$$=\frac{4x^2+x-3}{(4x-3)(2x-1)}=\frac{(4x-3)(x+1)}{(4x-3)(2x-1)}$$
$$=\frac{x+1}{2x-1}$$

71. $8x^3-1=(2x-1)(4x^2+2x+1)$

The LCM is $(2x-1)(4x^2+2x+1)$.

$$\frac{2x}{4x^2+2x+1}+\frac{4x+1}{8x^3-1}$$
$$=\frac{2x}{4x^2+2x+1}\cdot\frac{2x-1}{2x-1}+\frac{4x+1}{(2x-1)(4x^2+2x+1)}$$
$$=\frac{4x^2-2x+4x+1}{(2x-1)(4x^2+2x+1)}=\frac{4x^2+2x+1}{(2x-1)(4x^2+2x+1)}$$
$$=\frac{1}{2x-1}$$

73. $x^4-16=(x^2+4)(x^2-4)$

The LCM is $(x^2+4)(x^2-4)$.

$$\frac{x^2-12}{x^4-16}+\frac{1}{x^2-4}-\frac{1}{x^2+4}$$
$$=\frac{x^2-12}{(x^2+4)(x^2-4)}+\frac{1}{x^2-4}\cdot\frac{x^2+4}{x^2+4}-\frac{1}{x^2+4}\cdot\frac{x^2-4}{x^2-4}$$
$$=\frac{x^2-12+x^2+4-x^2+4}{(x^2+4)(x^2-4)}$$
$$=\frac{x^2-4}{(x^2+4)(x^2-4)}$$
$$=\frac{1}{x^2+4}$$

Critical Thinking

75. $$\frac{x^2-4x+4}{2x+1}\cdot\frac{2x^2+x}{x^3-4x}-\frac{3x-2}{x+1}$$
$$=\frac{(x-2)(x-2)}{2x+1}\cdot\frac{x(2x+1)}{x(x+2)(x-2)}-\frac{3x-2}{x+1}$$
$$=\frac{x-2}{x+2}-\frac{3x-2}{x+1}$$

The LCM is $(x+2)(x+1)$.

$$\frac{x-2}{x+2}-\frac{3x-2}{x+1}$$
$$=\frac{x-2}{x+2}\cdot\frac{x+1}{x+1}-\frac{3x-2}{x+1}\cdot\frac{x+2}{x+2}$$
$$=\frac{x^2-x-2-3x^2-4x+4}{x(x-2)9x+2)}=\frac{-2x^2-5x+2}{(x+2)(x+1)}$$
$$=-\frac{2x^2+5x-2}{(x+2)(x+1)}$$

77. The LCM is ab.

$$\left(\frac{a-2b}{b}+\frac{b}{a}\right)\div\left(\frac{b+a}{a}-\frac{2a}{b}\right)$$
$$=\left(\frac{a-2b}{b}\cdot\frac{a}{a}+\frac{b}{a}\cdot\frac{b}{b}\right)\div\left(\frac{b+a}{a}\cdot\frac{b}{b}-\frac{2a}{b}\cdot\frac{a}{a}\right)$$
$$=\left(\frac{a^2-2ab+b^2}{ab}\right)\div\left(\frac{b^2+ab-2a^2}{ab}\right)$$
$$=\frac{(a-b)(a-b)}{ab}\cdot\frac{ab}{(b+2a)(b-a)}$$
$$=\frac{a-b}{-(b+2a)}$$
$$=-\frac{a-b}{b+2a}=\frac{b-a}{b+2a}$$

Section 8.3

Concept Check

1. A complex fraction is a fraction whose numerator or denominator contains one or more fractions.

3. The LCM is 3.

$$\frac{2-\frac{1}{3}}{4+\frac{11}{3}}=\frac{2-\frac{1}{3}}{4+\frac{11}{3}}\cdot\frac{3}{3}=\frac{2\cdot 3-\frac{1}{3}\cdot 3}{4\cdot 3+\frac{11}{3}\cdot 3}$$
$$=\frac{6-1}{12+11}=\frac{5}{23}$$

5. The LCM is 6.

$$\frac{3-\frac{2}{3}}{5+\frac{5}{6}}=\frac{3-\frac{2}{3}}{5+\frac{5}{6}}\cdot\frac{6}{6}=\frac{3\cdot 6-\frac{2}{3}\cdot 6}{5\cdot 6+\frac{5}{6}\cdot 6}$$
$$=\frac{18-4}{30+5}=\frac{14}{35}=\frac{2}{5}$$

Objective A Exercises

7. The LCM is x^2.

$$\frac{1+\frac{1}{x}}{1-\frac{1}{x^2}}=\frac{1+\frac{1}{x}}{1-\frac{1}{x^2}}\cdot\frac{x^2}{x^2}=\frac{1\cdot x^2+\frac{1}{x}\cdot x^2}{1\cdot x^2-\frac{1}{x^2}\cdot x^2}$$
$$=\frac{x^2+x}{x^2-1}=\frac{x(x+1)}{(x-1)(x+1)}$$
$$=\frac{x}{x-1}$$

9. The LCM is a.

$$\frac{a-2}{\frac{4}{a}-a}=\frac{a-2}{\frac{4}{a}-a}\cdot\frac{a}{a}=\frac{a\cdot a-2\cdot a}{\frac{4}{a}\cdot a-a\cdot a}$$
$$=\frac{a^2-2a}{4-a^2}=\frac{a(a-2)}{(2-a)(2+a)}$$
$$=-\frac{a}{a+2}$$

11. The LCM is a^2.

$$\frac{2+\frac{1}{a}}{4-\frac{1}{a^2}}=\frac{2+\frac{1}{a}}{4-\frac{1}{a^2}}\cdot\frac{a^2}{a^2}=\frac{2a^2+a}{4a^2-1}$$
$$=\frac{a(2a+1)}{(2a+1)(2a-1)}=\frac{a}{2a-1}$$

13. The LCM is x.

$$\frac{x-\frac{1}{x}}{x+\frac{1}{x}}=\frac{x-\frac{1}{x}}{x+\frac{1}{x}}\cdot\frac{x}{x}$$
$$=\frac{x^2-1}{x^2+1}=\frac{(x-1)(x+1)}{x^2+1}$$

15. The LCM is a^2.

$$\frac{\frac{1}{a^2}-\frac{1}{a}}{\frac{1}{a^2}+\frac{1}{a}}=\frac{\frac{1}{a^2}-\frac{1}{a}}{\frac{1}{a^2}+\frac{1}{a}}\cdot\frac{a^2}{a^2}$$
$$=\frac{1-a}{1+a}=-\frac{a-1}{a+1}$$

17. The LCM is $x + 2$.

$$\frac{2-\frac{4}{x+2}}{5-\frac{10}{x+2}} = \frac{2-\frac{4}{x+2}}{5-\frac{10}{x+2}} \cdot \frac{x+2}{x+2}$$

$$= \frac{2(x+2)-4}{5(x+2)-10} = \frac{2x+4-4}{5x+10-10}$$

$$= \frac{2x}{5x} = \frac{2}{5}$$

19. The LCM is $2a - 3$.

$$\frac{\frac{3}{2a-3}+2}{\frac{-6}{2a-3}-4} = \frac{\frac{3}{2a-3}+2}{\frac{-6}{2a-3}-4} \cdot \frac{2a-3}{2a-3}$$

$$= \frac{3+2(2a-3)}{-6-4(2a-3)} = \frac{3+4a-6}{-6-8a+12}$$

$$= \frac{4a-3}{-8a+6} = \frac{4a-3}{-2(4a-3)}$$

$$= -\frac{1}{2}$$

21. The LCM is $(x - 4)(x + 1)$.

$$\frac{1-\frac{1}{x-4}}{1-\frac{6}{x+1}} = \frac{1-\frac{1}{x-4}}{1-\frac{6}{x+1}} \cdot \frac{(x-4)(x+1)}{(x-4)(x+1)}$$

$$= \frac{(x-4)(x+1)-(x+1)}{(x-4)(x+1)-6(x-4)}$$

$$= \frac{x^2-3x-4-x-1}{x^2-3x-4-6x+24}$$

$$= \frac{x^2-4x-5}{x^2-9x+20} = \frac{(x-5)(x+1)}{(x-5)(x-4)}$$

$$= \frac{x+1}{x-4}$$

23. The LCM is $(x - 3)(2 - x)$.

$$\frac{1-\frac{2}{x-3}}{1+\frac{3}{2-x}} = \frac{1-\frac{2}{x-3}}{1+\frac{3}{2-x}} \cdot \frac{(x-3)(2-x)}{(x-3)(2-x)}$$

$$= \frac{(x-3)(2-x)-2(2-x)}{(x-3)(2-x)+3(x-3)}$$

$$= \frac{2x-x^2-6+3x-4+2x}{2x-x^2-6+3x+3x-9}$$

$$= \frac{-x^2+7x-10}{-x^2+8x-15} = \frac{-(x^2-7x+10)}{-(x^2-8x+15)}$$

$$= \frac{(x-5)(x-2)}{(x-5)(x-3)}$$

$$= \frac{x-2}{x-3}$$

25. The LCM is $(2x + 3)$.

$$\frac{x-4+\frac{9}{2x+3}}{x+3-\frac{5}{2x+3}} = \frac{x-4+\frac{9}{2x+3}}{x+3-\frac{5}{2x+3}} \cdot \frac{2x+3}{2x+3}$$

$$= \frac{x(2x+3)-4(2x+3)+9}{x(2x+3)+3(2x+3)-5}$$

$$= \frac{2x^2+3x-8x-12+9}{2x^2+3x+6x+9-5}$$

$$= \frac{2x^2-5x-3}{2x^2+9x+4} = \frac{(2x+1)(x-3)}{(2x+1)(x+4)}$$

$$= \frac{x-3}{x+4}$$

27. The LCM is $(x+4)(x-3)$.

$$\frac{x-3+\frac{10}{x+4}}{x+7+\frac{16}{x-3}}=\frac{x-3+\frac{10}{x+4}}{x+7+\frac{16}{x-3}}\cdot\frac{(x+4)(x-3)}{(x+4)(x-3)}$$

$$=\frac{x(x+4)(x-3)-3(x+4)(x-3)+10(x-3)}{x(x+4)(x-3)+7(x+4)(x-3)+16(x+4)}$$

$$=\frac{(x-3)(x^2+4x-3x-12+10)}{(x+4)(x^2-3x+7x-21+16)}$$

$$=\frac{(x-3)(x^2+x-2)}{(x+4)(x^2+4x-5)}=\frac{(x-3)(x+2)(x-1)}{(x+4)(x+5)(x-1)}$$

$$=\frac{(x-3)(x+2)}{(x+4)(x+5)}$$

29. The LCM is x^2.

$$\frac{1-\frac{1}{x}-\frac{6}{x^2}}{1-\frac{4}{x}+\frac{3}{x^2}}=\frac{1-\frac{1}{x}-\frac{6}{x^2}}{1-\frac{4}{x}+\frac{3}{x^2}}\cdot\frac{x^2}{x^2}$$

$$=\frac{x^2-x-6}{x^2-4x+3}=\frac{(x-3)(x+2)}{(x-3)(x-1)}$$

$$=\frac{x+2}{x-1}$$

31. The LCM is x^2.

$$\frac{1+\frac{1}{x}-\frac{12}{x^2}}{\frac{9}{x^2}+\frac{3}{x}-2}=\frac{1+\frac{1}{x}-\frac{12}{x^2}}{\frac{9}{x^2}+\frac{3}{x}-2}\cdot\frac{x^2}{x^2}$$

$$=\frac{x^2+x-12}{9+3x-2x^2}=\frac{(x-3)(x+4)}{(3-x)(3+2x)}$$

$$=-\frac{x+4}{2x+3}$$

33. The LCM is x^2y^2.

$$\frac{\frac{1}{y^2}-\frac{1}{xy}-\frac{2}{x^2}}{\frac{1}{y^2}-\frac{3}{xy}+\frac{2}{x^2}}=\frac{\frac{1}{y^2}-\frac{1}{xy}-\frac{2}{x^2}}{\frac{1}{y^2}-\frac{3}{xy}+\frac{2}{x^2}}\cdot\frac{x^2y^2}{x^2y^2}$$

$$=\frac{x^2-xy-2y^2}{x^2-3xy+2y^2}=\frac{(x+y)(x-2y)}{(x-y)(x-2y)}$$

$$=\frac{x+y}{x-y}$$

35. The LCM is $x(x+1)$.

$$\frac{\frac{x}{x+1}-\frac{1}{x}}{\frac{x}{x+1}+\frac{1}{x}}=\frac{\frac{x}{x+1}-\frac{1}{x}}{\frac{x}{x+1}+\frac{1}{x}}\cdot\frac{x(x+1)}{x(x+1)}$$

$$=\frac{x^2-(x+1)}{x^2+(x+1)}=\frac{x^2-x-1}{x^2+x+1}$$

37. The LCM is $a(a-2)$.

$$\frac{\frac{1}{a}-\frac{3}{a-2}}{\frac{2}{a}+\frac{5}{a-2}}=\frac{\frac{1}{a}-\frac{3}{a-2}}{\frac{2}{a}+\frac{5}{a-2}}\cdot\frac{a(a-2)}{a(a-2)}$$

$$=\frac{a-2-3a}{2a-4+5a}=\frac{-2a-2}{7a-4}$$

$$=-\frac{2a+2}{7a-4}=-\frac{2(a+1)}{7a-4}$$

39. The LCM is $(x-1)(x+1)$.

$$\frac{\frac{x-1}{x+1}-\frac{x+1}{x-1}}{\frac{x-1}{x+1}+\frac{x+1}{x-1}}=\frac{\frac{x-1}{x+1}-\frac{x+1}{x-1}}{\frac{x-1}{x+1}+\frac{x+1}{x-1}}\cdot\frac{(x-1)(x+1)}{(x-1)(x+1)}$$

$$=\frac{(x-1)(x-1)-(x+1)(x+1)}{(x-1)(x-1)+(x+1)(x+1)}$$

$$=\frac{x^2-2x+1-x^2-2x-1}{x^2-2x+1+x^2+2x+1}=\frac{-4x}{2x^2+2}$$

$$=\frac{-4x}{2(x^2+1)}=-\frac{2x}{x^2+1}$$

41. The LCM is a.

$$a+\frac{a}{a+\frac{1}{a}}=a+\frac{a}{a+\frac{1}{a}}\cdot\frac{a}{a}$$

$$=a+\frac{a^2}{a^2+1}$$

The LCM is a^2+1.

$$a+\frac{a^2}{a^2+1}=a\cdot\frac{a^2+1}{a^2+1}+\frac{a^2}{a^2+1}$$

$$=\frac{a^3+a+a^2}{a^2+1}$$

$$=\frac{a^3+a^2+a}{a^2+1}=\frac{a(a^2+a+1)}{a^2+1}$$

43. $\dfrac{1}{1-\frac{1}{a}}=\dfrac{1}{1-\frac{1}{a}}\cdot\dfrac{a}{a}=\dfrac{a}{a-1}$

The reciprocal is $\dfrac{a-1}{a}$.

Critical Thinking

45. $\dfrac{x^{-1}}{y^{-1}}+\dfrac{y}{x}=\dfrac{\frac{1}{x}}{\frac{1}{y}}+\dfrac{y}{x}=\dfrac{\frac{1}{x}}{\frac{1}{y}}\cdot\dfrac{xy}{xy}+\dfrac{y}{x}=\dfrac{y}{x}+\dfrac{y}{x}=\dfrac{2y}{x}$

47. The LCM is $x(x+h)$.

$$\frac{\frac{1}{x+h}-\frac{1}{x}}{h}=\frac{\frac{1}{x+h}-\frac{1}{x}}{h}\cdot\frac{x(x+h)}{x(x+h)}$$

$$=\frac{x-(x+h)}{hx(x+h)}=\frac{-h}{hx(x+h)}$$

$$=-\frac{1}{x(x+h)}$$

Projects or Group Activities

49. a. $P(x)=\dfrac{Cx}{\left[1-\frac{1}{(x+1)^{60}}\right]}$

$$=\frac{Cx}{\left[1-\frac{1}{(x+1)^{60}}\right]}\cdot\frac{(x+1)^{60}}{(x+1)^{60}}$$

$$=\frac{Cx(x+1)^{60}}{(x+1)^{60}-1}$$

b. monthly interest rate $=\dfrac{0.08}{12}=0.006667$

$$P(0.006667)=\frac{20{,}000\cdot 0.006667(1.006667)^{60}}{1.00666y^{60}-1}$$

$$\approx 405.53$$

The monthly payment is \$405.53.

Section 8.4

Concept Check

1. Multiplication Property of Equations

3. We can clear the denominators in *equations*, as in (a), but not in *expressions*, as in (b).

Objective A Exercises

5. The following values would make a denominator 0.

$x+1=0$, $x=-1$ and $x-2=0$, $x=2$

7. The following values would make a denominator 0.

$x^2-9x=x(x-9)$

$x=0$ and $x-9=0$, $x=9$

9. The LCM is 12.

$$\frac{x}{3}-\frac{1}{4}=\frac{1}{12}$$
$$\frac{12}{1}\left(\frac{x}{3}-\frac{1}{4}\right)=\frac{12}{1}\left(\frac{1}{12}\right)$$
$$\frac{\overset{4}{\cancel{12}}}{1}\cdot\frac{x}{\underset{1}{\cancel{3}}}-\frac{\overset{3}{\cancel{12}}}{1}\cdot\frac{1}{\underset{1}{\cancel{4}}}=1$$
$$4x-3=1$$
$$4x=4$$
$$x=1$$

1 checks as a solution. The solution is 1.

11. The LCM is 18.

$$\frac{2y}{9}-\frac{1}{6}=\frac{y}{9}+\frac{1}{6}$$
$$\frac{18}{1}\left(\frac{2y}{9}-\frac{1}{6}\right)=\frac{18}{1}\left(\frac{y}{9}+\frac{1}{6}\right)$$
$$\frac{\overset{2}{\cancel{18}}}{1}\cdot\frac{2y}{\underset{1}{\cancel{9}}}-\frac{\overset{3}{\cancel{18}}}{1}\cdot\frac{1}{\underset{1}{\cancel{6}}}=\frac{\overset{2}{\cancel{18}}}{1}\cdot\frac{y}{\underset{1}{\cancel{9}}}+\frac{\overset{3}{\cancel{18}}}{1}\cdot\frac{1}{\underset{1}{\cancel{6}}}$$
$$4y-3=2y+3$$
$$2y-3=3$$
$$2y=6$$
$$y=3$$

3 checks as a solution. The solution is 3.

13. The LCM is 12.

$$\frac{3x+4}{12}-\frac{1}{3}=\frac{5x+2}{12}-\frac{1}{2}$$
$$\frac{12}{1}\left(\frac{3x+4}{12}-\frac{1}{3}\right)=\frac{12}{1}\left(\frac{5x+2}{12}-\frac{1}{2}\right)$$
$$\frac{\overset{1}{\cancel{12}}}{1}\cdot\frac{3x+4}{\underset{1}{\cancel{12}}}-\frac{\overset{4}{\cancel{12}}}{1}\cdot\frac{1}{\underset{1}{\cancel{3}}}=\frac{\overset{1}{\cancel{12}}}{1}\cdot\frac{5x+2}{\underset{1}{\cancel{12}}}-\frac{\overset{6}{\cancel{12}}}{1}\cdot\frac{1}{\underset{1}{\cancel{2}}}$$
$$3x+4-4=5x+2-6$$
$$3x=5x-4$$
$$-2x=-4$$
$$x=2$$

2 checks as a solution. The solution is 2.

15. The LCM is $3x-2$.

$$\frac{12}{3x-2}=3$$
$$\frac{\overset{1}{\cancel{3x-2}}}{1}\cdot\frac{12}{\underset{1}{\cancel{3x-2}}}=\frac{3x-2}{1}\cdot\frac{3}{1}$$
$$12=9x-6$$
$$18=9x$$
$$2=x$$

2 checks as a solution. The solution is 2.

17. The LCM is $4-3x$.

$$\frac{6}{4-3x}=3$$
$$\frac{\overset{1}{\cancel{4-3x}}}{1}\cdot\frac{6}{\underset{1}{\cancel{4-3x}}}=\frac{4-3x}{1}\cdot\frac{3}{1}$$
$$6=12-9x$$
$$-6=-9x$$
$$\frac{-6}{-9}=x$$
$$\frac{2}{3}=x$$

$\frac{2}{3}$ checks as a solution. The solution is $\frac{2}{3}$.

19. The LCM is n.

$$3+\frac{8}{n}=5$$
$$\frac{n}{1}\left(3+\frac{8}{n}\right)=\frac{n}{1}\cdot 5$$
$$\frac{n}{1}\cdot 3+\frac{n}{1}\cdot\frac{8}{n}=5n$$
$$3n+8=5n$$
$$8=2n$$
$$4=n$$

4 checks as a solution. The solution is 4.

21. The LCM is x.

$$3-\frac{12}{x}=7$$
$$\frac{x}{1}\left(3-\frac{12}{x}\right)=\frac{x}{1}\cdot 7$$
$$\frac{x}{1}.3-\frac{x}{1}\cdot\frac{12}{x}=7x$$
$$3x-12=7x$$
$$-12=4x$$
$$-3=x$$

-3 checks as a solution. The solution is -3.

23. The LCM is x.

$$\frac{6}{x}+3=11$$
$$\frac{x}{1}\left(\frac{6}{x}+3\right)=\frac{x}{1}\cdot 11$$
$$\frac{x}{1}\cdot\frac{6}{x}+\frac{x}{1}\cdot 3=11x$$
$$6+3x=11x$$
$$6=8x$$
$$\frac{6}{8}=x$$
$$\frac{3}{4}=x$$

$\frac{3}{4}$ checks as a solution. The solution is $\frac{3}{4}$.

25. The LCM is $(x+3)(x-1)$.

$$\frac{5}{x+3}=\frac{3}{x-1}$$
$$\frac{(x+3)(x-1)}{1}\cdot\frac{5}{x+3}=\frac{(x+3)(x-1)}{1}\cdot\frac{3}{x-1}$$
$$5x-5=3x+9$$
$$2x=14$$
$$x=7$$

7 checks as a solution. The solution is 7.

27. The LCM is $(3x-4)(1-2x)$.

$$\frac{5}{3x-4}=\frac{-3}{1-2x}$$
$$\frac{(3x-4)(1-2x)}{1}\cdot\frac{5}{3x-4}=\frac{(3x-4)(1-2x)}{1}\cdot\frac{-3}{1-2x}$$
$$5-10x=-9x+12$$
$$-x=7$$
$$x=-7$$

-7 checks as a solution. The solution is -7.

29. The LCM is $(5y-1)(2y-1)$.

$$\frac{4}{5y-1}=\frac{2}{2y-1}$$
$$\frac{(5y-1)(2y-1)}{1}\cdot\frac{4}{5y-1}=\frac{(5y-1)(2y-1)}{1}\cdot\frac{2}{2y-1}$$
$$8y-4=10y-2$$
$$-2y=2$$
$$y=-1$$

-1 checks as a solution. The solution is -1.

31. The LCM is $x+2$.

$$\frac{2x}{x+2}-5=\frac{7x}{x+2}$$
$$\frac{x+2}{1}\left(\frac{2x}{x+2}-5\right)=\frac{x+2}{1}\cdot\frac{7x}{x+2}$$
$$\frac{x+2}{1}\cdot\frac{2x}{x+2}-\frac{x+2}{1}\cdot 5=7x$$
$$2x-5x-10=7x$$
$$-3x-10=7x$$
$$-10x=10$$
$$x=-1$$

-1 checks as a solution. The solution is -1.

33. The LCM is $x+4$.

$$\frac{x}{x+4}=3-\frac{4}{x+4}$$
$$\frac{x+4}{1}\cdot\frac{x}{x+4}=\frac{x+4}{1}\left(3-\frac{4}{x+4}\right)$$
$$x=\frac{x+4}{1}\cdot 3-\frac{x+4}{1}\cdot\frac{4}{x+4}$$
$$x=3x+12-4$$
$$x=3x+8$$
$$-2x=8$$
$$x=-4$$

-4 does not check as a solution. The equation has no solution.

35. The LCM is $(x+12)(x+5)$.

$$\frac{x}{x+12}=\frac{1}{x+5}$$

$$\frac{\overset{1}{\cancel{(x+12)}}(x+5)}{1}\cdot\frac{x}{\underset{1}{\cancel{x+12}}}=\frac{(x+12)\overset{1}{\cancel{(x+5)}}}{1}\cdot\frac{1}{\underset{1}{\cancel{x+5}}}$$

$$x^2+5x=x+12$$

$$x^2+4x-12=0$$

$$(x+6)(x-2)=0$$

$$x+6=0 \qquad x-2=0$$

$$x=-6 \qquad x=2$$

Both −6 and 2 check as solutions. The solutions are −6 and 2.

37. The LCM is $(3n-8)(n+2)$.

$$\frac{5}{3n-8}=\frac{n}{n+2}$$

$$\frac{\overset{1}{\cancel{(3n-8)}}(n+2)}{1}\cdot\frac{5}{\underset{1}{\cancel{3n-8}}}=\frac{(3n-8)\overset{1}{\cancel{(n+2)}}}{1}\cdot\frac{n}{\underset{1}{\cancel{n+2}}}$$

$$5n+10=3n^2-8n$$

$$0=3n^2-13n-10$$

$$0=(3n+2)(n-5)$$

$$3n+2=0 \qquad n-5=0$$

$$3n=-2 \qquad n=5$$

$$n=-\frac{2}{3}$$

Both $-\frac{2}{3}$ and 5 check as solutions. The solutions are $-\frac{2}{3}$ and 5.

39. The LCM is $x-3$.

$$x-\frac{6}{x-3}=\frac{2x}{x-3}$$

$$\frac{x-3}{1}\cdot\left(x-\frac{6}{x-3}\right)=\frac{\overset{1}{\cancel{x-3}}}{1}\cdot\frac{2x}{\underset{1}{\cancel{x-3}}}$$

$$\frac{x-3}{1}\cdot x-\frac{\overset{1}{\cancel{x-3}}}{1}\cdot\frac{6}{\underset{1}{\cancel{x-3}}}=2x$$

$$x^2-3x-6=2x$$

$$x^2-5x-6=0$$

$$(x-6)(x+1)=0$$

$$x-6=0 \qquad x+1=0$$

$$x=6 \qquad x=-1$$

Both 6 and −1 check as solutions. The solutions are 6 and −1.

Critical Thinking

41. The LCM is 15.

$$\frac{3}{5}y-\frac{1}{3}(1-y)=\frac{2y-5}{15}$$

$$\frac{\overset{3}{\cancel{15}}}{1}\cdot\frac{3}{\cancel{5}}y-\frac{\overset{5}{\cancel{15}}}{1}\cdot\frac{1}{\underset{1}{\cancel{3}}}(1-y)=\frac{\overset{1}{\cancel{15}}}{1}\cdot\frac{2y-5}{\underset{1}{\cancel{15}}}$$

$$9y-5+5y=2y-5$$

$$14y-5=2y-5$$

$$12y=0$$

$$y=0$$

0 checks as a solution. The solution is 0.

43. The LCM is $(x+2)(x+1)(x-1)$.

$$\frac{x+1}{x^2+x-2}=\frac{x+2}{x^2-1}+\frac{3}{x+2}$$

$$\frac{x+1}{(x+2)(x-1)}=\frac{x+2}{(x-1)(x+1)}+\frac{3}{x+2}$$

$$\frac{\cancel{(x+2)}(x+1)\cancel{(x-1)}}{1}\cdot\frac{x+1}{\cancel{(x+2)}\cancel{(x-1)}}$$

$$=\frac{(x+2)\cancel{(x+1)}\cancel{(x-1)}}{1}\cdot\frac{x+2}{\cancel{(x-1)}\cancel{(x+1)}}$$

$$+\frac{\cancel{(x+2)}(x+1)(x-1)}{1}\cdot\frac{3}{\cancel{x+2}}$$

$$\begin{aligned}(x+1)(x+1)&=(x+2)(x+2)+(x+1)(x-1)3\\ x^2+2x+1&=x^2+4x+4+(x^2-1)3\\ x^2+2x+1&=x^2+4x+4+3x^2-3\\ x^2+2x+1&=4x^2+4x+1\\ 0&=3x^2+2x\\ 0&=x(3x+2)\end{aligned}$$

$$3x+2=0$$

$$x=0 \qquad 3x=-2$$

$$x=-\frac{2}{3}$$

Both 0 and $-\frac{2}{3}$ check as solutions. The solutions are 0 and $-\frac{2}{3}$.

Projects or Group Activities

45. Strategy To find the illumination, solve $I=\frac{s}{r^2}$ for I when $r=5$ and $s=100$.

Solution

$$\begin{aligned}I&=\frac{s}{r^2}\\ I&=\frac{100}{5^2}\\ I&=\frac{100}{25}\\ I&=4\end{aligned}$$

The illumination on the desk is 4 lm.

47. Strategy To find the candela, solve $I=\frac{s}{r^2}$ for s when $r=4$ and $I=20$.

Solution

$$\begin{aligned}I&=\frac{s}{r^2}\\ 20&=\frac{s}{4^2}\\ 20&=\frac{s}{16}\\ 16\cdot 20&=16\cdot\frac{s}{16}\\ 320&=s\end{aligned}$$

The lamp should have 320 candela.

49. Strategy To find the distance, solve $I=\frac{s}{r^2}$ for r when $s=40$ and $I=10$.

Solution

$$\begin{aligned}I&=\frac{s}{r^2}\\ 10&=\frac{40}{s^2}\\ s^2\cdot 10&=s^2\cdot\frac{40}{s^2}\\ 10s^2&=40\\ \frac{10s^2}{10}&=\frac{40}{10}\\ s^2&=4\\ x&=2\end{aligned}$$

The lamp should be placed 2 m above the desk.

Check Your Progress: Chapter 8

1. $$\frac{x^2-2x-8}{x^2-8x+16}=\frac{(x-4)(x+2)}{(x-4)(x-4)}=\frac{x+2}{x-4}$$

2. $$\frac{2x^2-11x-40}{6x^2-x-40}=\frac{(2x+5)(x-8)}{(2x+5)(3x-8)}=\frac{x-8}{3x-8}$$

3. $$\frac{x^2-3x-18}{x^2-5x-24}\cdot\frac{x^2-2x-15}{x^2+12x+27}$$

$$=\frac{(x-6)(x+3)}{(x-8)(x+3)}\cdot\frac{(x-5)(x+3)}{(x+3)(x+9)}=\frac{(x-6)(x-5)}{(x-8)(x+9)}$$

4.
$$\frac{x^2+x-72}{x^2+14x+45}\cdot\frac{2x^2+15x+25}{3x^2-15x-72}$$
$$=\frac{(x-8)(x+9)}{(x+5)(x+9)}\cdot\frac{(2x+5)(x+5)}{3(x-8)(x+3)}$$
$$=\frac{2x+5}{3x+9}$$

5.
$$\frac{6x^3y^2}{18a^4b}\div\frac{3xy}{9a^2b^5}=\frac{6x^3y^2}{18a^4b}\cdot\frac{9a^2b^5}{3xy}=\frac{b^4x^2y}{a^2}$$

6.
$$\frac{3x^2+17x-28}{x^2+2x-15}\div\frac{12x^2-13x-4}{x^2-6x+9}$$
$$=\frac{(3x-4)(x+7)}{(x+5)(x-3)}\cdot\frac{(x-3)(x-3)}{(4x+1)(3x-4)}$$
$$=\frac{(x+7)(x-3)}{(x+5)(4x+1)}$$

7.
$x^2+4x=x(x+4)$

$x^2+9x+20=(x+4)(x+5)$

LCM: $x(x+4)(x+5)$

8.
$x^2-4=(x+2)(x-2)$

$x^2+2x-8=(x+4)(x-2)$

LCM: $(x+2)(x-2)(x+4)$

9.
$$\frac{x+6}{x-1}+\frac{4x+5}{x-4}$$
$$=\frac{(x+6)}{(x-1)}\cdot\frac{(x-4)}{(x-4)}+\frac{(4x+5)}{(x-4)}\cdot\frac{(x-1)}{(x-1)}$$
$$=\frac{x^2+2x-24}{(x-1)(x-4)}+\frac{4x^2+x-5}{(x-1)(x-4)}$$
$$=\frac{5x^2+3x-29}{(x-1)(x-4)}$$

10.
$$\frac{x+9}{x-3}+\frac{3x+4}{x^2-12x+27}=\frac{x+9}{x-3}+\frac{3x+4}{(x-9)(x-3)}$$
$$=\frac{(x+9)}{(x-3)}\cdot\frac{(x-9)}{(x-9)}+\frac{3x+4}{(x-9)(x-3)}$$
$$=\frac{x^2-81}{(x-3)(x-9)}+\frac{3x+4}{(x-9)(x-3)}$$
$$=\frac{x^2+3x-77}{(x-3)(x-9)}$$

11.
$$\frac{x+9}{3x+4}-\frac{x+3}{x+1}$$
$$=\frac{(x+9)}{(3x+4)}\cdot\frac{(x+1)}{(x+1)}-\frac{(x+3)}{(x+1)}\cdot\frac{(3x+4)}{(3x+4)}$$
$$=\frac{x^2+10x+9}{(3x+4)(x+1)}-\frac{3x^2+13x+12}{(3x+4)(x+1)}$$
$$=\frac{x^2+10x+9-3x^2-13x-12}{(3x+4)(x+1)}$$
$$=\frac{-2x^2-3x-3}{(3x+4)(x+1)}=-\frac{2x^2+3x+3}{(3x+4)(x+1)}$$

12.
$$\frac{x-8}{3x^2+20x-63}-\frac{x+2}{3x-7}$$
$$=\frac{x-8}{(3x-7)(x+9)}-\frac{x+2}{3x-7}\cdot\frac{x+9}{x+9}$$
$$=\frac{x-8}{(3x-7)(x+9)}-\frac{x^2+11x+18}{(3x-7)(x+9)}$$
$$=\frac{x-8-x^2-11x-18}{(3x-7)(x+9)}$$
$$=\frac{-x^2-10x-26}{(3x-7)(x+6)}=-\frac{x^2+10x+26}{(3x-7)(x+6)}$$

13.
$$\frac{1+\frac{3}{x}}{1-\frac{9}{x^2}}=\frac{1+\frac{3}{x}}{1-\frac{9}{x^2}}\cdot\frac{x^2}{x^2}=\frac{x^2+3x}{x^2-9}$$
$$=\frac{x(x+3)}{(x+3)(x-3)}=\frac{x}{x-3}$$

14. $\dfrac{\dfrac{7}{x-3}-\dfrac{2}{3x}}{\dfrac{5}{3x}+\dfrac{1}{x-3}}=\dfrac{\dfrac{7}{x-3}-\dfrac{2}{3x}}{\dfrac{5}{3x}+\dfrac{1}{x-3}}\cdot\dfrac{3x(x-3)}{3x(x-3)}$

$$=\dfrac{21x-2x+6}{5x-15+3x}=\dfrac{19x+6}{8x-15}$$

15. $$\dfrac{5}{y+3}-2=\dfrac{7}{y+3}$$

$$\dfrac{5}{y+3}(y+3)-2(y+3)=\dfrac{7}{y+3}(y+3)$$

$$5-2y-6=7$$

$$-2y-1=7$$

$$-2y=8$$

$$y=-4$$

−4 checks as a solution. The solution is −4.

16. $$5+\dfrac{8}{a-2}=\dfrac{4a}{a-2}$$

$$5(a-2)+\dfrac{8}{a-2}(a-2)=\dfrac{4a}{a-2}(a-2)$$

$$5a-10+8=4a$$

$$5a-2=4a$$

$$a-2=0$$

$$a=2$$

2 does not check as a solution. There is no solution.

Section 8.5

Concept Check

1. A ratio is the quotient of two quantities that have the same units. A rate is the quotient of two quantities that have different units.

3. a. $\dfrac{50\text{ ft}}{4\text{ s}}$ Ratio

$$\dfrac{50\text{ ft}}{4\text{ s}}=\dfrac{25\text{ ft}}{2\text{ s}}$$

b. $\dfrac{28\text{ in}}{21\text{ in}}$ Ratio

$$\dfrac{28\text{ in}}{21\text{ in}}=\dfrac{4}{3}$$

c. $\dfrac{20\text{ mi}}{2\text{ hr}}$ Rate

$$\dfrac{20\text{ mi}}{2\text{ hr}}=\dfrac{10\text{ mi}}{1\text{ hr}}$$

d. $\dfrac{3\text{ gal}}{18\text{ gal}}$ Ratio

$$\dfrac{3\text{ gal}}{18\text{ gal}}=\dfrac{1}{6}$$

5. a. $\overline{YZ}$ **b.** $\angle R$

Objective A Exercises

7. $$\dfrac{x}{12}=\dfrac{3}{4}$$

$$\overset{1}{\cancel{12}}\cdot\dfrac{x}{\cancel{12}}=\overset{3}{\cancel{12}}\cdot\dfrac{3}{\cancel{4}}$$

$$x=9$$

The solution is 9.

9. $$\dfrac{4}{9}=\dfrac{x}{27}$$

$$\overset{3}{\cancel{27}}\cdot\dfrac{4}{\cancel{9}}=\overset{1}{\cancel{27}}\cdot\dfrac{x}{\cancel{27}}$$

$$12=x$$

The solution is 12.

11. $$\dfrac{x+3}{12}=\dfrac{5}{6}$$

$$\overset{1}{\cancel{12}}\cdot\dfrac{x+3}{\cancel{12}}=\overset{2}{\cancel{12}}\cdot\dfrac{5}{\cancel{6}}$$

$$x+3=10$$

$$x=7$$

The solution is 7.

13. $$\dfrac{18}{x+4}=\dfrac{9}{5}$$

$$5\cancel{(x+4)}\cdot\dfrac{18}{\cancel{x+4}}=\overset{1}{\cancel{5}}(x+4)\cdot\dfrac{9}{\cancel{5}}$$

$$90=9x+36$$

$$54=9x$$

$$6=x$$

The solution is 6.

15. $\frac{2}{x}=\frac{4}{x+1}$

$x(x+1)\cdot\frac{2}{x}=x(x+1)\cdot\frac{4}{x+1}$

$2x+2=4x$

$2=2x$

$1=x$

The solution is 1.

17. $\frac{x+3}{4}=\frac{x}{8}$

$8\cdot\frac{x+3}{4}=8\cdot\frac{x}{8}$

$2x+6=x$

$6=-x$

$-6=x$

The solution is −6.

19. $\frac{2}{x-1}=\frac{6}{2x+1}$

$(x-1)(2x+1)\cdot\frac{2}{x-1}=(x-1)(2x+1)\cdot\frac{6}{2x+1}$

$4x+2=6x-6$

$8=2x$

$4=x$

The solution is 4.

21. $\frac{2x}{7}=\frac{x-2}{14}$

$14\cdot\frac{2x}{7}=14\cdot\frac{x-2}{14}$

$4x=x-2$

$3x=-2$

$x=-\frac{2}{3}$

The solution is $-\frac{2}{3}$.

23. Strategy To solve for the number of voters who favor the amendment, write and solve a proportion using x to represent the number of voters who favor the amendment.

Solution

$\frac{4}{7}=\frac{x}{35,000}$

$35,000\cdot\frac{4}{7}=35,000\cdot\frac{x}{35,000}$

$20,000=x$

There are approximately 20,000 voters who favor the amendment.

25. Strategy To solve for the number of Americans with no health insurance, write and solve a proportion using x to represent the number of millions of Americans with no health insurance.

Solution

$\frac{3}{20}=\frac{x}{300}$

$300\cdot\frac{3}{20}=300\cdot\frac{x}{300}$

$45=x$

There are approximately 45 million Americans with no health insurance.

27. Strategy To solve for the number vents needed for the office building, write and solve a proportion using x to represent vents needed for the office building.

Solution

$\frac{2}{300}=\frac{x}{21,000}$

$21,000\cdot\frac{2}{300}=21,000\cdot\frac{x}{21,000}$

$140=x$

The office building would need 140 air vents.

29. Strategy To find the length of the claw, write and solve a proportion using x to represent the longest previously known scorpion claw.
Change18 in. to feet. $18 \div 12 = 1.5$ ft

Solution

$$\frac{1.5}{8.2} = \frac{x}{6.7}$$
$$8.2 \cdot 6.7 \cdot \frac{1.5}{8.2} = 8.2 \cdot 6.7 \cdot \frac{x}{6.7}$$
$$10.05 = 8.2x$$
$$1.23 = x$$

The longest previously known scorpion claw was approximately 1.23 ft.

31. Strategy To find the number of fish in the lake, write and solve a proportion using x to represent the number of fish in the lake.

Solution

$$\frac{4}{80} = \frac{40}{x}$$
$$80x \cdot \frac{4}{80} = 80x \cdot \frac{40}{x}$$
$$4x = 3200$$
$$x = 800$$

There are approximately 800 fish in the lake.

33. Strategy To find the number of panels needed, write and solve a proportion using x to represent the number of panel..

Solution

$$\frac{3}{10} = \frac{x}{600}$$
$$600 \cdot \frac{3}{10} = 600 \cdot \frac{x}{600}$$
$$180 = x$$

180 panels will needed.

35. Strategy To determine if the shipment will be accepted:, compare the two ratios.

Solution

$$\frac{3}{100} < \frac{400}{20{,}000} = \frac{2}{100}$$

Since $\frac{2}{100}$ is less than the required amount of $\frac{3}{100}$, the shipment will be accepted.

37. Strategy To find the height, write and solve a proportion using x to represent the height of the person.

Solution

$$\frac{1}{54} = \frac{1.25}{x}$$
$$54x \cdot \frac{1}{54} = 54x \cdot \frac{1.25}{x}$$
$$x = 67.5$$

The height of the person is 67.5 in.

39. Strategy To find the distance, write and solve a proportion using d to represent the distance between the two cities.

Solution

$$\frac{\frac{3}{4}}{100} = \frac{5\frac{5}{8}}{d}$$
$$\frac{\frac{3}{4}}{100} \cdot \frac{4}{4} = \frac{\frac{45}{8}}{d} \cdot \frac{8}{8}$$
$$\frac{3}{400} = \frac{45}{8d}$$
$$400d \cdot \frac{3}{400} = 400d \cdot \frac{45}{8d}$$
$$3d = 2250$$
$$d = 750$$

The distance between the two cities is 750 mi.

41. Strategy To find the pounds of fuel, write and solve a proportion using x to represent the number of pounds of fuel.

Solution

$$\frac{170{,}000}{1} = \frac{x}{\frac{3}{4}}$$

$$170{,}000 = \frac{x}{\frac{3}{4}} \cdot \frac{4}{4}$$

$$170{,}000 = \frac{4x}{3}$$

$$3 \cdot 170{,}000 = \cancel{3} \cdot \frac{4x}{\cancel{3}}$$

$$510{,}000 = 4x$$

$$127{,}500 = x$$

In 45 s, 127,500 lb of fuel are burned.

43. Strategy To find the number of gallons of yellow paint needed, write and solve a proportion using x to represent the gallons of yellow paint and $60 - x$ to represent the number of gallons of blue paint.

Solution

$$\frac{3}{5} = \frac{x}{60-x}$$

$$\cancel{5}(60-x) \cdot \frac{3}{\cancel{5}} = 5\cancel{(60-x)} \cdot \frac{x}{\cancel{(60-x)}}$$

$$180 - 3x = 5x$$

$$180 - 3x + 3x = 5x + 3x$$

$$180 = 8x$$

$$22.5 = x$$

22.5 gal of yellow paint will be needed.

45. Strategy To find number of additional acres needed:

- find the number of acres needed for 1320 bushels by writing and solving a proportion using x to represent the number of acres needed.
- subtract 50 acres from the number of acres needed to get the additional number of acres needed.

Solution

$$\frac{50}{1100} = \frac{x}{1320}$$

$$\overset{12}{\cancel{13{,}200}} \cdot \frac{50}{\underset{1}{\cancel{1100}}} = \overset{10}{\cancel{13{,}200}} \cdot \frac{x}{\underset{1}{\cancel{1320}}}$$

$$600 = 10x$$

$$60 = x$$

$60 - 50 = 10$

10 additional acres are needed.

Objective B

47. Strategy To find AC, write a proportion using the fact that in similar triangles, the ratios of corresponding sides are equal. Solve the proportion for AC.

Solution

$$\frac{AC}{AB} = \frac{DF}{DE}$$

$$\frac{AC}{4} = \frac{15}{9}$$

$$\overset{9}{\cancel{36}} \cdot \frac{AC}{\underset{1}{\cancel{4}}} = \frac{15}{\underset{1}{\cancel{9}}} \cdot \overset{4}{\cancel{36}}$$

$$9AC = 60$$

$$AC \approx 6.7$$

The length of AC is 6.7 cm.

49. Strategy To find the height of triangle ABC, write a proportion using the fact that in similar triangles, the ratio of corresponding sides equals the ratio of corresponding heights. Solve the proportion of the height.

Solution

$$\frac{h_{ABC}}{BC} = \frac{h_{DFE}}{FE}$$

$$\frac{h}{5} = \frac{7}{12}$$

$$\overset{12}{\cancel{60}} \cdot \frac{h}{\cancel{5}} = \frac{7}{\cancel{12}} \cdot \overset{5}{\cancel{60}}$$

$$12h = 35$$

$$h \approx 2.9$$

The height of triangle ABC is 2.9 m.

51. Strategy To find the perimeter:

- Find side DF by writing a proportion using the fact that the ratios of corresponding sides of similar triangles are equal.
- Use the formula for the perimeter of a triangle.

Solution

$$\frac{AC}{BC} = \frac{DF}{EF}$$

$$\frac{5}{6} = \frac{DF}{9}$$

$$\overset{3}{\cancel{18}} \cdot \frac{5}{\cancel{6}} = \frac{DF}{\cancel{9}} \cdot \overset{2}{\cancel{18}}$$

$$15 = 2DF$$

$$7.5 = DF$$

$$P = a + b + c = 7.5 + 9 + 6 = 22.5$$

The perimeter of triangle DEF is 22.5 ft.

53. Strategy To find the area:

- Find the height of triangle ABC by writing a proportion using the fact that in similar triangles, the ratio of corresponding sides equals the ratio of corresponding heights. Solve the proportion for the height (h).
- Use the formula for the area of a triangle.

Solution

$$\frac{h_{ABC}}{AB} = \frac{h_{DEF}}{DE}$$

$$\frac{h}{12} = \frac{12}{18}$$

$$\overset{3}{\cancel{36}} \cdot \frac{h}{\cancel{12}} = \frac{12}{\cancel{18}} \cdot \overset{2}{\cancel{36}}$$

$$3h = 24$$

$$h = 8$$

$$A = \frac{1}{2}bh = \frac{1}{2}(12)(8) = 48$$

The area of triangle ABC is 48 m^2.

55. Strategy To find BC, write a proportion using the fact that in similar triangles, the ratios of corresponding sides are equal. Solve the proportion for BC.

Solution

$$\frac{BD}{BC} = \frac{AE}{AC}$$

$$\frac{5}{BC} = \frac{8}{10}$$

$$10(BC)\frac{5}{\cancel{BC}} = \frac{8}{\cancel{10}}(BC)\overset{1}{\cancel{10}}$$

$$50 = 8BC$$

$$BC = 6.25$$

The length of BC is 6.25 cm.

57. Strategy To find *DA*,

• write a proportion using the fact that in similar triangles, the ratios of corresponding sides are equal. Solve the proportion for *BD*.

• subtract the length of *BD* from the length of *BA* (15 in) to get the length of *DA*.

Solution

$$\frac{AB}{BD}=\frac{AC}{DE}$$

$$\frac{15}{BD}=\frac{10}{6}$$

$$6(BD)\frac{15}{BD}=\frac{10}{6}6(BD)$$

$$90=10BD$$

$$9=BD$$

$15-9=6$

The length of *DA* is 6 in.

59. Strategy Triangle *MNO* is similar to triangle *PQD*. Solve a proportion to find the length of *OP*. Let *x* represent the length of *OP* and $39-x$ represent the length of *OM*.

Solution

$$\frac{NO}{OQ}=\frac{OM}{OP}$$

$$\frac{24}{12}=\frac{39-x}{x}$$

$$12x\cdot\frac{24}{12}=\frac{39-x}{x}\cdot 12x$$

$$24x=468-12x$$

$$36x=468$$

$$x=13$$

The length of *OP* is 13 cm.

61. True

63. Strategy To find the width of the river, write a proportion using the fact that in similar triangles, the ratios of corresponding sides are equal. Solve the proportion for the width *CD*.

Solution

$$\frac{BO}{AB}=\frac{OC}{CD}$$

$$\frac{8}{14}=\frac{20}{CD}$$

$$14(CD)\frac{8}{14}=\frac{20}{CD}CD14$$

$$8CD=280$$

$$CD=35$$

The width of the river is 35 m.

Critical Thinking

65. Strategy Let *x* equal the number and $\frac{1}{x}$ be the reciprocal. The sum of a number and its reciprocal is $\frac{26}{5}$.

Solution

$$x+\frac{1}{x}=\frac{26}{5}$$

$$5x\cdot x+5x\cdot\frac{1}{x}=5x\cdot\frac{26}{5}$$

$$5x^2+5=26x$$

$$5x^2-26x+5=0$$

$$(5x-1)(x-5)=0$$

$$5x-1=0 \qquad x-5=0$$

$$5x=1 \qquad x=5$$

$$x=\frac{1}{5}$$

The number is 5 or $\frac{1}{5}$.

67. **Strategy** Write and solve a proportion to find the number of foul shots made. Let x be the number of shots attempted and $x - 42$ be the number of foul shots made.

Solution

$$\frac{5}{6} = \frac{x-42}{x}$$
$$6x \cdot \frac{5}{6} = 6x \cdot \frac{x-42}{x}$$
$$5x = 6x - 252$$
$$5x - 6x = 6x - 6x - 252$$
$$-x = -252$$
$$x = 252$$

$252 - 42 = 210$

The player made 210 foul shots.

Projects or Group Activities

69. a. **Strategy** Write and solve a proportion to find the circumference of the earth using x for the circumference.

Solution

$$\frac{7.5°}{520} = \frac{360°}{x}$$
$$520x \cdot \frac{7.5}{520} = 520x \cdot \frac{360}{x}$$
$$7.5x = 187{,}200$$
$$x = 24960$$

Eratosthenes' calculation would be 24,960 mi.

b. **Strategy** Subtract the accepted value (24,800) from Eratosthenes' estimate.

Solution

$24{,}960 - 24{,}874 = 86$

Eratosthenes' estimate is 86 mi different from the accepted value.

Section 8.6

1. True

3. R

Objective A Exercises

5.
$$d = rt$$
$$\frac{d}{r} = \frac{rt}{r}$$
$$\frac{d}{r} = t$$

7.
$$PV = nRT$$
$$\frac{PV}{nR} = \frac{nRT}{nR}$$
$$\frac{PV}{nR} = T$$

9.
$$P = 2l + 2w$$
$$P - 2w = 2l + 2w - 2w$$
$$P - 2w = 2l$$
$$\frac{P-2w}{2} = \frac{2l}{2}$$
$$\frac{P-2w}{2} = l$$

11.
$$A = \frac{1}{2}h(b_1 + b_2)$$
$$2 \cdot A = 2 \cdot \frac{1}{2}h(b_1 + b_2)$$
$$2A = h(b_1 + b_2)$$
$$2A = hb_1 + hb_2$$
$$2A - hb_2 = hb_1 + hb_2 - hb_2$$
$$2A - hb_2 = hb_1$$
$$\frac{2A - hb_2}{h} = \frac{hb_1}{h}$$
$$\frac{2A - hb_2}{h} = b_1$$

13.
$$V = \frac{1}{3}Ah$$
$$3 \cdot V = 3 \cdot \frac{1}{3}Ah$$
$$3V = Ah$$
$$\frac{3V}{A} = \frac{Ah}{A}$$
$$\frac{3V}{A} = h$$

15.
$$R=\frac{C-S}{t}$$
$$t\cdot R=t\cdot\frac{C-S}{t}$$
$$Rt=C-S$$
$$Rt-C=C-C-S$$
$$Rt-C=-S$$
$$-1(Rt-C)=-1(-S)$$
$$C-Rt=S$$

17.
$$A=P+Prt$$
$$A=P(1+rt)$$
$$\frac{A}{1+rt}=\frac{P(1+rt)}{1+rt}$$
$$\frac{A}{1+rt}=P$$

19.
$$A=Sw+w$$
$$A=w(S+1)$$
$$\frac{A}{S+1}=\frac{w(S+1)}{S+1}$$
$$\frac{A}{S+1}=w$$

21.
$$3x+y=10$$
$$3x-3x+y=-3x+10$$
$$y=-3x+10$$

23.
$$4x-y=3$$
$$4x-4x-y=-4x+3$$
$$-y=-4x+3$$
$$-1(-y)=-1(-4x+3)$$
$$y=4x-3$$

25.
$$3x+2y=6$$
$$3x-3x+2y=-3x+6$$
$$2y=-3x+6$$
$$\frac{2y}{2}=\frac{-3x+6}{2}$$
$$y=-\frac{3}{2}x+3$$

27.
$$2x-5y=10$$
$$2x-2x-5y=-2x+10$$
$$-5y=-2x+10$$
$$\frac{-5y}{-5}=\frac{-2x+10}{-5}$$
$$y=\frac{2}{5}x-2$$

29.
$$2x+7y=14$$
$$2x-2x+7y=-2x+14$$
$$7y=-2x+14$$
$$\frac{7y}{7}=\frac{-2x+14}{7}$$
$$y=-\frac{2}{7}x+2$$

31.
$$x+3y=6$$
$$x-x+3y=-x+6$$
$$3y=-x+6$$
$$\frac{3y}{3}=\frac{-x+6}{3}$$
$$y=-\frac{1}{3}x+2$$

33.
$$x+3y=6$$
$$x+3y-3y=-3y+6$$
$$x=-3y+6$$

35.
$$3x-y=3$$
$$3x-y+y=y+3$$
$$3x=y+3$$
$$\frac{3x}{3}=\frac{y+3}{3}$$
$$x=\frac{1}{3}y+1$$

37.
$$2x+5y=10$$
$$2x+5y-5y=-5y+10$$
$$2x=-5y+10$$
$$\frac{2x}{2}=\frac{-5y+10}{2}$$
$$x=-\frac{5}{2}y+5$$

39.
$$x-2y+1=0$$
$$x-2y+1-1=0-1$$
$$x-2y=-1$$
$$x-2y+2y=2y-1$$
$$x=2y-1$$

41. a.

$$A = P(1+i)$$
$$A = P + Pi$$
$$A - P = P - P + Pi$$
$$A - P = Pi$$
$$\frac{A-P}{P} = \frac{Pi}{P}$$
$$\frac{A-P}{P} = i$$

$$A = P(1+i)$$
$$\frac{A}{P} = \frac{P(1+i)}{P}$$
$$\frac{A}{P} = 1+i$$
$$\frac{A}{P} - 1 = 1 - 1 + i$$
$$\frac{A}{P} - 1 = i$$

Yes

b.

$$A = P(1+i)$$
$$A = P + Pi$$
$$A - A - Pi = P + Pi - Pi - A$$
$$-Pi = P - A$$
$$\frac{-Pi}{-P} = \frac{P-A}{-P}$$
$$i = -\frac{P-A}{P}$$

$$A = P(1+i)$$
$$A = P + Pi$$
$$A - P = P - P + Pi$$
$$A - P = Pi$$
$$\frac{A-P}{P} = \frac{Pi}{P}$$
$$\frac{A-P}{P} = i$$

Yes

Critical Thinking

43.

$$\frac{1}{R_1} + \frac{1}{R_2} = \frac{1}{R}$$
$$\cancel{R_1} \cdot R_2 \cdot R \cdot \frac{1}{\cancel{R_1}} + R_1 \cdot \cancel{R_2} \cdot R \cdot \frac{1}{\cancel{R_2}} = R_1 \cdot R_2 \cdot \cancel{R} \cdot \frac{1}{\cancel{R}}$$
$$R_2R + R_1R = R_1R_2$$
$$R_2R - R_2R + R_1R = R_1R_2 - R_2R$$
$$R_1R = R_2(R_1 - R)$$
$$\frac{RR_1}{R_1 - R} = \frac{R_2(R_1 - R)}{(R_1 - R)}$$
$$\frac{RR_1}{R_1 - R} = R_2$$

Section 8.7

Concept Check

1. The rate of work is the amount of a task that is competed per unit of time.

3. $\frac{1}{x}$

5. Jen

7. **a.** $\frac{1}{x}$ of the job per hour

   b. $\frac{3}{x}$

9. $\frac{5}{h}$

11. **a.** 8 + 4 = 12 mph

    b. 8 – 4 = 4 mph

Objective A Exercises

13.

	Rate	•	Time	=	Part of Job
Elect.	$\frac{1}{10}$	•	t	=	$\frac{t}{10}$
Assist	$\frac{1}{12}$	•	t	=	$\frac{t}{12}$

15. Strategy • time to fill the fountain together: t

	Rate	Time	Part
First	$\frac{1}{3}$	t	$\frac{t}{3}$
Second	$\frac{1}{6}$	t	$\frac{t}{6}$

Solution

$$\frac{t}{3} + \frac{t}{6} = 1$$
$$6 \cdot \left(\frac{t}{3} + \frac{t}{6}\right) = 6 \cdot 1$$
$$2t + t = 6$$
$$3t = 6$$
$$t = 2$$

It would take 2 h to fill the fountain room.

17. Strategy • Time to remove the earth with both skiploaders working together: t

	Rate	Time	Part
First	$\frac{1}{12}$	t	$\frac{t}{12}$
Second	$\frac{1}{4}$	t	$\frac{t}{4}$

• The sum of the parts of the task completed by each must equal 1.

Solution

$$\frac{t}{12}+\frac{t}{4}=1$$
$$12\left(\frac{t}{12}+\frac{t}{4}\right)=12\cdot 1$$
$$t+3t=12$$
$$4t=12$$
$$t=3$$

With both skiploaders, it would take 3 h to remove the earth.

19. Strategy • Time to solve with both computers working: t

	Rate	Time	Part
Small	$\frac{1}{75}$	t	$\frac{t}{75}$
Large	$\frac{1}{50}$	t	$\frac{t}{50}$

• The sum of the parts completed by each air conditioner must equal 1.

Solution

$$\frac{t}{75}+\frac{t}{50}=1$$
$$150\left(\frac{t}{75}+\frac{t}{50}\right)=150\cdot 1$$
$$2t+3t=150$$
$$5t=150$$
$$t=30$$

It would take 30 h to solve the problem with both computers working.

21. Strategy • Time cool the room with both air conditioners: t

	Rate	Time	Part
Smaller	$\frac{1}{60}$	t	$\frac{t}{60}$
Larger	$\frac{1}{40}$	t	$\frac{t}{40}$

• The sum of the parts printed by each press must equal 1.

Solution

$$\frac{t}{60}+\frac{t}{40}=1$$
$$240\left(\frac{t}{60}+\frac{t}{40}\right)=240\cdot 1$$
$$4t+6t=240$$
$$10t=240$$
$$t=24$$

It would take 24 min for both air conditioners.

23. Strategy • Time for the second welder to compete the job: t

	Rate	Time	Part
First welder	$\frac{1}{10}$	6	$\frac{6}{10}$
Second welder	$\frac{1}{t}$	6	$\frac{6}{t}$

• The sum of the parts printed by each press must equal 1.

Solution

$$\frac{6}{10}+\frac{6}{t}=1$$
$$10t\left(\frac{6}{10}+\frac{6}{t}\right)=10t\cdot 1$$
$$6t+60=10t$$
$$6t-6t+60=10t-6t$$
$$60=4t$$
$$15=t$$

It would take 15 h for the second welder working alone.

25. Strategy • Time for the second pipeline to fill the tank: t

	Rate	Time	Part
First pipeline	$\frac{1}{45}$	30	$\frac{30}{45}$
Second pipeline	$\frac{1}{t}$	30	$\frac{30}{t}$

• The sum of the parts completed by both machines must equal 1.

Solution

$$\frac{30}{45}+\frac{30}{t}=1$$
$$45t\left(\frac{30}{45}+\frac{30}{t}\right)=45t\cdot 1$$
$$30t+1350=45t$$
$$1350=15t$$
$$90=t$$

It would take the second pipeline 90 min to fill the tank.

27. Strategy • Time for the old machine working alone to complete the wall: t

	Rate	Time	Part
New reaper	$\frac{1}{1.5}$	1	$\frac{1}{1.5}$
Old reaper	$\frac{1}{t}$	1	$\frac{1}{t}$

• The sum of the parts completed by both must equal 1.

Solution

$$\frac{1}{1.5}+\frac{1}{t}=1$$
$$1.5t\left(\frac{1}{1.5}+\frac{1}{t}\right)=1.5t\cdot 1$$
$$t+1.5=1.5t$$
$$t-t+1.5=1.5t-t$$
$$1.5=0.5t$$
$$3=t$$

It would take the old reaper 3 h alone.

29. Strategy • Time for the second technician to complete the task: t

	Rate	Time	Part
First tech	$\frac{1}{4}$	2	$\frac{2}{4}$
Second tech	$\frac{1}{6}$	t	$\frac{t}{6}$

• The sum completed by each must equal 1.

Solution

$$\frac{2}{4}+\frac{t}{6}=1$$
$$12\left(\frac{2}{4}+\frac{t}{6}\right)=12\cdot 1$$
$$6+2t=12$$
$$2t=6$$
$$t=3$$

It would take the second technician 3 h to complete the wiring.

31. Strategy • Time for the smaller heating unit: t

	Rate	Time	Part
Large heating unit	$\frac{1}{8}$	2	$\frac{2}{8}$
Second heating unit	$\frac{1}{t}$	11	$\frac{11}{t}$

• The sum of the parts of the task must equal 1.

Solution

$$\frac{2}{8}+\frac{11}{t}=1$$
$$8t\left(\frac{2}{8}+\frac{11}{t}\right)=8t\cdot 1$$
$$2t+88=8t$$
$$88=6t$$
$$14\frac{2}{3}=t$$

The small heating unit would take $14\frac{2}{3}$ h to heat the pool.

33. Strategy • Time for the apprentice to complete the repairs: t

	Rate	Time	Part
Mechanic	$\frac{1}{2}$	1	$\frac{1}{2}$
Apprentice	$\frac{1}{6}$	t	$\frac{t}{6}$

• The sum of the parts completed by each worker must equal 1.

Solution

$$\frac{1}{2}+\frac{t}{6}=1$$
$$6\left(\frac{1}{2}+\frac{t}{6}\right)=6\cdot 1$$
$$3+t=6$$
$$t=3$$

It would take 3 h to complete the repairs.

35. Less than k

Objective B Exercises

37.

a.	*Dist.*	÷	*Rate*	=	*Time*
Against	1440	÷	$380-r$	=	$\frac{1440}{380-r}$
With	1600	÷	$380+r$	=	$\frac{1600}{380+r}$

b. $\frac{1440}{3880-r}=\frac{1600}{380+r}$

39. Strategy • Rate walking: r

• Rate driving: $10r$

	Distance	Rate	Time
Driving	80	$10r$	$\frac{80}{10r}$
Hiking	4	r	$\frac{4}{r}$

• The total time of the trip was 3 h.

Solution

$$\frac{80}{10r}+\frac{4}{r}=3$$
$$10r\left(\frac{80}{10r}+\frac{4}{r}\right)=10r(3)$$
$$80+40=30r$$
$$120=30r$$
$$4=r$$

The hiking was 4 mph.

41. Strategy • Rate of helicopter: r

• Rate of jet: $4r$

	Distance	Rate	Time
Helicopter	180	r	$\frac{180}{r}$
Jet	1080	$4r$	$\frac{1080}{4r}$

• The total time of the trip was 5 h.

Solution

$$\frac{180}{r}+\frac{1080}{4r}=5$$
$$4r\left(\frac{180}{r}+\frac{1080}{4r}\right)=4r\cdot 5$$
$$720+1080=20r$$
$$1800=20r$$
$$90=r$$
$$4r=90\cdot 4=360$$

The rate of the jet was 360 mph.

43. Strategy • First rate: r

• Second rate: $r + 2$

	Distance	Rate	Time
1st rate	15	r	$\frac{15}{r}$
2nd rate	19	$r+2$	$\frac{19}{r+2}$

• The total time of the trip was 4 h.

Solution

$$\frac{15}{r}+\frac{19}{r+2}=4$$
$$r(r+2)\left(\frac{15}{r}+\frac{19}{r+2}\right)=r(r+2)(4)$$
$$15r+30+19r=4r^2+8r$$
$$0=4r^2-26r-30$$
$$0=2(2r^2-13r-15)$$
$$0=2(2r-15)(r+1)$$

$2r-15=0 \qquad r+1=0$

$2r=15 \qquad r=-1$

$r=\frac{15}{2}$

Since the rate cannot be −1, the rate for the first 15 mi was 7.5 mph.

45. Strategy • Rate of the technician through congested traffic: r

	Distance	Rate	Time
Congested traffic	10	r	$\frac{10}{r}$
Expressway	20	$r+$	$\frac{20}{r+20}$

• The total time of the trip was 1 h.

Solution

$$\frac{10}{r}+\frac{20}{r+20}=1$$
$$r(r+20)\left(\frac{10}{r}+\frac{20}{r+20}\right)=r(r+20)\cdot 1$$
$$10(r+20)+20r=r(r+20)$$
$$10r+200+20r=r^2+20r$$
$$30r+200=r^2+20r$$
$$0=r^2-10r-200$$
$$0=(r-20)(r+10)$$

$r-20=0 \qquad r+10=0$

$r=20 \qquad r=-10$

The solution −10 is not possible because the rate cannot be negative. The rate of travel in congested traffic was 20 mph.

47. Strategy • Distance on way: d

	Distance	Rate	Time
With current	d	$15+3$	$\frac{d}{18}$
Against current	d	$15-3$	$\frac{d}{12}$

• The total time traveled is 3 h.

Solution

$$\frac{d}{18}+\frac{d}{12}=3$$
$$36\left(\frac{d}{18}+\frac{d}{12}\right)=36(3)$$
$$2d+3d=108$$
$$5d=108$$
$$d=21.6$$

The family can travel 21.6 mi downstream.

49. Strategy • Rate of the current: r

	Distance	Rate	Time
With current	25	$20+r$	$\frac{25}{20+r}$
Against current	15	$20-r$	$\frac{15}{20-r}$

• The time traveled with the current equals the time traveled against the current.

Solution

$$\frac{25}{20+r}=\frac{15}{20-r}$$
$$(20+r)(20-r)\frac{25}{20+r}=(20+r)(20-r)\frac{15}{20-r}$$
$$500-25r=300+15r$$
$$200=40r$$
$$5=r$$

The rate of the current is 5 mph.

51. Strategy • Rate of the freight train: r

• Rate of the express train: $r+20$

	Distance	Rate	Time
Freight train	360	r	$\frac{360}{r}$
Express train	600	$r+20$	$\frac{600}{r+20}$

• The time of the freight train equals the time of the express train.

Solution

$$\frac{360}{r}=\frac{600}{r+20}$$
$$r(r+20)\cdot\frac{360}{r}=r(r+20)\cdot\frac{600}{r+20}$$
$$360(r+20)=600r$$
$$360r+7200=600r$$
$$7200=240r$$
$$30=r$$
$$50=r+20$$

The rate of the freight train is 30 mph.
The rate of the express train is 50 mph.

53. Strategy • Rate of the current: r

	Distance	Rate	Time
With current	24	$6+r$	$\frac{24}{6+r}$
Against gulf current	12	$6-r$	$\frac{12}{6-r}$

• The time traveled with the current equals the time traveled against the gulf current.

Solution

$$\frac{24}{6+r}=\frac{12}{6-r}$$
$$(6+r)(6-r)\left(\frac{24}{6+r}\right)=(6+r)(6-r)\left(\frac{12}{6-r}\right)$$
$$144-24r=72+12r$$
$$72=36r$$
$$2=r$$

The rate of the current is 2 mph.

55. Strategy • Rate of the jet stream: r

	Distance	Rate	Time
With jet stream	2400	$550+r$	$\frac{2400}{550+r}$
Against jet stream	2000	$550-r$	$\frac{2000}{550-r}$

• The time traveled with the jet stream equals the time traveled against the jet stream.

Solution

$$\frac{2400}{550+r}=\frac{2000}{550-r}$$
$$\overset{1}{\cancel{(550+r)}}(550-r)\frac{2400}{\underset{1}{\cancel{550+r}}}=(550+r)\overset{1}{\cancel{(550-r)}}\frac{2000}{\underset{1}{\cancel{550-r}}}$$
$$(550-r)2400=2000(550+r)$$
$$1{,}320{,}000-2400r=1{,}100{,}000+2000r$$
$$220{,}000=4400r$$
$$50=r$$

The rate of the jet stream is 50 mph.

57. Strategy • Rate of the current: r

	Distance	Rate	Time
With current	25	$20+r$	$\frac{25}{20+r}$
Against current	15	$20-r$	$\frac{15}{20-r}$

• The time traveled with the current equals the time traveled against the current.

Solution

$$\frac{25}{20+r}=\frac{15}{20-r}$$

$$(20+r)(20-r)\frac{25}{20+r}=(20+r)(20-r)\frac{15}{20-r}$$

$$500-25r=300+15r$$

$$200=40r$$

$$5=r$$

The rate of the current is 5 mph.

Critical Thinking

59. Strategy • Time to fill the tank with all three pipes working: t

	Rate	Time	Part
First pipe	$\frac{1}{2}$	t	$\frac{t}{2}$
Second pipe	$\frac{1}{4}$	t	$\frac{t}{4}$
Third pipe	$\frac{1}{5}$	t	$\frac{t}{5}$

• The sum of the parts completed by each pipe must equal 1.

Solution

$$\frac{t}{2}+\frac{t}{4}+\frac{t}{5}=1$$

$$20\left(\frac{t}{2}+\frac{t}{4}+\frac{t}{5}\right)=20\cdot 1$$

$$10t+5t+4t=20$$

$$19t=20$$

$$t=\frac{20}{19}=1\frac{1}{19}$$

It would take $1\frac{1}{19}$ h to fill the tank with all three pipes working.

61. Strategy To find the canoeing time

• Hiking rate: r

• Canoe rate: $3r$

	Distance	Rate	Time
Canoe	18	$3r$	$\frac{18}{3r}$
Hike	3	r	$\frac{3}{r}$

• The time hiking was 1 h less than the time canoeing.

• Find the time canoeing.

Solution

$$\frac{3}{r}=\frac{18}{3r}-1$$

$$3r\left(\frac{3}{r}\right)=3r\left(\frac{18}{3r}-1\right)$$

$$9=18-3r$$

$$-9=-3r$$

$$3=r$$

Time canoeing: $\frac{18}{3r}=\frac{18}{3(3)}=\frac{18}{9}=2$

The time canoeing was 2 h.

Projects or Group Activities

63. Strategy • Usual speed: r

• Speed in bad weather: $r - 10$

	Distance	Rate	Time
Usual	150	r	$\frac{150}{r}$
Bad weather	150	$r - 10$	$\frac{150}{r-10}$

• The time during bad weather is $\frac{1}{2}$ h more than the usual time.

Solution

$$\frac{150}{r-10} = \frac{150}{r} + \frac{1}{2}$$
$$2r(r-10)\cdot\frac{150}{r-10} = 2r(r-10)\left(\frac{150}{r} + \frac{1}{2}\right)$$
$$300r = 300(r-10) + r(r-10)$$
$$300r = 300r - 3000 + r^2 - 10r$$
$$0 = r^2 - 10r - 3000$$
$$0 = (r-60)(r+50)$$

$r - 60 = 0$ $\quad$ $r + 50 = 0$

$r = 60$ $\quad$ $r = -50$

The solution −50 is not possible because the rate cannot be negative. The bus usually travels 60 mph.

Section 8.8

Concept Check

1. $y = kx$

3. $z = kxy$

Objective A Exercises

5. Strategy: Write the basic direct variation equation replacing the variable with the given values. Solve for k.
Write the direct variation equation replacing k with its value. Substitute 5000 for s and solve for P.

Solution:

$P = ks$

$4000 = 250k$

$16 = k$

$P = 16s = 16(5000) = 80{,}000$

When the company sells 5000 products its profit will be $80,000.

7. Strategy: Write the basic direct variation equation replacing the variable with the given values. Solve for k.
Write the direct variation equation replacing k with its value. Substitute 15 for d and solve for p.

Solution:

$p = kd$

$4.5 = 10k$

$0.45 = k$

$p = 0.45d = 0.45(15) = 6.75$

The pressure is 6.75 lb/in^2.

9. **Strategy:** Write the basic direct variation equation replacing the variable with the given values. Solve for *k*.
Write the direct variation equation replacing *k* with its value. Substitute 10 for *t* and solve for *d*.

Solution:

$d = kt^2$

$144 = k(3)^2$

$144 = 9k$

$16 = k$

$d = 16t^2 = 16(10^2) = 16(100) = 1600$

In 10 s the object will fall 1600 ft.

11. **Strategy:** Write the basic inverse variation equation replacing the variable with the given values. Solve for *k*.
Write the inverse variation equation replacing *k* with its value. Substitute 5 for *n* and solve for *T*.

Solution:

$T = \frac{k}{n}$

$500 = \frac{k}{1}$

$500 = k$

$T = \frac{500}{n} = \frac{500}{5} = 100$

It will take 5 computers 100 s to solve the same problem.

13. **Strategy:** Write the basic direct variation equation replacing the variable with the given values. Solve for *k*.
Write the direct variation equation replacing *k* with its value. Substitute 15 for *v* and solve for *L*.

Solution:

$L = kv^2$

$640 = k(20)^2$

$640 = 400k$

$1.6 = k$

$L = 1.6v^2 = 1.6(15)^2 = 360$

The load on the sail will be 360 lbs.

15. **Strategy:** Write the basic direct variation equation replacing the variable with the given values. Solve for *k*.
Write the direct variation equation replacing *k* with its value. Substitute 86 for *d* and solve for *b*.

Solution:

$b = kd$

$3.1 = k56$

$0.055357 = k$

$b = 0.055357d = 0.055357(86) = 4.8$

4.8 million barrels were spilled.

17. Strategy: Write the basic combined variation equation replacing the variable with the given values. Solve for *k*. Write the combined variation equation replacing *k* with its value. Substitute 24 for *r*, 180 for *v* and solve for *I*.

Solution:

$I = \frac{kv}{r}$

$10 = \frac{110k}{11}$

$110 = 110k$

$1 = k$

$I = \frac{1v}{r} = \frac{180}{24} = 7.5$

The current is 7.5 amps.

19. Strategy: Write the basic inverse variation equation replacing the variable with the given values. Solve for *k*. Write the inverse variation equation replacing *k* with its value. Substitute 5 for *d* and solve for *I*.

Solution:

$I = \frac{k}{d^2}$

$12 = \frac{k}{10^2}$

$12 = \frac{k}{100}$

$1200 = k$

$I = \frac{1200}{d^2} = \frac{1200}{5^2} = \frac{1200}{25} = 48$

The intensity is 48 foot-candles when the distance is 5 ft.

Critical Thinking

21. inversely

Projects or Group Activities

23. If x is doubled than y is doubled.

25. y is divided by 4.

Chapter 8 Review Exercises

1. $\frac{6a^2b^7}{25x^3y} \div \frac{12a^3b^4}{5x^2y^2} = \frac{6a^2b^7}{25x^3y}\cdot\frac{5x^2y^2}{12a^3b^4}$

$= \frac{\overset{1}{\cancel{6}}\cdot\overset{1}{\cancel{5}}a^2b^7x^2y^2}{\underset{5}{\cancel{25}}\cdot\underset{2}{\cancel{12}}a^3b^4x^3y} = \frac{b^3y}{10ax}$

2. $\frac{x+7}{15x} + \frac{x-2}{20x} = \frac{4}{4}\left(\frac{x+7}{15x}\right) + \frac{3}{3}\left(\frac{x-2}{20x}\right)$

$= \frac{4x+28}{60x} + \frac{3x-6}{60x} = \frac{7x+22}{60x}$

3. $\frac{3x^3+9x^2}{6xy^2-18y^2}\cdot\frac{4xy^3-12y^3}{5x^2+15x}$

$\frac{3x^2\overset{1}{\cancel{(x+3)}}}{6y^2\underset{1}{\cancel{(x-3)}}}\cdot\frac{4y^3\overset{1}{\cancel{(x-3)}}}{5x\underset{1}{\cancel{(x+3)}}} = \frac{\overset{1}{\cancel{3}}x^{2-1}\overset{1}{\cancel{2}}\cdot 2y^{3-1}}{\underset{1}{\cancel{2}}\cdot\underset{1}{\cancel{3}}\cdot 5}$

$= \frac{2xy}{5}$

4. $\frac{2x(x-y)}{x^2y(x+y)} \div \frac{3(x-y)}{x^2y^2} = \frac{2x\overset{1}{\cancel{(x-y)}}}{x^2y(x+y)}\cdot\frac{x^2y^2}{3\underset{1}{\cancel{(x-y)}}}$

$= \frac{2xy}{3(x+y)}$

5. $\frac{x-\frac{16}{5x-2}}{3x-4-\frac{88}{5x-2}} = \frac{x-\frac{16}{5x-2}}{3x-4-\frac{88}{5x-2}}\cdot\frac{5x-2}{5x-2}$

$= \frac{x(5x-2)-\frac{16}{\underset{1}{\cancel{5x-2}}}\cdot\overset{1}{\cancel{(5x-2)}}}{3x(5x-2)-4(5x-2)-\frac{88}{\underset{1}{\cancel{5x-2}}}\cdot\overset{1}{\cancel{(5x-2)}}}$

$= \frac{5x^2-2x-16}{15x^2-6x-20x+8-88} = \frac{5x^2-2x-16}{15x^2-26x-80}$

$= \frac{\overset{1}{\cancel{(5x+8)}}(x-2)}{\underset{1}{\cancel{(5x+8)}}(3x-10)} = \frac{x-2}{3x-10}$

6. $\frac{x^2+x-30}{15+2x-x^2} = \frac{(x+6)\overset{1}{\cancel{(x-5)}}}{(3+x)\underset{-1}{\cancel{(5-x)}}} = -\frac{x+6}{x+3}$

7. $$\frac{16x^5y^3}{24xy^{10}}=\frac{\cancel{2}\cdot\cancel{2}\cdot\cancel{2}\cdot 2x^{5-1}}{\cancel{2}\cdot\cancel{2}\cdot\cancel{2}\cdot 3y^{10-3}}=\frac{2x^4}{3y^7}$$

8.
$$\frac{20}{x+2}=\frac{5}{16}$$
$$16\cancel{(x+2)}\cdot\frac{20}{\cancel{x+2}}=\cancel{16}(x+2)\cdot\frac{5}{\cancel{16}}$$
$$320=5x+10$$
$$310=5x$$
$$62=x$$

62 checks as a solution. 62 is the solution.

9.
$$\frac{10-23x+12y^2}{6y^2-y-5}\div\frac{4y^2-13y+10}{18y^2+3y-10}$$
$$=\frac{12y^2-23y+10}{6y^2-y-5}\cdot\frac{18y^2+3y-10}{4y^2-13y+10}$$
$$=\frac{(3y-2)\cancel{(4y-5)}}{\cancel{(6y+5)}(y-1)}\cdot\frac{(3y-2)\cancel{(6y+5)}}{\cancel{(4y-5)}(y-2)}$$
$$=\frac{(3y-2)^2}{(y-1)(y-2)}$$

10.
$$3ax-x=5$$
$$x(3a-1)=5$$
$$\frac{x(3a-1)}{3a-1}=\frac{5}{3a-1}$$
$$x=\frac{5}{3a-1}$$

11.
$$\frac{2}{x}+\frac{3}{4}=1$$
$$4x\cdot\frac{2}{x}+4x\cdot\frac{3}{4}=4x\cdot 1$$
$$8+3x=4x$$
$$8=x$$

8 checks as a solution. The solution is 8.

12.
$$\frac{x}{y}+\frac{3}{x}=\frac{x}{y}\cdot\frac{x}{x}+\frac{3}{x}\cdot\frac{y}{y}$$
$$=\frac{x^2}{xy}+\frac{3y}{xy}=\frac{x^2+3y}{xy}$$

13.
$$5x+4y=20$$
$$5x-5x+4y=-5x+20$$
$$4y=-5x+20$$
$$\frac{4y}{4}=\frac{-5x+20}{4}$$
$$y=-\frac{5}{4}x+5$$

14.
$$\frac{8ab^2}{15x^3y}\cdot\frac{5xy^4}{16a^2b}$$
$$\frac{8\cdot 5ab^2xy^4}{15\cdot 16a^2bx^3y}=\frac{by^3}{6ax^2}$$

15.
$$\frac{1-\frac{1}{x}}{1-\frac{8x-7}{x^2}}=\frac{1-\frac{1}{x}}{1-\frac{8x-7}{x^2}}\cdot\frac{x^2}{x^2}$$
$$=\frac{1\cdot x^2-\frac{1}{x}\cdot x^2}{1\cdot x^2-\frac{8x-7}{x^2}\cdot x^2}=\frac{x^2-x}{x^2-8x+7}$$
$$=\frac{x\cancel{(x-1)}}{\cancel{(x-1)}(x-7)}=\frac{x}{x-7}$$

16.
$$\frac{x}{12x^2+16x-3},\ \frac{4x^2}{6x^2+7x-3}$$
$$\frac{x}{(6x-1)(2x+3)},\ \frac{4x^2}{(2x+3)(3x-1)}$$

The LCM is $(6x-1)(2x+3)(3x-1)$.

$$\frac{x}{(6x-1)(2x+3)}\cdot\frac{3x-1}{3x-1}=\frac{3x^2-x}{(6x-1)(2x+3)(3x-1)}$$
$$\frac{4x^2}{(2x+3)(3x-1)}\cdot\frac{6x-1}{6x-1}=\frac{24x^3-4x^2}{(6x-1)(2x+3)(3x-1)}$$

17.
$$T=2(ab+bc+ca)$$
$$T=2ab+2bc+2ca$$
$$T-2bc=2a(b+c)$$
$$\frac{T-2bc}{2(b+c)}=a$$
$$\frac{T-2bc}{2b+2C}=a$$

18.
$$\frac{5}{7}+\frac{x}{2}=2-\frac{x}{7}$$
$$14\left(\frac{5}{7}+\frac{x}{2}\right)=14\left(2-\frac{x}{7}\right)$$
$$\overset{2}{\cancel{14}}\cdot\frac{5}{\underset{1}{\cancel{7}}}+\overset{7}{\cancel{14}}\cdot\frac{x}{\underset{1}{\cancel{2}}}=14\cdot 2-\overset{2}{\cancel{14}}\cdot\frac{x}{\underset{1}{\cancel{7}}}$$
$$10+7x=28-2x$$
$$9x=18$$
$$x=2$$

2 checks as a solution. The solution is 2.

19.
$$\frac{2+\frac{1}{x}}{3-\frac{2}{x}}=\frac{2+\frac{1}{x}}{3-\frac{2}{x}}\cdot\frac{x}{x}$$
$$=\frac{2\cdot x+\frac{1}{x}\cdot x}{3\cdot x-\frac{2}{x}\cdot x}=\frac{2x+1}{3x-2}$$

20.
$$\frac{2x}{x-5}-\frac{x+1}{x-2}=\frac{2x}{x-5}\cdot\frac{x-2}{x-2}-\frac{x+1}{x-2}\cdot\frac{x-5}{x-5}$$
$$=\frac{2x^2-4x}{(x-5)(x-2)}-\frac{x^2-4x-5}{(x-5)(x-2)}$$
$$=\frac{2x^2-4x-(x^2-4x-5)}{(x-5)(x-2)}=\frac{2x^2-4x-x^2+4x+5}{(x-5)(x-2)}$$
$$=\frac{x^2+5}{(x-5)(x-2)}$$

21.
$$i=\frac{100m}{c}$$
$$c(i)=\frac{100m}{c}(c)$$
$$ci=100m$$
$$c=\frac{100m}{i}$$

22.
$$\frac{x+8}{x+4}=1+\frac{5}{x+4}$$
$$\overset{1}{\cancel{(x+4)}}\left(\frac{x+8}{\underset{1}{\cancel{x+4}}}\right)=(x+4)\cdot 1+\overset{1}{\cancel{(x+4)}}\left(\frac{5}{\underset{1}{\cancel{x+4}}}\right)$$
$$x+8=x+4+5$$
$$0=1$$

There is no solution.

23.
$$\frac{20x^2-45x}{6x^3+4x^2}\div\frac{40x^3-90x^2}{12x^2+8x}$$
$$=\frac{20x^2-45x}{6x^3+4x^2}\cdot\frac{12x^2+8x}{40x^3-90x^2}$$
$$=\frac{5x(4x-9)}{2x^2(3x+2)}\cdot\frac{4x(3x+2)}{10x^2(4x-9)}$$
$$=\frac{\overset{1}{\cancel{5}}x\overset{1}{\cancel{(4x-9)}}}{\underset{1}{\cancel{2}}x^2\underset{1}{\cancel{(3x+2)}}}\cdot\frac{\overset{1}{\cancel{2}}\cdot\overset{1}{\cancel{2}}\cdot x\overset{1}{\cancel{(3x+2)}}}{\underset{1}{\cancel{5}}\cdot\underset{1}{\cancel{2}}x^2\underset{1}{\cancel{(4x-9)}}}=\frac{1}{x^2}$$

24.
$$\frac{2y}{5y-7}+\frac{3}{7-5y}=\frac{2y}{5y-7}+\frac{3}{-1(5y-7)}$$
$$=\frac{2y}{5y-7}-\frac{3}{5y-7}=\frac{2y-3}{5y-7}$$

25.
$$\frac{5x+3}{2x^2+5x-3}-\frac{3x+4}{2x^2+5x-3}$$
$$=\frac{5x+3-(3x+4)}{2x^2+5x-3}=\frac{5x+3-3x-4}{2x^2+5x-3}$$
$$=\frac{\overset{1}{\cancel{2x-1}}}{\underset{1}{\cancel{(2x-1)}}(x+3)}=\frac{1}{x+3}$$

26.
$$10x^2-11x+3=(5x-3)(2x-1)$$
$$20x^2-17x+3=(5x-3)(4x-1)$$
$$\text{LCM}=(5x-3)(2x-1)(4x-1)$$

27.
$$4x+9y=18$$
$$4x-4x+9y=-4x+18$$
$$9y=-4x+18$$
$$\frac{9y}{9}=\frac{-4x+18}{9}$$
$$y=-\frac{4}{9}x+2$$

28.
$$\frac{2x^2-5x-3}{3x^2-7x-6}\cdot\frac{3x^2+8x+4}{x^2+4x+4}$$
$$=\frac{(2x+1)\overset{1}{\cancel{(x-3)}}}{\underset{1}{\cancel{(3x+2)}}\underset{1}{\cancel{(x-3)}}}\cdot\frac{\overset{1}{\cancel{(3x+2)}}\overset{1}{\cancel{(x+2)}}}{\underset{1}{\cancel{(x+2)}}(x+2)}$$
$$=\frac{2x+1}{x+2}$$

29.

$$\frac{20}{2x+3}=\frac{17x}{2x+3}-5$$

$$(2x+3)\cdot\frac{20}{2x+3}=(2x+3)\cdot\frac{17x}{2x+3}-5(2x+3)$$

$$20=17x-10x-15$$

$$20=7x-15$$

$$35=7x$$

$$5=x$$

5 checks as a solution. The solution is 5.

30.

$$\frac{x-1}{x+2}+\frac{3x-2}{5-x}+\frac{5x^2+15x-11}{x^2-3x-10}$$

LCM $=(x-5)(x+2)$

$$\frac{x-1}{x+2}+\frac{3x-2}{5-x}+\frac{5x^2+15x-11}{x^2-3x-10}$$

$$=\frac{x-1}{x+2}-\frac{3x-2}{x-5}+\frac{5x^2+15x-11}{(x+2)(x-5)}$$

$$=\frac{x-1}{x+2}\cdot\frac{x-5}{x-5}-\frac{3x-2}{x-5}\cdot\frac{x+2}{x+2}+\frac{5x^2+15x-11}{(x+2)(x-5)}$$

$$=\frac{x^2-6x+5}{(x+2)(x-5)}-\frac{3x^2+4x-4}{(x+2)(x-5)}+\frac{5x^2+15x-11}{(x+2)(x-5)}$$

$$=\frac{x^2-6x+5-(3x^2+4x-4)+5x^2+15x-11}{(x+2)(x-5)}$$

$$=\frac{x^2-6x+5-3x^2-4x+4+5x^2+15x-11}{(x+2)(x-5)}$$

$$=\frac{3x^2+5x-2}{(x+2)(x-5)}=\frac{(3x-1)\overset{1}{\cancel{(x+2)}}}{\underset{1}{\cancel{(x+2)}}(x-5)}=\frac{3x-1}{x-5}$$

31.

$$\frac{6}{x-7}=\frac{8}{x-6}$$

$$(x-6)\overset{1}{\cancel{(x-7)}}\frac{6}{\underset{1}{\cancel{x-7}}}=\overset{1}{\cancel{(x-6)}}(x-7)\frac{8}{\underset{1}{\cancel{x-6}}}$$

$$6x-36=8x-56$$

$$20=2x$$

$$10=x$$

10 checks as a solution. The solution is 10.

32.

$$\frac{3}{20}=\frac{x}{80}$$

$$80\cdot\frac{3}{20}=\frac{x}{80}\cdot80$$

$$12=x$$

12 checks as a solution. The solution is 12.

33. Strategy Triangle *NMO* is similar to triangle *QPD*. Solve a proportion to find the length of *QO*. Let *x* represent the length of *QO* and $25-x$ represent the length of *NO*.

Solution

$$\frac{MO}{PO}=\frac{NO}{QO}$$

$$\frac{6}{9}=\frac{25-x}{x}$$

$$\overset{1}{\cancel{9}}x\cdot\frac{6}{\underset{1}{\cancel{9}}}=9\overset{1}{\cancel{x}}\cdot\frac{25-x}{\underset{1}{\cancel{x}}}$$

$$6x=225-9x$$

$$15x=225$$

$$x=15$$

The length of *QO* is 15 cm.

34. Strategy To find the area:

- Find the height of triangle *DEF* by writing a proportion using the fact that in similar triangles, the ratio of corresponding sides equals the ratio of corresponding heights. Solve the proportion for the height (*h*).
- Write a proportion and solve to find the base, *DF*.
- Use the formula for the area of a triangle.

Solution

$$\frac{AB}{DE}=\frac{8}{h}$$

$$\frac{9}{12}=\frac{8}{h}$$

$$\overset{1}{\cancel{12}}h\cdot\frac{9}{\underset{1}{\cancel{12}}}=\frac{8}{\underset{1}{\cancel{h}}}\cdot12\overset{1}{\cancel{h}}$$

$$9h=96$$

$$h=10\frac{2}{3}$$

$$\frac{AB}{DE}=\frac{AC}{DF}$$

$$\frac{9}{12}=\frac{12}{DF}$$

$$\overset{1}{\cancel{12}}DF\cdot\frac{9}{\underset{1}{\cancel{12}}}=12\overset{1}{\cancel{DF}}\cdot\frac{12}{\underset{1}{\cancel{DF}}}$$

$$9DF=144$$

$$DF=16$$

$$A=\frac{1}{2}bh=\frac{1}{2}(16)\left(10\frac{2}{3}\right)=8\cdot\frac{32}{3}=\frac{256}{3}$$

The area of triangle *DEF* is $\frac{256}{3}$ in^2.

35. Strategy •Time to fill the pool using both hoses: t

	Rate	Time	Part
First hose	$\frac{1}{15}$	t	$\frac{t}{15}$
Second hose	$\frac{1}{10}$	t	$\frac{t}{10}$

• The sum of the parts of the task completed by each hose must equal 1.

Solution

$$\frac{t}{15}+\frac{t}{10}=1$$
$$30\cdot\frac{t}{15}+30\cdot\frac{t}{10}=30\cdot 1$$
$$2t+3t=30$$
$$5t=30$$
$$t=6$$

It would take 6 hours to fill the pool.

36. Strategy •Rate of the bus: r
•Rate of the car: $r + 10$

	Distance	Rate	Time
Bus	245	r	$\frac{245}{r}$
Car	315	$r+10$	$\frac{315}{r+10}$

• The time of the bus equals the time of the car.

Solution

$$\frac{315}{r+10}=\frac{245}{r}$$
$$r(r+10)\frac{315}{r+10}=r(r+10)\frac{245}{r}$$
$$315r=245r+2450$$
$$70r=2450$$
$$r=35$$
$$r+10=45$$

The rate of the car is 45 mph.

37. Strategy •Rate of the wind: r

	Distance	Rate	Time
With the wind	2100	$400+r$	$\frac{2100}{400+r}$
Against the wind	1900	$400-r$	$\frac{1900}{400-r}$

• The time traveled with the wind equals the time traveled against the wind.

Solution

$$\frac{2100}{400+r}=\frac{1900}{400-r}$$
$$(400+r)(400-r)\frac{2100}{400+r}=(400+r)(400-r)\frac{1900}{400-r}$$
$$840{,}000-2100r==760{,}000+1900r$$
$$80{,}000=4000r$$
$$20=r$$

The rate of the wind is 20 mph.

38. Strategy • Unknown runs: x
Write and solve a proportion.

Solution

$$\frac{\text{ERA}}{9}=\frac{15}{100}$$
$$900\cdot\frac{\text{ERA}}{9}=900\cdot\frac{15}{100}$$
$$100\text{ERA}=135$$
$$\text{ERA}=1.35$$

The pitcher's ERA is 1.35.

39. Strategy: • Write the basic inverse variation equation replacing the variable with the given values. Solve for *k*.
• Write the inverse variation equation replacing *k* with its value. Substitute 100 for *R* and solve for *I*.

Solution:

$$I=\frac{k}{R}$$
$$4=\frac{k}{50}$$
$$200=k$$

$$I=\frac{200}{R}$$
$$I=\frac{200}{100}$$
$$I=2$$

The current is 2 amps.

40. Strategy Write a ratio using *m* for the number of miles. Solve for *m*.

Solution

$$\frac{2.5\text{ in}}{10\text{ mi}}=\frac{12\text{ in}}{m}$$
$$10m\left(\frac{2.5}{10}\right)=10m\left(\frac{12}{m}\right)$$
$$2.5m=120$$
$$m=48$$

12 in represents 48 mi.

41. Strategy Write a direct variation using *v* and *s*. Use 50 for *v* and 170 for *s*. Solve for *k*. Use *k* and 65 for *v* and solve for *s*.

Solution

$$s=kv^2$$
$$170=k50^2$$
$$170=2500k$$
$$0.068=k$$

$$s=0.068v^2$$
$$s=0.068\cdot 65^2$$
$$s=287.3$$

The stopping distance at 65 mph is 287.3 ft.

42. Strategy Time for the apprentice: *t*

	Time	Part
Electrician	65	$\frac{1}{65}$
Apprentice	t	$\frac{1}{t}$
Together	40	$\frac{1}{40}$

The sum of the electrician and the apprentice must equal the part together.

Solution

$$\frac{1}{65}+\frac{1}{t}=\frac{1}{40}$$
$$520t\left(\frac{1}{65}\right)+520t\left(\frac{1}{t}\right)=520t\left(\frac{1}{40}\right)$$
$$8t+520=13t$$
$$520=5t$$
$$104=t$$

The apprentice could install the fan in 104 min.

Chapter 8 Test

1. $\dfrac{x}{x+3}-\dfrac{2x-5}{x^2+x-6}$

$$\frac{x}{x+3}-\frac{2x-5}{(x+3)(x-2)}$$

$$\text{LCM}=(x+3)(x-2)$$

$$\frac{x}{x+3}\cdot\frac{x-2}{x-2}-\frac{2x-5}{(x+3)(x-2)}$$
$$=\frac{x^2-2x}{(x+3)(x-2)}-\frac{2x-5}{(x+3)(x-2)}$$
$$=\frac{x^2-2x-(2x-5)}{(x+3)(x-2)}=\frac{x^2-2x-2x+5}{(x+3)(x-2)}$$
$$=\frac{x^2-4x+5}{(x+3)(x-2)}$$

2.
$$\frac{3}{x+4}=\frac{5}{x+6}$$
$$(x+6)\cancel{(x+4)}\cdot\frac{3}{\cancel{x+4}}=\cancel{(x+6)}(x+4)\frac{5}{\cancel{x+6}}$$
$$3x+18=5x+20$$
$$-2=2x$$
$$-1=x$$

-1 checks as a solution. The solution is -1.

3. $\dfrac{x^2+2x-3}{x^2+6x+9}\cdot\dfrac{2x^2-11x+5}{2x^2+3x-5}$

$$\frac{\cancel{(x+3)}\cancel{(x-1)}}{\cancel{(x+3)}(x+3)}\cdot\frac{(2x-1)(x-5)}{(2x+5)\cancel{(x-1)}}=\frac{(2x-1)(x-5)}{(x+3)(2x+5)}$$

4. $\dfrac{16x^5y}{24x^2y^4}=\dfrac{\overset{2}{\cancel{16}}x^5y}{\underset{3}{\cancel{24}}x^2y^4}=\dfrac{2x^3}{3y^3}$

5.
$$d=s+rt$$
$$d-s=rt$$
$$\frac{d-s}{r}=\frac{rt}{r}$$
$$\frac{d-s}{r}=t$$

6.
$$\frac{6}{x}-2=1$$
$$\cancel{x}\cdot\frac{6}{\cancel{x}}-2\cdot x=1\cdot x$$
$$6-2x=x$$
$$6=3x$$
$$2=x$$

2 checks as a solution. The solution is 2.

7. $\dfrac{x^2+4x-5}{1-x^2}=\dfrac{(x+5)\cancel{(x-1)}}{(1+x)\cancel{(1-x)}}=-\dfrac{x+5}{x+1}$

8. $6x-3=3(2x-1)$

$2x^2+x-1=(2x-1)(x+1)$

$\text{LCM}=3(2x-1)(x+1)$

9. $\dfrac{2}{2x-1}-\dfrac{3}{3x+1}=\dfrac{2}{2x-1}\cdot\dfrac{3x+1}{3x+1}-\dfrac{3}{3x+1}\cdot\dfrac{2x-1}{2x-1}$

$$=\frac{6x+2}{(2x-1)(3x+1)}-\frac{6x-3}{(2x-1)(3x+1)}$$
$$=\frac{6x+2-(6x-3)}{(2x-1)(3x+1)}$$
$$=\frac{6x+2-6x+3}{(2x-1)(3x+1)}=\frac{5}{(2x-1)(3x+1)}$$

10. $\dfrac{x^2+3x+2}{x^2+5x+4}\div\dfrac{x^2-x-6}{x^2+2x-15}$

$$=\frac{\cancel{(x+2)}\cancel{(x+1)}}{\cancel{(x+1)}(x+4)}\cdot\frac{(x+5)\cancel{(x-3)}}{\cancel{(x-3)}\cancel{(x+2)}}=\frac{x+5}{x+4}$$

11. $\dfrac{1+\frac{1}{x}-\frac{12}{x^2}}{1+\frac{2}{x}-\frac{8}{x^2}}=\dfrac{1+\frac{1}{x}-\frac{12}{x^2}}{1+\frac{2}{x}-\frac{8}{x^2}}\cdot\dfrac{x^2}{x^2}$

$$=\frac{x^2\cdot1+x^2\cdot\frac{1}{x}-x^2\cdot\frac{12}{x^2}}{x^2\cdot1+x^2\cdot\frac{2}{x}-x^2\cdot\frac{8}{x^2}}=\frac{x^2+x-12}{x^2+2x-8}$$
$$=\frac{\cancel{(x+4)}(x-3)}{\cancel{(x+4)}(x-2)}=\frac{x-3}{x-2}$$

12. $\dfrac{3}{x^2-2x}=\dfrac{3}{x(x-2)}$ $\qquad\dfrac{x}{x^2-4}=\dfrac{x}{(x-2)(x+2)}$

$\text{LCM}=x(x-2)(x+2)$

$$\frac{3}{x(x-2)}\cdot\frac{x+2}{x+2}=\frac{3x+6}{x(x-2)(x+2)}$$
$$\frac{x}{(x-2)(x+2)}\cdot\frac{x}{x}=\frac{x^2}{x(x-2)(x+2)}$$

13. $\dfrac{2x}{x^2+3x-10}-\dfrac{4}{x^2+3x-10}=\dfrac{2x-4}{x^2+3x-10}$

$$=\frac{2\cancel{(x-2)}}{(x+5)\cancel{(x-2)}}=\frac{2}{x+5}$$

14. $$3x - 8y = 16$$
$$3x - 3x - 8y = -3x + 16$$
$$-8y = -3x + 16$$
$$\frac{-8y}{-8} = \frac{-3x+16}{-8}$$
$$y = \frac{3}{8}x - 2$$

15. $$\frac{2x}{x+1} - 3 = \frac{-2}{x+1}$$
$$\cancel{(x+1)}\left(\frac{2x}{\cancel{x+1}}\right) - 3(x+1) = \cancel{(x+1)}\left(\frac{-2}{\cancel{x+1}}\right)$$
$$2x - 3x - 3 = -2$$
$$-x = 1$$
$$x = -1$$

-1 does not check as a solution. There is no solution.

16. $$\frac{x^3y^4}{x^2-4x+4} \cdot \frac{x^2-x-2}{x^6y^4}$$
$$= \frac{x^3y^4}{(x-2)(x-2)} \cdot \frac{(x-2)(x+1)}{x^6y^4} = \frac{x+1}{x^3(x-2)}$$

17. Strategy $\angle CAE = \angle CBD$ and $\angle C = \angle C$, thus triangle CAE is similar to triangle CBD. Write and solve a proportion to find the length of CE. Let x represent the length of CD. Then
$CE = CD + DE = x + 8$ and
$AC = AB + BC = 5 + 3 = 8$.

Solution
$$\frac{AC}{BC} = \frac{CE}{CD}$$
$$\frac{8}{3} = \frac{8+x}{x}$$
$$\cancel{3}x \cdot \frac{8}{\cancel{3}} = 3\cancel{x} \cdot \frac{8+x}{\cancel{x}}$$
$$8x = 24 + 3x$$
$$5x = 24$$
$$x = 4.8$$

$CE = x + 8 = 4.8 + 8 = 12.8$

The length of CE is 12.8 ft.

18. Strategy To find the amount of salt needed, write and solve a proportion using x to represent the additional salt.

Solution
$$\frac{4}{10} = \frac{x+4}{15}$$
$$\overset{3}{\cancel{30}} \cdot \frac{4}{\underset{1}{\cancel{10}}} = \overset{2}{\cancel{30}} \cdot \frac{x+4}{\underset{1}{\cancel{15}}}$$
$$12 = 2x + 8$$
$$4 = 2x$$
$$2 = x$$

$x + 4 = 2 + 4 = 6$

6 lb of salt are needed.

19. Strategy • Time to fill the pool with both pipes: t

	Rate	Time	Part
First pipe	$\frac{1}{6}$	t	$\frac{t}{6}$
Second pipe	$\frac{1}{12}$	t	$\frac{t}{12}$

• The sum of the parts completed by each
pipe must equal 1.

Solution
$$\frac{t}{6} + \frac{t}{12} = 1$$
$$\overset{2}{\cancel{12}} \cdot \frac{t}{\underset{1}{\cancel{6}}} + \overset{1}{\cancel{12}} \cdot \frac{t}{\underset{1}{\cancel{12}}} = 12 \cdot 1$$
$$2t + t = 12$$
$$3t = 12$$
$$t = 4$$

It would take both pipes 4 h to fill the pool.

20. Strategy • Rate of the wind: r

	Distance	Rate	Time
With the wind	260	$110 + r$	$\dfrac{260}{110 + r}$
Against the wind	180	$110 - r$	$\dfrac{180}{110 - r}$

• The time traveled with the wind equals the time traveled against the wind.

Solution

$$\frac{260}{110 + r} = \frac{180}{110 - r}$$

$$(110 - r)\,\cancel{(110 + r)}\,\frac{260}{\cancel{110 + r}} = \cancel{(110 - r)}\,(110 + r)\,\frac{180}{\cancel{110 - r}}$$

$$28,600 - 260r = 19,800 + 180r$$

$$8800 = 440r$$

$$20 = r$$

The rate of the wind is 20 mph.

21. Strategy Write and solve a proportion using x to represent the number of sprinklers needed for 3600 ft² of lawn.

Solution

$$\frac{3}{200} = \frac{x}{3600}$$

$$\overset{18}{\cancel{3600}} \cdot \frac{3}{\underset{1}{\cancel{200}}} = \overset{1}{\cancel{3600}} \cdot \frac{x}{\underset{1}{\cancel{3600}}}$$

$$54 = x$$

54 sprinklers are needed for 3600 ft² of lawn.

22. Strategy: Write the inverse variation equation replacing the variable with the given values. Solve for k.
Write the inverse variation equation replacing k with its value. Substitute 5 for d. Then solve for I.

Solution:

$$I = \frac{k}{d^2}$$

$$50 = \frac{k}{8^2}$$

$$50 = \frac{k}{64}$$

$$3200 = k$$

$$I = \frac{3200}{d^2}$$

$$I = \frac{3200}{5^2}$$

$$I = 128$$

The intensity is 128 decibels.

Cumulative Review Exercises

1. $\left(\dfrac{2}{3}\right)^2 \div \left(\dfrac{3}{2} - \dfrac{2}{3}\right) + \dfrac{1}{2} = \left(\dfrac{2}{3}\right)^2 \div \left(\dfrac{9}{6} - \dfrac{4}{6}\right) + \dfrac{1}{2}$

$$= \frac{4}{9} \div \left(\frac{5}{6}\right) + \frac{1}{2} = \frac{4}{9} \cdot \frac{6}{5} + \frac{1}{2}$$

$$= \frac{2 \cdot 2 \cdot 2 \cdot \cancel{3}}{\cancel{3} \cdot 3 \cdot 5} + \frac{1}{2} = \frac{8}{15} + \frac{1}{2}$$

$$= \frac{16}{30} + \frac{15}{30} = \frac{31}{30}$$

2. $-a^2 + (a - b)^2$

$$-(-2)^2 + (-2 - 3)^2 = -4 + (-5)^2 = -4 + 25 = 21$$

3. $-2x - (-3y) + 7x - 5y = -2x + 3y + 7x - 5y$

$$= -2x + 7x + 3y - 5y$$

$$= 5x - 2y$$

4. $2\left[3x - 7(x - 3) - 8\right] = 2\left[3x - 7x + 21 - 8\right]$

$$= 2\left[-4x + 13\right] = -8x + 26$$

5.
$$4-\frac{2}{3}x=7$$
$$-\frac{2}{3}x=3$$
$$-\frac{3}{2}\left(-\frac{2}{3}x\right)=3\left(-\frac{3}{2}\right)$$
$$x=-\frac{9}{2}$$

The solution is $-\frac{9}{2}$.

6.
$$3\left[x-2(x-3)\right]=2(3-2x)$$
$$3\left[x-2x+6\right]=6-4x$$
$$3\left[-x+6\right]=6-4x$$
$$-3x+18=6-4x$$
$$x=-12$$

The solution is -12.

7.
$$P\cdot B=A$$
$$16\frac{2}{3}\%\cdot 60=A$$
$$\frac{1}{6}\cdot 60=A$$
$$10=A$$

8.
$$x-3(1-2x)\geq 1-4(3-2x)$$
$$x-3+6x\geq 1-12+8x$$
$$7x-3\geq -11+8x$$
$$-x-3\geq -11$$
$$-x\geq -8$$
$$x\leq 8$$
$$\{x\,|\,x\leq 8\}$$

9.
$$3x-2y=6$$
$$-2y=-3x+6$$
$$\frac{-2y}{-2}=\frac{-3x}{-2}+\frac{6}{-2}$$
$$y=\frac{3}{2}x-3$$
$$m=\frac{3}{2}$$
$$y-y_1=m(x-x_1)$$
$$y-(-1)=\frac{3}{2}(x-(-2))$$
$$y+1=\frac{3}{2}(x+2)$$
$$y+1=\frac{3}{2}x+3$$
$$y=\frac{3}{2}x+2$$

10.
(1) $2x-y+z=2$
(2) $3x+y-2z=9$
(3) $x-y+z=0$

Multiply equation (3) by –2 and add to equation (1).

$$2x-y+z=2$$
$$-2x+2y-2z=0$$
(4) $y-z=2$

Multiply equation (3) by –3 and add to equation (2).

$$3x+y-2z=9$$
$$-3x+3y-3z=0$$
(5) $4y-5z=9$

Use equations (4) and (5) to solve for y and z.

Multiply equation (4) by –4

$$-4y+4z=-8$$
$$4y-5z=9$$
$$-z=1$$
$$z=-1$$

Replace z with –1 in equation (4).

$$y-z=2$$
$$y-(-1)=2$$
$$y+1=2$$
$$y=1$$

Replace place y with 1 and z with –1 in equation (3).

$$x-1-1=0$$
$$x-2=0$$
$$x=2$$

The solution is (2, 1, –1).

11. $\left(a^2b^5\right)\left(ab^2\right)=a^{2+1}b^{5+2}=a^3b^7$

12.
$$(a-3b)(a+4b)=a^2+4ab-3ab-12b^2$$
$$=a^2+ab-12b^2$$

13.

$$x-2\overline{)x^3+0x^2+0x-8}\quad\text{quotient: } x^2+2x+4$$

$$\begin{array}{l} x^3-2x^2 \\ \hline 2x^2+0x \\ 2x^2-4x \\ \hline 4x-8 \\ 4x-8 \\ \hline 0 \end{array}$$

$$\left(x^3-8\right)\div\left(x-2\right)=x^2+2x+4$$

14. $12x^2-x-1$

$12\cdot(-1)=-12$

Factors of –12 whose sum is –1: –4 and 3

$12x^2-4x+3x-1$

$4x(3x-1)+1(3x-1)=(3x-1)(4x+1)$

15.

$$P(x)=\frac{x-1}{2x-3}$$

$$P(-2)=\frac{-2-1}{2(-2)-3}$$

$$P(-2)=\frac{-3}{-4-3}$$

$$P(-2)=\frac{-3}{-7}$$

$$P(-2)=\frac{3}{7}$$

16. $2a^3+7a^2-15a$

The GCF is a: $a(2a^2+7a-15)$

$2\cdot(-15)=-30$

Factors of –30 whose sum is 7: 10 and –3

$a(2a^2+10a-3a-15)$

$a[2a(a+5)-3(a+5)]=a(a+5)(2a-3)$

17. $4b^2-100$

The GCF is 4: $4(b^2-25)$

$4\left[b^2-(5)^2\right]=4(b+5)(b-5)$

18. $(x+3)(2x-5)=0$

$x+3=0$

$x=-3$

$$\begin{array}{ll} x+3=0 & 2x-5=0 \\ x=-3 & 2x=5 \\ & x=\frac{5}{2} \end{array}$$

The solutions are -3 and $\frac{5}{2}$.

19. $$\frac{12x^4y^2}{18xy^7}=\frac{\cancel{2}\cdot 2\cdot\cancel{3}x^{4-1}}{\cancel{2}\cdot\cancel{3}\cdot 3y^{7-2}}=\frac{2x^3}{3y^5}$$

20. $$\frac{x^2-7x+10}{25-x^2}=\frac{\cancel{(x-5)}(x-2)}{(5+x)\cancel{(5-x)}}=-\frac{x-2}{x+5}$$

21. $$\frac{x^2-x-56}{x^2+8x+7}\div\frac{x^2-13x+40}{x^2-4x-5}$$

$$=\frac{\cancel{(x-8)}\cancel{(x+7)}}{\cancel{(x+7)}\cancel{(x+1)}}\cdot\frac{\cancel{(x-5)}\cancel{(x+1)}}{\cancel{(x-8)}\cancel{(x-5)}}=1$$

22. The GCF is $(2x-1)(x+1)$.

$$\frac{2}{2x-1}-\frac{1}{x+1}=\frac{2}{2x-1}\cdot\frac{x+1}{x+1}-\frac{1}{x+1}\cdot\frac{2x-1}{2x-1}$$

$$=\frac{2x+2}{(2x-1)(x+1)}-\frac{2x-1}{(2x-1)(x+1)}$$

$$=\frac{2x+2-(2x-1)}{(2x-1)(x+1)}$$

$$=\frac{2x+2-2x+1}{(2x-1)(x+1)}=\frac{3}{(2x-1)(x+1)}$$

23. $$\frac{1-\frac{2}{x}-\frac{15}{x^2}}{1-\frac{25}{x^2}}=\frac{1-\frac{2}{x}-\frac{15}{x^2}}{1-\frac{25}{x^2}}\cdot\frac{x^2}{x^2}$$

$$\frac{1\cdot x^2-\frac{2}{x}\cdot x^2-\frac{15}{x^2}\cdot x^2}{1\cdot x^2-\frac{25}{x^2}\cdot x^2}=\frac{x^2-2x-15}{x^2-25}$$

$$=\frac{\cancel{(x-5)}(x+3)}{\cancel{(x-5)}(x+5)}=\frac{x+3}{x+5}$$

24. $$\frac{3x}{x-3}-2=\frac{10}{x-3}$$

$$\cancel{(x-3)}\left(\frac{3x}{\cancel{x-3}}\right)-2(x-3)=\cancel{(x-3)}\left(\frac{10}{\cancel{x-3}}\right)$$

$$3x-2(x-3)=10$$

$$3x-2x+6=10$$

$$x+6=10$$

$$x=4$$

4 checks as a solution. The solution is 4.

25.

$$\frac{2}{x-2}=\frac{12}{x+3}$$

$$(x+3)\overset{1}{\cancel{(x-2)}}\frac{2}{\underset{1}{\cancel{x-2}}}=\overset{1}{\cancel{(x+3)}}(x-2)\frac{12}{\underset{1}{\cancel{x+3}}}$$

$$2x+6=12x-24$$

$$30=10x$$

$$3=x$$

3 checks as a solution. The solution is 3.

26.

$$f=v+at$$

$$f-v=v-v+at$$

$$f-v=at$$

$$\frac{f-v}{a}=\frac{at}{a}$$

$$\frac{f-v}{a}=t$$

27. Strategy Let x be the number. Write an equation and solve.

Solution

$$5x-13=-8$$

$$5x=5$$

$$x=1$$

The number is 1.

28. Strategy • Percent of silver in the alloy: x

	Amount	Percent	Quantity
40% silver	60	0.40	0.40(60)
Silver alloy	120	x	$120x$
Mixture	180	0.60	0.60(180)

• The sum of the quantities before mixing is equal to the quantity after mixing.

Solution

$$0.40(60)+120x=0.60(180)$$

$$24+120x=108$$

$$120x=84$$

$$x=0.70$$

The silver alloy is 70% silver.

29. Strategy To find the cost of a \$5000 policy, write and solve a proportion using x to represent the cost for a \$5000 policy.

Solution

$$\frac{16}{1000}=\frac{x}{5000}$$

$$\overset{5}{\cancel{5000}}\cdot\frac{16}{\underset{1}{\cancel{1000}}}=\overset{1}{\cancel{5000}}\cdot\frac{x}{\underset{1}{\cancel{5000}}}$$

$$80=x$$

The cost of a \$5000 policy is \$80.

30. Strategy Write the combined variation equation replacing the variable with the given values. Solve for k.
Write the inverse variation equation replacing k with its value. Substitute 8000 for l and $\frac{1}{2}$ for d. Then solve for r.

Solution

$$r=\frac{lk}{d^2}$$

$$3.2=\frac{16{,}000k}{\left(\frac{1}{4}\right)^2}$$

$$3.2=256{,}000$$

$$0.0000125=k$$

$$r=\frac{0.0000125l}{d^2}$$

$$r=\frac{0.0000125\cdot 8000}{\left(\frac{1}{2}\right)^2}$$

$$r=0.4$$

The resistance is 0.4 ohms.

31. Strategy •Time to fill the pool with both pipes: t

	Rate	Time	Part
First pipe	$\frac{1}{9}$	t	$\frac{t}{9}$
Second pipe	$\frac{1}{18}$	t	$\frac{t}{18}$

• The sum of the parts completed by each pipe must equal 1.

Solution

$$\frac{t}{9}+\frac{t}{18}=1$$
$$\overset{2}{\cancel{18}}\cdot\frac{t}{\underset{1}{\cancel{9}}}+\overset{1}{\cancel{18}}\cdot\frac{t}{\underset{1}{\cancel{18}}}=18\cdot 1$$
$$2t+t=18$$
$$3t=18$$
$$t=6$$

It would take both pipes 6 min to fill the tank.

32. Strategy • Rate of current: r

	Distance	Rate	Time
With current	14	$5+r$	$\frac{14}{5+r}$
Against current	6	$5-r$	$\frac{6}{5-r}$

• The two times are equal.

Solution

$$\frac{14}{5+r}=\frac{6}{5-r}$$
$$(5+r)(5-r)\left(\frac{14}{5+r}\right)=(5+r)(5-r)\left(\frac{6}{5-r}\right)$$
$$70-14r=30+6r$$
$$40=20r$$
$$2=r$$

The rate of the current is 2 mph.

Chapter 9: Exponents and Radicals

Prep Test

1. $48 = ? \cdot 3$

$\left(\frac{1}{3}\right)48 = ? \cdot 3\left(\frac{1}{3}\right)$

$16 = ?$

2. $2^5 = 2 \cdot 2 \cdot 2 \cdot 2 \cdot 2 = 32$

3. $6\left(\frac{3}{2}\right) = \frac{6}{1}\left(\frac{3}{2}\right) = \frac{3 \cdot 2}{1}\left(\frac{3}{2}\right) = \frac{3 \cdot 3}{1 \cdot 1} = 9$

4. $\frac{1}{2} - \frac{2}{3} + \frac{1}{4} = \frac{6}{12} - \frac{8}{12} + \frac{3}{12}$

$= \frac{6 - 8 + 3}{12}$

$= \frac{1}{12}$

5. $(3 - 7x) - (4 - 2x)$

$= 3 - 7x - 4 + 2x$

$= -5x - 1$

6. $\frac{3x^5y^6}{12x^4y} = \frac{xy^5}{4}$

7. $(3x - 2)^2 = (3x - 2)(3x - 2)$

$= 9x^2 - 6x - 6x + 4$

$= 9x^2 - 12x + 4$

8. $(2 + 4x)(5 - 3x)$

$= 10 - 6x + 20x - 12x^2$

$= -12x^2 + 14x + 10$

9. $(6x - 1)(6x + 1)$

$= 36x^2 + 6x - 6x - 1$

$= 36x^2 - 1$

10. $x^2 - 14x - 5 = 10$

$x^2 - 14x - 15 = 0$

$(x - 15)(x + 1) = 0$

$x - 15 = 0 \quad x + 1 = 0$

$x = 15 \quad x = -1$

The solutions are −1 and 15.

Section 9.1

Concept Check

1. 125

3. −32

5. 9

7. 3

Objective A Exercises

9. (i) and (iii) are not real numbers.

11. $8^{1/3} = (2^3)^{1/3} = 2$

13. $9^{3/2} = (3^2)^{3/2} = 3^3 = 27$

15. $27^{-2/3} = (3^3)^{-2/3} = 3^{-2} = \frac{1}{3^2} = \frac{1}{9}$

17. $32^{2/5} = (2^5)^{2/5} = 2^2 = 4$

19. $(-25)^{5/2}$

Not a real number.

The base of the exponential expression is a negative number and the denominator of the exponent is a positive even number.

21. $\left(\dfrac{25}{49}\right)^{-3/2}$

$= \left(\dfrac{5^2}{7^2}\right)^{-3/2} = \left(\left(\dfrac{5}{7}\right)^2\right)^{-3/2}$

$= \left(\dfrac{5}{7}\right)^{-3} = \dfrac{5^{-3}}{7^{-3}} = \dfrac{7^3}{5^3}$

$= \dfrac{343}{125}$

23. $x^{1/2}x^{1/2} = x$

25. $y^{-1/4}y^{3/4} = y^{1/2}$

27. $x^{-2/3} \cdot x^{3/4} = x^{1/12}$

29. $a^{1/3} \cdot a^{3/4} \cdot a^{-1/2} = a^{7/12}$

31. $\dfrac{a^{1/2}}{a^{3/2}} = a^{-1} = \dfrac{1}{a}$

33. $\dfrac{y^{-3/4}}{y^{1/4}} = y^{-1} = \dfrac{1}{y}$

35. $\dfrac{y^{2/3}}{y^{-5/6}} = y^{9/6} = y^{3/2}$

37. $(x^2)^{-1/2} = x^{-1} = \dfrac{1}{x}$

39. $(x^{-2/3})^6 = x^{-4} = \dfrac{1}{x^4}$

41. $(a^{-1/2})^{-2} = a$

43. $(x^{-3/8})^{-4/5} = x^{3/10}$

45. $(a^{1/2} \cdot a)^2 = (a^{3/2})^2 = a^3$

47. $(x^{-1/2} \cdot x^{3/4})^{-2} = (x^{1/4})^{-2} = x^{-1/2} = \dfrac{1}{x^{1/2}}$

49. $(y^{-1/2} \cdot y^{2/3})^{2/3} = (y^{1/6})^{2/3} = y^{1/9}$

51. $(x^{-3}y^6)^{-1/3} = xy^{-2} = \dfrac{x}{y^2}$

53. $(x^{-2}y^{1/3})^{-3/4} = x^{3/2}y^{-1/4} = \dfrac{x^{3/2}}{y^{1/4}}$

55. $\left(\dfrac{x^{1/2}}{y^2}\right)^4 = \dfrac{x^2}{y^8}$

57. $\dfrac{x^{1/4} \cdot x^{-1/2}}{x^{2/3}} = \dfrac{x^{-1/4}}{x^{2/3}} = x^{-11/12} = \dfrac{1}{x^{11/12}}$

59. $\left(\dfrac{y^{2/3} \cdot y^{-5/6}}{y^{1/9}}\right)^9 = \left(\dfrac{y^{-1/6}}{y^{1/9}}\right)^9 = \left(y^{-5/18}\right)^9$

$= y^{-5/2} = \dfrac{1}{y^{5/2}}$

61. $\left(\dfrac{b^2 \cdot b^{-3/4}}{b^{-1/2}}\right)^{-1/2} = \left(\dfrac{b^{5/4}}{b^{-1/2}}\right)^{-1/2} = \left(b^{7/4}\right)^{-1/2}$

$= b^{-7/8} = \dfrac{1}{b^{7/8}}$

63. $(a^{2/3}b^2)^6(a^3b^3)^{1/3} = (a^4b^{12})(ab) = a^5b^{13}$

65. $(16x^{-2}y^4)^{-1/2}(xy^{1/2}) = (16)^{-1/2}(xy^{-2})(xy^{1/2})$

$= (16)^{-1/2}x^2y^{-3/2}$

$= \dfrac{x^2}{16^{1/2}y^{3/2}}$

$= \dfrac{x^2}{4y^{3/2}}$

67. $(x^{-2/3}y^{-3})^3(27x^{-3}y^6)^{-1/3} = (x^{-2}y^{-9})(27)^{-1/3}(xy^{-2})$

$= (27)^{-1/3}x^{-1}y^{-11}$

$= \dfrac{1}{27^{1/3}xy^{11}}$

$= \dfrac{1}{3xy^{11}}$

69. $\dfrac{\left(4a^{4/3}b^{-2}\right)^{-1/2}}{(a^{1/6}b^{-3/2})^2} = \dfrac{\left((4)^{-1/2}a^{-2/3}b\right)}{a^{1/3}b^{-3}} = \dfrac{b^4}{2a}$

71. $\left(\dfrac{x^{1/2}y^{-3/4}}{y^{2/3}}\right)^{-6} = \left(x^{1/2}y^{-17/12}\right)^{-6}$

$= x^{-3}y^{17/2} = \dfrac{y^{17/2}}{x^3}$

73. $\left(\dfrac{b^{-3}}{64a^{-1/2}}\right)^{-2/3} = \dfrac{b^2}{64^{-2/3}a^{1/3}}$

$= \dfrac{64^{2/3}b^2}{a^{1/3}} = \dfrac{16b^2}{a^{1/3}}$

75. $y^{3/2}(y^{1/2} - y^{1/2}) = y^{4/2} - y^{2/2} = y^2 - y$

77. $a^{-1/4}(a^{5/4} - a^{9/4}) = a^{4/4} - a^{8/4} = a - a^2$

Objective B Exercises

79. False

81. $3^{1/4} = \sqrt[4]{3}$

83. $a^{3/2} = (a^3)^{1/2} = \sqrt{a^3}$

85. $(2t)^{5/2} = \sqrt{(2t)^5} = \sqrt{32t^5}$

87. $-2x^{2/3} = -2(x)^{2/3} = -2\sqrt[3]{x^2}$

89. $(a^2b)^{2/3} = \sqrt[3]{(a^2b)^2} = \sqrt[3]{a^4b^2}$

91. $(a^2b^4)^{3/5} = \sqrt[5]{(a^2b^4)^3} = \sqrt[5]{a^6b^{12}}$

93. $(4x-3)^{3/4} = \sqrt[4]{(4x-3)^3}$

95. $x^{-2/3} = \dfrac{1}{x^{2/3}} = \dfrac{1}{\sqrt[3]{x^2}}$

97. $\sqrt{14} = 14^{1/2}$

99. $\sqrt[3]{x} = x^{1/3}$

101. $\sqrt[3]{x^4} = x^{4/3}$

103. $\sqrt[5]{b^3} = b^{3/5}$

105. $\sqrt[3]{2x^2} = (2x^2)^{1/3}$

107. $-\sqrt{3x^5} = -(3x^5)^{1/2}$

109. $3x\sqrt[3]{y^2} = 3xy^{2/3}$

111. $\sqrt{a^2 - 2} = (a^2 - 2)^{1/2}$

Objective C Exercises

113. Positive

115. Not a real number

117. $\sqrt{x^{16}} = x^8$

119. $-\sqrt{x^8} = -x^4$

121. $\sqrt[3]{x^3y^9} = xy^3$

123. $-\sqrt[3]{x^{15}y^3} = -x^5y$

125. $\sqrt{16a^4b^{12}} = 4a^2b^6$

127. The square root of a negative number is not a real number.

129. $\sqrt[3]{27x^9} = 3x^3$

131. $\sqrt[3]{-64x^9y^{12}} = -4x^3y^4$

133. $-\sqrt[4]{x^8y^{12}} = -x^2y^3$

135. $\sqrt[5]{x^{20}y^{10}} = x^4y^2$

137. $\sqrt[4]{81x^4y^{20}} = 3xy^5$

139. $\sqrt[5]{32a^5b^{10}} = 2ab^2$

Critical Thinking

141. No. If $x \geq 0$, the statement is true. However, if $x < 0$ then $\sqrt{x^2} = |x|$.

Projects or Groups Activities

143.
$$\sqrt{2} \approx 1 + \cfrac{1}{2 + \cfrac{1}{2 + \cfrac{1}{2 + \cfrac{1}{2}}}}$$
$$= 1 + \cfrac{1}{2 + \cfrac{1}{2 + \cfrac{1}{2.5}}} = 1 + \cfrac{1}{2 + \cfrac{1}{2.4}}$$
$$\approx 1 + \frac{1}{2.417} \approx 1.414$$

Section 9.2

Concept Check

1. Neither

3. Perfect square

5. Perfect cube

7. Perfect cube

9. No

11. No

Objective A Exercises

13. $\sqrt{x^4y^3z^5} = \sqrt{x^4y^2z^4(yz)}$
$= \sqrt{x^4y^2z^4}\sqrt{yz}$
$= x^2yz^2\sqrt{yz}$

15. $\sqrt{8a^3b^8} = \sqrt{4a^2b^8(2a)}$
$= \sqrt{4a^2b^8}\sqrt{2a}$
$= 2ab^4\sqrt{2a}$

17. $\sqrt{45x^2y^3z^5} = \sqrt{9x^2y^2z^4(5yz)}$
$= \sqrt{9x^2y^2z^4}\sqrt{5yz}$
$= 3xyz^2\sqrt{5yz}$

19. $\sqrt[4]{48x^4y^5z^6} = \sqrt[4]{16x^4y^4z^4(3yz^2)}$
$= \sqrt[4]{16x^4y^4z^4}\sqrt[4]{3yz^2}$
$= 2xyz\sqrt[4]{3yz^2}$

21. $\sqrt[3]{a^{16}b^8} = \sqrt[3]{a^{15}b^6(ab^2)}$
$= \sqrt[3]{a^{15}b^6}\sqrt[3]{ab^2}$
$= a^5b^2\sqrt[3]{ab^2}$

23. $\sqrt[3]{-125x^2y^4} = \sqrt[3]{-125y^3(x^2y)}$
$= \sqrt[3]{-125y^3}\sqrt[3]{x^2y}$
$= -5y\sqrt[3]{x^2y}$

25. $\sqrt[3]{a^4b^5c^6} = \sqrt[3]{a^3b^3c^6(ab^2)}$
$= \sqrt[3]{a^3b^3c^6}\sqrt[3]{ab^2}$
$= abc^2\sqrt[3]{ab^2}$

27. $\sqrt[4]{16x^9y^5} = \sqrt[4]{16x^8y^4(xy)}$
$= \sqrt[4]{16x^8y^4}\sqrt[4]{xy}$
$= 2x^2y\sqrt[4]{xy}$

Objective B Exercises

29. True

31. $2\sqrt{x} - 8\sqrt{x} = -6\sqrt{x}$

33. $\sqrt{8x} - \sqrt{32x} = \sqrt{4 \cdot 2x} - \sqrt{16 \cdot 2x}$
$= \sqrt{4}\sqrt{2x} - \sqrt{16}\sqrt{2x}$
$= 2\sqrt{2x} - 4\sqrt{2x}$
$= -2\sqrt{2x}$

35. $\sqrt{18b} + \sqrt{75b} = \sqrt{9 \cdot 2b} + \sqrt{25 \cdot 3b}$
$= \sqrt{9}\sqrt{2b} + \sqrt{25}\sqrt{3b}$
$= 3\sqrt{2b} + 5\sqrt{3b}$

37. $3\sqrt{8x^2y^3} - 2x\sqrt{32y^3} = 3\sqrt{4 \cdot 2x^2y^3} - 2x\sqrt{16 \cdot 2y^3}$
$= 3\sqrt{4x^2y^2}\sqrt{2y} - 2x\sqrt{16y^2}\sqrt{2y}$
$= 3 \cdot 2xy\sqrt{2y} - 2x \cdot 4y\sqrt{2y}$
$= 6xy\sqrt{2y} - 8xy\sqrt{2y}$
$= -2xy\sqrt{2y}$

39. $2a\sqrt{27ab^5} + 3b\sqrt{3a^3b} = 2a\sqrt{3^3ab^5} + 3b\sqrt{3a^3b}$
$= 2a\sqrt{3^2b^4}\sqrt{3ab} + 3b\sqrt{a^2}\sqrt{3ab}$
$= 2a \cdot 3b^2\sqrt{3ab} + 3ab\sqrt{3ab}$
$= 6ab^2\sqrt{3ab} + 3ab\sqrt{3ab}$

41. $\sqrt[3]{16} - \sqrt[3]{54} = \sqrt[3]{8 \cdot 2} - \sqrt[3]{27 \cdot 2}$
$= \sqrt[3]{8}\sqrt[3]{2} - \sqrt[3]{27}\sqrt[3]{2}$
$= 2\sqrt[3]{2} - 3\sqrt[3]{2}$
$= -\sqrt[3]{2}$

43. $2b\sqrt[3]{16b^2} + \sqrt[3]{128b^5} = 2b\sqrt[3]{8 \cdot 2b^2} + \sqrt[3]{64b^3 \cdot 2b^2}$
$= 2b\sqrt[3]{8}\sqrt[3]{2b^2} + \sqrt[3]{64b^3}\sqrt[3]{2b^2}$
$= 4b\sqrt[3]{2b^2} + 4b\sqrt[3]{2b^2}$
$= 8b\sqrt[3]{2b^2}$

45. $3\sqrt[4]{32a^5} - a\sqrt[4]{162a} = 3\sqrt[4]{16a^4 \cdot 2a} - a\sqrt[4]{81 \cdot 2a}$
$= 3\sqrt[4]{16a^4}\sqrt[4]{2a} - a\sqrt[4]{81}\sqrt[4]{2a}$
$= 3 \cdot 2a\sqrt[4]{2a} - a \cdot 3\sqrt[4]{2a}$
$= 6a\sqrt[4]{2a} - 3a\sqrt[4]{2a}$
$= 3a\sqrt[4]{2a}$

47. $2\sqrt{50} - 3\sqrt{125} + \sqrt{98}$
$= 2\sqrt{25 \cdot 2} - 3\sqrt{25 \cdot 5} + \sqrt{49 \cdot 2}$
$= 2\sqrt{25}\sqrt{2} - 3\sqrt{25}\sqrt{5} + \sqrt{49}\sqrt{2}$
$= 10\sqrt{2} - 15\sqrt{5} + 7\sqrt{2}$
$= 17\sqrt{2} - 15\sqrt{5}$

49. $\sqrt{9b^3} - \sqrt{25b^3} + \sqrt{49b^3}$
$= \sqrt{9b^2 \cdot b} - \sqrt{25b^2 \cdot b} + \sqrt{49b^2 \cdot b}$
$= \sqrt{9b^2}\sqrt{b} - \sqrt{25b^2}\sqrt{b} + \sqrt{49b^2}\sqrt{b}$
$= 3b\sqrt{b} - 5b\sqrt{b} + 7b\sqrt{b}$
$= 5b\sqrt{b}$

51. $2x\sqrt{8xy^2} - 3y\sqrt{32x^3} + \sqrt{4x^3y^3}$
$= 2x\sqrt{4y^2 \cdot 2x} - 3y\sqrt{16x^2 \cdot 2x} + \sqrt{4x^2y^2 \cdot xy}$
$= 2x\sqrt{4y^2}\sqrt{2x} - 3y\sqrt{16x^2}\sqrt{2x} + \sqrt{4x^2y^2}\sqrt{xy}$
$= 4xy\sqrt{2x} - 12xy\sqrt{2x} + 2xy\sqrt{xy}$
$= -8xy\sqrt{2x} + 2xy\sqrt{xy}$

Critical Thinking

53. $\sqrt[3]{54xy^3} - 5\sqrt[3]{2xy^3} + y\sqrt[3]{128x}$
$= \sqrt[3]{27y^3 \cdot 2x} - 5\sqrt[3]{y^3 \cdot 2x} + y\sqrt[3]{64 \cdot 2x}$
$= \sqrt[3]{27y^3}\sqrt[3]{2x} - 5\sqrt[3]{y^3}\sqrt[3]{2x} + y\sqrt[3]{64}\sqrt[3]{2x}$
$= 3y\sqrt[3]{2x} - 5y\sqrt[3]{2x} + 4y\sqrt[3]{2x}$
$= 2y\sqrt[3]{2x}$

55. $2a\sqrt[4]{32b^5} - 3b\sqrt[4]{162a^4b} + \sqrt[4]{2a^4b^5}$
$= 2a\sqrt[4]{16b^4 \cdot 2b} - 3b\sqrt[4]{81a^4 \cdot 2b} + \sqrt[4]{a^4b^4 \cdot 2b}$
$= 2a\sqrt[4]{16b^4}\sqrt[4]{2b} - 3b\sqrt[4]{81a^4}\sqrt[4]{2b} + \sqrt[4]{a^4b^4}\sqrt[4]{2b}$
$= 4ab\sqrt[4]{2b} - 9ab\sqrt[4]{2b} + ab\sqrt[4]{2b}$
$= -4ab\sqrt[4]{2b}$

Projects or Group Activities

57. Domain: $(-\infty, 3]$

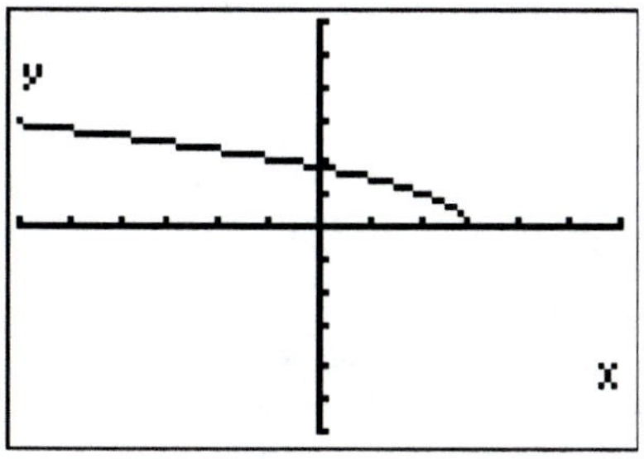

59. Domain: $(-\infty, \infty)$

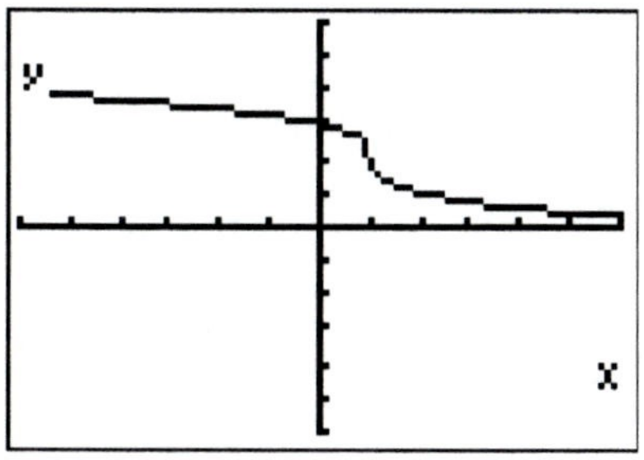

Section 9.3

Concept Check

1. If $\sqrt[n]{a}$ and $\sqrt[n]{b}$ are real numbers, then $\sqrt[n]{a} \cdot \sqrt[n]{b} = \sqrt[n]{ab}$.

3. $3 - \sqrt{5}$

5. $4 + 3\sqrt{11}$

7. (i) and (iii) can be simplified using the Product Property of Radicals.

Objective A Exercises

9. $\sqrt{8}\sqrt{32} = \sqrt{256} = 16$

11. $\sqrt[3]{4}\sqrt[3]{8} = 2\sqrt[3]{4}$

13. $\sqrt{x^2y^5}\sqrt{xy} = \sqrt{x^3y^6}$
$= \sqrt{x^2y^6 \cdot x} = xy^3\sqrt{x}$

15. $\sqrt{2x^2y}\sqrt{32xy} = \sqrt{64x^3y^2}$
$= \sqrt{64x^2y^2 \cdot x} = 8xy\sqrt{x}$

17. $\sqrt[3]{x^2y}\sqrt[3]{16x^4y^2} = \sqrt[3]{16x^6y^3}$
$= \sqrt[3]{8x^6y^3 \cdot 2} = 2x^2y\sqrt[3]{2}$

19. $\sqrt[4]{12ab^3}\sqrt[4]{4a^5b^2} = \sqrt[4]{48a^6b^5}$
$= \sqrt[4]{16a^4b^4 \cdot 3a^2b} = 2ab\sqrt[4]{3a^2b}$

21. $\sqrt{3}(\sqrt{27} - \sqrt{3}) = \sqrt{81} - \sqrt{9} = 9 - 3 = 6$

23. $\sqrt{x}(\sqrt{x} - \sqrt{2}) = \sqrt{x^2} - \sqrt{2x} = x - \sqrt{2x}$

25. $\sqrt{2x}(\sqrt{8x} - \sqrt{32}) = \sqrt{16x^2} - \sqrt{64x}$
$= \sqrt{16x^2} - \sqrt{64 \cdot x} = 4x - 8\sqrt{x}$

27. $(3-2\sqrt{5})(2+\sqrt{5})=6+3\sqrt{5}-4\sqrt{5}-2(\sqrt{5})^2$
$=6+3\sqrt{5}-4\sqrt{5}-10=-4-\sqrt{5}$

29. $(-2+\sqrt{7})(3+5\sqrt{7})=-6-10\sqrt{7}+3\sqrt{7}+5(\sqrt{7})^2$
$=-6-10\sqrt{7}+3\sqrt{7}+35=29-7\sqrt{7}$

31. $(6+3\sqrt{2})(4-2\sqrt{2})=24-12\sqrt{2}+12\sqrt{2}-6(\sqrt{2})^2$
$=24-12\sqrt{2}+12\sqrt{2}-12=12$

33. $(5-2\sqrt{7})(5+2\sqrt{7})=25+10\sqrt{7}-10\sqrt{7}-4(\sqrt{7})^2$
$=25+10\sqrt{7}-10\sqrt{7}-28=-3$

35. $(3-\sqrt{2x})(1+5\sqrt{2x})$
$=3+15\sqrt{2x}-\sqrt{2x}-5(\sqrt{2x})^2$
$=3+15\sqrt{2x}-\sqrt{2x}-10x$
$=-10x+14\sqrt{2x}+3$

37. $(2+\sqrt{x})^2=(2+\sqrt{x})(2+\sqrt{x})$
$=4+2\sqrt{x}+2\sqrt{x}+(\sqrt{x})^2$
$=x+4\sqrt{x}+4$

39. $(\sqrt{3x}-5)^2=(\sqrt{3x}-5)(\sqrt{3x}-5)$
$=(\sqrt{3x})^2-5\sqrt{3x}-5\sqrt{3x}+25$
$=3x-10\sqrt{3x}+25$

41. $(4-\sqrt{2x+1})^2=(4-\sqrt{2x+1})(4-\sqrt{2x+1})$
$=16-4\sqrt{2x+1}-4\sqrt{2x+1}+(\sqrt{2x+1})^2$
$=2x+1-8\sqrt{2x+1}+16$
$=2x-8\sqrt{2x+1}+17$

43. True

Objective B Exercises

45. To rationalize the denominator of a radical expression means to rewrite the expression with no radicals in the denominator. To do this, multiply both the numerator and the denominator by the same expression, one that removes the radicals(s) from the denominator of the original expression.

47. $\sqrt[3]{4x}$

49. $\sqrt{3}+x$

51. $\dfrac{\sqrt{60y^4}}{\sqrt{12y}}=\sqrt{\dfrac{60y^4}{12y}}$
$=\sqrt{5y^3}=\sqrt{y^2\cdot 5y}$
$=y\sqrt{5y}$

53. $\dfrac{\sqrt{65ab^4}}{\sqrt{5ab}}=\sqrt{\dfrac{65ab^4}{5ab}}$
$=\sqrt{13b^3}=\sqrt{b^2\cdot 13b}$
$=b\sqrt{13b}$

55. $\dfrac{1}{\sqrt{2}}=\dfrac{1}{\sqrt{2}}\cdot\dfrac{\sqrt{2}}{\sqrt{2}}=\dfrac{\sqrt{2}}{\sqrt{2^2}}=\dfrac{\sqrt{2}}{2}$

57. $\dfrac{2}{\sqrt{3y}}=\dfrac{2}{\sqrt{3y}}\cdot\dfrac{\sqrt{3y}}{\sqrt{3y}}=\dfrac{2\sqrt{3y}}{\sqrt{9y^2}}=\dfrac{2\sqrt{3y}}{3y}$

59. $\dfrac{9}{\sqrt{3a}}=\dfrac{9}{\sqrt{3a}}\cdot\dfrac{\sqrt{3a}}{\sqrt{3a}}=\dfrac{9\sqrt{3a}}{\sqrt{9a^2}}$
$=\dfrac{9\sqrt{3a}}{3a}=\dfrac{3\sqrt{3a}}{a}$

61. $\sqrt{\dfrac{y}{2}}=\dfrac{\sqrt{y}}{\sqrt{2}}=\dfrac{\sqrt{y}}{\sqrt{2}}\cdot\dfrac{\sqrt{2}}{\sqrt{2}}=\dfrac{\sqrt{2y}}{\sqrt{4}}=\dfrac{\sqrt{2y}}{2}$

63. $\dfrac{5}{\sqrt[3]{9}}=\dfrac{5}{\sqrt[3]{9}}\cdot\dfrac{\sqrt[3]{3}}{\sqrt[3]{3}}=\dfrac{5\sqrt[3]{3}}{\sqrt[3]{27}}=\dfrac{5\sqrt[3]{3}}{3}$

65. $\dfrac{5}{\sqrt[3]{3y}}=\dfrac{5}{\sqrt[3]{3y}}\cdot\dfrac{\sqrt[3]{9y^2}}{\sqrt[3]{9y^2}}=\dfrac{5\sqrt[3]{9y^2}}{\sqrt[3]{27y^3}}=\dfrac{5\sqrt[3]{9y^2}}{3y}$

67. $\frac{6x}{\sqrt[4]{9x}}=\frac{6x}{\sqrt[4]{9x}}\cdot\frac{\sqrt[4]{9x^3}}{\sqrt[4]{9x^3}}=\frac{6x\sqrt[4]{9x^3}}{\sqrt[4]{81x^4}}$

$=\frac{6x\sqrt[4]{9x^3}}{3x}=2\sqrt[4]{9x^3}$

69. $\frac{9x^2}{\sqrt[5]{27x}}=\frac{9x^2}{\sqrt[5]{27x}}\cdot\frac{\sqrt[5]{9x^4}}{\sqrt[5]{9x^4}}=\frac{9x^2\sqrt[5]{9x^4}}{\sqrt[5]{243x^5}}$

$=\frac{9x^2\sqrt[5]{9x^4}}{3x}=3x\sqrt[5]{9x^4}$

71. $\frac{\sqrt{15a^2b^5}}{\sqrt{30a^5b^3}}=\sqrt{\frac{15a^2b^5}{30a^5b^3}}=\sqrt{\frac{b^2}{2a^3}}=\frac{\sqrt{b^2}}{\sqrt{a^2\cdot 2a}}$

$=\frac{b}{a\sqrt{2a}}=\frac{b}{a\sqrt{2a}}\cdot\frac{\sqrt{2a}}{\sqrt{2a}}=\frac{b\sqrt{2a}}{a\sqrt{4a^2}}$

$=\frac{b\sqrt{2a}}{2a^2}$

73. $\frac{\sqrt{12x^3y}}{\sqrt{20x^4y}}=\sqrt{\frac{12x^3y}{20x^4y}}=\sqrt{\frac{3}{5x}}$

$=\frac{\sqrt{3}}{\sqrt{5x}}=\frac{\sqrt{3}}{\sqrt{5x}}\cdot\frac{\sqrt{5x}}{\sqrt{5x}}=\frac{\sqrt{15x}}{\sqrt{25x^2}}$

$=\frac{\sqrt{15x}}{5x}$

75. $\frac{-2}{1-\sqrt{2}}=\frac{-2}{1-\sqrt{2}}\cdot\frac{1+\sqrt{2}}{1+\sqrt{2}}$

$=\frac{-2(1+\sqrt{2})}{1^2-(\sqrt{2})^2}=\frac{-2-2\sqrt{2}}{1-2}=\frac{-2-2\sqrt{2}}{-1}$

$=2+2\sqrt{2}$

77. $\frac{-4}{3-\sqrt{2}}=\frac{-4}{3-\sqrt{2}}\cdot\frac{3+\sqrt{2}}{3+\sqrt{2}}$

$=\frac{-4(3+\sqrt{2})}{3^2-(\sqrt{2})^2}=\frac{-12-4\sqrt{2}}{9-2}=\frac{-12-4\sqrt{2}}{7}$

79. $\frac{5}{2-\sqrt{7}}=\frac{5}{2-\sqrt{7}}\cdot\frac{2+\sqrt{7}}{2+\sqrt{7}}$

$=\frac{5(2+\sqrt{7})}{2^2-(\sqrt{7})^2}=\frac{10+5\sqrt{7}}{4-7}=\frac{10+5\sqrt{7}}{-3}$

$=-\frac{10+5\sqrt{7}}{3}$

81. $\frac{-7}{\sqrt{x}-3}=\frac{-7}{\sqrt{x}-3}\cdot\frac{\sqrt{x}+3}{\sqrt{x}+3}$

$=\frac{-7(\sqrt{x}+3)}{(\sqrt{x})^2-3^2}=\frac{-(7\sqrt{x}+21)}{x-9}$

$=-\frac{7\sqrt{x}+21}{x-9}$

83. $\frac{\sqrt{3}+\sqrt{4}}{\sqrt{2}+\sqrt{3}}=\frac{\sqrt{3}+\sqrt{2^2}}{\sqrt{2}+\sqrt{3}}=\frac{\sqrt{3}+2}{\sqrt{2}+\sqrt{3}}\cdot\frac{\sqrt{2}-\sqrt{3}}{\sqrt{2}-\sqrt{3}}$

$=\frac{\sqrt{6}-(\sqrt{3})^2+2\sqrt{2}-2\sqrt{3}}{(\sqrt{2})^2-(\sqrt{3})^2}$

$=\frac{\sqrt{6}-3+2\sqrt{2}-2\sqrt{3}}{2-3}$

$=\frac{\sqrt{6}-3+2\sqrt{2}-2\sqrt{3}}{-1}$

$=-\sqrt{6}+3-2\sqrt{2}+2\sqrt{3}$

85. $\frac{2+3\sqrt{5}}{1-\sqrt{5}}=\frac{2+3\sqrt{5}}{1-\sqrt{5}}\cdot\frac{1+\sqrt{5}}{1+\sqrt{5}}$

$=\frac{2+2\sqrt{5}+3\sqrt{5}+3(\sqrt{5})^2}{1^2-(\sqrt{5})^2}$

$=\frac{2+5\sqrt{5}+15}{1-5}$

$=\frac{17+5\sqrt{4}}{-4}$

$=-\frac{17+5\sqrt{4}}{4}$

87. $\dfrac{2\sqrt{a}-\sqrt{b}}{4\sqrt{a}+3\sqrt{b}} = \dfrac{2\sqrt{a}-\sqrt{b}}{4\sqrt{a}+3\sqrt{b}} \cdot \dfrac{4\sqrt{a}-3\sqrt{b}}{4\sqrt{a}-3\sqrt{b}}$

$= \dfrac{8\sqrt{a^2}-6\sqrt{ab}-4\sqrt{ab}+3\sqrt{b^2}}{(4\sqrt{a})^2-(3\sqrt{b})^2}$

$= \dfrac{8a-10\sqrt{ab}+3b}{16a-9b}$

89. $\dfrac{3\sqrt{y}-y}{\sqrt{y}+2y} = \dfrac{3\sqrt{y}-y}{\sqrt{y}+2y} \cdot \dfrac{\sqrt{y}-2y}{\sqrt{y}-2y}$

$= \dfrac{3(\sqrt{y})^2-6y\sqrt{y}-y\sqrt{y}+2y^2}{(\sqrt{y})^2-(2y)^2}$

$= \dfrac{3y-7y\sqrt{y}+2y^2}{y-4y^2}$

$= \dfrac{3-7\sqrt{y}+2y}{1-4y}$

Critical Thinking

91.

$(\sqrt{8}-\sqrt{2})^3 = (2\sqrt{2}-\sqrt{2})^3 = (\sqrt{2}(2-1))^3$

$= (\sqrt{2})^3 = \sqrt{8} = 2\sqrt{2}$

93. $(\sqrt{2}-\sqrt{3})^2 = (\sqrt{2}-\sqrt{3})(\sqrt{2}-\sqrt{3})$

$= 2-\sqrt{6}-\sqrt{6}+3 = 5-2\sqrt{6}$

95. $\dfrac{\sqrt{9+h}-3}{h} \cdot \dfrac{\sqrt{9+h}+3}{\sqrt{9+h}+3} = \dfrac{9+h-9}{h(\sqrt{9+h}+3)}$

$= \dfrac{1}{\sqrt{9+h}+3}$

Projects or Group Activities

97.

$\dfrac{1}{(3+\sqrt[3]{2})} \cdot \dfrac{9-3\sqrt[3]{2}+\sqrt[3]{4}}{9-3\sqrt[3]{2}+\sqrt[3]{4}} = \dfrac{9-3\sqrt[3]{2}+\sqrt[3]{4}}{29}$

Check Your Progress: Chapter 9

1. $(32)^{4/5} = (32^{1/5})^4 = 2^4 = 16$

2. $(16)^{-3/4} = (16^{1/4})^{-3} = 2^{-3} = \dfrac{1}{2^3} = \dfrac{1}{8}$

3. $\left(\dfrac{27}{8}\right)^{-1/3} = \left(\dfrac{27^{1/3}}{8^{1/3}}\right)^{-1} = \left(\dfrac{3}{2}\right)^{-1} = \dfrac{2}{3}$

4. $\left(\dfrac{64}{81}\right)^{-3/2} = \left(\dfrac{64^{1/2}}{81^{1/2}}\right)^{-3} = \left(\dfrac{8}{9}\right)^{-3} = \left(\dfrac{9}{8}\right)^{3}$

$= \dfrac{729}{512}$

5. $x^{3/4} \cdot x^{-1/2} = x^{3/4+(-1/2)} = x^{1/4}$

6. $(8x^6)^{2/3} = 8^{2/3} \cdot (x^6)^{2/3} = 4x^4$

7. $\dfrac{z^{5/6}}{z^{3/4}} = z^{5/6-3/4} = z^{1/12}$

8. $\left(\dfrac{a^{-1/3}b^{3/2}}{c^{2/3}}\right)^{-6} = \dfrac{(a^{-1/3})^{-6}(b^{3/2})^{-6}}{(c^{2/3})^{-6}} = \dfrac{a^2b^{-9}}{c^{-4}}$

$= \dfrac{a^2c^4}{b^9}$

9. $\sqrt{9x^{12}} = (9x^{12})^{1/2} = 3x^6$

10. $\sqrt[5]{-32a^5b^{15}} = (-32a^5b^{15})^{1/5} = -2ab^3$

11. $\sqrt{72a^3b^{10}} = (72a^3b^{10})^{1/2} = (36 \cdot 2 \cdot a^2 \cdot a \cdot b^{10})^{1/2}$

$= 3ab^5(2a)^{1/2} = 6ab^5\sqrt{2a}$

12. $\sqrt[3]{16x^7y^3z^{11}} = (16x^7y^3z^{11})^{1/3}$

$= (8 \cdot 2 \cdot x^6 \cdot x \cdot y^3 \cdot z^9 \cdot z^2)^{1/3} = 2x^2yz^3(2xz^2)^{1/3}$

$= 2x^2yz^3\sqrt[3]{2xz^2}$

13. $6\sqrt{8a^2b^3} - 4a\sqrt{32b^3} = 6\sqrt{4 \cdot 2a^2b^3} - 4a\sqrt{16 \cdot 2b^3}$
$= 6\sqrt{4a^2b^2}\sqrt{2b} - 4a\sqrt{16b^2}\sqrt{2b}$
$= 6 \cdot 2ab\sqrt{2b} - 4a \cdot 4b\sqrt{2b}$
$= 12ab\sqrt{2b} - 16ab\sqrt{2b}$
$= -4ab\sqrt{2b}$

14. $3\sqrt{50} - 9\sqrt{72} + 6\sqrt{98}$
$= 3\sqrt{25 \cdot 2} - 9\sqrt{36 \cdot 2} + 6\sqrt{49 \cdot 2}$
$= 3 \cdot 5\sqrt{2} - 9 \cdot 6\sqrt{2} + 6 \cdot 7\sqrt{2}$
$= 15\sqrt{2} - 54\sqrt{2} + 42\sqrt{2}$
$= 3\sqrt{2}$

15. $x\sqrt{3x^3} + 2x^2\sqrt{27x} - \sqrt{75x^5}$
$= x\sqrt{3x^2x} + 2x^2\sqrt{9 \cdot 3x} - \sqrt{25 \cdot 3x^4x}$
$= x \cdot x\sqrt{3x} + 2 \cdot 3x^2\sqrt{3x} - 5x^2\sqrt{3x}$
$= x^2\sqrt{3x} + 6x^2\sqrt{3x} - 5x^2\sqrt{3x}$
$= 2x^2\sqrt{3x}$

16. $\left(\sqrt[3]{4x^2y}\right)\left(\sqrt[3]{6xy^4}\right) = \sqrt[3]{24x^3y^5}$
$= \sqrt[3]{8x^3y^3 \cdot 3y^2} = 2xy\sqrt[3]{3y^2}$

17. $\sqrt{3x}(2\sqrt{6x^3} - \sqrt{12x})$
$= 2\sqrt{18x^4} - \sqrt{36x^2} = 2\sqrt{9 \cdot 2 \cdot x^4} - \sqrt{36x^2}$
$= 6x^2\sqrt{2} - 6x$

18. $(2\sqrt{5} + 7)(3\sqrt{5} - 1)$
$= 6\sqrt{25} - 2\sqrt{5} + 21\sqrt{5} - 7$
$= 30 + 19\sqrt{5} - 7$
$= 23 + 19\sqrt{5}$

19. $(2\sqrt{x} - 3)^2 = (2\sqrt{x} - 3)(2\sqrt{x} - 3)$
$= 4\sqrt{x^2} - 6\sqrt{x} - 6\sqrt{x} + 9$
$= 4x - 12\sqrt{x} + 9$

20. $\frac{6}{\sqrt{8}} \cdot \frac{\sqrt{8}}{\sqrt{8}} = \frac{6\sqrt{8}}{\sqrt{64}} = \frac{6\sqrt{4 \cdot 2}}{8}$
$= \frac{12\sqrt{2}}{8} = \frac{3\sqrt{2}}{2}$

21. $\sqrt[3]{\frac{3}{4}} = \frac{\sqrt[3]{3}}{\sqrt[3]{4}} \cdot \frac{\sqrt[3]{2}}{\sqrt[3]{2}} = \frac{\sqrt[3]{6}}{\sqrt[3]{8}} = \frac{\sqrt[3]{6}}{2}$

22. $\frac{7}{2\sqrt{3} + 3} \cdot \frac{2\sqrt{3} - 3}{2\sqrt{3} - 3} = \frac{7(2\sqrt{3} - 3)}{4\sqrt{9} - 9}$
$\frac{14\sqrt{3} - 21}{12 - 9} = \frac{14\sqrt{3} - 21}{3}$

23. $\frac{2\sqrt{x}}{\sqrt{x} - 2} \cdot \frac{\sqrt{x} + 2}{\sqrt{x} + 2} = \frac{2\sqrt{x}(\sqrt{x} + 2)}{(\sqrt{x} - 2)(\sqrt{x} + 2)}$
$= \frac{2\sqrt{x^2} + 4\sqrt{x}}{\sqrt{x^2} - 4} = \frac{2x + 4\sqrt{x}}{x - 4}$

24. $\frac{3\sqrt{2} + 5}{2\sqrt{2} - 1} \cdot \frac{2\sqrt{2} + 1}{2\sqrt{2} + 1} = \frac{(3\sqrt{2} + 5)(2\sqrt{2} + 1)}{(2\sqrt{2} - 1)(2\sqrt{2} + 1)}$
$= \frac{6\sqrt{4} + 3\sqrt{2} + 10\sqrt{2} + 5}{4\sqrt{4} - 1} = \frac{12 + 13\sqrt{2} + 5}{8 - 1}$
$= \frac{17 + 13\sqrt{2}}{7}$

Section 9.4

Concept Check

1. Sometimes true

3. No

Objective A Exercises

5. $\sqrt[3]{4x} = -2$
$(\sqrt[3]{4x})^3 = (-2)^3$
$4x = -8$
$x = -2$
The solution is −2.

7. $\sqrt{3x-2}=5$

$(\sqrt{3x-2})^2=(5)^2$

$3x-2=25$

$3x=27$

$x=9$

The solution is 9.

9. No solution

11. $\sqrt[3]{2x-6}=4$

$(\sqrt[3]{2x-6})^3=(4)^3$

$2x-6=64$

$2x=70$

$x=35$

The solution is 35.

13. $\sqrt[4]{3x}+2=5$

$\sqrt[4]{3x}=3$

$(\sqrt[4]{3x})^4=(3)^4$

$3x=81$

$x=27$

The solution is 27.

15. $\sqrt[3]{2x-3}+5=2$

$\sqrt[3]{2x-3}=-3$

$(\sqrt[3]{2x-3})^3=(-3)^3$

$2x-3=-27$

$2x=-24$

$x=-12$

The solution is −12.

17. $4\sqrt{x-2}+2=x+3$

$4\sqrt{x-2}=x+1$

$\left(4\sqrt{x-2}\right)^2=\left(x+1\right)^2$

$16(x-2)=x^2+2x+1$

$16x-32=x^2+2x+1$

$x^2-14x+33=0$

$(x-3)(x-11)=0$

$x-3=0 \quad x-11=0$

$x=3 \qquad x=11$

Check both solutions in the original equation

$4\sqrt{x-2}+2=x+3$

$4\sqrt{3-2}+2=3+3$

$4\sqrt{1}+2=6$

$6=6$

$4\sqrt{x-2}+2=x+3$

$4\sqrt{11-2}+2=11+3$

$4\sqrt{9}+2=14$

$14=14$

The solution is 3 and 11.

19. $\sqrt{x}+\sqrt{x-5}=5$

$\sqrt{x}=5-\sqrt{x-5}$

$(\sqrt{x})^2=(5-\sqrt{x-5})^2$

$x=25-10\sqrt{x-5}+x-5$

$0=20-10\sqrt{x-5}$

$-20=-10\sqrt{x-5}$

$2=\sqrt{x-5}$

$(2)^2=(\sqrt{x-5})^2$

$4=x-5$

$x=9$

The solution is 9.

21. $\sqrt{2x+5}-\sqrt{2x}=1$

$\sqrt{2x+5}=1+\sqrt{2x}$

$(\sqrt{2x+5})^2=(1+\sqrt{2x})^2$

$2x+5=1+2\sqrt{2x}+2x$

$5=1+2\sqrt{2x}$

$4=2\sqrt{2x}$

$2=\sqrt{2x}$

$(2)^2=(\sqrt{2x})^2$

$4=2x$

$x=2$

The solution is 2.

23. $\sqrt{2x}-\sqrt{x-1}=1$

$\sqrt{2x}=1+\sqrt{x-1}$

$(\sqrt{2x})^2=(1+\sqrt{x-1})^2$

$2x=1+2\sqrt{x-1}+x-1$

$x=2\sqrt{x-1}$

$(x)^2=(2\sqrt{x-1})^2$

$x^2=4(x-1)$

$x^2=4x-4$

$x^2-4x+4=0$

$(x-2)(x-2)=0$

$x-2=0 \quad x-2=0$

$x=2 \qquad x=2$

The solution is 2.

25. $\sqrt{2x+2}+\sqrt{x}=3$

$\sqrt{2x+2}=3-\sqrt{x}$

$(\sqrt{2x+2})^2=(3-\sqrt{x})^2$

$2x+2=9-6\sqrt{x}+x$

$x+2=9-6\sqrt{x}$

$x-7=-6\sqrt{x}$

$(x-7)^2=(-6\sqrt{x})^2$

$x^2-14x+49=36x$

$x^2-50x+49=0$

$(x-49)(x-1)=0$

$x-49=0 \quad x-1=0$

$x=49 \qquad x=1$

Check both solutions in the original equation

$\sqrt{2x+2}+\sqrt{x}=3$

$\sqrt{2(49)+2}+\sqrt{49}=3$

$\sqrt{100}+\sqrt{49}=3$

$10+7=3$

$17\neq 3$

$\sqrt{2(1)+2}+\sqrt{1}=3$

$\sqrt{4}+\sqrt{1}=3$

$2+1=3$

$3=3$

The solution is 1.

27. $\sqrt{x}<\sqrt{x+5}$.

Therefore $\sqrt{x}-\sqrt{x+5}<0$ and cannot equal a positive number.

Objective B Exercises

29. Strategy: To find the distance the object will fall substitute the given values for t and g in the equation and solve for d.

Solution: $t = \sqrt{\frac{2d}{g}}$

$3 = \sqrt{\frac{2d}{5.5}}$

$(3)^2 = (\sqrt{\frac{2d}{5.5}})^2$

$9 = \frac{2d}{5.5}$

$49.5 = 2d$

$24.75 = d$

On the moon, the object will fall 24.75 ft in 3 s.

31. a) **Strategy:** To find the height of the water evaluate the function for $t = 10$.

Solution: $h(t) = (88.18 - 3.18t)^{2/5}$

$h(10) = (88.18 - 3.18(10))^{2/5}$

$h(10) = (88.18 - 31.8)^{2/5}$

$h(10) = (56.38)^{2/5}$

$h(10) = 5.0$

The height of the water is 5.0 ft.

b) Strategy: To find how long it will take to empty the take substitute the given value for h in the equation and solve for t.

Solution: $h(t) = (88.18 - 3.18t)^{2/5}$

$0 = (88.18 - 3.18t)^{2/5}$

$0 = ((88.18 - 3.18t)^{2/5})^{5/2}$

$0 = 88.18 - 3.18t$

$3.18t = 88.16$

$t = 27.7$

The tank will empty in 27.7 s.

33. Strategy: To find the length of the pendulum substitute the given value for T in the equation and solve for L.

Solution: $T = 2\pi\sqrt{\frac{L}{32}}$

$3 = 2\pi\sqrt{\frac{L}{32}}$

$\left(\frac{3}{2\pi}\right)^2 = \left(\sqrt{\frac{L}{32}}\right)^2$

$\left(\frac{3}{2\pi}\right)^2 = \frac{L}{32}$

$32\left(\frac{3}{2\pi}\right)^2 = L$

$7.3 = L$

The length of the pendulum is 7.30 ft.

35. Strategy: Find the difference in the widths. Use the Pythagorean Theorem to find the width of the screen of a regular TV and then repeat the process to find the width of the screen for HDTV. Subtract the width of the regular TV from the width of the HDTV.

Solution: $c^2 = a^2 + b^2$

for the regular TV:

$27^2 = 16.2^2 + b^2$

$729 = 262.44 + b^2$

$466.56 = b^2$

$(466.56)^{1/2} = (b^2)^{1/2}$

$21.6 = b$

for the HDTV

$33^2 = 16.2^2 + b^2$

$1089 = 262.44 + b^2$

$826.56 = b^2$

$(826.56)^{1/2} = (b^2)^{1/2}$

$28.75 = b$

$28.75 - 21.6 = 7.15$

The HDTV is approximately 7.15 in. wider.

Critical Thinking

37. Strategy: Use the Pythagorean Theorem to find out the longest pole that can be placed in the box.

Solution:
Find the length diagonal of the bottom of the box.
$c^2 = a^2 + b^2$
$c^2 = 2^2 + 3^2$
$c^2 = 4 + 9$
$c^2 = 13$

This represents the value one of the legs of the right triangle needed to find the length of the pole.

$c^2 = a^2 + b^2$
$c^2 = 13 + 4^2$
$c^2 = 13 + 16$
$c^2 = 29$
$c = 5.4$

The longest pole can be 5.4 ft.

Projects or Group Activities

39. No. One way to see this is to calculate the distance of the bottom of the ladder from the wall at one second intervals. A second way is to note that it takes 4 s for the top of the ladder to reach the ground. In that 4 s, the bottom of the ladder has moved 4 ft. Thus the average speed is 1 ft/s.

Section 9.5

Concept Check

1. An imaginary number is a number whose square is a negative number.
A complex number is a number of the form $a + bi$ where a and b are real numbers and $i = \sqrt{-1}$.

3. 3; 7

5. 7; 0

Objective A Exercises

7. $\sqrt{-25} = i\sqrt{25} = 5i$

9. $\sqrt{-98} = i\sqrt{98} = i\sqrt{49 \cdot 2} = 7i\sqrt{2}$

11. $\dfrac{6+\sqrt{-4}}{2} = \dfrac{6+i\sqrt{4}}{2} = \dfrac{6+2i}{2} = 3+i$

13. $\dfrac{6-5\sqrt{-8}}{4} = \dfrac{6-5i\sqrt{4\cdot 2}}{4} = \dfrac{6-10i\sqrt{2}}{4}$
$= \dfrac{3}{2} - \dfrac{5i\sqrt{2}}{2}$

15. $-b + \sqrt{b^2 - 4ac}$
$-4 + \sqrt{(4)^2 - 4(1)(5)} = -4 + \sqrt{16 - 20}$
$= -4 + \sqrt{-4} = -4 + i\sqrt{4}$
$= -4 + 2i$

17. $-b + \sqrt{b^2 - 4ac}$
$-(-4) + \sqrt{(-4)^2 - 4(2)(10)} = 4 + \sqrt{16 - 80}$
$= 4 + \sqrt{-64} = 4 + i\sqrt{64}$
$= 4 + 8i$

19. $-b + \sqrt{b^2 - 4ac}$
$-(-8) + \sqrt{(-8)^2 - 4(3)(6)} = 8 + \sqrt{64 - 72}$
$= 8 + \sqrt{-8} = 8 + i\sqrt{4 \cdot 2}$
$= 8 + 2i\sqrt{2}$

21. $-b + \sqrt{b^2 - 4ac}$
$-2 + \sqrt{(2)^2 - 4(4)(7)} = -2 + \sqrt{4 - 112}$
$= -2 + \sqrt{-108} = -2 + i\sqrt{36 \cdot 3}$
$= -2 + 6i\sqrt{3}$

23. $-b+\sqrt{b^2-4ac}$

$-5+\sqrt{(5)^2-4(-2)(-6)}=-5+\sqrt{25-48}$

$=-5+\sqrt{-23}$

$=-5+i\sqrt{23}$

25. $-b+\sqrt{b^2-4ac}$

$-4+\sqrt{(4)^2-4(-3)(-6)}=-4+\sqrt{16-72}$

$=-4+\sqrt{-56}=-4+i\sqrt{4\cdot 14}$

$=-4+2i\sqrt{14}$

Objective B Exercises

27. $(2+4i)+(6-5i)=8-i$

29. $(-2-4i)-(6-8i)=-8+4i$

31. $(8-2i)-(2+4i)=6-6i$

33. $5+(6-4i)=11-4i$

35. $3i-(6+5i)=-6-2i$

37. The real parts of the complex numbers are additive inverses.

Objective C Exercises

39. $(7i)(-9i)=-63i^2=-63(-1)=63$

41. $\sqrt{-2}\sqrt{-8}=i\sqrt{2}\cdot i\sqrt{8}=i^2\sqrt{16}=-\sqrt{16}=-4$

43. $(5+2i)(5-2i)=25-10i+10i-4i^2$

$=25-4i^2=25-4(-1)=29$

45. $2i(6+2i)=12i+4i^2=12i+4(-1)$

$=-4+12i$

47. $-i(4-3i)=-4i+3i^2=-4i+3(-1)$

$=-3-4i$

49. $(5-2i)(3+i)=15+5i-6i-2i^2$

$=15-i-2i^2$

$=15-i-2(-1)$

$=17-i$

51. $(6+5i)(3+2i)=18+12i+15i+10i^2$

$=18+27i+10i^2$

$=18+27i+10(-1)$

$=8+27i$

53. $(2+5i)^2=4+20i+25i^2$

$=4+20i+25(-1)$

$=-21+20i$

55. $\left(\frac{6}{5}+\frac{3}{5}i\right)\left(\frac{2}{3}-\frac{1}{3}i\right)=\frac{4}{5}-\frac{2}{5}i+\frac{2}{5}i-\frac{1}{5}i^2$

$=\frac{4}{5}-\frac{1}{5}i^2$

$=\frac{4}{5}-\frac{1}{5}(-1)$

$=\frac{4}{5}+\frac{1}{5}=1$

57. True

Objective D Exercises

59. $\frac{3}{i}=\frac{3}{i}\cdot\frac{i}{i}=\frac{3i}{i^2}=\frac{3i}{-1}=-3i$

61. $\frac{2-3i}{-4i}=\frac{2-3i}{-4i}\cdot\frac{i}{i}=\frac{2i-3i^2}{-4i^2}$

$=\frac{2i-3(-1)}{-4(-1)}=\frac{2i+3}{4}$

$=\frac{3}{4}+\frac{1}{2}i$

63. $\frac{4}{5+i}=\frac{4}{5+i}\cdot\frac{5-i}{5-i}=\frac{20-4i}{25-i^2}$

$=\frac{20-4i}{25-(-1)}=\frac{20-4i}{26}$

$=\frac{10}{13}-\frac{2}{13}i$

65. $\dfrac{2}{2-i}=\dfrac{2}{2-i}\cdot\dfrac{2+i}{2+i}=\dfrac{4+2i}{4-i^2}$

$=\dfrac{4+2i}{4-(-1)}=\dfrac{4+2i}{5}$

$=\dfrac{4}{5}+\dfrac{2}{5}i$

67. $\dfrac{1-3i}{3+i}=\dfrac{1-3i}{3+i}\cdot\dfrac{3-i}{3-i}=\dfrac{3-i-9i+3i^2}{9-i^2}$

$=\dfrac{3-10i+3(-1)}{9-(-1)}=\dfrac{-10i}{10}$

$=-i$

69. $\dfrac{3i}{1+4i}=\dfrac{3i}{1+4i}\cdot\dfrac{1-4i}{1-4i}=\dfrac{3i-12i^2}{1-16i^2}$

$=\dfrac{3i-12(-1)}{1-16(-1)}=\dfrac{3i+12}{17}$

$=\dfrac{12}{17}+\dfrac{3}{17}i$

71. $\dfrac{2-3i}{3+i}=\dfrac{2-3i}{3+i}\cdot\dfrac{3-i}{3-i}=\dfrac{6-2i-9i+3i^2}{9-i^2}$

$=\dfrac{6-11i+3(-1)}{9-(-1)}=\dfrac{3-11i}{10}$

$=\dfrac{3}{10}-\dfrac{11}{10}i$

73. $\dfrac{5+3i}{3-i}=\dfrac{5+3i}{3-i}\cdot\dfrac{3+i}{3+i}=\dfrac{15+5i+9i+3i^2}{9-i^2}$

$=\dfrac{15+14i+3(-1)}{9-(-1)}=\dfrac{12+14i}{10}$

$=\dfrac{6+7i}{5}$

$=\dfrac{6}{5}+\dfrac{7}{5}i$

75. True

Critical Thinking

77. $x^2-10x+29=0$

$(5-3i)^2-10(5-3i)+29=0$

$25-30i-9-50+30i+29=0$

$-5=0$

No, $5-3i$ is not a solution.

79. $x^2-2x+4=0$

$(1-i\sqrt{3})^2-2(1-i\sqrt{3})+4=0$

$1-2i\sqrt{3}-3-2+2i\sqrt{3}+4=0$

$0=0$

Yes, $1-i\sqrt{3}$ is a solution.

Projects or Group Activities

81. 1

83. i

Chapter 9 Review Exercises

1. $(16x^{-4}y^{12})^{1/4}(100x^6y^{-2})^{1/2}$

$=(16)^{1/4}x^{-1}y^3\cdot 100^{1/2}x^3y^{-1}$

$=20x^2y^2$

2. $\sqrt[4]{3x-5}=2$

$(\sqrt[4]{3x-5})^4=2^4$

$3x-5=16$

$3x=21$

$x=7$

3. $(6-5i)(4+3i)=24+18i-20i-15i^2$

$=24-2i-15(-1)$

$=39-2i$

4. $7y\sqrt[3]{x^2}=7x^{2/3}y$

5. $(\sqrt{3}+8)(\sqrt{3}-2)=\sqrt{3^2}-2\sqrt{3}+8\sqrt{3}-16$

$=3+6\sqrt{3}-16$

$=6\sqrt{3}-13$

6. $\sqrt{4x+9}+10=11$

$\sqrt{4x+9}=1$

$(\sqrt{4x+9})^2=1^2$

$4x+9=1$

$4x=-8$

$x=-2$

7. $\dfrac{x^{-3/2}}{x^{7/2}}=x^{-10/2}=x^{-5}=\dfrac{1}{x^5}$

8. $\dfrac{8}{\sqrt{3y}}=\dfrac{8}{\sqrt{3y}}\cdot\dfrac{\sqrt{3y}}{\sqrt{3y}}=\dfrac{8\sqrt{3y}}{\sqrt{3^2y^2}}=\dfrac{8\sqrt{3y}}{3y}$

9. $\sqrt[3]{-8a^6b^{12}}=-2a^2b^4$

10. $\sqrt{50a^4b^3}-ab\sqrt{18a^2b}$

$=\sqrt{25a^4b^2\cdot 2b}-ab\sqrt{9a^2\cdot 2b}$

$=5a^2b\sqrt{2b}-3a^2b\sqrt{2b}$

$=2a^2b\sqrt{2b}$

11. $\dfrac{14}{4-\sqrt{2}}=\dfrac{14}{4-\sqrt{2}}\cdot\dfrac{4+\sqrt{2}}{4+\sqrt{2}}=\dfrac{56+14\sqrt{2}}{16-\sqrt{2^2}}$

$=\dfrac{56+14\sqrt{2}}{16-2}=\dfrac{56+14\sqrt{2}}{14}$

$=4+\sqrt{2}$

12. $\dfrac{5+2i}{3i}=\dfrac{5+2i}{3i}\cdot\dfrac{-3i}{-3i}=\dfrac{-15i-6i^2}{-9i^2}$

$=\dfrac{-15i-6(-1)}{-9(-1)}=\dfrac{-15i+6}{9}$

$=\dfrac{2}{3}-\dfrac{5}{3}i$

13. $\sqrt{18a^3b^6}=\sqrt{9a^2b^6\cdot 2a}=3ab^3\sqrt{2a}$

14. $(17+8i)-(15-4i)=2+12i$

15. $3x\sqrt[3]{54x^8y^{10}}-2x^2y\sqrt[3]{16x^5y^7}$

$=3x\sqrt[3]{27x^6y^9\cdot 2x^2y}-2x^2y\sqrt[3]{8x^3y^6\cdot 2x^2y}$

$=9x^3y^3\sqrt[3]{2x^2y}-4x^3y^3\sqrt[3]{2x^2y}$

$=5x^3y^3\sqrt[3]{2x^2y}$

16. $\sqrt[3]{16x^4y}\sqrt[3]{4xy^5}=\sqrt[3]{64x^5y^6}$

$=\sqrt[3]{64x^3y^6\cdot x^2}=4xy^2\sqrt[3]{x^2}$

17. $i(3-7i)=3i-7i^2=3i-7(-1)=7+3i$

18. $\dfrac{(4a^{-2/3}b^4)^{-1/2}}{(a^{-1/6}b^{3/2})^2}=\dfrac{(4)^{-1/2}a^{1/3}b^{-2}}{a^{-1/3}b^3}$

$=\dfrac{a^{2/3}}{2b^5}$

19. $\sqrt[5]{-64a^8b^{12}}=\sqrt[5]{-32a^5b^{10}\cdot 2a^3b^2}$

$=-2ab^2\sqrt[5]{2a^3b^2}$

20. $\dfrac{5+9i}{1-i}=\dfrac{5+9i}{1-i}\cdot\dfrac{1+i}{1+i}=\dfrac{5+5i+9i+9i^2}{1-i^2}$

$=\dfrac{5+14i+9(-1)}{1-(-1)}=\dfrac{-4+14i}{2}$

$=-2+7i$

21. $\sqrt{-12}\sqrt{-6}=i\sqrt{12}\cdot i\sqrt{6}=i^2\sqrt{72}$

$=-1\sqrt{36\cdot 2}=-6\sqrt{2}$

22. $\sqrt{x-5}+\sqrt{x+6}=11$
$\sqrt{x-5}=11-\sqrt{x+6}$
$(\sqrt{x-5})^2=(11-\sqrt{x+6})^2$
$x-5=121-22\sqrt{x+6}+x+6$
$-5=127-22\sqrt{x+6}$
$-132=-22\sqrt{x+6}$
$6=\sqrt{x+6}$
$(6)^2=(\sqrt{x+6})^2$
$36=x+6$
$30=x$
The solution is 30.

23. $\sqrt[4]{81a^8b^{12}}=3a^2b^3$

24. $\frac{9}{\sqrt[3]{3x}}=\frac{9}{\sqrt[3]{3x}}\cdot\frac{\sqrt[3]{9x^2}}{\sqrt[3]{9x^2}}=\frac{9\sqrt[3]{9x^2}}{\sqrt[3]{27x^3}}$
$=\frac{9\sqrt[3]{9x^2}}{3x}=\frac{3\sqrt[3]{9x^2}}{x}$

25. $(-8+3i)-(4-7i)=-12+10i$

26. $(2+\sqrt{2x-1})^2=4+4\sqrt{2x-1}+2x-1$
$=2x+4\sqrt{2x-1}+3$

27. $4x\sqrt{12x^2y}+\sqrt{3x^4y}-x^2\sqrt{27y}$
$=4x\sqrt{4x^2\cdot 3y}+\sqrt{x^4\cdot 3y}-x^2\sqrt{9\cdot 3y}$
$=8x^2\sqrt{3y}+x^2\sqrt{3y}-3x^2\sqrt{3y}$
$=6x^2\sqrt{3y}$

28. $81^{-1/4}=(3^4)^{-1/4}=3^{-1}=\frac{1}{3}$

29. $(a^{16})^{-5/8}=a^{-10}=\frac{1}{a^{10}}$

30. $-\sqrt{49x^6y^{16}}=-7x^3y^8$

31. $4a^{2/3}=4\sqrt[3]{a^2}$

32. $(9x^2y^4)^{-1/2}(x^6y^6)^{1/3}=9^{-1/2}x^{-1}y^{-2}x^2y^2$
$=9^{-1/2}x^1y^0=9^{-1/2}x$
$=\frac{x}{9^{1/2}}=\frac{x}{3}$

33. $\sqrt[4]{x^6y^8z^{10}}=\sqrt[4]{x^4y^8z^8\cdot x^2z^2}$
$=xy^2z^2\sqrt[4]{x^2z^2}$

34. $\sqrt{54}+\sqrt{24}=\sqrt{9\cdot 6}+\sqrt{4\cdot 6}$
$=3\sqrt{6}+2\sqrt{6}=5\sqrt{6}$

35. $\sqrt{48x^5y}-x\sqrt{80x^2y}=\sqrt{16x^4\cdot 3xy}-x\sqrt{16x^2\cdot y}$
$=4x^2\sqrt{3xy}-4x^2\sqrt{y}$

36. $\sqrt{32}\sqrt{50}=\sqrt{1600}=40$

37. $\sqrt{3x}(3+\sqrt{3x})=3\sqrt{3x}+\sqrt{(3x)^2}$
$=3\sqrt{3x}+3x$

38. $\frac{\sqrt{125x^6}}{\sqrt{5x^3}}=\sqrt{\frac{125x^6}{5x^3}}=\sqrt{25x^3}=5x\sqrt{x}$

39. $\frac{2-3\sqrt{7}}{6-\sqrt{7}}=\frac{2-3\sqrt{7}}{6-\sqrt{7}}\cdot\frac{6+\sqrt{7}}{6+\sqrt{7}}$
$=\frac{12+2\sqrt{7}-18\sqrt{7}-3(\sqrt{7})^2}{6^2-(\sqrt{7})^2}$
$=\frac{12-16\sqrt{7}-21}{36-7}$
$=\frac{-9-16\sqrt{7}}{29}$

40. $\sqrt{-36}=i\sqrt{36}=6i$

41. $-b+\sqrt{b^2-4ac}$

$-(-8)+\sqrt{(-8)^2-4(1)(25)}=8+\sqrt{64-100}$

$=8+\sqrt{-36}=8+i\sqrt{36}$

$=8+6i$

42. $-b+\sqrt{b^2-4ac}$

$-2+\sqrt{2^2-4(1)(9)}=-2+\sqrt{4-36}$

$=-2+\sqrt{-32}=-2+i\sqrt{16\cdot 2}$

$=-2+4i\sqrt{2}$

43. $(5+2i)+(4-3i)=9-i$

44. $(3+2\sqrt{5})(3-2\sqrt{5})$

$=9-6\sqrt{5}+6\sqrt{5}-4(\sqrt{5})^2$

$=9-20=-11$

45. $(3-9i)-7=-4-9i$

46. $(4-i)^2=(4-i)(4-i)$

$=16-8i+i^2=16-8i+(-1)$

$=15-8i$

47. $\frac{-6}{i}=\frac{-6}{i}\cdot\frac{i}{i}=\frac{-6i}{i^2}=\frac{-6i}{-1}=6i$

48. $\frac{7}{2-i}=\frac{7}{2-i}\cdot\frac{2+i}{2+i}=\frac{14+7i}{4-i^2}$

$=\frac{14+7i}{4-(-1)}=\frac{14+7i}{5}$

$=\frac{14}{5}+\frac{7}{5}i$

49. $\sqrt{2x-7}+2=5$

$\sqrt{2x-7}=3$

$\left(\sqrt{2x-7}\right)^2=3^2$

$2x-7=9$

$2x=16$

$x=8$

The solution is 8.

50. $\sqrt[3]{9x}=-6$

$\left(\sqrt[3]{9x}\right)^3=(-6)^3$

$9x=-216$

$x=-24$

The solution is −24.

51. Strategy: Use the Pythagorean Theorem to find the width of the rectangle.

Solution: $c^2=a^2+b^2$

$13^2=12^2+b^2$

$169=144+b^2$

$25=b^2$

$(25)^{1/2}=(b^2)^{1/2}$

$5=b$

The width of the rectangle is 5 in.

52. Strategy: To find the amount of power substitute the given value for v in the equation and solve for P.

Solution: $v=4.05\sqrt[3]{P}$

$20=4.05\sqrt[3]{P}$

$4.94=\sqrt[3]{P}$

$(4.94)^3=(\sqrt[3]{P})^3$

$120\approx P$

The amount of power is 120 watts.

53. Strategy: To find the distance required substitute the given values for v and a in the equation and solve for s.

Solution: $v=\sqrt{2as}$

$88=\sqrt{2(16s)}$

$(88)^2=(\sqrt{32s})^2$

$7744=32s$

$242=s$

The distance required is 242 ft.

54. Strategy: To find the distance use the Pythagorean Theorem.

Solution: $c^2 = a^2 + b^2$
$12^2 = 10^2 + b^2$
$144 = 100 + b^2$
$44 = b^2$
$(44)^{1/2} = (b^2)^{1/2}$
$6.63 = b$
The distance is 6.63 ft.

Chapter 9 Test

1. $\frac{1}{2}\sqrt[4]{x^3} = \frac{1}{2}x^{3/4}$

2. $\sqrt[3]{54x^7y^3} - x\sqrt[3]{128x^4y^3} - x^2\sqrt[3]{2xy^3}$
$= \sqrt[3]{27x^6y^3 \cdot 2x} - x\sqrt[3]{64x^3y^3 \cdot 2x} - x^2\sqrt[3]{y^3 \cdot 2x}$
$= 3x^2y\sqrt[3]{2x} - 4x^2y\sqrt[3]{2x} - x^2y\sqrt[3]{2x}$
$= -2x^2y\sqrt[3]{2x}$

3. $3y^{2/5} = 3\sqrt[5]{y^2}$

4. $(2 + 5i)(4 - 2i) = 8 - 4i + 20i - 10i^2$
$= 8 + 16i - 10(-1)$
$= 18 + 16i$

5. $(3 - 2\sqrt{x})^2 = (3 - 2\sqrt{x})(3 - 2\sqrt{x})$
$= 9 - 12\sqrt{x} + 4(\sqrt{x})^2$
$= 4x - 12\sqrt{x} + 9$

6. $\frac{r^{2/3}r^{-1}}{r^{-1/2}} = \frac{r^{-1/3}}{r^{-1/2}} = r^{1/6}$

7. $\sqrt{x+12} - \sqrt{x} = 2$
$\sqrt{x+12} = 2 + \sqrt{x}$
$(\sqrt{x+12})^2 = (2 + \sqrt{x})^2$
$x + 12 = 4 + 4\sqrt{x} + x$
$12 = 4 + 4\sqrt{x}$
$8 = 4\sqrt{x}$
$2 = \sqrt{x}$
$(2)^2 = (\sqrt{x})^2$
$4 = x$
The solution is 4.

8. $\sqrt[4]{4a^5b^3}\sqrt[4]{8a^3b^7} = \sqrt[4]{32a^8b^{10}}$
$= \sqrt[4]{16a^8b^8 \cdot 2b^2} = 2a^2b^2\sqrt[4]{2b^2}$

9. $\sqrt{3x}(\sqrt{x} - \sqrt{25x}) = \sqrt{3x^2} - \sqrt{75x^2}$
$= \sqrt{x^2 \cdot 3} - \sqrt{25x^2 \cdot 3} = x\sqrt{3} - 5x\sqrt{3}$
$= -4x\sqrt{3}$

10. $(5 - 2i) - (8 - 4i) = -3 + 2i$

11. $\sqrt{32x^4y^7} = \sqrt{16x^4y^6 \cdot 2y} = 4x^2y^3\sqrt{2y}$

12. $(2\sqrt{3} + 4)(3\sqrt{3} - 1)$
$= 6\sqrt{3^2} - 2\sqrt{3} + 12\sqrt{3} - 4$
$= 18 + 10\sqrt{3} - 4$
$= 14 + 10\sqrt{3}$

13. $\sqrt{-5} \cdot \sqrt{-20} = i\sqrt{5} \cdot i\sqrt{20} = i^2\sqrt{100}$
$= -1\sqrt{100} = -10$

14. $\frac{4 - 2\sqrt{5}}{2 - \sqrt{5}} = \frac{4 - 2\sqrt{5}}{2 - \sqrt{5}} \cdot \frac{2 + \sqrt{5}}{2 + \sqrt{5}}$
$= \frac{8 + 4\sqrt{5} - 4\sqrt{5} - 2(\sqrt{5})^2}{2^2 - (\sqrt{5})^2}$
$= \frac{8 - 10}{4 - 5} = \frac{-2}{-1} = 2$

15. $\sqrt{18a^3}+a\sqrt{50a}=\sqrt{9a^2\cdot 2a}+a\sqrt{25\cdot 2a}$
$=3a\sqrt{2a}+5a\sqrt{2a}$
$=8a\sqrt{2a}$

16. $(\sqrt{a}-3\sqrt{b})(2\sqrt{a}+5\sqrt{b})$
$=2\sqrt{a^2}+5\sqrt{ab}-6\sqrt{ab}-15\sqrt{b^2}$
$=2a-\sqrt{ab}-15b$

17. $\dfrac{(2x^{1/3}y^{-2/3})^6}{(x^{-4}y^8)^{1/4}}=\dfrac{2^6x^2y^{-4}}{x^{-1}y^2}=\dfrac{64x^3}{y^6}$

18. $\dfrac{10x}{\sqrt[3]{5x^2}}=\dfrac{10x}{\sqrt[3]{5x^2}}\cdot\dfrac{\sqrt[3]{25x}}{\sqrt[3]{25x}}=\dfrac{10x\sqrt[3]{25x}}{\sqrt[3]{125x^3}}$
$=\dfrac{10x\sqrt[3]{25x}}{5x}=2\sqrt[3]{25x}$

19. $\dfrac{2+3i}{1-2i}=\dfrac{2+3i}{1-2i}\cdot\dfrac{1+2i}{1+2i}=\dfrac{2+4i+3i+6i^2}{1-4i^2}$
$=\dfrac{2+7i+6(-1)}{1-4(-1)}=\dfrac{-4+7i}{5}$
$=-\dfrac{4}{5}+\dfrac{7}{5}i$

20. $\sqrt[3]{2x-2}+4=2$
$\sqrt[3]{2x-2}=-2$
$(\sqrt[3]{2x-2})^3=(-2)^3$
$2x-2=-8$
$2x=-6$
$x=-3$
The solution is −3.

21. $\left(\dfrac{4a^4}{b^2}\right)^{-3/2}=\dfrac{(4a^4)^{-3/2}}{(b^2)^{-3/2}}=\dfrac{4^{-3/2}a^{-6}}{b^{-3}}=\dfrac{b^3}{8a^6}$

22. $\sqrt[3]{27a^4b^3c^7}=\sqrt[3]{27a^3b^3c^6\cdot ac}$
$=3abc^2\sqrt[3]{ac}$

23. $\dfrac{\sqrt{32x^5y}}{\sqrt{2xy^3}}=\sqrt{\dfrac{32x^5y}{2xy^3}}=\sqrt{\dfrac{16x^4}{y^2}}=\dfrac{4x^2}{y}$

24. $\dfrac{5x}{\sqrt{5x}}=\dfrac{5x}{\sqrt{5x}}\cdot\dfrac{\sqrt{5x}}{\sqrt{5x}}=\dfrac{5x\sqrt{5x}}{(\sqrt{5x})^2}$
$=\dfrac{5x\sqrt{5x}}{5x}=\sqrt{5x}$

25. Strategy: To find the distance the object has fallen substitute the value for v in the equation and solve for d.

Solution: $v=\sqrt{64d}$
$192=\sqrt{64d}$
$(192)^2=(\sqrt{64d})^2$
$36{,}864=64d$
$576=d$
The object has fallen 576 ft.

Cumulative Review Exercises

1. $2^3\cdot 3-4(3-4\cdot 5)=2^3\cdot 3-4(3-20)$
$=2^3\cdot 3-4(-17)$
$=8\cdot 3-4(-17)$
$=24+68$
$=92$

2. $4a^2b-a^3=4(-2)^2(3)-(-2)^3$
$=4(4)(3)-(-8)$
$=16(3)+8$
$=48+8$
$=56$

3. $-3(4x-1)-2(1-x)=-12x+3-2+2x$
$=-10x+1$

4. $5-\frac{2}{3}x=4$

$5-\frac{2}{3}x-5=4-5$

$-\frac{2}{3}x=-1$

$\left(-\frac{3}{2}\right)\left(-\frac{2}{3}x\right)=-1\left(-\frac{3}{2}\right)$

$x=\frac{3}{2}$

The solution is $\frac{3}{2}$.

5. $2[4-2(3-2x)]=4(1-x)$

$2[4-6+4x]=4-4x$

$2[-2+4x]=4-4x$

$-4+8x=4-4x$

$-4+12x=4$

$12x=8$

$x=\frac{2}{3}$

The solution is $\frac{2}{3}$.

6. $6x-3(2x+2)>3-3(x+2)$

$6x-6x-6>3-3x-6$

$-6>-3x-3$

$-3>-3x$

$1<x$

$\{x\,|\,x>1\}$

7. $2+|4-3x|=5$

$|4-3x|=3$

$4-3x=-3$ $\quad$ $4-3x=3$

$-3x=-7$ $\quad$ $-3x=-1$

$x=\frac{7}{3}$ $\quad$ $x=\frac{1}{3}$

The solutions are $\frac{1}{3}$ and $\frac{7}{3}$.

8. $|2x+3|\le 9$

$-9\le 2x+3\le 9$

$-9-3\le 2x+3-3\le 9-3$

$-12\le 2x\le 6$

$-6\le x\le 3$

$\{x|\,-6\le x\le 3\}$

9. $A=\frac{1}{2}bh=\frac{1}{2}(25)(15)=12.5(15)=187.5$

The area is 187.5 cm^2.

10. $V=L\times W\times H=3.5\times 2\times 2=14\text{ ft}^3$

The volume is 14 ft^3.

11. Solve $3x-2y=-6$ for y.

$3x-2y=-6$

$-2y=-3x-6$

$y=\frac{3}{2}x+3$

The y-intercept is (0, 3)

The slope is $\frac{3}{2}$.

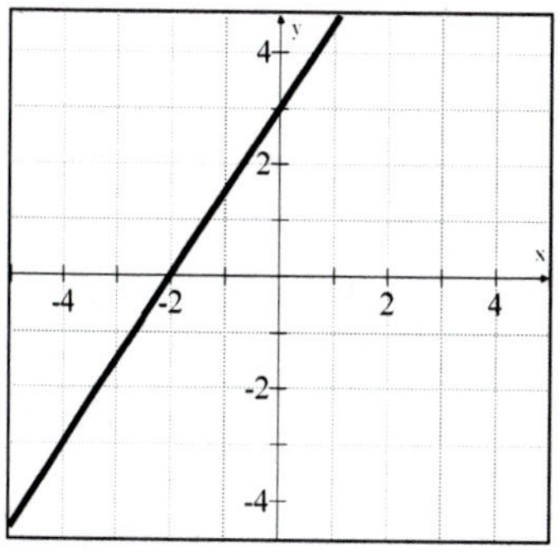

12. $3x+2y\le 4$

$2y\le -3x+4$

$y\le -\frac{3}{2}x+2$

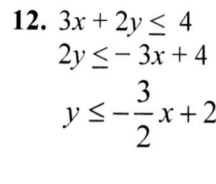

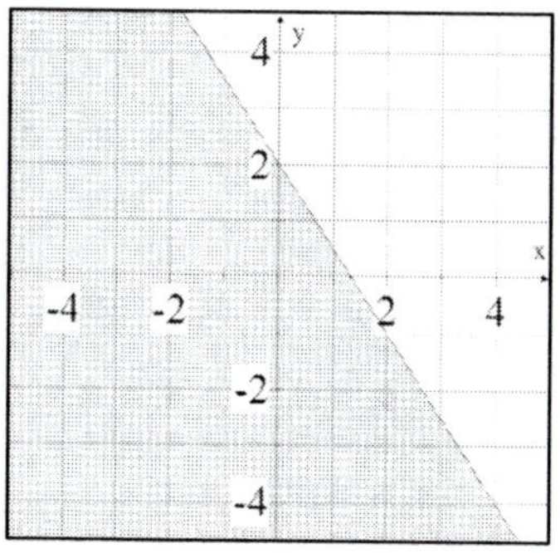

13. Find the slope of the line.

$m=\frac{y_2-y_1}{x_2-x_1}=\frac{2-3}{-1-2}=\frac{-1}{-3}=\frac{1}{3}$

Use the point-slope formula to find the equation of the line.

$y-y_1=m(x-x_1)$

$y-3=\frac{1}{3}(x-2)$

$$y-3=\frac{1}{3}x-\frac{2}{3}$$
$$y=\frac{1}{3}x+\frac{7}{3}$$

The equation of the line is $y=\frac{1}{3}x+\frac{7}{3}$.

14.
$$\begin{aligned} 2x-y&=4\\ -2x+3y&=0\\ \hline 2y&=4\\ y&=2\\ 2x-2&=4\\ 2x&=6\\ x&=3 \end{aligned}$$

The solution is (3, 2).

15. $(2^{-1}x^2y^{-6})(2^{-1}y^{-4})^{-2}$
$=(2^{-1}x^2y^{-6})(2^2y^8)$
$=2x^2y^2$

16. $81x^2-y^2=(9x+y)(9x-y)$

17. $x^5+2x^3-3x=x(x^4+2x^2-3)$
$=x(x^2+3)(x^2-1)$
$=x(x^2+3)(x+1)(x-1)$

18. $P=\frac{R-C}{n}$
$P\cdot n=\frac{R-C}{n}\cdot n$
$nP=R-C$
$nP+C=R$
$C=R-nP$

19. $\left(\frac{x^{-2/3}y^{1/2}}{y^{-1/3}}\right)^6=\frac{x^{-4}y^3}{y^{-2}}=\frac{y^5}{x^4}$

20. $\sqrt{40x^3}-x\sqrt{90x}=\sqrt{4x^2\cdot 10x}-x\sqrt{9\cdot 10x}$
$=2x\sqrt{10x}-3x\sqrt{10x}$
$=-x\sqrt{10x}$

21. $(\sqrt{3}-2)(\sqrt{3}-5)$
$=\sqrt{3^2}-5\sqrt{3}-2\sqrt{3}+10$
$=3-7\sqrt{3}+10$
$=13-7\sqrt{3}$

22. $\frac{4}{\sqrt{6}-\sqrt{2}}=\frac{4}{\sqrt{6}-\sqrt{2}}\cdot\frac{\sqrt{6}+\sqrt{2}}{\sqrt{6}+\sqrt{2}}$
$=\frac{4\sqrt{6}+4\sqrt{2}}{\sqrt{6^2}-\sqrt{2^2}}=\frac{4\sqrt{6}+4\sqrt{2}}{6-2}=\frac{4\sqrt{6}+4\sqrt{2}}{4}$
$=\sqrt{6}+\sqrt{2}$

23. $\frac{2i}{3-i}=\frac{2i}{3-i}\cdot\frac{3+i}{3+i}=\frac{6i+2i^2}{9-i^2}$
$=\frac{6i+2(-1)}{9-(-1)}=\frac{6i-2}{10}$
$=\frac{-1+3i}{5}$
$=-\frac{1}{5}+\frac{3}{5}i$

24. $\sqrt[3]{3x-4}+5=1$
$\sqrt[3]{3x-4}=-4$
$(\sqrt[3]{3x-4})^3=(-4)^3$
$3x-4=-64$
$3x=-60$
$x=-20$

25. $\frac{BC}{EF}=\frac{AC}{DE}$
$\frac{8}{12}=\frac{18}{DE}$
$8(DE)=12(18)$
$8(DE)=216$
$DE=27$

The length of side DE is 27 m.

26. **Strategy**: Let x represent amount invested at 3.5%.

	Principal	Rate	Interest
Amount invested at 3.5%	x	0.035	$(0.035)x$
Amount invested at 4.5%	$10{,}000 - x$	0.045	$0.045(10{,}000 - x)$

The total amount of interest earned is \$425.
$0.035x + 0.045(10{,}000 - x) = 425$

Solution:
$$\begin{aligned} 0.035x + 0.045(10{,}000 - x) &= 425 \\ 0.035x + 450 - 0.045x &= 425 \\ 450 - 0.010x &= 425 \\ -0.010x &= -25 \\ x &= 2500 \end{aligned}$$

\$2500 must be invested at 3.5%.

27. **Strategy:** Let x represent the unknown rate of the car.
The unknown rate of the plane is $5x$.

	Distance	Rate	Time
car	25	x	$\frac{25}{x}$
Plane	625	$5x$	$\frac{625}{5x}$

The total time of the trip was 3 h.
$$\frac{25}{x} + \frac{625}{5x} = 3$$

Solution: $\frac{25}{x} + \frac{625}{5x} = 3$
$$5x\left(\frac{25}{x} + \frac{625}{5x}\right) = 5x(3)$$
$$125 + 625 = 15x$$
$$750 = 15x$$
$$50 = x$$
$$250 = 5x$$
The rate of the plane is 250 mph.

28. **Strategy:** To find the time it takes light to travel from the earth to the moon use the formula $D = RT$. Substitute in the given values for R and D and solve for T.

Solution: $D = RT$
$$1.86 \times 10^5 \cdot T = 232{,}500$$
$$T = 1.25 \times 10^0$$
$$T = 1.25$$
The time is 1.25 s.

29. **Strategy:** To find the height if the periscope substitute in the given value for d and solve for h.

Solution: $d = \sqrt{1.5h}$
$$7 = \sqrt{1.5h}$$
$$(7)^2 = (\sqrt{1.5h})^2$$
$$49 = 1.5h$$
$$32.7 \approx h$$
The height of the periscope is 32.7 ft.

30. $m = \dfrac{y_2 - y_1}{x_2 - x_1} = \dfrac{400 - 0}{5000 - 0} = \dfrac{400}{5000} = 0.08$

The annual income is 8% of the investment.

Chapter 10: Quadratic Equations

Prep Test

1. $\sqrt{18} = \sqrt{9 \cdot 2} = 3\sqrt{2}$

2. $3i$

3. $\dfrac{3x-2}{x-1} - 1 = \dfrac{3x-2}{x-1} - \dfrac{x-1}{x-1}$

$= \dfrac{3x-2-x+1}{x-1}$

$= \dfrac{2x-1}{x-1}$

4. $b^2 - 4ac$

$(-4)^2 - 4(2)(1) = 16 - 8$

$= 8$

5. $4x^2 + 28x + 49 = (2x+7)(2x+7)$

Yes, it is a perfect square trinomial.

6. $4x^2 - 4x + 1 = (2x-1)(2x-1) = (2x-1)^2$

7. $9x^2 - 4 = (3x+2)(3x-2)$

8. $\{x \mid x < -1\} \cap \{x \mid x < 4\}$

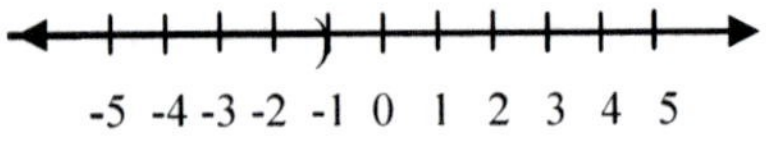

9. $x(x-1) = x + 15$

$x^2 - x = x + 15$

$x^2 - 2x - 15 = 0$

$(x-5)(x+3) = 0$

$x - 5 = 0 \quad x + 3 = 0$

$x = 5 \qquad x = -3$

The solutions are −3 and 5.

10. $\dfrac{4}{x-3} = \dfrac{16}{x}$

$x(x-3)\left(\dfrac{4}{x-3}\right) = x(x-3)\left(\dfrac{16}{x}\right)$

$4x = 16(x-3)$

$4x = 16x - 48$

$-12x = -48$

$x = 4$

The solution is 4.

Section 10.1

Concept Check

1. (i) and (iii) are not quadratic equations.

3. The Principle of Zero Products states that if the product of two factors is zero, then at least one of the factors must be zero. This principle is used to solve quadratic equations when they are written as a product of factors.

5. $5(x+4) = 0$

$x + 4 = 0$

$x = -4$

The solution is −4.

7. $2x(x-4) = 0$

$2x = 0 \quad x - 4 = 0$

$x = 0 \qquad x = 4$

The solutions are 0 and 4.

Objective A Exercises

9. $x^2 - 4x = 0$

$x(x-4) = 0$

$x = 0 \quad x - 4 = 0$

$x = 4$

The solutions are 0 and 4.

11. $t^2 - 25 = 0$
$(t+5)(t-5) = 0$
$t+5 = 0 \quad t-5 = 0$
$t = -5 \qquad t = 5$
The solutions are −5 and 5.

13. $s^2 - s - 6 = 0$
$(s-3)(s+2) = 0$
$s-3 = 0 \quad s+2 = 0$
$s = 3 \qquad s = -2$
The solutions are −2 and 3.

15. $y^2 - 6y + 9 = 0$
$(y-3)(y-3) = 0$
$y-3 = 0 \quad y-3 = 0$
$y = 3 \qquad y = 3$
The solution is 3.

17. $9z^2 - 18z = 0$
$9z(z-2) = 0$
$9z = 0 \quad z-2 = 0$
$z = 0 \qquad z = 2$
The solutions are 0 and 2.

19. $r^2 - 3r = 10$
$r^2 - 3r - 10 = 0$
$(r-5)(r+2) = 0$
$r-5 = 0 \quad r+2 = 0$
$r = 5 \qquad r = -2$
The solutions are −2 and 5.

21. $v^2 + 10 = 7v$
$v^2 - 7v + 10 = 0$
$(v-5)(v-2) = 0$
$v-5 = 0 \quad v-2 = 0$
$v = 5 \qquad v = 2$
The solutions are 2 and 5.

23. $2x^2 - 9x - 18 = 0$
$(2x+3)(x-6) = 0$
$2x+3 = 0 \quad x-6 = 0$
$2x = -3 \qquad x = 6$
$x = -\frac{3}{2}$
The solutions are $-\frac{3}{2}$ and 6.

25. $4z^2 - 9z + 2 = 0$
$(4z-1)(z-2) = 0$
$4z-1 = 0 \quad z-2 = 0$
$4z = 1 \qquad z = 2$
$z = \frac{1}{4}$
The solutions are $\frac{1}{4}$ and 2.

27. $3w^2 + 11w = 4$
$3w^2 + 11w - 4 = 0$
$(3w-1)(w+4) = 0$
$3w-1 = 0 \quad w+4 = 0$
$3w = 1 \qquad w = -4$
$w = \frac{1}{3}$
The solutions are −4 and $\frac{1}{3}$.

29. $6x^2 = 23x + 18$
$6x^2 - 23x - 18 = 0$
$(2x-9)(3x+2) = 0$
$2x-9 = 0 \quad 3x+2 = 0$
$2x = 9 \qquad 3x = -2$
$x = \frac{9}{2} \qquad x = -\frac{2}{3}$
The solutions are $-\frac{2}{3}$ and $\frac{9}{2}$.

31. $4-15u-4u^2=0$

$(1-4u)(4+u)=0$

$1-4u=0 \quad 4+u=0$

$-4u=-1 \qquad u=-4$

$u=\frac{1}{4}$

The solutions are $\frac{1}{4}$ and -4.

33. $x+18=x(x-6)$

$x+18=x^2-6x$

$0=x^2-7x-18$

$0=(x-9)(x+2)$

$x-9=0 \quad x+2=0$

$x=9 \qquad x=-2$

The solutions are -2 and 9.

35. $4s(s+3)=s-6$

$4s^2+12s=s-6$

$4s^2+11s+6=0$

$(4s+3)(s+2)=0$

$4s+3=0 \quad s+2=0$

$4s=-3 \qquad s=-2$

$s=-\frac{3}{4}$

The solutions are -2 and $-\frac{3}{4}$.

37. $u^2-2u+4=(2u-3)(u+2)$

$u^2-2u+4=2u^2+u-6$

$0=u^2+3u-10$

$0=(u+5)(u-2)$

$u+5=0 \quad u-2=0$

$u=-5 \qquad u=2$

The solutions are -5 and 2.

39. $(3x-4)(x+4)=x^2-3x-28$

$3x^2+8x-16=x^2-3x-28$

$2x^2+11x+12=0$

$(2x+3)(x+4)=0$

$2x+3=0 \quad x+4=0$

$2x=-3 \qquad x=-4$

$x=-\frac{3}{2}$

The solutions are -4 and $-\frac{3}{2}$.

41. $(x-r_1)(x-r_2)=0$

$(x-2)(x-5)=0$

$x^2-7x+10=0$

43. $(x-r_1)(x-r_2)=0$

$[x-(-2)][x-(-4)]=0$

$(x+2)(x+4)=0$

$x^2+6x+8=0$

45. $(x-r_1)(x-r_2)=0$

$(x-6)[x-(-1)]=0$

$(x-6)(x+1)=0$

$x^2-5x-6=0$

47. $(x-r_1)(x-r_2)=0$

$(x-3)[x-(-3)]=0$

$(x-3)(x+3)=0$

$x^2-9=0$

49. $(x-r_1)(x-r_2)=0$

$(x-4)(x-4)=0$

$x^2-8x+16=0$

51. $(x-r_1)(x-r_2)=0$

$(x-0)(x-5)=0$

$x^2-5x=0$

53. $(x-r_1)(x-r_2)=0$

$(x-3)\left(x-\frac{1}{2}\right)=0$

$x^2-\frac{7}{2}x+\frac{3}{2}=0$

$2\left(x^2-\frac{7}{2}x+\frac{3}{2}\right)=0\cdot 2$

$2x^2-7x+3=0$

55. $(x-r_1)(x-r_2)=0$

$\left(x-\left(-\frac{3}{4}\right)\right)(x-2)=0$

$\left(x+\frac{3}{4}\right)(x-2)=0$

$x^2-\frac{5}{4}x-\frac{3}{2}=0$

$4\left(x^2-\frac{5}{4}x-\frac{3}{2}\right)=0\cdot 4$

$4x^2-5x-6=0$

57. $(x-r_1)(x-r_2)=0$

$\left(x-\left(-\frac{5}{3}\right)\right)[x-(-2)]=0$

$\left(x+\frac{5}{3}\right)(x+2)=0$

$x^2+\frac{11}{3}x+\frac{10}{3}=0$

$3\left(x^2+\frac{11}{3}x+\frac{10}{3}\right)=0\cdot 3$

$3x^2+11x+10=0$

59. $(x-r_1)(x-r_2)=0$

$\left(x-\frac{1}{2}\right)\left(x-\frac{1}{3}\right)=0$

$x^2-\frac{5}{6}x+\frac{1}{6}=0$

$6\left(x^2-\frac{5}{6}x+\frac{1}{6}\right)=0\cdot 6$

$6x^2-5x+1=0$

61. $(x-r_1)(x-r_2)=0$

$\left(x-\frac{6}{5}\right)\left(x-\left(-\frac{1}{2}\right)\right)=0$

$\left(x-\frac{6}{5}\right)\left(x+\frac{1}{2}\right)=0$

$x^2-\frac{7}{10}x-\frac{3}{5}=0$

$10\left(x^2-\frac{7}{10}x-\frac{3}{5}\right)=0\cdot 10$

$10x^2-7x-6=0$

63. $(x-r_1)(x-r_2)=0$

$\left(x-\left(-\frac{1}{4}\right)\right)\left(x-\left(-\frac{1}{2}\right)\right)=0$

$\left(x+\frac{1}{4}\right)\left(x+\frac{1}{2}\right)=0$

$x^2+\frac{3}{4}x+\frac{1}{8}=0$

$8\left(x^2+\frac{3}{4}x+\frac{1}{8}\right)=0\cdot 8$

$8x^2+6x+1=0$

65. $c=0$

Objective C Exercises

67. $y^2 = 49$

$\sqrt{y^2} = \sqrt{49}$

$y = \pm\sqrt{49} = \pm 7$

The solutions are −7 and 7.

69. $z^2 = -4$

$\sqrt{z^2} = \sqrt{-4}$

$z = \pm\sqrt{-4} = \pm 2i$

The solutions are $-2i$ and $2i$.

71. $s^2 - 4 = 0$

$s^2 = 4$

$\sqrt{s^2} = \sqrt{4}$

$s = \pm\sqrt{4} = \pm 2$

The solutions are −2 and 2.

73. $4x^2 - 81 = 0$

$4x^2 = 81$

$x^2 = \frac{81}{4}$

$\sqrt{x^2} = \sqrt{\frac{81}{4}}$

$x = \pm\sqrt{\frac{81}{4}} = \pm\frac{9}{2}$

The solutions are $-\frac{9}{2}$ and $\frac{9}{2}$.

75. $y^2 + 49 = 0$

$y^2 = -49$

$\sqrt{y^2} = \sqrt{-49}$

$x = \pm\sqrt{-49} = \pm 7i$

The solutions are $-7i$ and $7i$.

77. $v^2 - 48 = 0$

$v^2 = 48$

$\sqrt{v^2} = \sqrt{48}$

$v = \pm\sqrt{48} = \pm 4\sqrt{3}$

The solutions are $-4\sqrt{3}$ and $4\sqrt{3}$.

79. $z^2 + 18 = 0$

$z^2 = -18$

$\sqrt{z^2} = \sqrt{-18}$

$z = \pm\sqrt{-18} = \pm 3i\sqrt{2}$

The solutions are $-3i\sqrt{2}$ and $3i\sqrt{2}$.

81. $(x-1)^2 = 36$

$\sqrt{(x-1)^2} = \sqrt{36}$

$x - 1 = \pm\sqrt{36} = \pm 6$

$x - 1 = 6 \quad x - 1 = -6$

$x = 7 \qquad x = -5$

The solutions are −5 and 7.

83. $5(z+2)^2 = 125$

$(z+2)^2 = 25$

$\sqrt{(z+2)^2} = \sqrt{25}$

$z + 2 = \pm\sqrt{25} = \pm 5$

$z + 2 = 5 \quad z + 2 = -5$

$z = 3 \qquad z = -7$

The solutions are −7 and 3.

85. $\left(v - \frac{1}{2}\right)^2 = \frac{1}{4}$

$\sqrt{\left(v - \frac{1}{2}\right)^2} = \sqrt{\frac{1}{4}}$

$v - \frac{1}{2} = \pm\sqrt{\frac{1}{4}} = \pm\frac{1}{2}$

$v - \frac{1}{2} = \frac{1}{2} \quad v - \frac{1}{2} = -\frac{1}{2}$

$v = 1 \qquad v = 0$

The solutions are 1 and 0.

87. $(x+5)^2 - 6 = 0$

$(x+5)^2 = 6$

$\sqrt{(x+5)^2} = \sqrt{6}$

$x+5 = \pm\sqrt{6}$

$x+5 = \sqrt{6} \qquad x+5 = -\sqrt{6}$

$x = -5+\sqrt{6} \qquad x = -5-\sqrt{6}$

The solutions are $-5-\sqrt{6}$ and $-5+\sqrt{6}$.

89. $(v-3)^2 + 45 = 0$

$(v-3)^2 = -45$

$\sqrt{(v-3)^2} = \sqrt{-45}$

$v-3 = \pm\sqrt{-45} = \pm 3i\sqrt{5}$

$v-3 = 3i\sqrt{5} \qquad v-3 = -3i\sqrt{5}$

$v = 3+3i\sqrt{5} \qquad v = 3-3i\sqrt{5}$

The solutions are $3-3i\sqrt{5}$ and $3+3i\sqrt{5}$.

91. $\left(u+\frac{2}{3}\right)^2 - 18 = 0$

$\left(u+\frac{2}{3}\right)^2 = 18$

$\sqrt{\left(u+\frac{2}{3}\right)^2} = \sqrt{18}$

$u+\frac{2}{3} = \pm\sqrt{18} = \pm 3\sqrt{2}$

$u+\frac{2}{3} = 3\sqrt{2} \qquad u+\frac{2}{3} = -3\sqrt{2}$

$u = -\frac{2}{3}+3\sqrt{2} \qquad u = -\frac{2}{3}-3\sqrt{2}$

$u = -\frac{2+9\sqrt{2}}{3} \qquad u = -\frac{2-9\sqrt{2}}{3}$

The solutions are $-\frac{2+9\sqrt{2}}{3}$ and $-\frac{2-9\sqrt{2}}{3}$.

93. Two complex solutions

95. Two equal real solutions

Critical Thinking

97. $(x-r_1)(x-r_2) = 0$

$(x-\sqrt{2})[x-(-\sqrt{2})] = 0$

$(x-\sqrt{2})(x+\sqrt{2}) = 0$

$x^2 - 2 = 0$

99. $(x-r_1)(x-r_2) = 0$

$(x-i)[x-(-i)] = 0$

$(x-i)(x+i) = 0$

$x^2 + 1 = 0$

101. $(x-r_1)(x-r_2) = 0$

$[x-(3-\sqrt{2})][x-(3+\sqrt{2})] = 0$

$x^2 - 6x + 7 = 0$

103. $(x-r_1)(x-r_2) = 0$

$[x-(5-i)][x-(5+i)] = 0$

$x^2 - 10x + 26 = 0$

105. $x^2 = \sqrt{7}$

$\sqrt{x^2} = \sqrt{\sqrt{7}}$

$x = \pm\sqrt[4]{7}$

The solutions are $-\sqrt[4]{7}$ and $\sqrt[4]{7}$.

107. $x^2 - \sqrt[3]{2} = 0$

$x^2 = \sqrt[3]{2}$

$\sqrt{x^2} = \sqrt{\sqrt[3]{2}}$

$x = \pm\sqrt[6]{2}$

The solutions are $-\sqrt[6]{2}$ and $\sqrt[6]{2}$.

Projects or Group Activities

109. No

$b = -(1+5) = -6$

$c = (1)(5) = 5$

111. Yes

$b = -(-1-2\sqrt{3}) + (-1+2\sqrt{3}) = -(-2) = 2$

$c = (-1-2\sqrt{3})(-1+2\sqrt{3}) = -11$

113. Yes

$b = -(2-i\sqrt{3}) + (2+i\sqrt{3}) = -(4) = -4$

$c = (2-i\sqrt{3})(2+i\sqrt{3}) = 7$

Section 10.2

Concept Check

1. a) No
b) Yes
c) Yes
d) No

3. Yes

Objective A Exercises

5. $x^2 - 4x - 5 = 0$

$x^2 - 4x = 5$

$x^2 - 4x + 4 = 5 + 4$

$(x-2)^2 = 9$

$\sqrt{(x-2)^2} = \sqrt{9}$

$x - 2 = \pm\sqrt{9} = \pm 3$

$x - 2 = 3 \quad x - 2 = -3$

$x = 5 \quad x = -1$

The solutions are −1 and 5.

7. $z^2 - 6z + 9 = 0$

$z^2 - 6z = -9$

$z^2 - 6z + 9 = -9 + 9$

$(z-3)^2 = 0$

$\sqrt{(z-3)^2} = \sqrt{0}$

$z - 3 = 0$

$z = 3$

The solution is 3.

9. $r^2 + 4r - 7 = 0$

$r^2 + 4r = 7$

$r^2 + 4r + 4 = 7 + 4$

$(r+2)^2 = 11$

$\sqrt{(r+2)^2} = \sqrt{11}$

$r + 2 = \pm\sqrt{11}$

$r + 2 = \sqrt{11} \quad r + 2 = -\sqrt{11}$

$r = -2 + \sqrt{11} \quad r = -2 - \sqrt{11}$

The solutions are $-2-\sqrt{11}$ and $-2+\sqrt{11}$

11. $x^2 - 6x + 7 = 0$

$x^2 - 6x = -7$

$x^2 - 6x + 9 = -7 + 9$

$(x-3)^2 = 2$

$\sqrt{(x-3)^2} = \sqrt{2}$

$x - 3 = \pm\sqrt{2}$

$x - 3 = \sqrt{2} \quad x - 3 = -\sqrt{2}$

$x = 3 + \sqrt{2} \quad x = 3 - \sqrt{2}$

The solutions are $3-\sqrt{2}$ and $3+\sqrt{2}$.

13. $p^2 - 3p + 1 = 0$

$p^2 - 3p = -1$

$p^2 - 3p + \frac{9}{4} = -1 + \frac{9}{4}$

$\left(p - \frac{3}{2}\right)^2 = \frac{5}{4}$

$\sqrt{\left(p - \frac{3}{2}\right)^2} = \sqrt{\frac{5}{4}}$

$p-\frac{3}{2}=\pm\frac{\sqrt{5}}{2}$

$p-\frac{3}{2}=\frac{\sqrt{5}}{2}$ $\quad p-\frac{3}{2}=-\frac{\sqrt{5}}{2}$

$p=\frac{3}{2}+\frac{\sqrt{5}}{2}$ $\quad p=\frac{3}{2}-\frac{\sqrt{5}}{2}$

The solutions are $\frac{3+\sqrt{5}}{2}$ and $\frac{3-\sqrt{5}}{2}$.

15. $y^2-6y=4$

$y^2-6y+9=4+9$

$(y-3)^2=13$

$\sqrt{(y-3)^2}=\sqrt{13}$

$y-3=\pm\sqrt{13}$

$y-3=\sqrt{13}$ $\quad y-3=-\sqrt{13}$

$y=3+\sqrt{13}$ $\quad y=3-\sqrt{13}$

The solutions are $3-\sqrt{13}$ and $3+\sqrt{13}$.

17. $z^2=z+4$

$z^2-z=4$

$z^2-z+\frac{1}{4}=4+\frac{1}{4}$

$\left(z-\frac{1}{2}\right)^2=\frac{17}{4}$

$\sqrt{\left(z-\frac{1}{2}\right)^2}=\sqrt{\frac{17}{4}}$

$z-\frac{1}{2}=\pm\frac{\sqrt{17}}{2}$

$z-\frac{1}{2}=\frac{\sqrt{17}}{2}$ $\quad z-\frac{1}{2}=-\frac{\sqrt{17}}{2}$

$z=\frac{1}{2}+\frac{\sqrt{17}}{2}$ $\quad z=\frac{1}{2}-\frac{\sqrt{17}}{2}$

The solutions are $\frac{1+\sqrt{17}}{2}$ and $\frac{1-\sqrt{17}}{2}$.

19. $z^2-2z+2=0$

$z^2-2z=-2$

$z^2-2z+1=-2+1$

$(z-1)^2=-1$

$\sqrt{(z-1)^2}=\sqrt{-1}$

$z-1=\pm i$

$z-1=i$ $\quad z-1=-i$

$z=1+i$ $\quad z=1-i$

The solutions are $1+i$ and $1-i$.

21. $v^2=4v-13$

$v^2-4v=-13$

$v^2-4v+4=-13+4$

$(v-2)^2=-9$

$\sqrt{(v-2)^2}=\sqrt{-9}$

$v-2=\pm 3i$

$v-2=3i$ $\quad v-2=-3i$

$v=2+3i$ $\quad v=2-3i$

The solutions are $2+3i$ and $2-3i$.

23. $p^2+6p=-13$

$p^2+6p+9=-13+9$

$(p+3)^2=-4$

$\sqrt{(p+3)^2}=\sqrt{-4}$

$p+3=\pm 2i$

$p+3=2i$ $\quad p+3=-2i$

$p=-3+2i$ $\quad p=-3-2i$

The solutions are $-3-2i$ and $-3+2i$.

25. $2s^2 = 4s + 5$

$2s^2 - 4s = 5$

$\frac{1}{2}\left(2s^2 - 4s\right) = \frac{1}{2}(5)$

$s^2 - 2s = \frac{5}{2}$

$s^2 - 2s + 1 = \frac{5}{2} + 1$

$(s-1)^2 = \frac{7}{2}$

$\sqrt{(s-1)^2} = \sqrt{\frac{7}{2}}$

$s - 1 = \pm\sqrt{\frac{7}{2}} = \pm\frac{\sqrt{14}}{2}$

$s - 1 = \frac{\sqrt{14}}{2} \qquad s - 1 = -\frac{\sqrt{14}}{2}$

$s = 1 + \frac{\sqrt{14}}{2} \qquad s = 1 - \frac{\sqrt{14}}{2}$

The solutions are $\frac{2-\sqrt{14}}{2}$ and $\frac{2+\sqrt{14}}{2}$.

27. $4x^2 - 4x + 5 = 0$

$4x^2 - 4x = -5$

$\frac{1}{4}\left(4x^2 - 4x\right) = \frac{1}{4}(-5)$

$x^2 - x = -\frac{5}{4}$

$x^2 - x + \frac{1}{4} = -\frac{5}{4} + \frac{1}{4}$

$\left(x - \frac{1}{2}\right)^2 = -1$

$\sqrt{\left(x - \frac{1}{2}\right)^2} = \sqrt{-1}$

$x - \frac{1}{2} = \pm i$

$x - \frac{1}{2} = i \qquad x - \frac{1}{2} = -i$

$x = \frac{1}{2} + i \qquad x = \frac{1}{2} - i$

The solutions are $\frac{1}{2} - i$ and $\frac{1}{2} + i$.

29. $9x^2 - 6x + 2 = 0$

$9x^2 - 6x = -2$

$\frac{1}{9}\left(9x^2 - 6x\right) = \frac{1}{9}(-2)$

$x^2 - \frac{2}{3}x = -\frac{2}{9}$

$x^2 - \frac{2}{3}x + \frac{1}{9} = -\frac{2}{9} + \frac{1}{9}$

$\left(x - \frac{1}{3}\right)^2 = -\frac{1}{9}$

$\sqrt{\left(x - \frac{1}{3}\right)^2} = \sqrt{-\frac{1}{9}}$

$x - \frac{1}{3} = \pm\frac{1}{3}i$

$x - \frac{1}{3} = \frac{1}{3}i \qquad x - \frac{1}{3} = -\frac{1}{3}i$

$x = \frac{1}{3} + \frac{1}{3}i \qquad x = \frac{1}{3} - \frac{1}{3}i$

The solutions are $\frac{1}{3} - \frac{1}{3}i$ and $\frac{1}{3} + \frac{1}{3}i$.

31. $y-2=(y-3)(y+2)$

$y-2=y^2-y-6$

$y^2-2y=4$

$y^2-2y+1=4+1$

$(y-1)^2=5$

$\sqrt{(y-1)^2}=\sqrt{5}$

$y-1=\pm\sqrt{5}$

$y-1=\sqrt{5} \qquad y-1=-\sqrt{5}$

$y=1+\sqrt{5} \qquad y=1-\sqrt{5}$

The solutions are $1-\sqrt{5}$ and $1+\sqrt{5}$.

33. $6t-2=(2t-3)(t-1)$

$6t-2=2t^2-5t+3$

$2t^2-11t=-5$

$\frac{1}{2}\left(2t^2-11t\right)=\frac{1}{2}(-5)$

$t^2-\frac{11}{2}t=-\frac{5}{2}$

$t^2-\frac{11}{2}t+\frac{121}{16}=-\frac{5}{2}+\frac{121}{16}$

$\left(t-\frac{11}{4}\right)^2=\frac{81}{16}$

$\sqrt{\left(t-\frac{11}{4}\right)^2}=\sqrt{\frac{81}{16}}$

$t-\frac{11}{4}=\pm\frac{9}{4}$

$t-\frac{11}{4}=\frac{9}{4} \qquad t-\frac{11}{4}=-\frac{9}{4}$

$t=\frac{20}{4}=5 \qquad t=\frac{2}{4}=\frac{1}{2}$

The solutions are $\frac{1}{2}$ and 5.

35. $(x-4)(x+1)=x-3$

$x^2-3x-4=x-3$

$x^2-4x=1$

$x^2-4x+4=1+4$

$(x-2)^2=5$

$\sqrt{(x-2)^2}=\sqrt{5}$

$x-2=\pm\sqrt{5}$

$x-2=\sqrt{5} \qquad x-2=-\sqrt{5}$

$x=2+\sqrt{5} \qquad x=2-\sqrt{5}$

The solutions are $2+\sqrt{5}$ and $2-\sqrt{5}$.

37. $z^2+2z=4$

$z^2+2z+1=4+1$

$(z+1)^2=5$

$\sqrt{(z+1)^2}=\sqrt{5}$

$z+1=\pm\sqrt{5}$

$z+1=\sqrt{5} \qquad z+1=-\sqrt{5}$

$z=\sqrt{5}-1 \qquad z=-\sqrt{5}-1$

$z=1.236 \qquad z=-3.236$

The solutions are -3.326 and 1.236.

39. $2x^2=4x-1$

$2x^2-4x=-1$

$\frac{1}{2}\left(2x^2-4x\right)=\frac{1}{2}(-1)$

$x^2-2x=-\frac{1}{2}$

$x^2-2x+1=-\frac{1}{2}+1$

$(x-1)^2=\frac{1}{2}$

$$\sqrt{(x-1)^2} = \sqrt{\frac{1}{2}}$$

$$x-1 = \pm\sqrt{\frac{1}{2}}$$

$$x-1 = \sqrt{\frac{1}{2}} \qquad x-1 = -\sqrt{\frac{1}{2}}$$

$$x = \sqrt{\frac{1}{2}}+1 \qquad x = -\sqrt{\frac{1}{2}}+1$$

$$x = 1.707 \qquad x = 0.293$$

The solutions are 0.293 and 1.707.

41. $c \le 4$

Objective B Exercises

43. The quadratic formula:

$$x = \frac{-b \pm \sqrt{b^2-4ac}}{2a}$$

a is the coefficient of x^2; b is the coefficient of x, and c is the constant term in the quadratic equation $ax^2+bx+c,\ a \ne 0$.

45. $x^2-3x-10=0$

$a=1,\ b=-3,\ c=-10$

$$x = \frac{-b \pm \sqrt{b^2-4ac}}{2a}$$

$$x = \frac{-(-3) \pm \sqrt{(-3)^2-4(1)(-10)}}{2(1)}$$

$$x = \frac{3 \pm \sqrt{9+40}}{2} = \frac{3 \pm \sqrt{49}}{2}$$

$$x = \frac{3 \pm 7}{2}$$

$$x = \frac{3+7}{2} \qquad x = \frac{3-7}{2}$$

$$x = \frac{10}{2} = 5 \qquad x = -\frac{4}{2} = -2$$

The solutions are −2 and 5.

47. $x^2-8x+9=0$

$a=1,\ b=-8,\ c=9$

$$x = \frac{-b \pm \sqrt{b^2-4ac}}{2a}$$

$$x = \frac{-(-8) \pm \sqrt{(-8)^2-4(1)(9)}}{2(1)}$$

$$x = \frac{8 \pm \sqrt{64-36}}{2} = \frac{8 \pm \sqrt{28}}{2}$$

$$x = \frac{8 \pm 2\sqrt{7}}{2} = 4 \pm \sqrt{7}$$

$$x = 4+\sqrt{7} \qquad x = 4-\sqrt{7}$$

The solutions are $4+\sqrt{7}$ and $4-\sqrt{7}$.

49. $v^2 = 6v + 19$

$v^2-6v-19=0$

$a=1,\ b=-6,\ c=-19$

$$x = \frac{-b \pm \sqrt{b^2-4ac}}{2a}$$

$$v = \frac{-(-6) \pm \sqrt{(-6)^2-4(1)(-19)}}{2(1)}$$

$$v = \frac{6 \pm \sqrt{36+76}}{2} = \frac{6 \pm \sqrt{112}}{2}$$

$$v = \frac{6 \pm 4\sqrt{7}}{2} = 3 \pm 2\sqrt{7}$$

$$v = 3+2\sqrt{7} \qquad v = 3-2\sqrt{7}$$

The solutions are $3+2\sqrt{7}$ and $3-2\sqrt{7}$.

51. $x^2 = 14x - 4$

$x^2 - 14x + 4 = 0$

$a = 1, b = -14, c = 4$

$$x = \frac{-b \pm \sqrt{b^2 - 4ac}}{2a}$$

$$x = \frac{-(-14) \pm \sqrt{(-14)^2 - 4(1)(4)}}{2(1)}$$

$$x = \frac{14 \pm \sqrt{196 - 16}}{2} = \frac{14 \pm \sqrt{180}}{2}$$

$$x = \frac{14 \pm 6\sqrt{5}}{2}$$

$$x = \frac{14 + 6\sqrt{5}}{2} \qquad x = \frac{14 - 6\sqrt{5}}{2}$$

$$x = 7 + 3\sqrt{5} \qquad x = 7 - 3\sqrt{5}$$

The solutions are $7 - 3\sqrt{5}$ and $7 + 3\sqrt{5}$.

53. $2z^2 - 2z - 1 = 0$

$a = 2, b = -2, c = -1$

$$z = \frac{-b \pm \sqrt{b^2 - 4ac}}{2a}$$

$$z = \frac{-(-2) \pm \sqrt{(-2)^2 - 4(2)(-1)}}{2(2)}$$

$$z = \frac{2 \pm \sqrt{4 + 8}}{4} = \frac{2 \pm \sqrt{12}}{4}$$

$$z = \frac{2 \pm 2\sqrt{3}}{4} = \frac{1 \pm \sqrt{3}}{2}$$

$$z = \frac{1 + \sqrt{3}}{2} \qquad z = \frac{1 - \sqrt{3}}{2}$$

The solutions are $\frac{1 - \sqrt{3}}{2}$ and $\frac{1 + \sqrt{3}}{2}$.

55. $4r^2 = 20r - 17$

$4r^2 - 20r + 17 = 0$

$a = 4, b = -20, c = 17$

$$r = \frac{-b \pm \sqrt{b^2 - 4ac}}{2a}$$

$$r = \frac{-(-20) \pm \sqrt{(-20)^2 - 4(4)(17)}}{2(4)}$$

$$r = \frac{-(-20) \pm \sqrt{400 - 272}}{8} = \frac{20 \pm \sqrt{128}}{8}$$

$$r = \frac{20 \pm \sqrt{128}}{8} = \frac{20 \pm 8\sqrt{2}}{8} = \frac{5 \pm 2\sqrt{2}}{2}$$

$$r = \frac{5 + 2\sqrt{2}}{2} \qquad r = \frac{5 - 2\sqrt{2}}{2}$$

The solutions are $\frac{5 - 2\sqrt{2}}{2}$ and $\frac{5 + 2\sqrt{2}}{2}$.

57. $z^2 + 2z + 2 = 0$

$a = 1, b = 2, c = 2$

$$z = \frac{-b \pm \sqrt{b^2 - 4ac}}{2a}$$

$$z = \frac{-2 \pm \sqrt{2^2 - 4(1)(2)}}{2(1)}$$

$$z = \frac{-2 \pm \sqrt{4 - 8}}{2} = \frac{-2 \pm \sqrt{-4}}{2}$$

$$z = \frac{-2 \pm 2i}{2} = -1 \pm i$$

$$z = -1 + i \qquad z = -1 - i$$

The solutions are $-1 - i$ and $-1 + i$.

59. $y^2 - 2y + 5 = 0$

$a = 1, b = -2, c = 5$

$$y = \frac{-b \pm \sqrt{b^2 - 4ac}}{2a}$$

$$y = \frac{-(-2) \pm \sqrt{(-2)^2 - 4(1)(5)}}{2(1)}$$

$$y = \frac{2 \pm \sqrt{4 - 20}}{2} = \frac{2 \pm \sqrt{-16}}{2}$$

$$y = \frac{2 \pm 4i}{2} = 1 \pm 2i$$

$y = 1 + 2i \qquad y = 1 - 2i$

The solutions are $1 - 2i$ and $1 + 2i$.

61. $s^2 - 4s + 13 = 0$

$a = 1, b = -4, c = 13$

$$s = \frac{-b \pm \sqrt{b^2 - 4ac}}{2a}$$

$$s = \frac{-(-4) \pm \sqrt{(-4)^2 - 4(1)(13)}}{2(1)}$$

$$s = \frac{4 \pm \sqrt{16 - 52}}{2} = \frac{4 \pm \sqrt{-36}}{2}$$

$$s = \frac{4 \pm 6i}{2} = 2 \pm 3i$$

$s = 2 + 3i \qquad s = 2 - 3i$

The solutions are $2 - 3i$ and $2 + 3i$.

63. $4x^2 - 4x + 33 = 0$

$a = 4, b = -4, c = 33$

$$x = \frac{-b \pm \sqrt{b^2 - 4ac}}{2a}$$

$$x = \frac{-(-4) \pm \sqrt{(-4)^2 - 4(4)(33)}}{2(4)}$$

$$x = \frac{4 \pm \sqrt{16 - 528}}{8} = \frac{4 \pm \sqrt{-512}}{8}$$

$$x = \frac{4 \pm 16i\sqrt{2}}{8} = \frac{1 \pm 4i\sqrt{2}}{2} = \frac{1}{2} \pm 2i\sqrt{2}$$

$x = \frac{1}{2} + 2i\sqrt{2} \qquad x = \frac{1}{2} - 2i\sqrt{2}$

The solutions are $\frac{1}{2} - 2i\sqrt{2}$ and $\frac{1}{2} + 2i\sqrt{2}$.

65. $9v^2 - 6v - 71 = 0$

$a = 9, b = -6, c = -71$

$$v = \frac{-b \pm \sqrt{b^2 - 4ac}}{2a}$$

$$v = \frac{-(-6) \pm \sqrt{(-6)^2 - 4(9)(-71)}}{2(9)}$$

$$v = \frac{6 \pm \sqrt{36 + 2556}}{18} = \frac{6 \pm \sqrt{2592}}{18}$$

$$v = \frac{6 \pm 36\sqrt{2}}{18} = \frac{1 \pm 6\sqrt{2}}{3}$$

$v = \frac{1 + 6\sqrt{2}}{3} \qquad v = \frac{1 - 6\sqrt{2}}{3}$

The solutions are $\frac{1 - 6\sqrt{2}}{3}$ and $\frac{1 + 6\sqrt{2}}{3}$.

67. $2w^2 - 2w - 5 = 0$

$a = 2, b = -2, c = -5$

$$w = \frac{-b \pm \sqrt{b^2 - 4ac}}{2a}$$

$$w = \frac{-(-2) \pm \sqrt{(-2)^2 - 4(2)(-5)}}{2(2)}$$

$$w = \frac{2 \pm \sqrt{4 + 40}}{4} = \frac{2 \pm \sqrt{44}}{4}$$

$$w = \frac{2 \pm 2\sqrt{11}}{4} = \frac{1 \pm \sqrt{11}}{2}$$

$w = \frac{1 + \sqrt{11}}{2} \qquad w = \frac{1 - \sqrt{11}}{2}$

The solutions are $\frac{1 - \sqrt{11}}{2}$ and $\frac{1 + \sqrt{11}}{2}$.

69. $2x^2+4x-6=0$

$a=2, b=4, c=-6$

$$x=\frac{-b\pm\sqrt{b^2-4ac}}{2a}$$

$$x=\frac{-(4)\pm\sqrt{(4)^2-4(2)(-6)}}{2(2)}$$

$$x=\frac{-4\pm\sqrt{16+48}}{4}=\frac{-4\pm\sqrt{64}}{4}$$

$$x=\frac{-4\pm8}{4}$$

$$x=\frac{-4+8}{4}=1 \qquad x=\frac{-4-8}{4}=-3$$

The solutions are –3 and 1.

71. $2x^2+x=(x-4)(x-2)$

$x^2+7x-8=0$

$a=1, b=7\ c=-8$

$$x=\frac{-b\pm\sqrt{b^2-4ac}}{2a}$$

$$x=\frac{-(7)\pm\sqrt{(7)^2-4(1)(-8)}}{2(1)}$$

$$x=\frac{-7\pm\sqrt{49+32}}{2}=\frac{-7\pm\sqrt{81}}{2}$$

$$x=\frac{-7\pm9}{2}$$

$$x=\frac{-7+9}{2}=1 \qquad x=\frac{-7-9}{2}=-8$$

The solutions are 1 and –8.

73. $(2x+1)(x+2)=(x-4)(x+3)$

$x^2+6x+14=0$

$a=1, b=6, c=14$

$$x=\frac{-b\pm\sqrt{b^2-4ac}}{2a}$$

$$x=\frac{-(6)\pm\sqrt{(6)^2-4(1)(14)}}{2(1)}$$

$$x=\frac{-6\pm\sqrt{36-56}}{2}=\frac{-6\pm\sqrt{-20}}{2}$$

$$x=\frac{-6\pm2i\sqrt{5}}{2}=-3\pm i\sqrt{5}$$

$$x=-3+i\sqrt{5} \qquad x=-3-i\sqrt{5}$$

The solutions are $-3-i\sqrt{5}$ and $-3+i\sqrt{5}$.

75. $2x^2-x=(x+3)(x-2)$

$x^2-2x+6=0$

$a=1, b=-2, c=6$

$$x=\frac{-b\pm\sqrt{b^2-4ac}}{2a}$$

$$x=\frac{-(-2)\pm\sqrt{(-2)^2-4(1)(6)}}{2(1)}$$

$$x=\frac{2\pm\sqrt{4-24}}{2}=\frac{2\pm\sqrt{-20}}{2}$$

$$x=\frac{2\pm2i\sqrt{5}}{2}=1\pm i\sqrt{5}$$

$$x=1+i\sqrt{5} \qquad x=1-i\sqrt{5}$$

The solutions are $1-i\sqrt{5}$ and $1+i\sqrt{5}$.

77. $5t^2-5t+7=(t-1)(t-2)$

$4t^2-2t+5=0$

$a=4, b=-2, c=5$

$$t=\frac{-b\pm\sqrt{b^2-4ac}}{2a}$$

$$t=\frac{-(-2)\pm\sqrt{(-2)^2-4(4)(5)}}{2(4)}$$

$$t=\frac{2\pm\sqrt{4-80}}{8}=\frac{2\pm\sqrt{-76}}{8}$$

$$t=\frac{2\pm2i\sqrt{19}}{8}=\frac{1}{4}\pm\frac{i\sqrt{19}}{4}$$

$$t=\frac{1}{4}+\frac{i\sqrt{19}}{4} \qquad t=\frac{1}{4}-\frac{i\sqrt{19}}{4}$$

The solutions are $\frac{1}{4}+\frac{i\sqrt{19}}{4}$ and

$\frac{1}{4}-\frac{i\sqrt{19}}{4}$.

79. $p^2 - 8p + 3 = 0$

$a = 1, b = -8, c = 3$

$$p = \frac{-b \pm \sqrt{b^2 - 4ac}}{2a}$$

$$p = \frac{-(-8) \pm \sqrt{(-8)^2 - 4(1)(3)}}{2(1)}$$

$$p = \frac{8 \pm \sqrt{64 - 12}}{2} = \frac{8 \pm \sqrt{52}}{2}$$

$$p = \frac{8 \pm 2\sqrt{13}}{2} = 4 \pm \sqrt{13}$$

$p = 4 + \sqrt{13} \qquad p = 4 - \sqrt{13}$

The solutions are 0.394 and 7.606.

81. $w^2 + 4w = 1$

$w^2 + 4w - 1 = 0$

$a = 1, b = 4, c = -1$

$$w = \frac{-b \pm \sqrt{b^2 - 4ac}}{2a}$$

$$w = \frac{-4 \pm \sqrt{4^2 - 4(1)(-1)}}{2(1)}$$

$$w = \frac{-4 \pm \sqrt{16 + 4}}{2} = \frac{-4 \pm \sqrt{20}}{2}$$

$$w = \frac{-4 \pm 2\sqrt{5}}{2} = -2 \pm \sqrt{5}$$

$w = -2 + \sqrt{5} \qquad w = -2 - \sqrt{5}$

The solutions are -4.236 and 0.236.

83. $2y^2 = y + 5$

$2y^2 - y - 5 = 0$

$a = 2, b = -1, c = -5$

$$y = \frac{-b \pm \sqrt{b^2 - 4ac}}{2a}$$

$$y = \frac{-(-1) \pm \sqrt{(-1)^2 - 4(2)(-5)}}{2(2)}$$

$$y = \frac{1 \pm \sqrt{1 + 40}}{4} = \frac{1 \pm \sqrt{41}}{4}$$

$$y = \frac{1 + \sqrt{41}}{4} \qquad y = \frac{1 - \sqrt{41}}{4}$$

The solutions are -1.351 and 1.851.

85. $3y^2 + y + 1 = 0$

$a = 3, b = 1, c = 1$

$b^2 - 4ac = 1^2 - 4(3)(1)$

$= 1 - 12 = -11$

$-11 < 0$

Since the discriminant is less than zero, the equation has two complex number solutions.

87. $4x^2 + 20x + 25 = 0$

$a = 4, b = 20, c = 25$

$b^2 - 4ac = (20)^2 - 4(4)(25)$

$= 400 - 400 = 0$

Since the discriminant is equal to zero, the equation has two equal real number solutions.

89. $3w^2 + 3w - 2 = 0$

$a = 3, b = 3, c = -2$

$b^2 - 4ac = 3^2 - 4(3)(-2)$

$= 9 + 24 = 33$

$33 > 0$

Since the discriminant is greater than zero, the equation has two unequal real number solutions.

91. $\sqrt{4ac}$

Critical Thinking

93. $x^2 - 6x + p = 0$

$$x^2 - 6x = -p$$

$$x^2 - 6x + 9 = -p + 9$$

$$(x-3)^2 = -p + 9$$

$$\sqrt{(x-3)^2} = \sqrt{9-p}$$

$$x - 3 = \pm\sqrt{9-p}$$

$$x = 3 \pm \sqrt{9-p}$$

x will have two real solutions if $9 - p > 0$.
Solving this inequality gives $p < 9$.
The values of p are $\{p | p < 9\}$.

95. $x^2 - 2x + p = 0$

$a = 1, b = -2, c = p$

$$x = \frac{-b \pm \sqrt{b^2 - 4ac}}{2a}$$

$$x = \frac{-(-2) \pm \sqrt{(-2)^2 - 4(1)(p)}}{2(1)}$$

$$x = \frac{2 \pm \sqrt{4-4p}}{2}$$

$$x = 1 \pm \sqrt{1-p}$$

x will have two complex solutions when $1 - p < 0$.
Solving the inequality gives $p > 1$.
The values of p are $(1, \infty)$.

97. $x^2 - 2ix + 15 = 0$

$a = 1, b = -2i, c = 15$

$$x = \frac{-b \pm \sqrt{b^2 - 4ac}}{2a}$$

$$x = \frac{-(-2i) \pm \sqrt{(-2i)^2 - 4(1)(15)}}{2(1)}$$

$$x = \frac{2i \pm \sqrt{-4-60}}{2} = \frac{2i \pm \sqrt{-64}}{2}$$

$$x = \frac{2i \pm 8i}{2}$$

$$x = \frac{2i + 8i}{2} = 5i \qquad x = \frac{2i - 8i}{2} = -3i$$

The solutions are $-3i$ and $5i$.

Projects or Group Activities

99. $h = -16t^2 + 70t + 4$

Find the time t it takes for the ball to hit the ground. When the ball hits the ground h is zero.

$$0 = -16t^2 + 70t + 4$$

$a = -16, b = 70, c = 4$

$$x = \frac{-b \pm \sqrt{b^2 - 4ac}}{2a}$$

$$x = \frac{-(70) \pm \sqrt{(70)^2 - 4(-16)(4)}}{2(-16)}$$

$$x = \frac{-70 \pm \sqrt{5156}}{-32}$$

$$x = \frac{-70 + \sqrt{5156}}{-32} = -0.056$$

$$x = \frac{-70 - \sqrt{5156}}{-32} = 4.431$$

The ball hits the ground 4.431 s after it is struck by the batter.
$s = 44.5t = 44.5(4.431) = 197.2$ ft
No, the ball will not clear a fence 325 ft from home plate. It will only have gone 197.2 ft when it hits the ground.

Section 10.3

Concept Check

1. (i), (ii), (iii), (iv) and (v)

Objective A Exercises

3. Yes

5. No

7. $x^4 - 13x^2 + 36 = 0$

$(x^2)^2 - 13(x^2) + 36 = 0$

$u^2 - 13u + 36 = 0$

$(u-4)(u-9) = 0$

$u - 4 = 0 \quad u - 9 = 0$

$u = 4 \qquad u = 9$

Replace u with x^2.

$x^2 = 4 \qquad x^2 = 9$

$\sqrt{x^2} = \sqrt{4} \quad \sqrt{x^2} = \sqrt{9}$

$x = \pm 2 \qquad x = \pm 3$

The solutions are −2, 2, −3 and 3.

9. $z^4 - 6z^2 + 8 = 0$

$(z^2)^2 - 6(z^2) + 8 = 0$

$u^2 - 6u + 8 = 0$

$(u-4)(u-2) = 0$

$u - 4 = 0 \quad u - 2 = 0$

$u = 4 \qquad u = 2$

Replace u with z^2.

$z^2 = 4 \qquad z^2 = 2$

$\sqrt{z^2} = \sqrt{4} \quad \sqrt{z^2} = \sqrt{2}$

$z = \pm 2 \qquad z = \pm\sqrt{2}$

The solutions are −2, 2, $-\sqrt{2}$ and $\sqrt{2}$.

11. $p - 3p^{1/2} + 2 = 0$

$(p^{1/2})^2 - 3(p^{1/2}) + 2 = 0$

$u^2 - 3u + 2 = 0$

$(u-2)(u-1) = 0$

$u - 2 = 0 \quad u - 1 = 0$

$u = 2 \qquad u = 1$

Replace u with $p^{1/2}$.

$p^{1/2} = 2 \qquad p^{1/2} = 1$

$\left(p^{1/2}\right)^2 = 2^2 \quad \left(p^{1/2}\right)^2 = 1^2$

$p = 4 \qquad p = 1$

The solutions are 1 and 4.

13. $x - x^{1/2} - 12 = 0$

$(x^{1/2})^2 - (x^{1/2}) - 12 = 0$

$u^2 - u - 12 = 0$

$(u-4)(u+3) = 0$

$u - 4 = 0 \quad u + 3 = 0$

$u = 4 \qquad u = -3$

Replace u with $x^{1/2}$.

$x^{1/2} = 4 \qquad x^{1/2} = -3$

$\left(x^{1/2}\right)^2 = 4^2 \quad \left(x^{1/2}\right)^2 = (-3)^2$

$x = 16 \qquad x = 9$

9 does not check as a solution. The solution is 16.

15. $z^4 + 3z^2 - 4 = 0$

$(z^2)^2 + 3(z^2) - 4 = 0$

$u^2 + 3u - 4 = 0$

$(u+4)(u-1) = 0$

$u + 4 = 0 \quad u - 1 = 0$

$u = -4 \qquad u = 1$

Replace u with z^2.

$z^2 = -4 \qquad z^2 = 1$

$\sqrt{z^2} = \sqrt{-4} \quad \sqrt{z^2} = \sqrt{1}$

$z = \pm 2i \qquad z = \pm 1$

The solutions are −1, 1, $-2i$ and $2i$.

17. $x^4 + 12x^2 - 64 = 0$

$(x^2)^2 + 12(x^2) - 64 = 0$

$u^2 + 12u - 64 = 0$

$(u + 16)(u - 4) = 0$

$u + 16 = 0 \quad u - 4 = 0$

$u = -16 \qquad u = 4$

Replace u with x^2.

$x^2 = -16 \qquad x^2 = 4$

$\sqrt{x^2} = \sqrt{-16} \quad \sqrt{x^2} = \sqrt{4}$

$x = \pm 4i \qquad x = \pm 2$

The solutions are −2, 2, $-4i$ and $4i$.

19. $p + 2p^{1/2} - 24 = 0$

$(p^{1/2})^2 + 2(p^{1/2}) - 24 = 0$

$u^2 + 2u - 24 = 0$

$(u + 6)(u - 4) = 0$

$u + 6 = 0 \quad u - 4 = 0$

$u = -6 \qquad u = 4$

Replace u with $p^{1/2}$.

$p^{1/2} = -6 \qquad p^{1/2} = 4$

$\left(p^{1/2}\right)^2 = (-6)^2 \quad \left(p^{1/2}\right)^2 = 4^2$

$p = 36 \qquad p = 16$

36 does not check as a solution. The solution is 16.

21. $y^{2/3} - 9y^{1/3} + 8 = 0$

$(y^{1/3})^2 - 9(y^{1/3}) + 8 = 0$

$u^2 - 9u + 8 = 0$

$(u - 8)(u - 1) = 0$

$u - 8 = 0 \quad u - 1 = 0$

$u = 8 \qquad u = 1$

Replace u with $y^{1/3}$.

$y^{1/3} = 8 \qquad y^{1/3} = 1$

$(y^{1/3})^3 = 8^3 \quad (y^{1/3})^3 = 1^3$

$y = 512 \qquad y = 1$

The solutions are 1 and 512.

23. $9w^4 - 13w^2 + 4 = 0$

$9(w^2)^2 - 13(w^2) + 4 = 0$

$9u^2 - 13u + 4 = 0$

$(9u - 4)(u - 1) = 0$

$9u - 4 = 0 \quad u - 1 = 0$

$9u = 4 \qquad u = 1$

$u = \dfrac{4}{9}$

Replace u with w^2.

$w^2 = \dfrac{4}{9} \qquad w^2 = 1$

$\sqrt{w^2} = \sqrt{\dfrac{4}{9}} \quad \sqrt{w^2} = \sqrt{1}$

$w = \pm\dfrac{2}{3} \qquad w = \pm 1$

The solutions are −1, 1, $-\dfrac{2}{3}$ and $\dfrac{2}{3}$.

Objective B Exercises

25. Exercises 30, 31, 32, 36, 37, 38, 40, 41, 44

27. $\sqrt{x+1} + x = 5$

$\sqrt{x+1} = 5 - x$

$\left(\sqrt{x+1}\right)^2 = (5 - x)^2$

$x + 1 = 25 - 10x + x^2$

$0 = x^2 - 11x + 24$

$(x - 3)(x - 8) = 0$

$x - 3 = 0 \quad x - 8 = 0$

$x = 3 \qquad x = 8$

8 does not check as a solution.
The solution is 3.

29. $x = \sqrt{x} + 6$

$x - 6 = \sqrt{x}$

$(x-6)^2 = (\sqrt{x})^2$

$x^2 - 12x + 36 = x$

$x^2 - 13x + 36 = 0$

$(x-9)(x-4) = 0$

$x - 9 = 0 \quad x - 4 = 0$

$x = 9 \qquad x = 4$

4 does not check as a solution.
The solution is 9.

31. $\sqrt{3w+3} = w+1$

$\left(\sqrt{3w+3}\right)^2 = (w+1)^2$

$3w + 3 = w^2 + 2w + 1$

$0 = w^2 - w - 2$

$(w-2)(w+1) = 0$

$w - 2 = 0 \quad w + 1 = 0$

$w = 2 \qquad w = -1$

The solutions are −1 and 2.

33. $\sqrt{4y+1} - y = 1$

$\sqrt{4y+1} = 1 + y$

$\left(\sqrt{4y+1}\right)^2 = (1+y)^2$

$4y + 1 = 1 + 2y + y^2$

$0 = y^2 - 2y$

$y(y-2) = 0$

$y = 0 \quad y - 2 = 0$

$y = 2$

The solutions are 0 and 2.

35. $\sqrt{10x+5} - 2x = 1$

$\sqrt{10x+5} = 1 + 2x$

$\left(\sqrt{10x+5}\right)^2 = (1+2x)^2$

$10x + 5 = 1 + 4x + 4x^2$

$0 = 4x^2 - 6x - 4$

$2(2x+1)(x-2) = 0$

$2x + 1 = 0 \quad x - 2 = 0$

$2x = -1 \qquad x = 2$

$x = -\frac{1}{2}$

The solutions are $-\frac{1}{2}$ and 2.

37. $\sqrt{p+11} = 1 - p$

$\left(\sqrt{p+11}\right)^2 = (1-p)^2$

$p + 11 = 1 - 2p + p^2$

$0 = p^2 - 3p - 10$

$(p-5)(p+2) = 0$

$p - 5 = 0 \quad p + 2 = 0$

$p = 5 \qquad p = -2$

5 does not check as a solution.
The solution is −2.

39. $\sqrt{x-1} - \sqrt{x} = -1$

$\sqrt{x-1} = \sqrt{x} - 1$

$\left(\sqrt{x-1}\right)^2 = (\sqrt{x} - 1)^2$

$x - 1 = x - 2\sqrt{x} + 1$

$2\sqrt{x} = 2$

$\sqrt{x} = 1$

$\left(\sqrt{x}\right)^2 = 1^2$

$x = 1$

The solution is 1.

41. $\sqrt{2x-1}=1-\sqrt{x-1}$

$$\left(\sqrt{2x-1}\right)^2=(1-\sqrt{x-1})^2$$
$$2x-1=1-2\sqrt{x-1}+x-1$$
$$2\sqrt{x-1}=-x+1$$
$$\left(2\sqrt{x-1}\right)^2=(-x+1)^2$$
$$4(x-1)=x^2-2x+1$$
$$4x-4=x^2-2x+1$$
$$0=x^2-6x+5$$
$$(x-5)(x-1)=0$$
$$x-5=0 \quad x-1=0$$
$$x=5 \qquad x=1$$

5 does not check as a solution.
The solution is 1.

43. $\sqrt{t+3}+\sqrt{2t+7}=1$

$$\sqrt{t+3}=1-\sqrt{2t+7}$$
$$\left(\sqrt{t+3}\right)^2=(1-\sqrt{2t+7})^2$$
$$t+3=1-2\sqrt{2t+7}+2t+7$$
$$2\sqrt{2t+7}=t+5$$
$$\left(2\sqrt{2t+7}\right)^2=(t+5)^2$$
$$4(2t+7)=t^2+10t+25$$
$$8t+28=t^2+10t+25$$
$$0=t^2+2t-3$$
$$(t+3)(t-1)=0$$
$$t+3=0 \quad t-1=0$$
$$t=-3 \qquad t=1$$

1 does not check as a solution.
The solution is −3.

Objective C Exercises

45. $y+2$

47. $x=\dfrac{10}{x-9}$

$$(x-9)x=\left(\frac{10}{x-9}\right)(x-9)$$
$$x^2-9x=10$$
$$x^2-9x-10=0$$
$$(x-10)(x+1)=0$$
$$x-10=0 \quad x+1=0$$
$$x=10 \qquad x=-1$$

The solutions are −1 and 10.

49. $\dfrac{y-1}{y+2}+y=1$

$$(y+2)\left(\frac{y-1}{y+2}+y\right)=1(y+2)$$
$$y-1+y(y+2)=y+2$$
$$y-1+y^2+2y=y+2$$
$$y^2+2y-3=0$$
$$(y+3)(y-1)=0$$
$$y+3=0 \quad y-1=0$$
$$y=-3 \qquad y=1$$

The solutions are −3 and 1.

51. $\dfrac{3r+2}{r+2}-2r=1$

$$(r+2)\left(\frac{3r+2}{r+2}-2r\right)=1(r+2)$$
$$3r+2-2r(r+2)=r+2$$
$$3r+2-2r^2-4r=r+2$$
$$-2r^2-2r=0$$
$$-2r(r+1)=0$$
$$-2r=0 \quad r+1=0$$
$$r=0 \qquad r=-1$$

The solutions are −1 and 0.

53. $\frac{1}{x+2}+\frac{x}{x-2}=\frac{x+6}{x^2-4}$

$(x+2)(x-2)\left(\frac{1}{x+2}+\frac{x}{x-2}\right)=\left(\frac{x+6}{x^2-4}\right)(x+2)(x-2)$

$(x-2)+x(x+2)=x+6$

$x-2+x^2+2x=x+6$

$x^2+2x-8=0$

$(x+4)(x-2)=0$

$x+4=0 \quad x-2=0$

$x=-4 \qquad x=2$

The solution is –4.

55. $\frac{16}{z-2}+\frac{16}{z+2}=6$

$(z-2)(z+2)\left(\frac{16}{z-2}+\frac{16}{z+2}\right)=6(z-2)(z+2)$

$16(z+2)+16(z-2)=6(z^2-4)$

$16z+32+16z-32=6z^2-24$

$0=6z^2-32z-24$

$2(3z^2-16z-12)=0$

$2(3z+2)(z-6)=0$

$3z+2=0 \quad z-6=0$

$3z=-2 \qquad z=6$

$z=-\frac{2}{3}$

The solutions are $-\frac{2}{3}$ and 6.

57. $\frac{t}{t-2}+\frac{2}{t-1}=4$

$(t-2)(t-1)\left(\frac{t}{t-2}+\frac{2}{t-1}\right)=4(t-2)(t-1)$

$t(t-1)+2(t-2)=4(t^2-3t+2)$

$t^2-t+2t-4=4t^2-12t+8$

$0=3t^2-13t+12$

$(3t-4)(t-3)=0$

$3t-4=0 \quad t-3=0$

$3t=4 \qquad t=3$

$t=\frac{4}{3}$

The solutions are $\frac{4}{3}$ and 3.

Critical Thinking

59. $\left(\sqrt{x}+3\right)^2-4\sqrt{x}-17=0$

Let $u=\sqrt{x}+3$.

$u^2-4(u-3)-17=0$

$u^2-4u+12-17=0$

$u^2-4u-5=0$

$(u-5)(u+1)=0$

$u-5=0 \quad u+1=0$

$u=5 \qquad u=-1$

Replace u with $\sqrt{x}+3$.

$\sqrt{x}+3=5 \quad \sqrt{x}+3=-1$

$\sqrt{x}=2 \qquad \sqrt{x}=-4$

$\left(\sqrt{x}\right)^2=2^2 \quad \left(\sqrt{x}\right)^2=(-4)^2$

$x=4 \qquad x=16$

The solution is 4.

Projects or Group Activities

61. a) $\{x \mid -\sqrt{29.7366} \le x \le \sqrt{29.7366}\}$

b)

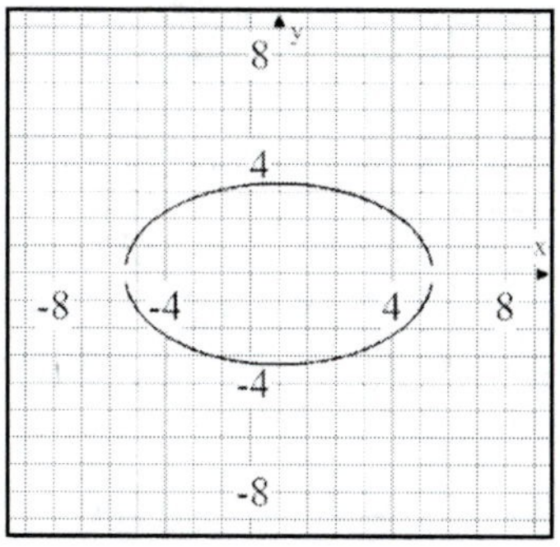

The $\pm$ symbol occurs in the equation so that the graph pictures the entire shape of the football.

c)

$$y = 3.3041\sqrt{1-\frac{x^2}{29.7336}} = 3.3041\sqrt{1-\frac{3^2}{29.7336}}$$

$y = 2.7592$ in

Check Your Progress: Chapter 10

1. $3x^2-10x-8=0$

$(3x+2)(x-4)=0$

$3x+2=0 \quad x-4=0$

$3x=-2 \qquad x=4$

$x=-\frac{2}{3}$

The solutions are $-\frac{2}{3}$ and 4.

2. $2x^2+3x=(x+3)(x+4)$

$x^2-4x-12=0$

$(x-6)(x+2)=0$

$x-6=0 \quad x+2=0$

$x=6 \qquad x=-2$

The solutions are –2 and 6.

3. $(x-r_1)(x-r_2)=0$

$(x-(-3))\left(x-\frac{3}{5}\right)=0$

$(x+3)\left(x-\frac{3}{5}\right)=0$

$x^2+\frac{12}{5}x-\frac{9}{5}=0$

$5\left(x^2+\frac{12}{5}x-\frac{9}{5}\right)=0\cdot 5$

$5x^2+12x-9=0$

4. $(x+3)^2=20$

$\sqrt{(x+3)^2}=\sqrt{20}$

$x+3=\pm\sqrt{20}=\pm 2\sqrt{5}$

$x+3=2\sqrt{5} \quad x+3=-2\sqrt{5}$

$x=-3+2\sqrt{5} \qquad x=-3-2\sqrt{5}$

The solutions are $-3+2\sqrt{5}$ and $-3-2\sqrt{5}$.

5. $(z-4)^2+9=5$

$(z-4)^2=-4$

$\sqrt{(z-4)^2}=\sqrt{-4}$

$z-4=\pm\sqrt{-4}=\pm 2i$

$z-4=2i \quad z-4=-2i$

$z=4+2i \qquad x=4-2i$

The solutions are $4+2i$ and $4-2i$.

6. $x^2 + 2x = 49$

$x^2 + 2x + 1 = 49 + 1$

$(x+1)^2 = 50$

$\sqrt{(x+1)^2} = \sqrt{50}$

$x + 1 = \pm 5\sqrt{2}$

$x + 1 = 5\sqrt{2} \qquad x + 1 = -5\sqrt{2}$

$x = -1 + 5\sqrt{2} \qquad x = -1 - 5\sqrt{2}$

The solutions are $-1 + 5\sqrt{2}$ and $-1 - 5\sqrt{2}$.

7. $4x^2 + 12x + 21 = 0$

$4x^2 + 12x = -21$

$\frac{1}{4}\left(4x^2 + 12x\right) = \frac{1}{4}(-21)$

$x^2 + 3x = -\frac{21}{4}$

$x^2 + 3x + \frac{9}{4} = -\frac{21}{4} + \frac{9}{4}$

$\left(x + \frac{3}{2}\right)^2 = -3$

$\sqrt{\left(x + \frac{3}{2}\right)^2} = \sqrt{-3}$

$x + \frac{3}{2} = \pm i\sqrt{3}$

$x + \frac{3}{2} = i\sqrt{3} \qquad x + \frac{3}{2} = -i\sqrt{3}$

$x = -\frac{3}{2} + i\sqrt{3} \qquad x = -\frac{3}{2} - i\sqrt{3}$

The solutions are $-\frac{3}{2} + i\sqrt{3}$ and $-\frac{3}{2} - i\sqrt{3}$.

8. $4x^2 - 4x - 31 = 0$

$a = 4, b = -4, c = -31$

$x = \frac{-b \pm \sqrt{b^2 - 4ac}}{2a}$

$x = \frac{-(-4) \pm \sqrt{(-4)^2 - 4(4)(-31)}}{2(4)}$

$x = \frac{4 \pm \sqrt{16 + 496}}{8} = \frac{4 \pm \sqrt{512}}{8}$

$x = \frac{4 \pm 16\sqrt{2}}{8} = \frac{1 \pm 4\sqrt{2}}{2}$

$x = \frac{1 + 4\sqrt{2}}{2} \qquad x = \frac{1 - 4\sqrt{2}}{2}$

The solutions are $\frac{1 + 4\sqrt{2}}{2}$ and $\frac{1 - 4\sqrt{2}}{2}$.

9. $x^2 + 8x + 25 = 0$

$a = 1, b = 8, c = 25$

$x = \frac{-b \pm \sqrt{b^2 - 4ac}}{2a}$

$x = \frac{-(8) \pm \sqrt{(8)^2 - 4(1)(25)}}{2(1)}$

$x = \frac{-8 \pm \sqrt{64 - 100}}{2} = \frac{-8 \pm \sqrt{-36}}{2}$

$x = \frac{-8 \pm 6i}{2} = -4 \pm 3i$

$x = -4 + 3i \qquad x = -4 - 3i$

The solutions are $-4 + 3i$ and $-4 - 3i$.

10. $x^4 + 8x^2 - 20 = 0$

$(x^2)^2 + 8(x^2) - 20 = 0$

$u^2 + 8u - 20 = 0$

$(u + 10)(u - 2) = 0$

$u + 10 = 0 \quad u - 2 = 0$

$u = -10 \qquad u = 2$

Replace u with x^2.

$$x^2 = -10 \qquad x^2 = 2$$
$$\sqrt{x^2} = \sqrt{-10} \qquad \sqrt{x^2} = \sqrt{2}$$
$$x = \pm i\sqrt{10} \qquad x = \pm\sqrt{2}$$

The solutions are $i\sqrt{10}$, $-i\sqrt{10}$, $\sqrt{2}$ and $-\sqrt{2}$.

11. $\sqrt{2x+1} - \sqrt{x+1} = 2$

$$\sqrt{2x+1} = \sqrt{x+1} + 2$$
$$\left(\sqrt{2x+1}\right)^2 = \left(\sqrt{x+1}+2\right)^2$$
$$2x+1 = x+1+4\sqrt{x+1}+4$$
$$x-4 = 4\sqrt{x+1}$$
$$(x-4)^2 = \left(4\sqrt{x+1}\right)^2$$
$$x^2 - 8x + 16 = 16x + 16$$
$$x^2 - 24x = 0$$
$$x(x-24) = 0$$
$$x = 0 \qquad x - 24 = 0$$
$$x = 24$$

The solution is 24.

12. $\dfrac{r}{r+1} - \dfrac{2}{r} = \dfrac{3}{10}$

$$(10r(r+1))\left(\frac{r}{r+1} - \frac{2}{r}\right) = \left(\frac{3}{10}\right)(10r(r+1))$$
$$10r^2 - 2(10r+10) = 3r(r+1)$$
$$10r^2 - 20r - 20 = 3r^2 + 3r$$
$$7r^2 - 23r - 20 = 0$$
$$(7r+5)(r-4) = 0$$
$$7r + 5 = 0 \qquad r - 4 = 0$$
$$7r = -5 \qquad r = 4$$
$$r = -\frac{5}{7}$$

The solution is $-\dfrac{5}{7}$ and 4.

Section 10.4

Concept Check

1. $\dfrac{1}{t}$

3. Down: $r+2$
Up: $r-2$

Objective A Exercises

5. **Strategy:** To find the maximum safe speed, substitute for d and solve for v.

Solution: $d = 0.04v^2 + 0.5v$
$$60 = 0.04v^2 + 0.5v$$
$$0 = 0.04v^2 + 0.5v - 60$$
$$v = \frac{-b \pm \sqrt{b^2 - 4ac}}{2a}$$
$$v = \frac{-0.5 \pm \sqrt{(0.5)^2 - 4(0.04)(-60)}}{2(0.04)}$$
$$v = \frac{-0.5 \pm \sqrt{9.85}}{0.08}$$
$$v \approx 33$$
$$v \approx -45$$

Since the speed cannot be a negative number the maximum speed is 33 mph.

7. **Strategy:** To find the time it takes for the projectile to return to Earth, substitute for s and v_0 and solve for t.

Solution: $s = v_0 t - 16t^2$
$$0 = 200t - 16t^2$$
$$0 = 8t(25 - 2t)$$
$$8t = 0 \qquad 25 - 2t = 0$$
$$t = 0 \qquad t = 12.5$$

The projectile will take 12.5 s to return to Earth.

9. Strategy: Substitute the given value for H and V and solve for the length of the side. In a square base, $L = W$.

Solution: $V = LWH$

Let x represent the length of the side of the square base. $V = x^2H$

$971{,}199 = x^2(31)$

$x^2 = 31{,}329$

$x = 177$

The length of a side of the square base is 177 m.

11. Strategy: Let t represent the time it takes the smaller pipe to fill the tank.
The time it takes the larger pipe to fill the tank is $t - 6$.

	Rate	Time	Part
Smaller pipe	$\frac{1}{t}$	4	$\frac{4}{t}$
Larger pipe	$\frac{1}{t-6}$	4	$\frac{4}{t-6}$

The sum of the parts of the task completed by each pipe equals 1.

Solution: $\frac{4}{t} + \frac{4}{t-6} = 1$

$t(t-6)\left(\frac{4}{t} + \frac{4}{t-6}\right) = 1t(t-6)$

$4(t-6) + 4t = t^2 - 6t$

$4t - 24 + 4t = t^2 - 6t$

$t^2 - 14t + 24 = 0$

$(t-12)(t-2) = 0$

$t - 12 = 0 \quad t - 2 = 0$

$t = 12 \qquad t = 2$

$t - 6 = 12 - 6 = 6$

$t - 6 = 2 - 6 = -4$

$t = 2$ is not possible since time cannot be a negative number. It will take the smaller pipe 12 min and the larger pipe 6 min.

13. Strategy: Let t represent the time it takes the faster computer working alone.

	Rate	Time	Part
Slower computer	$\frac{1}{t+4}$	3	$\frac{3}{t+4}$
Faster computer	$\frac{1}{t}$	1	$\frac{1}{t}$

The sum of the parts of the task completed by each computer equals 1.

Solution: $\frac{1}{t} + \frac{3}{t+4} = 1$

$t(t+4)\left(\frac{1}{t} + \frac{3}{t+4}\right) = 1t(t+4)$

$t + 4 + 3t = t^2 + 4t$

$4t + 4 = t^2 + 4t$

$t^2 - 4 = 0$

$(t+2)(t-2) = 0$

$t + 2 = 0 \quad t - 2 = 0$

$t = -2 \qquad t = 2$

Time cannot be a negative number. It will take the faster computer 2 h working alone.

15. Strategy: Let t represent the time it takes the experienced carpenter.
The time it takes the apprentice is $t + 2$.

	Rate	Time	Part
Experienced carpenter	$\frac{1}{t}$	2	$\frac{2}{t}$
Apprentice carpenter	$\frac{1}{t+2}$	4	$\frac{4}{t+2}$

The sum of the parts of the task completed by each carpenter equals 1.

Solution: $\frac{2}{t} + \frac{4}{t+2} = 1$

$t(t+2)\left(\frac{2}{t} + \frac{4}{t+2}\right) = 1t(t+2)$

$2(t+2) + 4t = t^2 + 2t$

$2t + 4 + 4t = t^2 + 2t$

$t^2 - 4t - 4 = 0$

$$t = \frac{-b \pm \sqrt{b^2 - 4ac}}{2a}$$

$$t = \frac{-(-4) \pm \sqrt{(-4)^2 - 4(1)(-4)}}{2(1)}$$

$$t = \frac{4 \pm \sqrt{32}}{2} = 2 \pm 2\sqrt{2}$$

$t \approx 4.8$

$t \approx -0.8$

Time cannot be a negative number. It will take the apprentice carpenter $t + 2 \approx 6.8$ h working alone.

17. Strategy: Let r represent the rate of the wind.

	Distance	Rate	Time
With wind	4000	$1320 + r$	$\frac{4000}{1320+r}$
Against wind	4000	$1320 - r$	$\frac{4000}{1320-r}$

It took 0.5 h less time to make the return trip.

Solution: $\dfrac{4000}{1320-r} - \dfrac{4000}{1320+r} = 0.5$

$$(1320-r)(1320+r)\left(\frac{4000}{1320-r} - \frac{4000}{1320+r}\right) = 0.5(1320-r)(1320+r)$$

$$4000(1320+r) - 4000(1320-r) = 0.5(1{,}742{,}400 - r^2)$$

$$5{,}280{,}000 + 4000r - 5{,}280{,}000 + 4000r = 871{,}200 - 0.5r^2$$

$$8000r = 871{,}200 - 0.5r^2$$

$$0.5r^2 + 8000r - 871{,}200 = 0$$

$$r = \frac{-b \pm \sqrt{b^2 - 4ac}}{2a}$$

$$r = \frac{-(8000) \pm \sqrt{(8000)^2 - 4(0.5)(-871{,}200)}}{2(0.5)}$$

$$r = \frac{-8000 \pm \sqrt{65{,}742{,}400}}{1}$$

$r \approx 108$

$r \cong -16{,}108$

Since the rate cannot be a negative number. The rate of the wind was approximately 108 mph.

19. Strategy: Let r represent the rate of the jet stream.

	Distance	Rate	Time
With jet stream	3660	$630+r$	$\frac{3660}{630+r}$
Against jet stream	3660	$630-r$	$\frac{3660}{630-r}$

It took 1.75 h less time to make the trip flying with the jet stream.

Solution: $\frac{3660}{630-r}-\frac{3660}{630+r}=1.75$

$$(630+r)(630-r)\left(\frac{3660}{630-r}-\frac{3660}{630+r}\right)=1.75(630+r)(630-r)$$

$$3660(630+r)-3660(630-r)=1.75(396{,}900-r^2)$$

$$2{,}305{,}800+3660r-2{,}305{,}800+3660r=694{,}575-1.75r^2$$

$$7320r=694{,}575-1.75r^2$$

$$1.75r^2+7320r-694{,}575=0$$

$$r=\frac{-b\pm\sqrt{b^2-4ac}}{2a}$$

$$r=\frac{-(7320)\pm\sqrt{(7320)^2-4(1.75)(-694{,}575)}}{2(1.75)}$$

$$r=\frac{-7320\pm\sqrt{58{,}444{,}425}}{3.5}$$

$r\approx 93$

$r\approx -4276$

The rate cannot be a negative number. The rate of the jet stream is 93 mph.

21. Strategy: Let x represent the number of apartments.
The monthly rent is $1200 + 100x$.
The number of units rented is $100 - x$.
Solution:

$$(1200+100x)(100-x)=153600$$

$$-100x^2+8800x+120000=153600$$

$$100x^2-8800x-33600=0$$

$$x^2-88x+336=0$$

$$(x-4)(x-84)=0$$

$x-4=0 \quad x-84=0$

$x=4 \qquad x=84$

$100-4=96 \quad 1200+100(4)=1600$
96 units at $1600/month

$100-84=16 \quad 1200+100(84)=9600$
16 units at $9600/month

23. Strategy: Let x represent a side of the square base of the box.
The volume of the box is 49,000 cm^3.
Solution: $V = LWH$

$$49{,}000=(x)(x)(10)$$

$$49{,}000=10x^2$$

$$x^2=4900$$

$$x=70$$

The side of the original square is 20 cm more than side x.
$x+20=70+20=90$
The dimensions of the original square base are 90 cm by 90 cm.

25. Strategy: Let x represent the width of the rectangle. The length of the rectangle is $40 - x$. The area of the rectangle is 300 ft^2.

Solution: $A = LW$

$300 = x(40 - x)$

$300 = 40x - x^2$

$x^2 - 40x + 300 = 0$

$(x - 10)(x - 30) = 0$

$x - 10 = 0 \quad x - 30 = 0$

$x = 10 \quad x = 30$

$40 - x = 40 - 10 = 30$

$40 - x = 40 - 30 = 10$

The dimensions of the rectangle are 10 ft by 30 ft.

Critical Thinking

27. Strategy: Use the Pythagorean formula $a^2 + b^2 = c^2$ with $a = 1.5$, $b = 3.5$ and $c = x + 1.5$.

Solution: $a^2 + b^2 = c^2$

$(1.5)^2 + (3.5)^2 = (x + 1.5)^2$

$14.5 = (x + 1.5)^2$

$\sqrt{14.5} = \sqrt{(x + 1.5)^2}$

$\pm 3.8 \approx x + 1.5$

$3.8 = x + 1.5 \quad -3.18 = x + 1.5$

$x = 2.3 \quad x = -5.3$

Distance cannot be a negative number. The bottom of the scoop of ice cream is 2.3 in. from the bottom of the cone.

Section 10.5

Concept Check

1. It must be true that $x - 3 > 0$ and $x - 5 > 0$ or the $x - 3 < 0$ and $x - 5 < 0$. In other words, either both factors are positive or both factors are negative.

3. a) $x = 2$ True
b) $x = -2$ False
c) $x = -3$ False

5. a) $x = 2$ True
b) $x = 3$ False
c) $x = -1$ True

Objective A Exercises

7. $(x - 4)(x + 2) > 0$

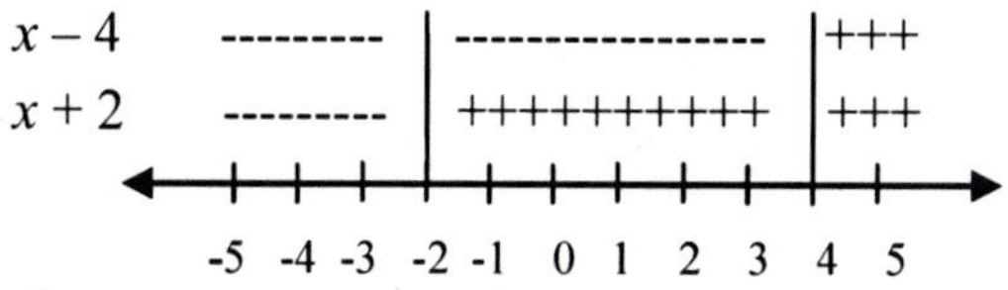

$\{x \mid x < -2 \text{ or } x > 4\}$

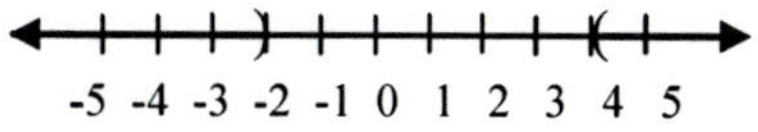

9. $x^2 - 3x + 2 \ge 0$

$(x - 2)(x - 1) \ge 0$

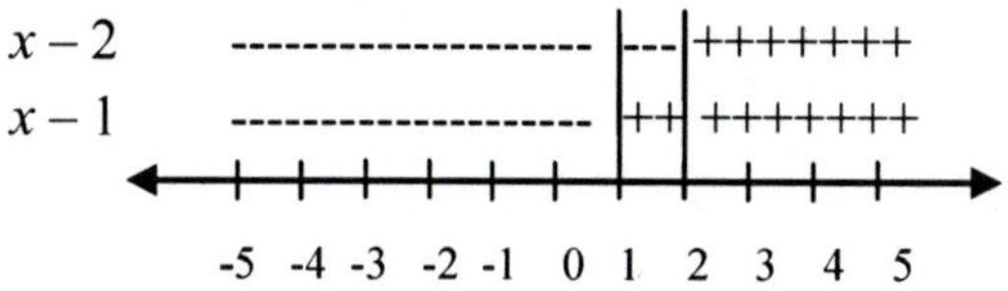

$\{x \mid x \le 1 \text{ or } x \ge 2\}$

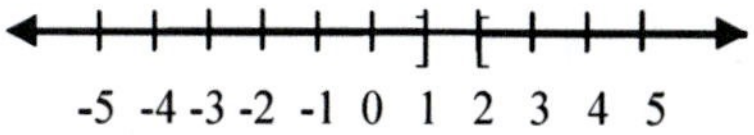

11. $x^2 - x - 12 < 0$

$(x - 4)(x + 3) < 0$

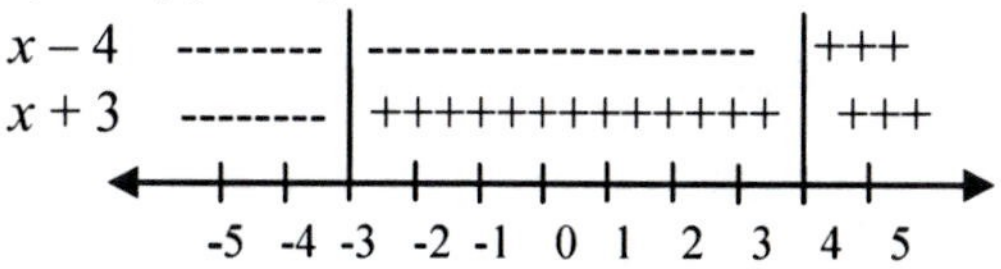

$\{x \mid -3 < x < 4\}$

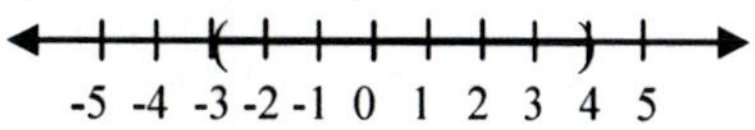

13. $(x - 1)(x + 2)(x - 3) < 0$

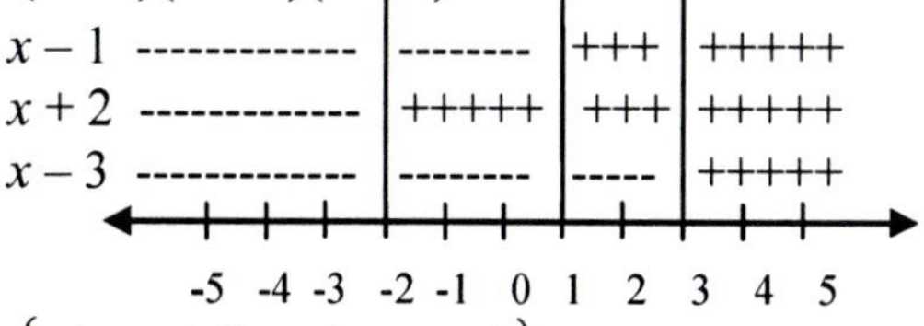

$\{x \mid x < -2 \text{ or } 1 < x < 3\}$

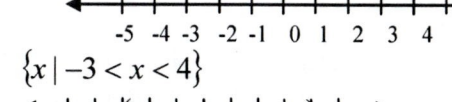

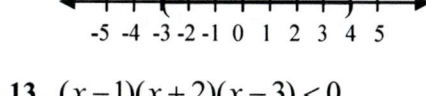

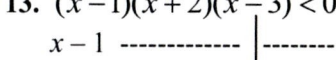

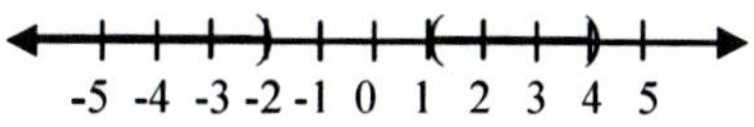

15. $(x+4)(x-2)(x-1) \geq 0$

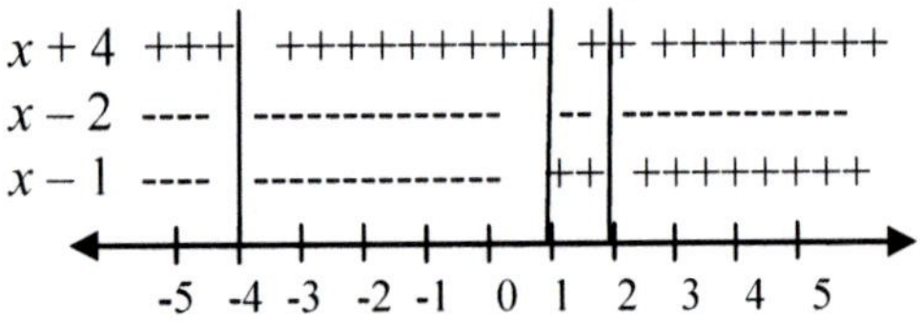

$\{x \mid -4 \leq x \leq 1 \text{ or } x \geq 2\}$

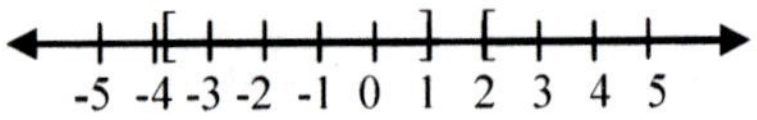

17. $\dfrac{x-4}{x+2} > 0$

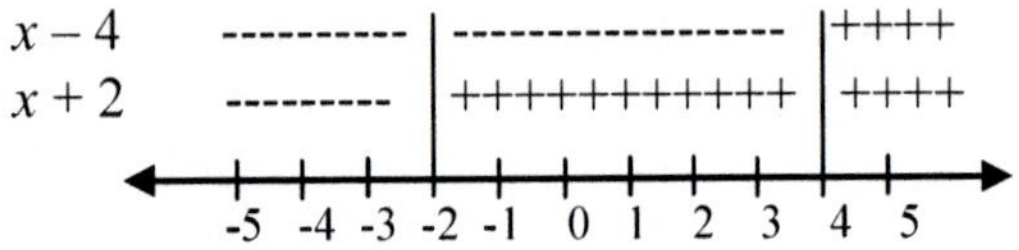

$\{x \mid x < -2 \text{ or } x > 4\}$

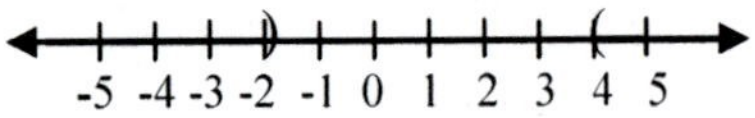

19. $\dfrac{x-3}{x+1} \leq 0$

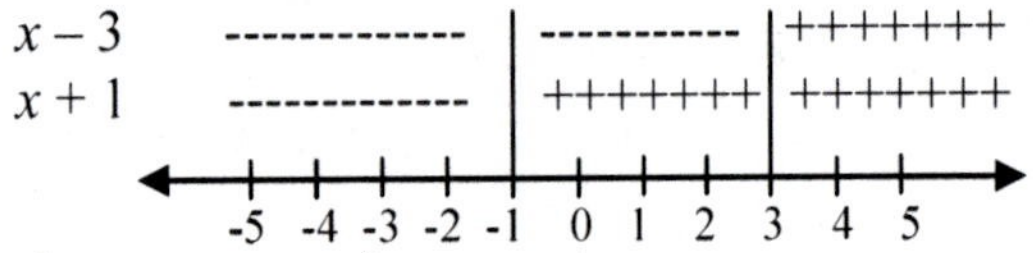

$\{x \mid -1 < x \leq 3\}$

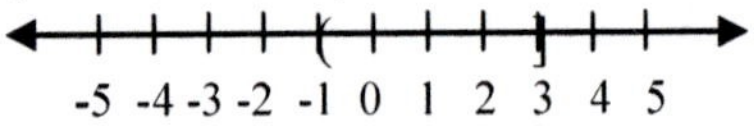

21. $\dfrac{(x-1)(x+2)}{x-3} \leq 0$

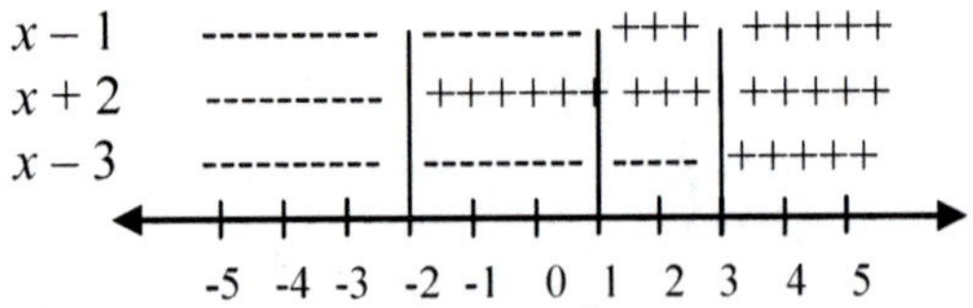

$\{x \mid x \leq -2 \text{ or } 1 \leq x < 3\}$

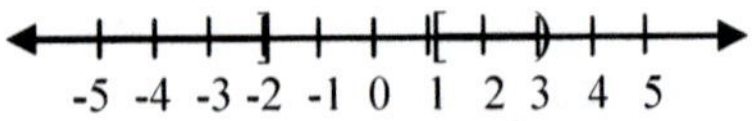

23. $x^2 - 16 > 0$

$(x+4)(x-4) > 0$

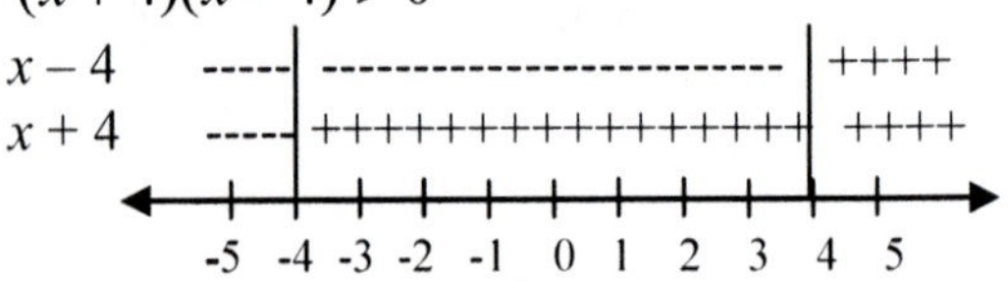

$\{x \mid x > 4 \text{ or } x < -4\}$

25. $x^2 - 9x \leq 36$

$x^2 - 9x - 36 \leq 0$

$(x-12)(x+3) \leq 0$

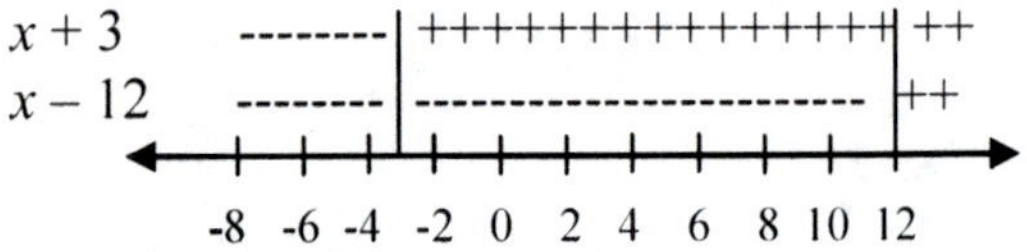

$\{x \mid -3 \leq x \leq 12\}$

27. $4x^2 - 8x + 3 < 0$

$(2x-3)(2x-1) < 0$

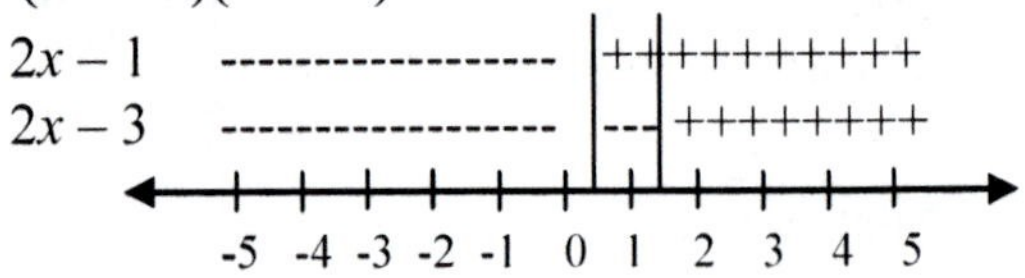

$\left\{x \mid \dfrac{1}{2} < x < \dfrac{3}{2}\right\}$

29. $\dfrac{3}{x-1} < 2$

$\dfrac{3}{x-1} - 2 < 0$

$\dfrac{3}{x-1} - \dfrac{2x-2}{x-1} < 0$

$\dfrac{-2x+5}{x-1} < 0$

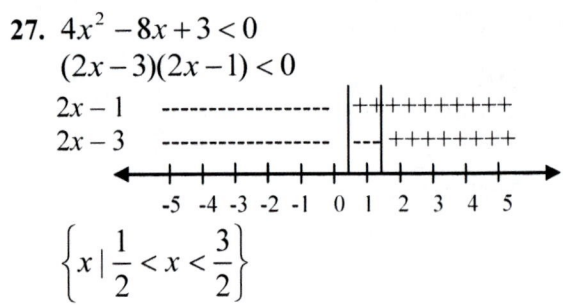

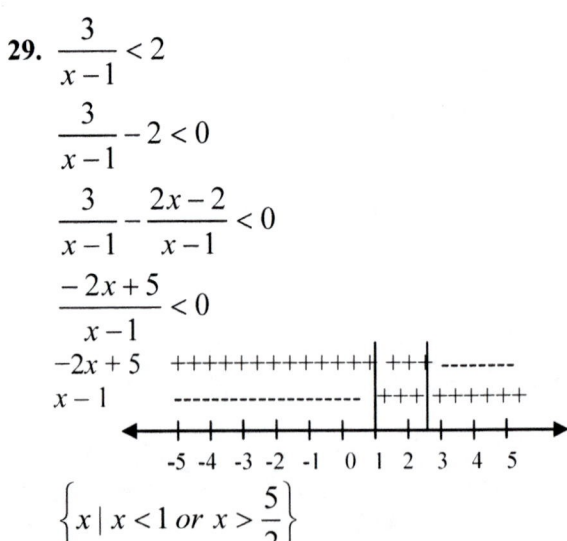

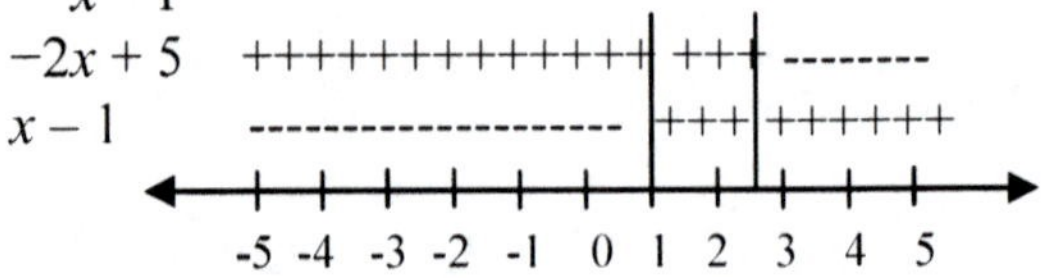

$\left\{x \mid x < 1 \text{ or } x > \dfrac{5}{2}\right\}$

31. $\dfrac{x-2}{(x+1)(x-1)} \le 0$

$x-2$ `--------------- |------|---| ++++++`
$x+1$ `--------------- |++++| ++| ++++++`
$x-1$ `--------------- |------|++| +++++++`

-5 -4 -3 -2 -1 0 1 2 3 4 5

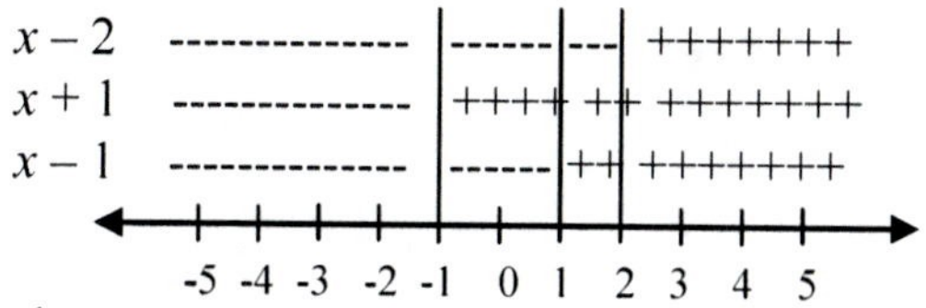

$\{x \mid x < -1 \text{ or } 1 < x \le 2\}$

33. $\dfrac{x}{2x-1} \ge 1$

$\dfrac{x}{2x-1} - 1 \ge 0$

$\dfrac{x}{2x-1} - \dfrac{2x-1}{2x-1} \ge 0$

$\dfrac{-x+1}{2x-1} \ge 0$

$-x+1$ `+++++++++++|+ ---------------`
$2x-1$ `---------------|+ ++++++++++`

-5 -4 -3 -2 -1 0 1 2 3 4 5

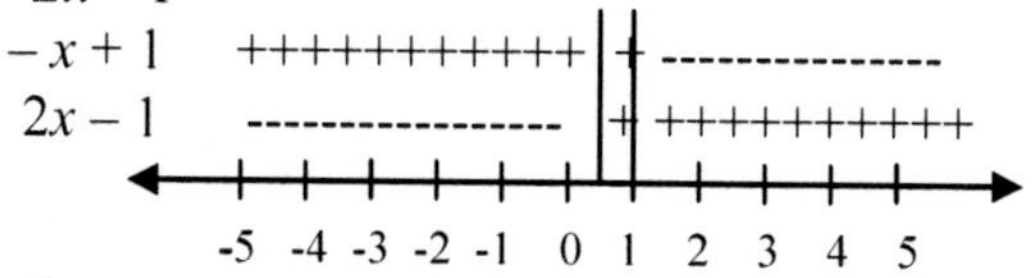

$\left\{x \mid \dfrac{1}{2} < x \le 1\right\}$

35. $\dfrac{3}{x-5} > \dfrac{1}{x+1}$

$\dfrac{3}{x-5} - \dfrac{1}{x+1} > 0$

$\dfrac{3(x+1)}{(x-5)(x+1)} - \dfrac{(x-5)}{(x-5)(x+1)} > 0$

$\dfrac{3x+3-x+5}{(x-5)(x+1)} > 0$

$\dfrac{2x+8}{(x-5)(x+1)} > 0$

$2x+8$ `---- | +++++|+++++++++++++| ++`
$x-5$ `---- | ---------------------| ++`
$x+1$ `---- | --------|++++++++++++| ++`

-5 -4 -3 -2 -1 0 1 2 3 4 5

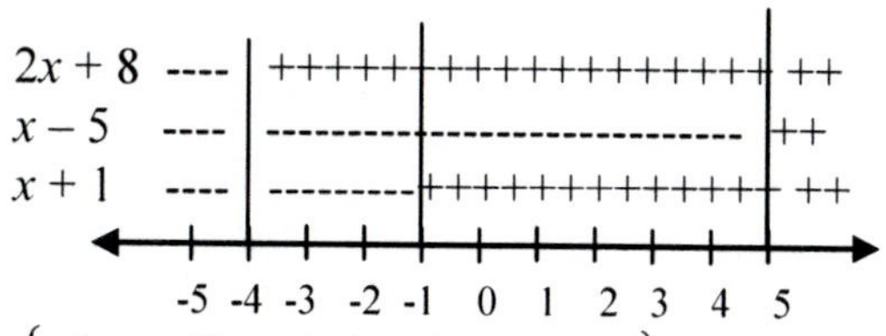

$\{x \mid x > 5\} \cup \{x \mid -4 < x < -1\}$

Critical Thinking

37. $(x-1)(x+3)(x-2)(x-4) \ge 0$

$x-1$ `-------- | ------------|++|++++| +++`
$x+3$ `-------- |++++++++|++|++++| +++`
$x-2$ `-------- |------------|---|++++| +++`
$x-4$ `-------- |------------|---|------| +++`

-5 -4 -3 -2 -1 0 1 2 3 4 5

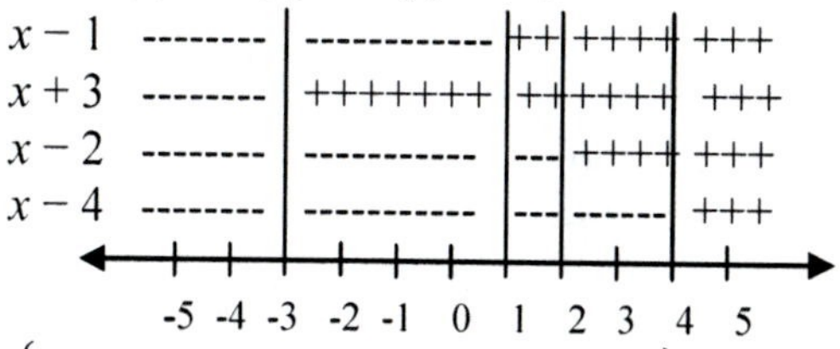

$\{x \mid x \le -3 \text{ or } 1 \le x \le 2 \text{ or } x \ge 4\}$

-5 -4 -3 -2 -1 0 1 2 3 4 5

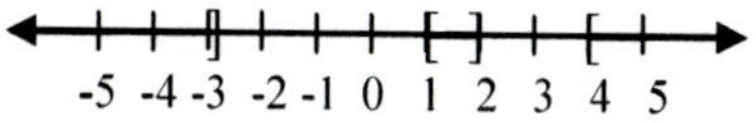

39. $(x^2 + 2x - 3)(x^2 + 3x + 2) \ge 0$

$(x-1)(x+3)(x+1)(x+2) \ge 0$

$x-1$ `------- |---|---|----- |+++++++++++`
$x+3$ `------- |++|++|+++|++++++++++++`
$x+2$ `------- |---|++|+++|++++++++++++`
$x+1$ `------- |---|---|+++|++++++++++`

-5 -4 -3 -2 -1 0 1 2 3 4 5

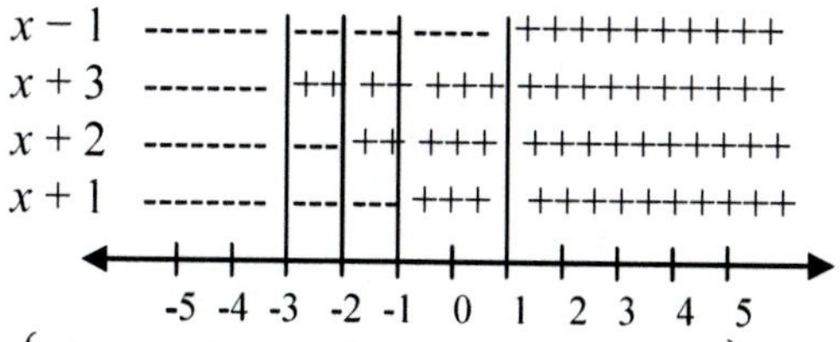

$\{x \mid x \le -3 \text{ or } -2 \le x \le -1 \text{ or } x \ge 1\}$

-5 -4 -3 -2 -1 0 1 2 3 4 5

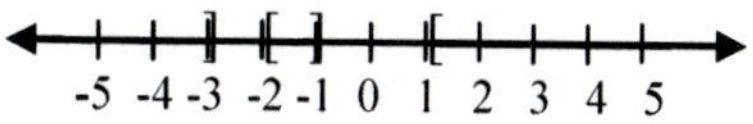

41. $\dfrac{x^2(3-x)(2x+1)}{(x+4)(x+2)} \ge 0$

x^2 `+++|++++|++ |+++++++|+++++`
$3-x$ `+++|+++|++ |+++++++| --------`
$2x+1$ `----|--------|---- |+++++++| +++++`
$x+4$ `---- |+++|+++|+++++++| ++++++`
$x+2$ `---- |-----|+++|+++++++| ++++++`

-5 -4 -3 -2 -1 0 1 2 3 4 5

$\left\{x \mid -4 < x < -2 \text{ or } -\dfrac{1}{2} \le x \le 3\right\}$

-5 -4 -3 -2 -1 0 1 2 3 4 5

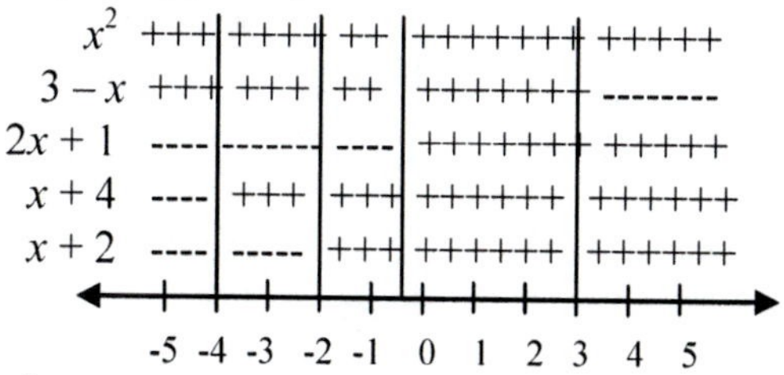

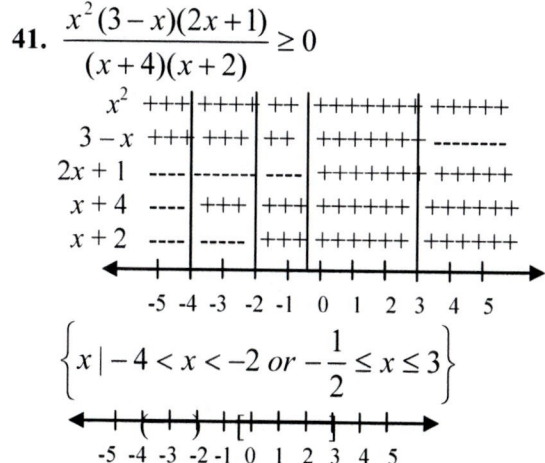

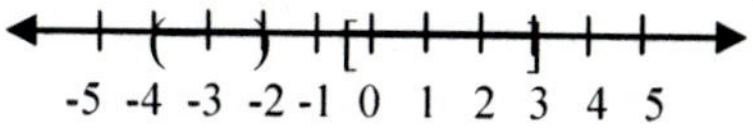

Projects or Group Activities

43. $d = rt - 5t^2$
$70t - 5t^2 > 200$
$-5t^2 + 70t - 200 > 0$
$t^2 - 14t + 40 < 0$
$(t-4)(t-10) < 0$
$t - 4 < 0 \quad t - 10 < 0$
$t < 4 \quad t < 10$
The arrow will be more than 200 m high between 4 s and 10 s.

Chapter 10 Review Exercises

1. $2x^2 - 3x = 0$
$x(2x-3) = 0$
$2x - 3 = 0 \quad x = 0$
$2x = 3$
$x = \frac{3}{2}$
The solutions are 0 and $\frac{3}{2}$.

2. $6x^2 + 9x = 6$
$6x^2 + 9x - 6 = 0$
$3(2x^2 + 3x - 2) = 0$
$3(2x-1)(x+2) = 0$
$2x - 1 = 0 \quad x + 2 = 0$
$2x = 1 \quad x = -2$
$x = \frac{1}{2}$
The solutions are $-2c$ and $\frac{c}{2}$.

3. $x^2 = 48$
$\sqrt{x^2} = \sqrt{48}$
$x = \pm\sqrt{48} = \pm 4\sqrt{3}$
The solutions are $4\sqrt{3}$ and $-4\sqrt{3}$.

4. $\left(x + \frac{1}{2}\right)^2 + 4 = 0$
$\left(x + \frac{1}{2}\right)^2 = -4$
$\sqrt{\left(x + \frac{1}{2}\right)^2} = \sqrt{-4}$
$x + \frac{1}{2} = \pm\sqrt{-4} = \pm 2i$
$x + \frac{1}{2} = 2i \quad x + \frac{1}{2} = -2i$
$x = 2i - \frac{1}{2} \quad x = -2i - \frac{1}{2}$
The solutions are $2i - \frac{1}{2}$ and $-2i - \frac{1}{2}$.

5. $x^2 + 4x + 3 = 0$
$x^2 + 4x = -3$
$x^2 + 4x + 4 = -3 + 4$
$(x+2)^2 = 1$
$\sqrt{(x+2)^2} = \sqrt{1}$
$x + 2 = \pm 1$
$x + 2 = 1 \quad x + 2 = -1$
$x = -1 \quad x = -3$
The solutions are −1 and −3.

6. $7x^2 - 14x + 3 = 0$

$7x^2 - 14x = -3$

$\frac{1}{7}(7x^2 - 14x) = \frac{1}{7}(-3)$

$x^2 - 2x = -\frac{3}{7}$

$x^2 - 2x + 1 = -\frac{3}{7} + 1$

$(x-1)^2 = \frac{4}{7}$

$\sqrt{(x-1)^2} = \sqrt{\frac{4}{7}}$

$x - 1 = \pm\sqrt{\frac{4}{7}} = \pm\frac{2\sqrt{7}}{7}$

$x - 1 = \frac{2\sqrt{7}}{7} \quad x - 1 = -\frac{2\sqrt{7}}{7}$

$x = 1 + \frac{2\sqrt{7}}{7} \quad x = 1 - \frac{2\sqrt{7}}{7}$

The solutions are $\frac{7+2\sqrt{7}}{7}$ and $\frac{7-2\sqrt{7}}{7}$.

7. $12x^2 - 25x + 12 = 0$

$a = 12, b = -25, c = 12$

$x = \frac{-b \pm \sqrt{b^2 - 4ac}}{2a}$

$x = \frac{-(-25) \pm \sqrt{(-25)^2 - 4(12)(12)}}{2(12)}$

$x = \frac{25 \pm \sqrt{625 - 576}}{24}$

$x = \frac{25 \pm \sqrt{49}}{24}$

$x = \frac{25 \pm 7}{24}$

$x = \frac{25+7}{24} = \frac{32}{24} = \frac{4}{3}$

$x = \frac{25-7}{24} = \frac{18}{24} = \frac{3}{4}$

The solutions are $\frac{4}{3}$ and $\frac{3}{4}$.

8. $x^2 - x + 8 = 0$

$a = 1, b = -1, c = 8$

$x = \frac{-b \pm \sqrt{b^2 - 4ac}}{2a}$

$x = \frac{-(-1) \pm \sqrt{(-1)^2 - 4(1)(8)}}{2(1)}$

$x = \frac{1 \pm \sqrt{1 - 32}}{2}$

$x = \frac{1 \pm \sqrt{-31}}{2}$

$x = \frac{1 \pm i\sqrt{31}}{2}$

The solutions are $\frac{1+i\sqrt{31}}{2}$ and $\frac{1-i\sqrt{31}}{2}$.

9. $(x - r_1)(x - r_2) = 0$

$(x - 0)(x - (-3)) = 0$

$x(x + 3) = 0$

$x^2 + 3x = 0$

10. $(x - r_1)(x - r_2) = 0$

$\left(x - \frac{3}{4}\right)\left(x - (-\frac{2}{3})\right) = 0$

$\left(x - \frac{3}{4}\right)\left(x + \frac{2}{3}\right) = 0$

$x^2 - \frac{1}{12}x - \frac{1}{2} = 0$

$12\left(x^2 - \frac{1}{12}x - \frac{1}{2}\right) = 0(12)$

$12x^2 - x - 6 = 0$

11. $x^2 - 2x + 8 = 0$

$x^2 - 2x = -8$

$x^2 - 2x + 1 = -8 + 1$

$(x-1)^2 = -7$

$\sqrt{(x-1)^2} = \sqrt{-7}$

$x - 1 = \pm\sqrt{-7} = \pm i\sqrt{7}$

$x = 1 \pm i\sqrt{7}$

The solutions are $1 + i\sqrt{7}$ and $1 - i\sqrt{7}$.

12. $(x-2)(x+3) = x - 10$

$x^2 + x - 6 = x - 10$

$x^2 = -4$

$\sqrt{x^2} = \sqrt{-4}$

$x = \pm\sqrt{-4}$

$x = \pm 2i$

The solutions are $2i$ and $-2i$.

13. $3x(x-3) = 2x - 4$

$3x^2 - 9x = 2x - 4$

$3x^2 - 11x + 4 = 0$

$a = 3, b = -11, c = 4$

$$x = \frac{-b \pm \sqrt{b^2 - 4ac}}{2a}$$

$$x = \frac{-(-11) \pm \sqrt{(-11)^2 - 4(3)(4)}}{2(3)}$$

$$x = \frac{11 \pm \sqrt{121 - 48}}{6}$$

$$x = \frac{11 \pm \sqrt{73}}{6}$$

The solutions are $\dfrac{11 + \sqrt{73}}{6}$ and $\dfrac{11 - \sqrt{73}}{6}$.

14. $3x^2 - 5x + 3 = 0$

$a = 3, b = -5, c = 3$

$b^2 - 4ac$

$(-5)^2 - 4(3)(3) = 25 - 36 = -11$

$-11 > 0$

Since the discriminant is less than zero the equation has two complex number solutions.

15. $(x+3)(2x-5) < 0$

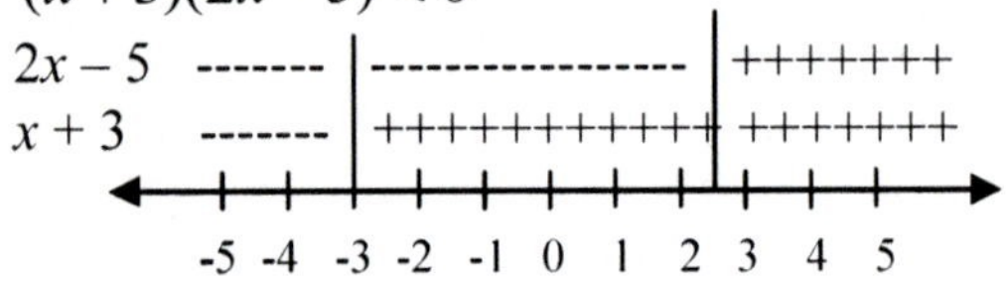

$$\left\{x \mid -3 < x < \frac{5}{2}\right\}$$

16. $(x-2)(x+4)(2x+3) \le 0$

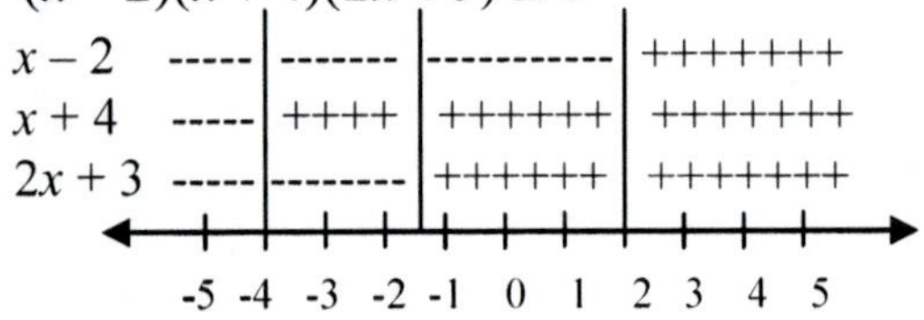

$$\left\{x \mid x \le -4 \text{ or } -\frac{3}{2} \le x \le 2\right\}$$

17. $x^{2/3} + x^{1/3} - 12 = 0$

$\left(x^{1/3}\right)^2 + x^{1/3} - 12 = 0$

$u^2 + u - 12 = 0$

$(u+4)(u-3) = 0$

$u + 4 = 0 \quad u - 3 = 0$

$u = -4 \qquad u = 3$

Replace u with $x^{1/3}$.

$x^{1/3} = -4 \qquad x^{1/3} = 3$

$\left(x^{1/3}\right)^3 = (-4)^3 \quad \left(x^{1/3}\right)^3 = 3^3$

$x = -64 \qquad x = 27$

The solutions are −64 and 27.

18. $2(x-1) + 3\sqrt{x-1} - 2 = 0$

$2(\sqrt{x-1})^2 + 3\sqrt{x-1} - 2 = 0$

$2u^2 + 3u - 2 = 0$

$(2u-1)(u+2) = 0$

$2u - 1 = 0 \qquad u + 2 = 0$

$2u = 1 \qquad u = -2$

$$u = \frac{1}{2}$$

Replace u with $\sqrt{x-1}$.

$\sqrt{x-1} = \frac{1}{2}$ $\quad\quad$ $\sqrt{x-1} = -2$

$\left(\sqrt{x-1}\right)^2 = \left(\frac{1}{2}\right)^2$ $\quad\quad$ $\left(\sqrt{x-1}\right) = (-2)^2$

$x-1 = \frac{1}{4}$ $\quad\quad$ $x-1 = 4$

$x = \frac{5}{4}$ $\quad\quad$ $x = 5$

5 does not check as a solution.

The solution is $\frac{5}{4}$.

19. $3x = \dfrac{9}{x-2}$

$3x(x-2) = \dfrac{9}{x-2}(x-2)$

$3x^2 - 6x = 9$

$3x^2 - 6x - 9 = 0$

$3(x^2 - 2x - 3) = 0$

$3(x-3)(x+1) = 0$

$x-3 = 0 \quad x+1 = 0$

$x = 3 \quad\quad x = -1$

The solutions are −1 and 3.

20. $\dfrac{3x+7}{x+2} + x = 3$

$\dfrac{3x+7}{x+2} = 3 - x$

$(x+2)\left(\dfrac{3x+7}{x+2}\right) = (3-x)(x+2)$

$3x + 7 = 3x + 6 - x^2 - 2x$

$x^2 + 2x + 1 = 0$

$(x+1)^2 = 0$

$\sqrt{(x+1)^2} = \sqrt{0}$

$x + 1 = 0$

$x = -1$

The solution is −1.

21. $\dfrac{x-2}{2x-3} \geq 0$

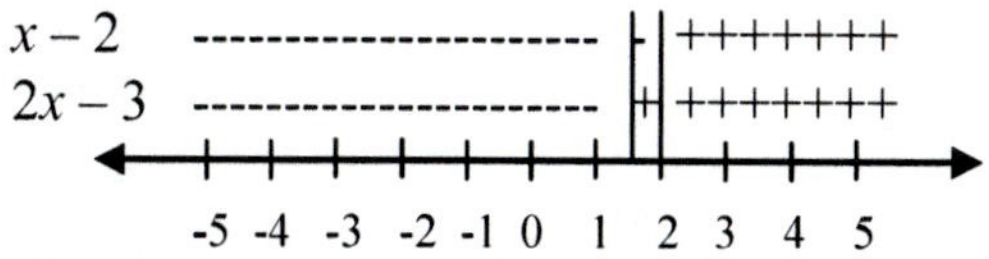

$\left\{x \mid x < \frac{3}{2} \text{ or } x \geq 2\right\}$

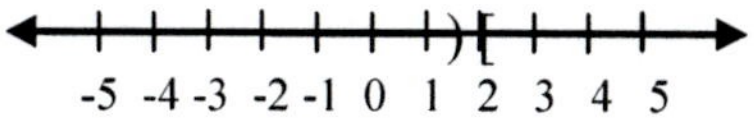

22. $\dfrac{(2x-1)(x+3)}{x-4} \leq 0$

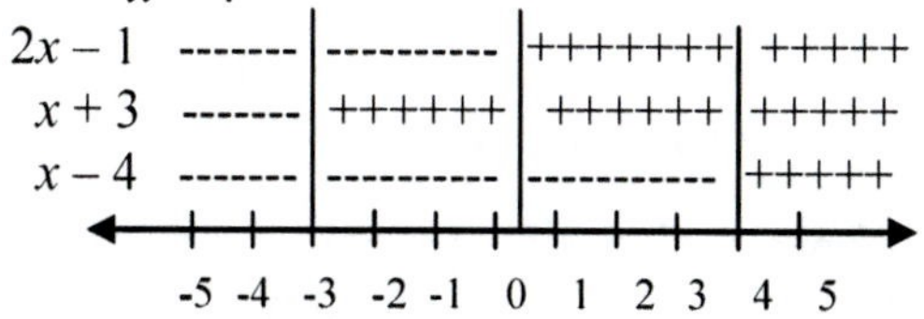

$\left\{x \mid x \leq -3 \text{ or } \frac{1}{2} \leq x < 4\right\}$

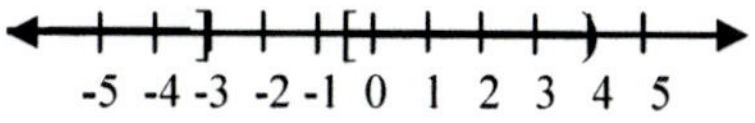

23. $x = \sqrt{x} + 2$

$x - 2 = \sqrt{x}$

$(x-2)^2 = \left(\sqrt{x}\right)^2$

$x^2 - 4x + 4 = x$

$x^2 - 5x + 4 = 0$

$(x-4)(x-1) = 0$

$x - 4 = 0 \quad x - 1 = 0$

$x = 4 \quad\quad x = 1$

1 does not check as a solution.

The solution is 4.

24. $2x = \sqrt{5x+24} + 3$

$2x - 3 = \sqrt{5x+24}$

$(2x-3)^2 = \left(\sqrt{5x+24}\right)^2$

$4x^2 - 12x + 9 = 5x + 24$

$4x^2 - 17x - 15 = 0$

$(4x+3)(x-5) = 0$

$4x + 3 = 0 \quad x - 5 = 0$

$4x = -3 \qquad x = 5$

$x = -\frac{3}{4}$

$-\frac{3}{4}$ does not check as a solution.

The solution is 5.

25. $\frac{x-2}{2x+3} - \frac{x-4}{x} = 2$

$(2x+3)(x)\left(\frac{x-2}{2x+3} - \frac{x-4}{x}\right) = 2(2x+3)(x)$

$x(x-2) - (x-4)(2x+3) = 2x(2x+3)$

$x^2 - 2x - 2x^2 - 3x + 8x + 12 = 4x^2 + 6x$

$0 = 5x^2 + 3x - 12$

$a = 5, b = 3, c = -12$

$x = \frac{-b \pm \sqrt{b^2 - 4ac}}{2a}$

$x = \frac{-3 \pm \sqrt{3^2 - 4(5)(-12)}}{2(5)}$

$x = \frac{-3 \pm \sqrt{9 + 240}}{10}$

$x = \frac{-3 \pm \sqrt{249}}{10}$

The solutions are $\frac{-3+\sqrt{249}}{10}$ and $\frac{-3-\sqrt{249}}{10}$.

26. $1 - \frac{x+4}{2-x} = \frac{x-3}{x+2}$

$(2-x)(x+2)\left(1 - \frac{x+4}{2-x}\right) = (2-x)(x+2)\frac{x-3}{x+2}$

$(2-x)(x+2) - (x+2)(x+4) = (2-x)(x-3)$

$4 - x^2 - x^2 - 6x - 8 = -x^2 + 5x - 6$

$x^2 + 11x - 2 = 0$

$a = 1, b = 11, c = -2$

$x = \frac{-b \pm \sqrt{b^2 - 4ac}}{2a}$

$x = \frac{-11 \pm \sqrt{11^2 - 4(1)(-2)}}{2(1)}$

$x = \frac{-11 \pm \sqrt{121 + 8}}{2}$

$x = \frac{-11 \pm \sqrt{129}}{2}$

The solutions are $\frac{-11+\sqrt{129}}{2}$ and $\frac{-11-\sqrt{129}}{2}$.

27. $(x - r_1)(x - r_2) = 0$

$\left(x - \frac{1}{3}\right)(x - (-3)) = 0$

$\left(x - \frac{1}{3}\right)(x + 3) = 0$

$x^2 + \frac{8}{3}x - 1 = 0$

$3\left(x^2 + \frac{8}{3}x - 1\right) = 0(3)$

$3x^2 + 8x - 3 = 0$

28. $2x^2 + 9x = 5$

$2x^2 + 9x - 5 = 0$

$(2x-1)(x+5) = 0$

$2x - 1 = 0 \quad x + 5 = 0$

$2x = 1 \quad x = -5$

$x = \frac{1}{2}$

The solutions are −5 and $\frac{1}{2}$.

29. $2(x+1)^2 - 36 = 0$

$2(x+1)^2 = 36$

$(x+1)^2 = 18$

$(x+1)^2 = \left(\sqrt{18}\right)^2$

$x + 1 = \pm\sqrt{18} = \pm 3\sqrt{2}$

$x = -1 \pm 3\sqrt{2}$

The solutions are $-1 + 3\sqrt{2}$ and $-1 - 3\sqrt{2}$.

30. $x^2 + 6x + 10 = 0$

$a = 1, b = 6, c = 10$

$x = \frac{-b \pm \sqrt{b^2 - 4ac}}{2a}$

$x = \frac{-6 \pm \sqrt{6^2 - 4(1)(10)}}{2(1)}$

$x = \frac{-6 \pm \sqrt{36 - 40}}{2} = \frac{-6 \pm \sqrt{-4}}{2}$

$x = \frac{-6 \pm 2i}{2}$

$x = -3 \pm i$

The solutions are $-3 + i$ and $-3 - i$.

31. $\frac{2}{x-4} + 3 = \frac{x}{2x-3}$

$(2x-3)(x-4)\left(\frac{2}{x-4} + 3\right) = (2x-3)(x-4)\frac{x}{2x-3}$

$2(2x-3) + 3(x-4)(2x-3) = x(x-4)$

$4x - 6 + 6x^2 - 33x + 36 = x^2 - 4x$

$5x^2 - 25x + 30 = 0$

$5(x^2 - 5x + 6) = 0$

$5(x-3)(x-2) = 0$

$x - 3 = 0 \quad x - 2 = 0$

$x = 3 \quad x = 2$

The solutions are 2 and 3.

32. $x^4 - 28x^2 + 75 = 0$

$\left(x^2\right)^2 - 28x^2 + 75 = 0$

$u^2 - 28u + 75 = 0$

$(u-25)(u-3) = 0$

$u - 25 = 0 \quad u - 3 = 0$

$u = 25 \quad u = 3$

Replace u with x^2.

$x^2 = 25 \quad x^2 = 3$

$\sqrt{x^2} = \sqrt{25} \quad \sqrt{x^2} = \sqrt{3}$

$x = \pm 5 \quad x = \pm\sqrt{3}$

The solutions are −5, 5, $\sqrt{3}$ and $-\sqrt{3}$.

33. $\sqrt{2x-1}+\sqrt{2x}=3$

$\sqrt{2x-1}=3-\sqrt{2x}$

$(\sqrt{2x-1})^2=\left(3-\sqrt{2x}\right)^2$

$2x-1=9-6\sqrt{2x}-2x$

$-10=-6\sqrt{2x}$

$5=3\sqrt{2x}$

$5^2=\left(3\sqrt{2x}\right)^2$

$25=18x$

$x=\dfrac{25}{18}$

The solution is $\dfrac{25}{18}$.

34. $2x^{2/3}+3x^{1/3}-2=0$

$2\left(x^{1/3}\right)^2+3x^{1/3}-2=0$

$2u^2+3u-2=0$

$(2u-1)(u+2)=0$

$2u-1=0 \quad u+2=0$

$2u=1 \qquad u=-2$

$u=\dfrac{1}{2}$

Replace u with $x^{1/3}$.

$x^{1/3}=\dfrac{1}{2} \qquad x^{1/3}=-2$

$\left(x^{1/3}\right)^3=\left(\dfrac{1}{2}\right)^3 \quad \left(x^{1/3}\right)^3=(-2)^3$

$x=\dfrac{1}{8} \qquad x=-8$

The solutions are -8 and $\dfrac{1}{8}$.

35. $\sqrt{3x-2}+4=3x$

$\sqrt{3x-2}=3x-4$

$\left(\sqrt{3x-2}\right)^2=(3x-4)^2$

$3x-2=9x^2-24x+16$

$9x^2-27x+18=0$

$9(x^2-3x+2)=0$

$(x-2)(x-1)=0$

$x-2=0 \quad x-1=0$

$x=2 \qquad x=1$

1 does not check as a solution.
The solution is 2.

36. $x^2-10x+7=0$

$x^2-10x=-7$

$x^2-10x+25=-7+25$

$(x-5)^2=18$

$\sqrt{(x-5)^2}=\sqrt{18}$

$x-5=\pm3\sqrt{2}$

$x=5\pm3\sqrt{2}$

The solutions are $5+3\sqrt{2}$ and $5-3\sqrt{2}$.

37. $\dfrac{2x}{x-4}+\dfrac{6}{x+1}=11$

$(x-4)(x+1)\left(\dfrac{2x}{x-4}+\dfrac{6}{x+1}\right)=11(x-4)(x+1)$

$2x(x+1)+6(x-4)=11(x-4)(x+1)$

$2x^2+2x+6x-24=11x^2-33x-44$

$0=9x^2-41x-20$

$0=(9x+4)(x-5)$

$9x+4=0 \quad x-5=0$

$9x=-4 \qquad x=5$

$x=-\dfrac{4}{9}$

The solutions are $-\dfrac{4}{9}$ and 5.

38. $9x^2 - 3x = 1$

$9x^2 - 3x - 1 = 0$

$a = 9, b = -3, c = -1$

$x = \dfrac{-b \pm \sqrt{b^2 - 4ac}}{2a}$

$x = \dfrac{-(-3) \pm \sqrt{(-3)^2 - 4(9)(-1)}}{2(9)}$

$x = \dfrac{3 \pm \sqrt{9+36}}{18} = \dfrac{3 \pm \sqrt{45}}{18}$

$x = \dfrac{3 \pm 3\sqrt{5}}{18}$

$x = \dfrac{1 \pm \sqrt{5}}{6}$

The solutions are $\dfrac{1+\sqrt{5}}{6}$ and $\dfrac{1-\sqrt{5}}{6}$.

39. $2x = 4 - 3\sqrt{x-1}$

$2x - 4 = 3\sqrt{x-1}$

$(2x-4)^2 = \left(3\sqrt{x-1}\right)^2$

$4x^2 - 16x + 16 = 9x - 9$

$4x^2 - 25x + 25 = 0$

$(4x-5)(x-5) = 0$

$4x - 5 = 0 \quad x - 5 = 0$

$4x = 5 \quad\quad x = 5$

$x = \dfrac{5}{4}$

$\dfrac{5}{4}$ does not check as a solution.

The solution is 5.

40. $1 - \dfrac{x+3}{3-x} = \dfrac{x-4}{x+3}$

$(3-x)(x+3)\left(1 - \dfrac{x+3}{3-x}\right) = (3-x)(x+3)\dfrac{x-4}{x+3}$

$(3-x)(x+3) - (x+3)(x+3) = (3-x)(x-4)$

$3x + 9 - x^2 - 3x - x^2 - 6x - 9 = -x^2 + 7x - 12$

$x^2 + 13x - 12 = 0$

$a = 1, b = 13, c = -12$

$x = \dfrac{-b \pm \sqrt{b^2 - 4ac}}{2a}$

$x = \dfrac{-13 \pm \sqrt{13^2 - 4(1)(-12)}}{2(1)}$

$x = \dfrac{-13 \pm \sqrt{169+48}}{2}$

$x = \dfrac{-13 \pm \sqrt{217}}{2}$

The solutions are $\dfrac{-13+\sqrt{217}}{2}$ and $-\dfrac{13-\sqrt{217}}{2}$.

41. $2x^2 - 5x = 6$

$2x^2 - 5x - 6 = 0$

$a = 2, b = -5, c = -6$

$b^2 - 4ac = (-5)^2 - 4(2)(-6) = 73$

$73 > 0$

Since the discriminant is greater than zero the equation has two unequal real number solutions.

42. $x^2 - 3x \le 10$

$x^2 - 3x - 10 \le 0$

$(x-5)(x+2) \le 0$

The zeros are −2 and 5. The factors have opposite signs between the zeros. The solution set is $\{x \mid -2 \le x \le 5\}$.

43. Strategy: Let r represent the rate of the rowing in calm water.

	Distance	Rate	Time
With current	16	$r+2$	$\frac{16}{r+2}$
Against current	16	$r-2$	$\frac{16}{r-2}$

The total time traveled was 6 h.

Solution: $\frac{16}{r+2}+\frac{16}{r-2}=6$

$(r-2)(r+2)\left(\frac{16}{r+2}+\frac{16}{r-2}\right)=6(r-2)(r+2)$

$16(r-2)+16(r+2)=6r^2-24$

$16r-32+16r+32=6r^2-24$

$0=6r^2-32r-24$

$0=2(3r^2-16r-12)$

$0=(3r+2)(r-6)$

$3r+2=0 \quad r-6=0$

$3r=-2 \qquad r=6$

$r=-\frac{2}{3}$

The rate cannot be a negative number. The rowing rate in calm water is 6 mph.

44. Strategy: Let x represent the width of the rectangle.
The length of the rectangle is $2x+2$.
The area of the rectangle is 60 cm^2.
Solution: $A=LW$

$60=x(2x+2)$

$60=2x^2+2x$

$0=2x^2+2x-60$

$0=2(x^2+x-30)$

$0=2(x+6)(x-5)$

$x+6=0 \quad x-5=0$

$x=-6 \qquad x=5$

The width cannot be a negative number.
$2x+2=2(5)+2=12$
The width is 5 cm.
The length is 12 cm.

45. Strategy: Let t represent the time it takes the new computer to print the payroll.
The time it takes the older computer to print the payroll is $t + 12$.

	Rate	Time	Part
New computer	$\frac{1}{t}$	8	$\frac{8}{t}$
Older computer	$\frac{1}{t+12}$	8	$\frac{8}{t+12}$

The sum of the parts of the task completed equals 1.

Solution: $\frac{8}{t}+\frac{8}{t+12}=1$

$t(t+12)\left(\frac{8}{t}+\frac{8}{t+12}\right)=1t(t+12)$

$8(t+12)+8t=t^2+12t$

$8t+96+8t=t^2+12t$

$t^2-4t-96=0$

$(t-12)(t+8)=0$

$t-12=0 \quad t+8=0$

$t=12 \qquad t=-8$

$t=-8$ is not possible since time cannot be a negative number. Working alone the new computer can print the payroll in 12 min.

46. Strategy: Let r represent the rate of the first car.
The rate of the second car is $r + 10$.

	Distance	Rate	Time
1st car	200	r	$\frac{200}{r}$
2nd car	200	$r+10$	$\frac{200}{r+10}$

The second car's time is one hour less than the first car's time.

Solution: $\frac{200}{r+10}=\frac{200}{r}-1$

$(r)(r+10)\left(\frac{200}{r+10}\right)=(r)(r+10)\left(\frac{200}{r}-1\right)$

$200r=200(r+10)-r(r+10)$

$200r=200r+2000-r^2-10r$

$0=r^2-10r-2000$

$0=(r-40)(r+50)$

$r-40=0 \quad r+50=0$

$r=40 \qquad r=-50$

The rate cannot be a negative number.
The rate of the first car is 40 mph.
The rate of the second car is 50 mph.

Chapter 10 Test

1. $3x^2+10x=8$

$3x^2+10x-8=0$

$(3x-2)(x+4)=0$

$(3x-2)(x+4)=0$

$3x-2=0 \quad x+4=0$

$3x=2 \qquad x=-4$

$x=\frac{2}{3}$

The solutions are -4 and $\frac{2}{3}$.

2. $6x^2-5x-6=0$

$(2x-3)(3x+2)=0$

$2x-3=0 \quad 3x+2=0$

$2x=3 \qquad 3x=-2$

$x=\frac{3}{2} \qquad x=-\frac{2}{3}$

The solutions are $\frac{3}{2}$ and $-\frac{2}{3}$.

3. $(x-r_1)(x-r_2)=0$

$(x-3)(x-(-3))=0$

$(x-3)(x+3)=0$

$x^2-9=0$

4. $(x-r_1)(x-r_2)=0$

$\left(x-\frac{1}{2}\right)(x-(-4))=0$

$\left(x-\frac{1}{2}\right)(x+4)=0$

$x^2+\frac{7}{2}x-2=0$

$2\left(x^2+\frac{7}{2}x-2\right)=0(2)$

$2x^2+7x-4=0$

5. $3(x-2)^2-24=0$

$3(x-2)^2=24$

$(x-2)^2=8$

$\sqrt{(x-2)^2}=\sqrt{8}$

$x-2=\pm 2\sqrt{2}$

$x=2\pm 2\sqrt{2}$

The solutions are $2+2\sqrt{2}$ and $2-2\sqrt{2}$.

6. $x^2-6x-2=0$

$x^2-6x=2$

$x^2-6x+9=2+9$

$(x-3)^2=11$

$\sqrt{(x-3)^2}=\sqrt{11}$

$x-3=\pm\sqrt{11}$

$x=3\pm\sqrt{11}$

The solutions are $3+\sqrt{11}$ and $3-\sqrt{11}$.

7. $3x^2-6x=2$

$\frac{1}{3}\left(3x^2-6x\right)=2\left(\frac{1}{3}\right)$

$x^2-2x=\frac{2}{3}$

$x^2-2x+1=\frac{2}{3}+1$

$(x-1)^2=\frac{5}{3}$

$\sqrt{(x-1)^2}=\sqrt{\frac{5}{3}}$

$x-1=\pm\frac{\sqrt{15}}{3}$

$x=1\pm\frac{\sqrt{15}}{3}$

$x=\frac{3\pm\sqrt{15}}{3}$

The solutions are $\frac{3+\sqrt{15}}{3}$ and $\frac{3-\sqrt{15}}{3}$.

8. $2x^2-2x=1$

$2x^2-2x-1=0$

$a=2, b=-2, c=-1$

$x=\frac{-b\pm\sqrt{b^2-4ac}}{2a}$

$x=\frac{-(-2)\pm\sqrt{(-2)^2-4(2)(-1)}}{2(2)}$

$x=\frac{2\pm\sqrt{4+8}}{4}=\frac{2\pm\sqrt{12}}{4}$

$x=\frac{2\pm 2\sqrt{3}}{4}$

$x=\frac{1\pm\sqrt{3}}{2}$

The solutions are $\frac{1+\sqrt{3}}{2}$ and $\frac{1-\sqrt{3}}{2}$.

9. $x^2+4x+12=0$

$a=1, b=4, c=12$

$$x=\frac{-b\pm\sqrt{b^2-4ac}}{2a}$$

$$x=\frac{-4\pm\sqrt{4^2-4(1)(12)}}{2(1)}$$

$$x=\frac{-4\pm\sqrt{16-48}}{2}=\frac{-4\pm\sqrt{-32}}{2}$$

$$x=\frac{-4\pm 4i\sqrt{2}}{2}$$

$$x=-2\pm 2i\sqrt{2}$$

The solutions are $-2+2i\sqrt{2}$ and $-2-2i\sqrt{2}$.

10. $2x+7x^{1/2}-4=0$

$2(x^{1/2})^2+7x^{1/2}-4=0$

$2u^2+7u-4=0$

$(2u-1)(u+4)=0$

$2u-1=0 \quad u+4=0$

$2u=1 \qquad u=-4$

$u=\frac{1}{2}$

Replace u with $x^{1/2}$.

$x^{1/2}=\frac{1}{2} \qquad x^{1/2}=-4$

$(x^{1/2})^2=\left(\frac{1}{2}\right)^2 \quad (x^{1/2})^2=(-4)^2$

$x=\frac{1}{4} \qquad x=16$

16 does not check as a solution.

The solution is $\frac{1}{4}$.

11. $x^4-4x^2+3=0$

$(x^2)^2-4x^2+3=0$

$u^2-4u+3=0$

$(u-1)(u-3)=0$

$u-1=0 \quad u-3=0$

$u=1 \qquad u=3$

Replace u with x^2.

$x^2=1 \qquad x^2=3$

$\sqrt{x^2}=\sqrt{1} \quad \sqrt{x^2}=\sqrt{3}$

$x=\pm 1 \qquad x=\pm\sqrt{3}$

The solutions are -1, 1, $\sqrt{3}$ and $-\sqrt{3}$.

12. $\sqrt{2x+1}+5=2x$

$\sqrt{2x+1}=2x-5$

$\left(\sqrt{2x+1}\right)^2=(2x-5)^2$

$2x+1=4x^2-20x+25$

$4x^2-22x+24=0$

$2(2x^2-11x+12)=0$

$2(2x-3)(x-4)=0$

$2x-3=0 \quad x-4=0$

$2x=3 \qquad x=4$

$x=\frac{3}{2}$

$\frac{3}{2}$ does not check as a solution.

The solution is 4.

13. $\sqrt{x-2}=\sqrt{x}-2$

$(\sqrt{x-2})^2=\left(\sqrt{x}-2\right)^2$

$x-2=x-4\sqrt{x}+4$

$-6=-4\sqrt{x}$

$3=2\sqrt{x}$

$3^2=\left(2\sqrt{x}\right)^2$

$9=4x$

$x=\frac{9}{4}$

$\frac{9}{4}$ does not check as a solution.

There is no solution.

14. $\frac{2x}{x-3}+\frac{5}{x-1}=1$

$(x-3)(x-1)\left(\frac{2x}{x-3}+\frac{5}{x-1}\right)=1(x-3)(x-1)$

$2x(x-1)+5(x-3)=1(x-3)(x-1)$

$2x^2-2x+5x-15=x^2-4x+3$

$x^2+7x-18=0$

$(x+9)(x-2)=0$

$x+9=0 \quad x-2=0$

$x=-9 \quad\quad x=2$

The solutions are −9 and 2.

15. $(x-2)(x+4)(x-4)<0$

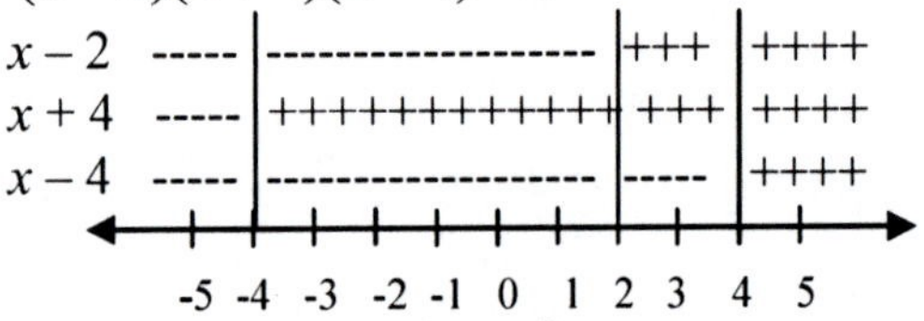

$\{x \mid x<-4 \text{ or } 2<x<4\}$

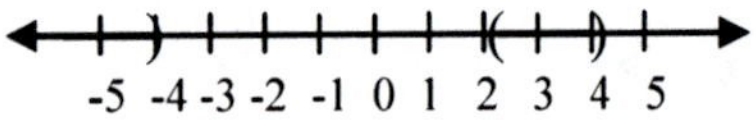

16. $\frac{2x-3}{x+4}\le 0$

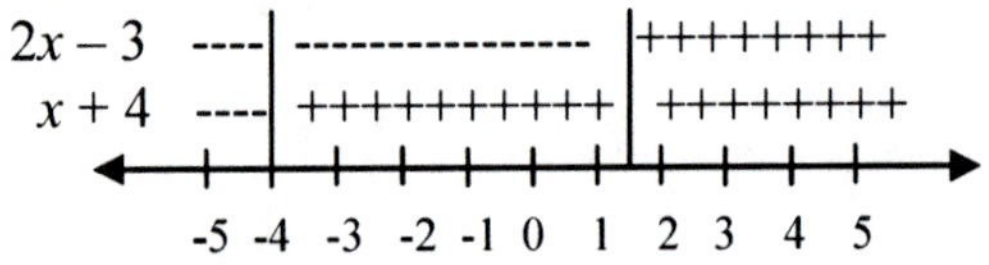

$\left\{x \mid -4<x\le\frac{3}{2}\right\}$

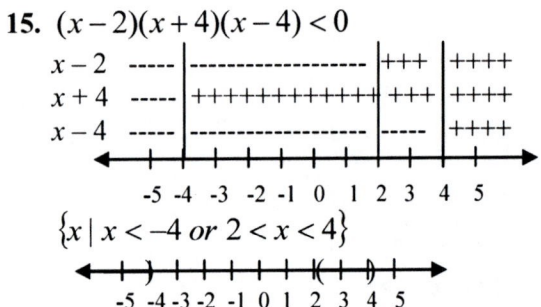

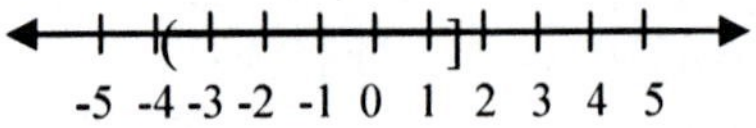

17. $9x^2+24x=-16$

$9x^2+24x+16=0$

$a=9, b=24, c=16$

$b^2-4ac=(24)^2-4(9)(16)=0$

Since the discriminant is equal to zero the equation has two equal real number solutions.

18. Strategy: To find the time when the ball hits the basket substitute 10 ft for h in the equation and solve for t.

Solution: $h=-16t^2+32t+6.5$

$10=-16t^2+32t+6.5$

$0=-16t^2+32t-3.5$

$0=16t^2-32t+3.5$

$a=16, b=-32, c=3.5$

$t=\frac{-b\pm\sqrt{b^2-4ac}}{2a}$

$t=\frac{-(-32)\pm\sqrt{(-32)^2-4(16)(3.5)}}{2(16)}$

$t=\frac{32\pm\sqrt{800}}{32}$

$t=1.88$

$t=0.12$

We need to find the time it takes to reach the basket after the ball has reached its peak. This occurs 1.88 s after the ball has been released.

19. Strategy: Let t represent the time it takes Cora to stain a bookcase.

The time it takes Clive to stain a bookcase is $t+6$.

	Rate	Time	Part
Cora	$\frac{1}{t}$	4	$\frac{4}{t}$
Clive	$\frac{1}{t+6}$	4	$\frac{4}{t+6}$

The sum of the parts of the task completed equals 1.

Solution: $\frac{4}{t}+\frac{4}{t+6}=1$

$t(t+6)\left(\frac{4}{t}+\frac{4}{t+6}\right)=1t(t+6)$

$4(t+6)+4t=t^2+6t$

$4t+24+4t=t^2+6t$

$t^2 - 2t - 24 = 0$

$(t-6)(t+4) = 0$

$t-6=0 \quad t+4=0$

$t=6 \quad\quad t=-4$

$t = -4$ is not possible since time cannot be a negative number. Working alone it will take Cora 6 h to stain the bookcase.

20. Strategy: Let r represent the rate of the canoe in calm water.

	Distance	Rate	Time
With current	6	$r+2$	$\frac{6}{r+2}$
Against current	6	$r-2$	$\frac{6}{r-2}$

The total time traveled was 4 h.

Solution: $\dfrac{6}{r+2} + \dfrac{6}{r-2} = 4$

$(r-2)(r+2)\left(\dfrac{6}{r+2} + \dfrac{6}{r-2}\right) = 4(r-2)(r+2)$

$6(r-2) + 6(r+2) = 4r^2 - 16$

$6r - 12 + 6r + 12 = 4r^2 - 16$

$0 = 4r^2 - 12r - 16$

$0 = 4(r^2 - 3r - 4)$

$0 = 4(r-4)(r+1)$

$r-4=0 \quad r+1=0$

$r=4 \quad\quad r=-1$

The rate cannot be a negative number. The rate of the canoe in calm water is 4 mph.

Cumulative Review Exercises

1. $2a^2 - b^2 \div c^2$

$2(3)^2 - (-4)^2 \div (-2)^2 = 2(9) - 16 \div 4$

$= 18 - 16 \div 4 = 18 - 4$

$= 14$

2. $V = \pi r^2 h = \pi(3)^2(6) = \pi(9)(6) = 54\pi$
The volume is 54π m^3.

3. (3, −4) and (−1, 2)

$m = \dfrac{y_2 - y_1}{x_2 - x_1} = \dfrac{2-(-4)}{-1-3} = \dfrac{2+4}{-4} = \dfrac{6}{-4}$

$m = -\dfrac{3}{2}$

4. $x - y = 1$

$y = x - 1$

$m = 1$ and (1, 2)

$y - y_1 = m(x - x_1)$

$y - 2 = 1(x-1)$

$y - 2 = x - 1$

$y = x + 1$

5. $-3x^3y + 6x^2y^2 - 9xy^3 = -3xy(x^2 - 2xy + 3y^2)$

6. $6x^2 - 7x - 20 = (2x-5)(3x+4)$

7. $x^2 + xy - 2x - 2y$
$= x(x+y) - 2(x+y)$
$= (x+y)(x-2)$

8.

$$\begin{array}{r} x^2 - 3x - 4 \\ 3x-4 \overline{\smash{)}\,3x^3 - 13x^2 + 0x + 10} \\ \underline{3x^3 - 4x^2} \qquad\qquad \\ -9x^2 + 0x \qquad \\ \underline{-9x^2 + 12x} \qquad \\ -12x + 10 \\ \underline{-12x + 16} \\ -6 \end{array}$$

$(3x^3 - 13x^2 + 10) \div (3x-4) = x^2 - 3x - 4 + \dfrac{-6}{3x-4}$

9. $\dfrac{x^2+2x+1}{8x^2+8x}\cdot\dfrac{4x^3-4x^2}{x^2-1}$

$=\dfrac{(x+1)(x+1)}{8x(x+1)}\cdot\dfrac{4x^2(x-1)}{(x+1)(x-1)}$

$=\dfrac{(x+1)(x+1)\cdot 4x^2(x-1)}{8x(x+1)\cdot(x+1)(x-1)}$

$=\dfrac{x}{2}$

10. $\dfrac{BC}{EF}=\dfrac{\textit{height of triangle ABC}}{\textit{height of triangle DEF}}$

$\dfrac{12}{24}=\dfrac{8}{h}$

$12h=24(8)$

$12h=192$

$h=16$

The height of triangle *DEF* is 16 cm.

11. $S=\dfrac{n}{2}(a+b)$

$2S=n(a+b)$

$2S=na+nb$

$2S-na=nb$

$\dfrac{2S-na}{n}=b$

12. $-2i(7-4i)=-14i+8i^2=-8-14i$

13. $a^{-1/2}(a^{1/2}-a^{3/2})=a^0-a^1$
$=1-a$

14. $\dfrac{\sqrt[3]{8x^4y^5}}{\sqrt[3]{16xy^6}}=\sqrt[3]{\dfrac{8x^4y^5}{16xy^6}}=\sqrt[3]{\dfrac{x^3}{2y}}$

$=\dfrac{\sqrt[3]{x^3}}{\sqrt[3]{2y}}$

$=\dfrac{x}{\sqrt[3]{2y}}\cdot\dfrac{\sqrt[3]{4y^2}}{\sqrt[3]{4y^2}}=\dfrac{x\sqrt[3]{4y^2}}{\sqrt[3]{8y^3}}$

$=\dfrac{x\sqrt[3]{4y^2}}{2y}$

15. $\dfrac{x}{x+2}-\dfrac{4x}{x+3}=1$

$(x+2)(x+3)\left(\dfrac{x}{x+2}-\dfrac{4x}{x+3}\right)=1(x+2)(x+3)$

$x(x+3)-4x(x+2)=1(x+2)(x+3)$

$x^2+3x-4x^2-8x=x^2+5x+6$

$0=4x^2+10x+6$

$0=2(2x^2+5x+3)$

$(2x+3)(x+1)=0$

$2x+3=0 \quad x+1=0$

$2x=-3 \quad x=-1$

$x=-\dfrac{3}{2}$

The solutions are -1 and $-\dfrac{3}{2}$.

16. $\dfrac{x}{2x+3}-\dfrac{3}{4x^2-9}=\dfrac{x}{2x-3}$

$(2x+3)(2x-3)\left(\dfrac{x}{2x+3}-\dfrac{3}{(2x+3)(2x-3)}\right)=(2x+3)(2x-3)\dfrac{x}{2x-3}$

$x(2x-3)-3=x(2x+3)$

$2x^2-3x-3=2x^2+3x$

$-3=6x$

$x=-\dfrac{1}{2}$

The solution is $-\dfrac{1}{2}$.

17. $x^4-6x^2+8=0$

$\left(x^2\right)^2-6x^2+8=0$

$u^2-6u+8=0$

$(u-4)(u-2)=0$

$u-4=0 \quad u-2=0$

$u=4 \qquad u=2$

Replace u with x^2.

$x^2=4 \qquad x^2=2$

$\sqrt{x^2}=\sqrt{4} \quad \sqrt{x^2}=\sqrt{2}$

$x=\pm 2 \qquad x=\pm\sqrt{2}$

The solutions are -2, 2, $\sqrt{2}$ and $-\sqrt{2}$.

18. $\sqrt{3x+1}-1=x$

$\sqrt{3x+1}=x+1$

$\left(\sqrt{3x+1}\right)^2=(x+1)^2$

$3x+1=x^2+2x+1$

$0=x^2-x$

$0=x(x-1)$

$x=0 \qquad x-1=0$

$x=1$

Both 0 and 1 check as solutions. The solutions are 0 and 1.

19. $|3x-2|<8$

$-8<3x-2<8$

$-8+2<3x-2+2<8+2$

$-6<3x<10$

$-2<x<\dfrac{10}{3}$

$\left\{x\,|\,-2<x<\dfrac{10}{3}\right\}$

20. $6x-5y=15$

$6x-5(0)=15$

$6x=15$

$x=\dfrac{15}{6}=\dfrac{5}{2}$

The x-intercept is $\left(\dfrac{5}{2},0\right)$.

$6(0)-5y=15$

$-5y=15$

$y=-3$

The y-intercept is $(0, -3)$.

21. Solve each inequality for y.

$x + y \le 3 \quad 2x - y < 4$

$y \le 3 - x \quad -y < 4 - 2x$

$y > 2x - 4$

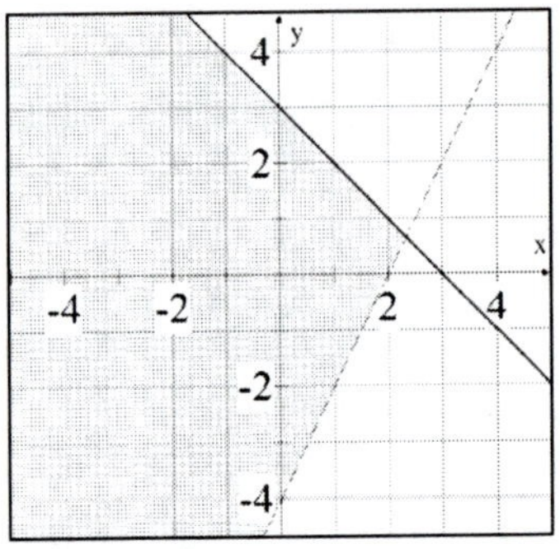

22. (1) $x + y + z = 2$
(2) $-x + 2y - 3z = -9$
(3) $x - 2y - 2z = -1$
Eliminate x and y. Add equations (2) and (3).

$$\begin{array}{r} -x + 2y - 3z = -9 \\ \underline{x - 2y - 2z = -1} \\ -5z = -10 \\ z = 2 \end{array}$$

Eliminate y and z. Multiply equation (1) by 2 and add to equation (3).

$2(x + y + z = 2)$

$2x + 2y + 2z = 4$

$$\begin{array}{r} 2x + 2y + 2z = 4 \\ \underline{x - 2y - 2z = -1} \\ 3x \qquad\quad = 3 \\ x = 1 \end{array}$$

Replace x with 1 and z with 2 in equation (1)

$x + y + z = 2$

$1 + y + 2 = 2$

$y + 3 = 2$

$y = -1$

The solution is $(1, -1, 2)$.

23. $f(x) = \dfrac{2x - 3}{x^2 - 1}$

$$f(-2) = \frac{2(-2) - 3}{(-2)^2 - 1} = \frac{-4 - 3}{4 - 1} = -\frac{7}{3}$$

24. $f(x) = \dfrac{x - 2}{x^2 - 2x - 15}$

$f(x) = \dfrac{x - 2}{(x - 5)(x + 3)}$

$x - 5 = 0 \quad x + 3 = 0$

$x = 5 \quad x = -3$

$\{x \mid x \ne -3, 5\}$

25. $x^3 + x^2 - 6x < 0$

$x(x^2 + x - 6) < 0$

$x(x + 3)(x - 2) < 0$

x	-------	--------	+++	++++++++
$x + 3$	-------	+++++	+++	++++++++
$x - 2$	-------	--------	------	++++++++

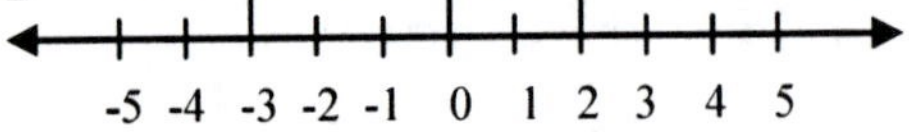

$\{x \mid x < -3\} \cup \{x \mid 0 < x < 2\}$

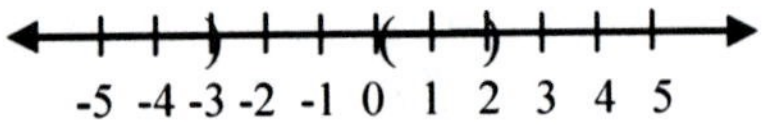

26. $\dfrac{(x - 1)(x - 5)}{x + 3} \ge 0$

$x - 1$	---------	------------	+++++++	++
$x - 5$	---------	------------	-----------	++
$x + 3$	---------	+++++++	+++++++	++

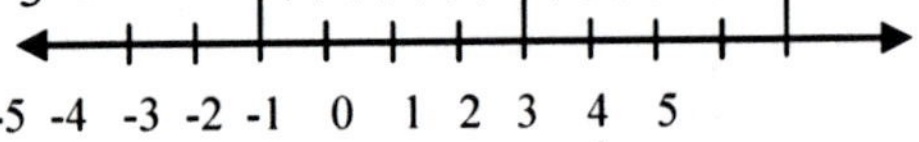

$\{x \mid -3 < x \le 1\} \cup \{x \mid x \ge 5\}$

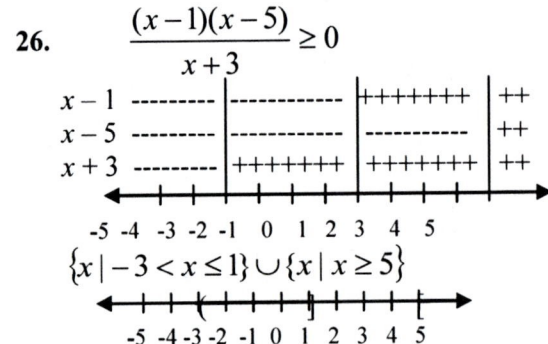

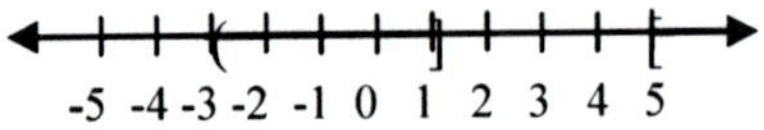

27. Strategy: Let P represent the length of the piston rod, T the tolerance and m the given length. Solve the absolute value inequality $|m-p| \le T$ for m.

Solution: $|m-p| \le T$

$$\left|m-9\frac{3}{8}\right| \le \frac{1}{64}$$

$$-\frac{1}{64} \le m-9\frac{3}{8} \le \frac{1}{64}$$

$$-\frac{1}{64}+9\frac{3}{8} \le m-9\frac{3}{8}+9\frac{3}{8} \le \frac{1}{64}+9\frac{3}{8}$$

$$9\frac{23}{64} \le m \le 9\frac{25}{64}$$

The lower limit is $9\frac{23}{64}$ in.

The upper limit is $9\frac{25}{64}$ in.

28. Strategy: The base of the triangle is $x + 8$. The height of the triangle is $2x - 4$.

Solution: $A = \frac{1}{2}bh$

$$A = \frac{1}{2}(x+8)(2x-4)$$

$$A = \frac{1}{2}(2x^2+12x-32)$$

$$A = x^2+6x-16 \ ft^2$$

The area is $(x^2 + 6x - 16)$ ft^2.

29. $2x^2+4x+3=0$

$a = 2, b = 4, c = 3$

$$b^2-4ac = 4^2-4(2)(3) = 16-24 = -8$$

$$-8 < 0$$

Since the discriminant is less than zero, the equation has two complex number solutions.

30. (0, 250,000) and (30, 0)

$$m = \frac{V_2-V_1}{t_2-t_1} = \frac{0-250{,}000}{30-0} = \frac{-25000}{3}$$

$$m = -\frac{25000}{3}$$

The building depreciates $\frac{\$25{,}000}{3}$, or about $8333 each year.

Chapter 11: Functions and Relations

Prep Test

1. $-\dfrac{b}{2a}$

$-\dfrac{(-4)}{2(2)} = -\dfrac{-4}{4} = -(-1) = 1$

2. $y = -x^2 + 2x + 1$
$y = -(-2)^2 + 2(-2) + 1 = -4 - 4 + 1 = -7$

3. $f(x) = x^2 - 3x + 2$
$f(-4) = (-4)^2 - 3(-4) + 2$
$f(-4) = 16 + 12 + 2$
$f(-4) = 30$

4. $p(r) = r^2 - 5$
$p(2 + h) = (2 + h)^2 - 5$
$= 4 + 4h + h^2 - 5$
$= h^2 + 4h - 1$

5. $0 = 3x^2 - 7x - 6$
$0 = (3x + 2)(x - 3)$
$0 = 3x + 2 \quad 0 = x - 3$
$-2 = 3x \quad 3 = x$
$-\dfrac{2}{3} = x$

The solutions are $-\dfrac{2}{3}$ and 3.

6. $0 = x^2 - 4x + 1$
$a = 1 \quad b = -4 \quad c = 1$

$x = \dfrac{-b \pm \sqrt{b^2 - 4ac}}{2a}$

$x = \dfrac{-(-4) \pm \sqrt{(-4)^2 - 4(1)(1)}}{2(1)}$

$x = \dfrac{4 \pm \sqrt{16 - 4}}{2} = \dfrac{4 \pm \sqrt{12}}{2} = \dfrac{4 \pm 2\sqrt{3}}{2}$

$x = \dfrac{4}{2} \pm \dfrac{2\sqrt{3}}{2} = 2 \pm \sqrt{3}$

The solutions are $2 + \sqrt{3}$ and $2 - \sqrt{3}$.

7. $x = 2y + 4$
$2y + 4 = x$
$2y = x - 4$
$\left(\dfrac{1}{2}\right)2y = \left(\dfrac{1}{2}\right)(x - 4)$
$y = \dfrac{1}{2}x - 2$

8. Domain: $\{-2, 3, 4, 6\}$
Range: $\{4, 5, 6\}$
Yes the relation is a function.

9. $\{x|x \neq 8\}$

10.

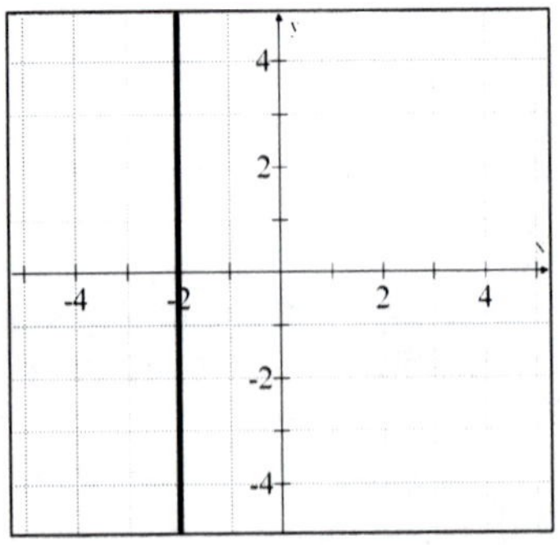

Section 11.1

Concept Check

1. (i) and (iii)

3. The axis of symmetry of the graph of a parabola is the vertical line that passes through the vertex of the parabola and is parallel to the y–axis.

5. Zero, one or two

7. $(-1,0)$ and $(3,0)$

9. When $a > 0$, the parabola opens up and the vertex of the parabola is the lowest point on the parabola, with the smallest y-coordinate. This point is called the minimum value of the function.
When $a < 0$, the parabola opens down and the vertex of the parabola is the highest point on the parabola, with the largest y-coordinate. This point is called the maximum value of the function.

Objective A Exercises

11. $y = x^2 - 2x - 4$

$$-\frac{b}{2a} = -\frac{-2}{2(1)} = -\frac{-2}{2} = -(-1) = 1$$

$$y = (1)^2 - 2(1) - 4 = -5$$

Vertex: $(1, -5)$
Axis of symmetry: $x = 1$

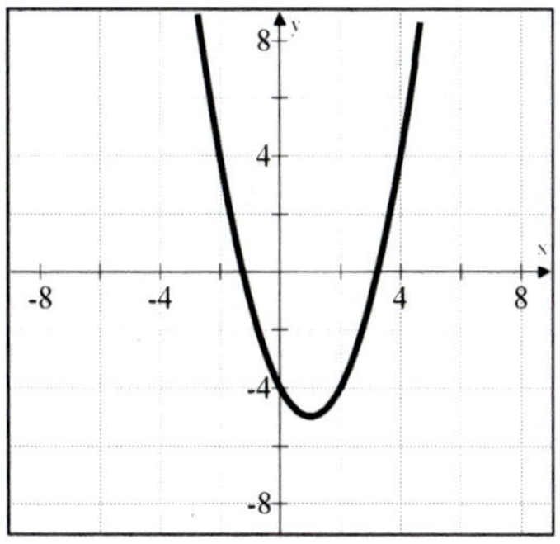

13. $y = -x^2 + 2x - 3$

$$-\frac{b}{2a} = -\frac{2}{2(-1)} = -\frac{2}{-2} = -(-1) = 1$$

$$y = -(1)^2 + 2(1) - 3 = -2$$

Vertex: $(1, -2)$
Axis of symmetry: $x = 1$

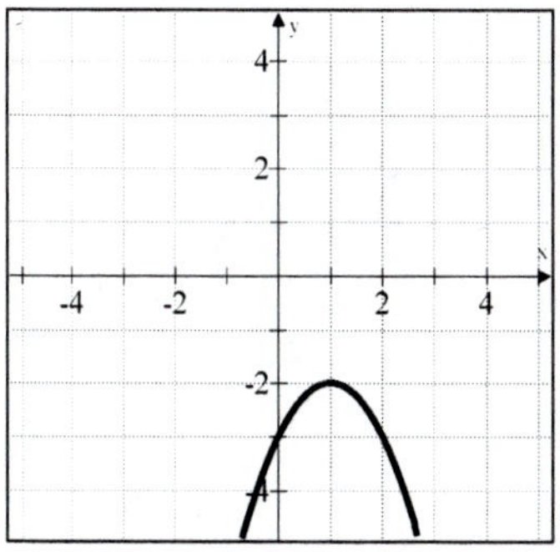

15. $f(x) = x^2 - x - 6$

$$-\frac{b}{2a} = -\frac{(-1)}{2(1)} = -\frac{-1}{2} = \frac{1}{2}$$

$$y = \left(\frac{1}{2}\right)^2 - \frac{1}{2} - 6 = -\frac{25}{4}$$

Vertex: $\left(\frac{1}{2}, -\frac{25}{4}\right)$

Axis of symmetry: $x = \frac{1}{2}$

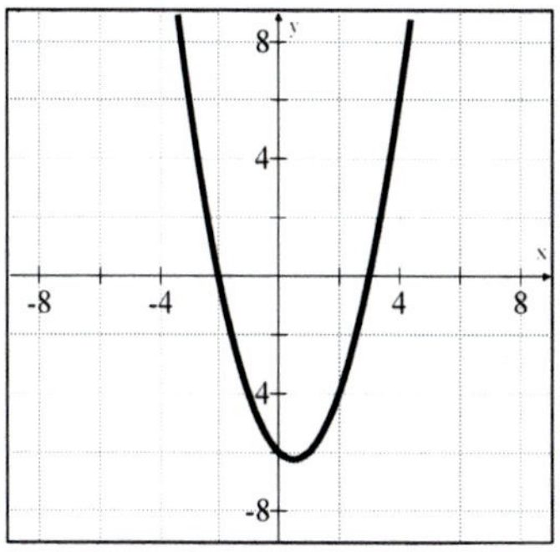

17. $F(x) = x^2 - 3x + 2$

$$-\frac{b}{2a} = -\frac{(-3)}{2(1)} = -\frac{-3}{2} = \frac{3}{2}$$

$$y = \left(\frac{3}{2}\right)^2 - 3\left(\frac{3}{2}\right) + 2 = -\frac{1}{4}$$

Vertex: $\left(\frac{3}{2}, -\frac{1}{4}\right)$

Axis of symmetry: $x = \frac{3}{2}$

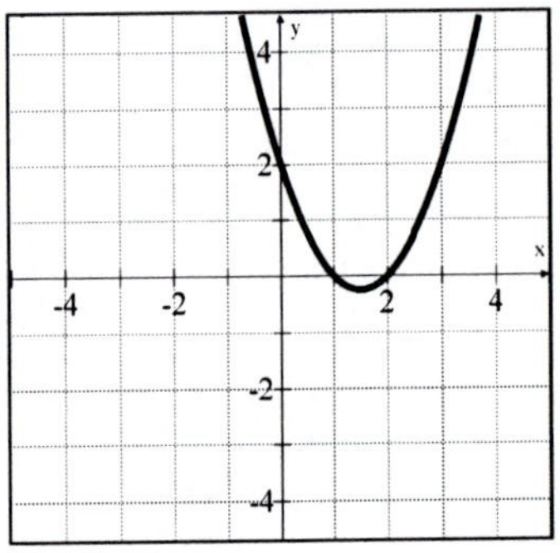

19. $y = -2x^2 + 6x$

$$-\frac{b}{2a} = -\frac{6}{2(-2)} = -\frac{6}{-4} = \frac{3}{2}$$

$$y = -2\left(\frac{3}{2}\right)^2 + 6\left(\frac{3}{2}\right) = \frac{9}{2}$$

Vertex: $\left(\frac{3}{2}, \frac{9}{2}\right)$

Axis of symmetry: $x = \frac{3}{2}$

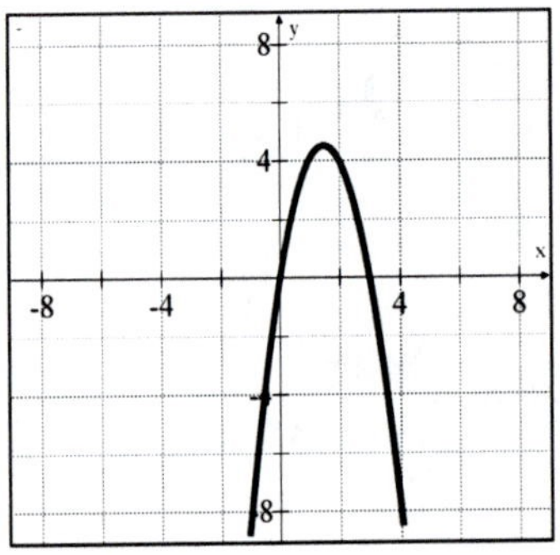

21. $y = -\frac{1}{4}x^2 - 1$

$$-\frac{b}{2a} = -\frac{(0)}{2\left(-\frac{1}{4}\right)} = -\frac{0}{-\frac{1}{2}} = 0$$

$$y = -\left(\frac{1}{4}\right)0^2 - 1 = -1$$

Vertex: $(0, -1)$

Axis of symmetry: $x = 0$

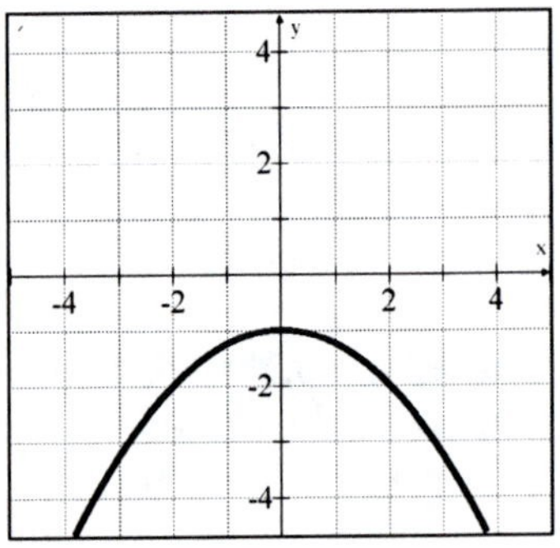

23. $P(x) = -\frac{1}{2}x^2 + 2x - 3$

$$-\frac{b}{2a} = -\frac{2}{2\left(-\frac{1}{2}\right)} = -\frac{2}{-1} = -(-2) = 2$$

$$y = \left(-\frac{1}{2}\right)2^2 + 2(2) - 3 = -1$$

Vertex: $(2, -1)$

Axis of symmetry: $x = 2$

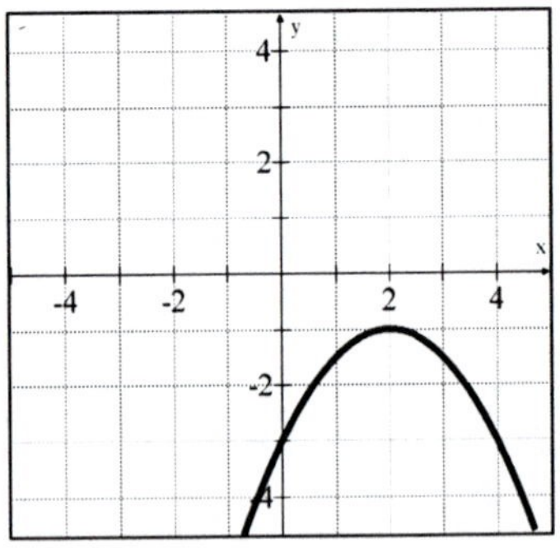

25. $y = -\frac{1}{2}x^2 + x - 3$

$-\frac{b}{2a} = -\frac{1}{2\left(-\frac{1}{2}\right)} = -\frac{1}{-1} = -(-1) = 1$

$y = \left(-\frac{1}{2}\right)1^2 + 1 - 3 = -\frac{5}{2}$

Vertex: $\left(1, -\frac{5}{2}\right)$

Axis of symmetry: $x = 1$

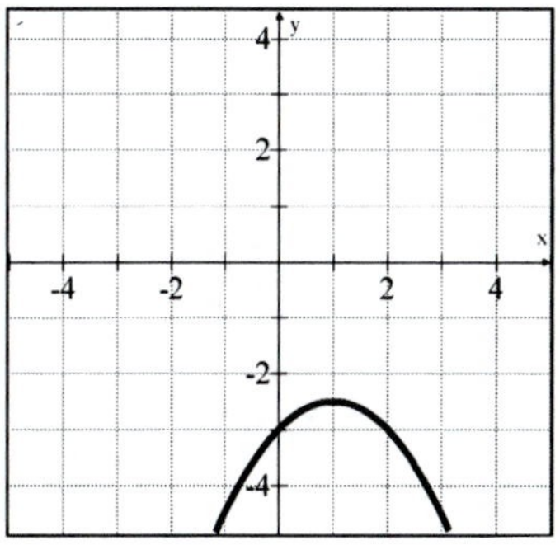

27. Domain: $\{x \mid -\infty < x < \infty\}$
Range: $\{y \mid y \geq 2\}$

29. Domain: $\{x \mid -\infty < x < \infty\}$
Range: $\{y \mid y \geq -5\}$

31. Domain: $\{x \mid -\infty < x < \infty\}$
Range: $\{y \mid y \leq 0\}$

33. Domain: $\{x \mid -\infty < x < \infty\}$
Range: $\{y \mid y \geq -7\}$

Objective B Exercises

35. $y = x^2 - 9$

$0 = x^2 - 9$

$0 = (x+3)(x-3)$

$x + 3 = 0 \qquad x - 3 = 0$

$x = -3 \qquad x = 3$

The x-intercepts are $(-3, 0)$ and $(3, 0)$.

37. $y = 3x^2 + 6x$

$0 = 3x^2 + 6x$

$0 = 3x(x+2)$

$3x = 0 \qquad x + 2 = 0$

$x = 0 \qquad x = -2$

The x-intercepts are $(0, 0)$ and $(-2, 0)$.

39. $y = x^2 - 2x - 8$

$0 = x^2 - 2x - 8$

$0 = (x-4)(x+2)$

$x - 4 = 0 \qquad x + 2 = 0$

$x = 4 \qquad x = -2$

The x-intercepts are $(4, 0)$ and $(-2, 0)$.

41. $y = 2x^2 - 5x - 3$

$0 = 2x^2 - 5x - 3$

$0 = (x-3)(2x+1)$

$x - 3 = 0 \qquad 2x + 1 = 0$

$x = 3 \qquad 2x = -1$

$x = -\frac{1}{2}$

The x-intercepts are $(3, 0)$ and $\left(-\frac{1}{2}, 0\right)$.

43. $y = x^2 + 4x - 3$

$0 = x^2 + 4x - 3$

$a = 1 \quad b = 4 \quad c = -3$

$x = \frac{-b \pm \sqrt{b^2 - 4ac}}{2a}$

$x = \frac{-4 \pm \sqrt{(4)^2 - 4(1)(-3)}}{2(1)}$

$x = \frac{-4 \pm \sqrt{16+12}}{2} = \frac{-4 \pm \sqrt{28}}{2} = \frac{-4 \pm 2\sqrt{7}}{2}$

$x = -2 \pm \sqrt{7}$

The x-intercepts are $(-2 + \sqrt{7}, 0)$ and $(-2 - \sqrt{7}, 0)$.

45. $y = -x^2 - 4x - 5$

$0 = -x^2 - 4x - 5$

$a = -1 \quad b = -4 \quad c = -5$

$$x = \frac{-b \pm \sqrt{b^2 - 4ac}}{2a}$$

$$x = \frac{-(-4) \pm \sqrt{(-4)^2 - 4(-1)(-5)}}{2(-1)}$$

$$x = \frac{4 \pm \sqrt{16 - 20}}{-2} = \frac{4 \pm \sqrt{-4}}{-2} = \frac{4 \pm 2i}{-2}$$

$x = -2 \pm i$

There are no real solutions. The parabola has no x-intercepts.

47. $f(x) = x^2 - 4x - 5$

$x^2 - 4x - 5 = 0$

$(x - 5)(x + 1) = 0$

$x - 5 = 0 \qquad x + 1 = 0$

$x = 5 \qquad x = -1$

The zeros are 5 and -1.

49. $f(x) = 3x^2 - 2x - 8$

$3x^2 - 2x - 8 = 0$

$(3x + 4)(x - 2) = 0$

$3x + 4 = 0 \qquad x - 2 = 0$

$3x = -4 \qquad x = 2$

$x = -\frac{4}{3}$

The zeros are $-\frac{4}{3}$ and 2.

51. $h(x) = 4x^2 - 4x + 1$

$4x^2 - 4x + 1 = 0$

$(2x - 1)(2x - 1) = 0$

$2x - 1 = 0 \qquad 2x - 1 = 0$

$2x = 1 \qquad 2x = 1$

$x = \frac{1}{2} \qquad x = \frac{1}{2}$

The zero is $\frac{1}{2}$.

53. $f(x) = -3x^2 + 4x$

$-3x^2 + 4x = 0$

$x(-3x + 4) = 0$

$-3x + 4 = 0 \qquad x = 0$

$-3x = -4$

$x = \frac{4}{3}$

The zeros are $\frac{4}{3}$ and 0.

55. $f(x) = -3x^2 + 12$

$-3x^2 + 12 = 0$

$-3(x^2 - 4) = 0$

$-3(x + 2)(x - 2) = 0$

$x + 2 = 0 \qquad x - 2 = 0$

$x = -2 \qquad x = 2$

The zeros are -2 and 2.

57. $f(x) = 2x^2 - 54$

$2x^2 - 54 = 0$

$2(x^2 - 27) = 0$

$x^2 - 27 = 0$

$x^2 = 27$

$x = \pm\sqrt{27} = \pm 3\sqrt{3}$

The zeros are $3\sqrt{3}$ and $-3\sqrt{3}$.

59. $f(x) = x^2 - 2x - 17$

$x^2 - 2x - 17 = 0$

$a = 1 \quad b = -2 \quad c = -17$

$$x = \frac{-b \pm \sqrt{b^2 - 4ac}}{2a}$$

$$x = \frac{-(-2) \pm \sqrt{(-2)^2 - 4(1)(-17)}}{2(1)}$$

$$x = \frac{2 \pm \sqrt{4 + 68}}{2} = \frac{2 \pm \sqrt{72}}{2} = \frac{2 \pm 6\sqrt{2}}{2}$$

$x = 1 \pm 3\sqrt{2}$

The zeros are $1 + 3\sqrt{2}$ and $1 - 3\sqrt{2}$.

61. $f(x) = x^2 + 4x + 5$

$x^2 + 4x + 5 = 0$

$a = 1 \quad b = 4 \quad c = 5$

$$x = \frac{-b \pm \sqrt{b^2 - 4ac}}{2a}$$

$$x = \frac{-4 \pm \sqrt{(4)^2 - 4(1)(5)}}{2(1)}$$

$$x = \frac{-4 \pm \sqrt{16 - 20}}{2} = \frac{-4 \pm \sqrt{-4}}{2} = \frac{-4 \pm 2i}{2}$$

$x = -2 \pm i$

The zeros are $-2 + i$ and $-2 - i$.

63. $f(x) = x^2 + 4x + 13$

$x^2 + 4x + 13 = 0$

$a = 1 \quad b = 4 \quad c = 13$

$$x = \frac{-b \pm \sqrt{b^2 - 4ac}}{2a}$$

$$x = \frac{-4 \pm \sqrt{(4)^2 - 4(1)(13)}}{2(1)}$$

$$x = \frac{-4 \pm \sqrt{16 - 52}}{2} = \frac{-4 \pm \sqrt{-36}}{2} = \frac{-4 \pm 6i}{2}$$

$x = -2 \pm 3i$

The zeros are $-2 + 3i$ and $-2 - 3i$.

65. $y = -x^2 - x + 3$

$a = -1 \quad b = -1 \quad c = 3$

$b^2 - 4ac$

$(-1)^2 - 4(-1)(3) = 1 + 12 = 13$

$13 > 0$

Since the discriminant is greater than zero the parabola has two x-intercepts.

67. $y = x^2 - 10x + 25$

$a = 1 \quad b = -10 \quad c = 25$

$b^2 - 4ac$

$(-10)^2 - 4(1)(25) = 100 - 100 = 0$

Since the discriminant is equal to zero the parabola has one x-intercept.

69. $y = -2x^2 + x - 1$

$a = -2 \quad b = 1 \quad c = -1$

$b^2 - 4ac$

$(1)^2 - 4(-2)(-1) = 1 - 8 = -7$

$-7 < 0$

Since the discriminant is less than zero the parabola has no x-intercepts.

71. $y = 4x^2 - x - 2$

$a = 4 \quad b = -1 \quad c = -2$

$b^2 - 4ac$

$(-1)^2 - 4(4)(-2) = 1 + 32 = 33$

$33 > 0$

Since the discriminant is greater than zero the parabola has two x-intercepts.

73. $y = 2x^2 + x + 4$

$a = 2 \quad b = 1 \quad c = 4$

$b^2 - 4ac$

$(1)^2 - 4(2)(4) = 1 - 32 = -31$

$-31 < 0$

Since the discriminant is less than zero the parabola has no x-intercepts.

75. $y = 4x^2 + 2x - 5$

$a = 4 \quad b = 2 \quad c = -5$

$b^2 - 4ac$

$(2)^2 - 4(4)(-5) = 4 + 80 = 84$

$84 > 0$

Since the discriminant is greater than zero the parabola has two x-intercepts.

77. a) $a > 0$
b) $a = 0$
c) $a < 0$

Objective C Exercises

79. $f(x) = 2x^2 + 4x$

$x = -\frac{b}{2a} = -\frac{4}{2(2)} = -1$

$f(x) = 2x^2 + 4x$

$f(-1) = 2(-1)^2 + 4(-1) = 2 - 4 = -2$

Since $a > 0$, the function has a minimum value. The minimum value of the function is −2.

81. $f(x) = -2x^2 + 4x - 5$

$x = -\frac{b}{2a} = -\frac{4}{2(-2)} = 1$

$f(x) = -2x^2 + 4x - 5$

$f(1) = -2(1)^2 + 4(1) - 5 = -2 + 4 - 5 = -3$

Since $a < 0$, the function has a maximum value. The maximum value of the function is −3.

83. $f(x) = -2x^2 - 3x$

$x = -\frac{b}{2a} = -\frac{-3}{2(-2)} = -\frac{3}{4}$

$f(x) = -2x^2 - 3x$

$f\left(-\frac{3}{4}\right) = -2\left(-\frac{3}{4}\right)^2 - 3\left(-\frac{3}{4}\right) = -\frac{9}{8} + \frac{9}{4}$

$= \frac{9}{8}$

Since $a < 0$, the function has a maximum value. The maximum value of the function is $\frac{9}{8}$.

85. $f(x) = 3x^2 + 3x - 2$

$x = -\frac{b}{2a} = -\frac{3}{2(3)} = -\frac{1}{2}$

$f(x) = 3x^2 + 3x - 2$

$f\left(-\frac{1}{2}\right) = 3\left(-\frac{1}{2}\right)^2 + 3\left(-\frac{1}{2}\right) - 2 = \frac{3}{4} - \frac{3}{2} - 2$

$= -\frac{11}{4}$

Since $a > 0$, the function has a minimum value. The minimum value of the function is $-\frac{11}{4}$.

87. $f(x) = -x^2 - x + 2$

$x = -\frac{b}{2a} = -\frac{-1}{2(-1)} = -\frac{1}{2}$

$f(x) = -x^2 - x + 2$

$f\left(-\frac{1}{2}\right) = -\left(-\frac{1}{2}\right)^2 - \left(-\frac{1}{2}\right) + 2 = -\frac{1}{4} + \frac{1}{2} + 2$

$= \frac{9}{4}$

Since $a < 0$, the function has a maximum value. The maximum value of the function is $\frac{9}{4}$.

89. $f(x) = 3x^2 + 5x + 2$

$x = -\frac{b}{2a} = -\frac{5}{2(3)} = -\frac{5}{6}$

$f(x) = 3x^2 + 5x + 2$

$f\left(-\frac{5}{6}\right) = 3\left(-\frac{5}{6}\right)^2 + 5\left(-\frac{5}{6}\right) + 2 = \frac{25}{12} - \frac{25}{6} + 2$

$= -\frac{1}{12}$

Since $a > 0$, the function has a minimum

value. The minimum value of the function is $-\frac{1}{12}$.

Objective D Exercises

91. **Strategy**: To find the price that will give the maximum revenue find the *P*-coordinate of the vertex.

Solution:

$$P=-\frac{b}{2a}=-\frac{125}{2\left(-\frac{1}{4}\right)}=250$$

A price of \$250 will give the maximum revenue.

93. **Strategy**: Let x represent one number. The other number is $20-x$.
Their product is $x(20-x)$.
To find one number find the x-coordinate of the vertex. To find the second number evaluate $20-x$ at the x-coordinate of the vertex.

Solution:

$$x(20-x)=20x-x^2$$

$$x=-\frac{b}{2a}=-\frac{20}{2(-1)}=10$$

$$20-x=20-10=10$$

The two numbers are 10 and 10.

95. **Strategy**: To find the time it takes the plane to reach its maximum height find the t-coordinate of the vertex. To find the maximum height evaluate the function at the t-coordinate of the vertex.

Solution:

$$t=-\frac{b}{2a}=-\frac{119}{2(-1.42)}\approx 42$$

$$h(t)=-1.42t^2+119t+6000$$

$$h(42)=-1.42(42)^2+119(42)+6000$$

$$=-2504.88+4998+6000=8493.12$$

The vertex of the parabola is approximately (42, 8500), so the maximum height of the plane is about 8500 m.

97. To find the time at which the stream reaches its maximum height, find the t-coordinate of the vertex. To find the maximum height, evaluate the function at the t-coordinate of the vertex.

Solution:

$$t=-\frac{b}{2a}=-\frac{30}{2(-16)}=0.9375$$

$$h(t)=-16t^2+30t$$

$$h(0.9375)=-16(0.9375)^2+30(0.9375)$$

$$=-14.0625+28.125=14.0625$$

The stream is at its maximum height in 0.9375 s.
The maximum height is 14.0625 ft.

99. **Strategy:** Let x represent the number of units. Find the maximum value of x in order to find the maximum revenue.
$R(x)=(1200+100x)(100-x)$
$R(x)=-100x^2+8800x+120000$

Solution:

$$x=-\frac{b}{2a}=-\frac{8800}{2(-100)}=44$$

$$R(x)=-100x^2+8800x+120000$$

$$R(44)=-100(44)^2+8800(44)+120000$$

$$=-193600+387200+120000$$

$$=313600$$

The maximum revenue of \$313,600 is reached when 44 units are rented.

101. **Strategy**: Let x represent width of the rectangular corral. The length is $200-2x$.
The area is $x(200-2x)$.
To find the width find the x-coordinate of the vertex. To find the length evaluate $200-2x$ at the x-coordinate of the vertex.

Solution:

$$x(200-2x)=200x-2x^2$$

$$x=-\frac{b}{2a}=-\frac{200}{2(-2)}=50$$

$$200-2x=200-2(50)=100$$

The width is 50 ft and the length is 100 ft.

103. **Strategy**: Let x represent width of the ball fields. The length is $\frac{2100-3x}{2}$.

The area is $x\left(\frac{2100-3x}{2}\right)$.

To find the width find the x-coordinate of the vertex. To find the length evaluate $\frac{2100-3x}{2}$ at the x-coordinate of the vertex.

Solution:

$$x\left(\frac{2100-3x}{2}\right) = x(1050-1.5x) = 1050x-1.5x^2$$

$$x = -\frac{b}{2a} = -\frac{1050}{2(-1.5)} = 350$$

$$\frac{2100-3x}{2} = \frac{2100-1050}{2} = \frac{1050}{2} = 525$$

The dimensions are 350 ft by 525 ft.

Critical Thinking

105. To find the root substitute $x = -2, y = 0$ and $x = 3, y = 0$ into $f(x) = mx^2 + nx + 1$. Use these two equations to find the relationship between m and n.

Solution:

$$0 = m(-2)^2 + n(-2) + 1 \quad 0 = m(3)^2 + n(3) + 1$$

$$0 = 4m - 2n + 1 \quad 0 = 9m + 3n + 1$$

$$4m - 2n + 1 = 9m + 3n + 1$$

$$4m - 2n = 9m + 3n$$

$$-5n = 5m$$

$$n = -m$$

n and m are opposites.

Substituting $-n$ for m and $-m$ for n we get

$$f(x) = mx^2 + nx + 1 = (-n)x^2 + (-m)x + 1$$

$$= -nx^2 - mx + 1$$

$$= nx^2 + mx - 1 = g(x)$$

Therefore since $f(x) = g(x)$ their roots are the same. $g(x)$ has roots -2 and 3.

Projects or Group Activities

107. **Strategy**: Let h be the height of the window, r is the radius of the semicircle, and $2r$ is the width of the window. Consider the perimeter and area of the rectangle added to the semicircle.

Solution:

$$P = (2h + 2r) + \pi r$$

$$50 = 2h + 2r + \pi r$$

$$h = \frac{50 - 2r - \pi r}{2} = 25 - r - \frac{1}{2}\pi r$$

$$A = (2rh) + \left(\frac{1}{2}\pi r^2\right)$$

$$A = 2r\left(25 - r - \frac{1}{2}\pi r\right) + \frac{1}{2}\pi r^2$$

$$= 50r - 2r^2 - \pi r^2 + \frac{1}{2}\pi r^2$$

$$= 50r - 3.57r^2$$

Find the maximum value for r.

$$r = -\frac{b}{2a} = -\frac{50}{2(-3.57)} \approx 7$$

$$h = 25 - r - \frac{1}{2}\pi r = 25 - 7 - \frac{1}{2}\pi(7) = 7$$

$$h \approx 7$$

Both r and h are approximately 7.

Section 11.2

Concept Check

1. Left

3. Up

5. $(-3, 5)$

Objective A Exercises

7.

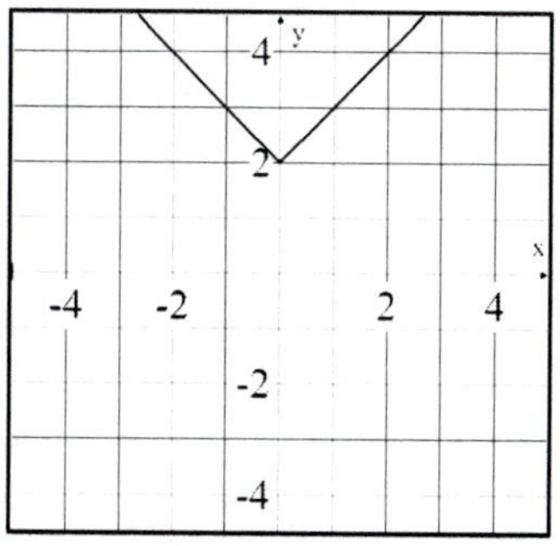

9.

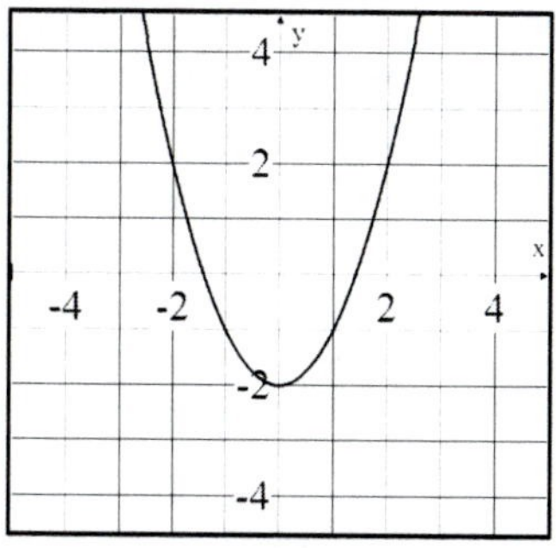

11.

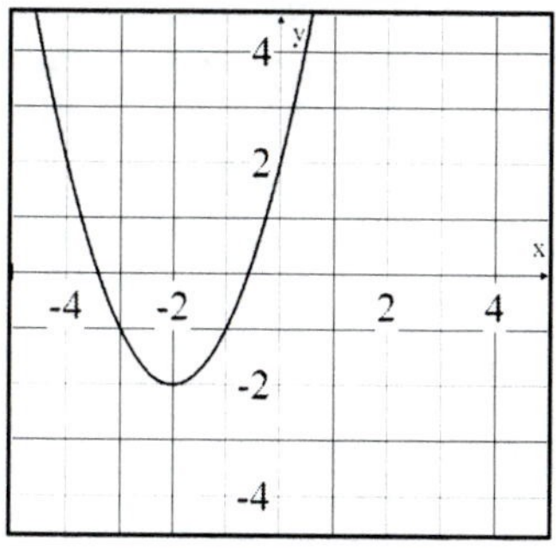

13.

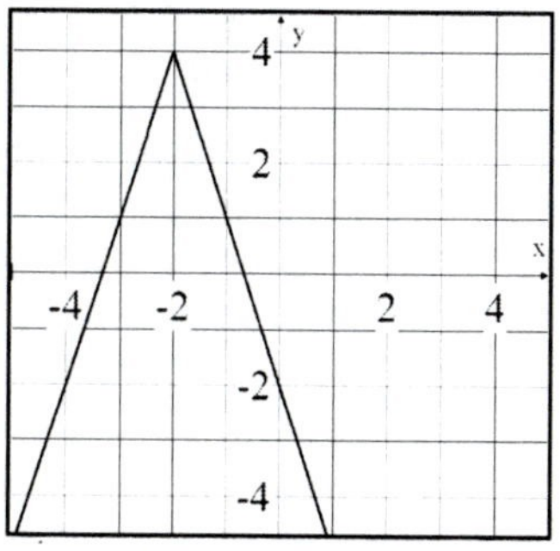

15.

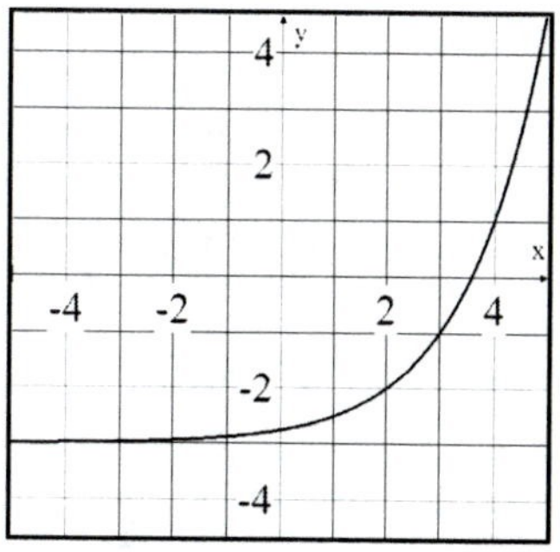

17.

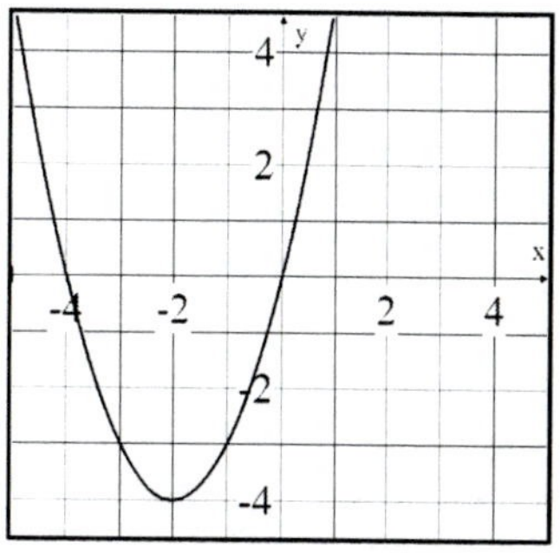

19.

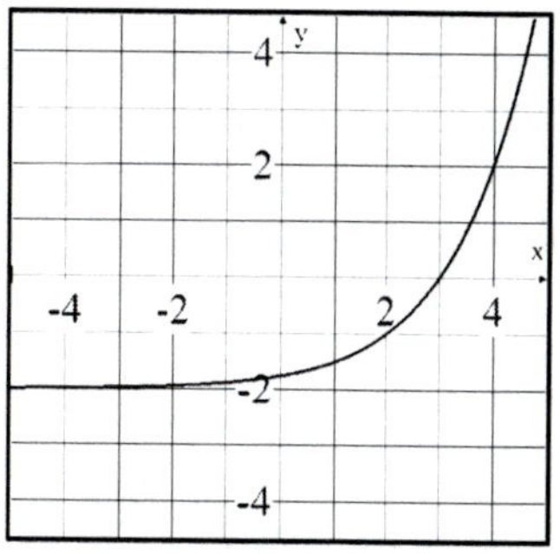

21.

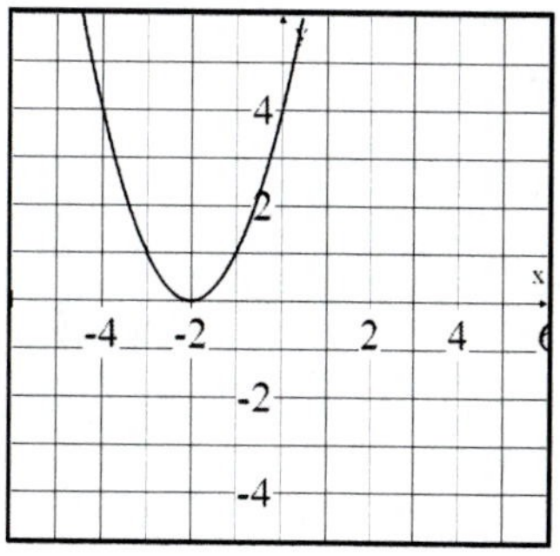

23. (0, 9)

Objective B Exercises

25.

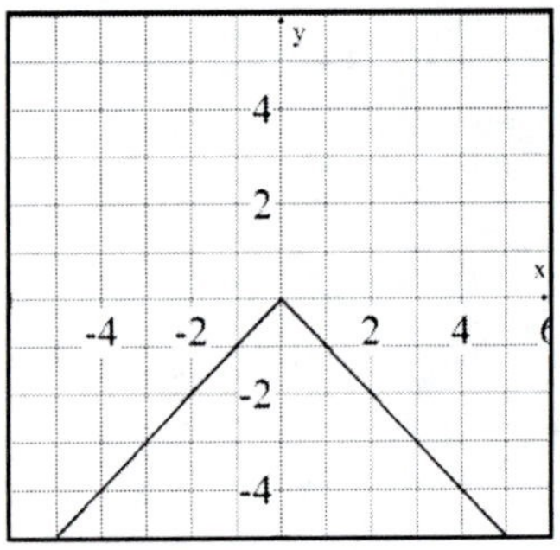

27.

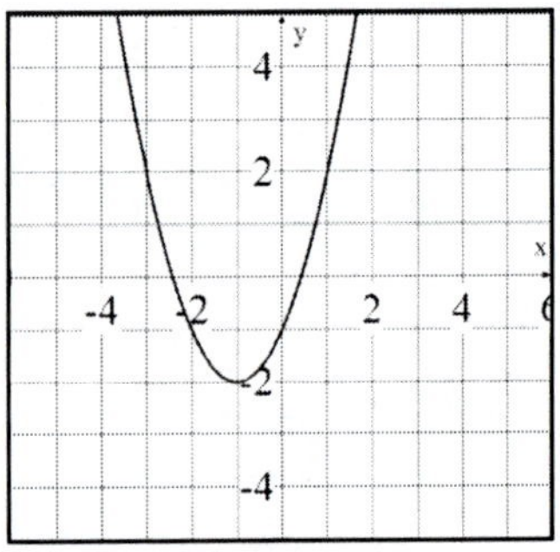

29.

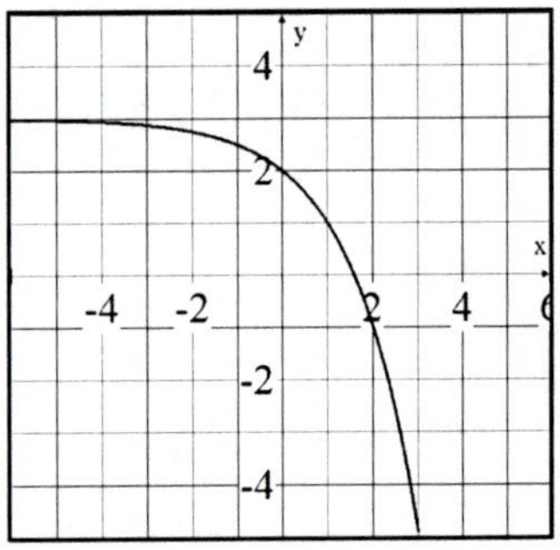

31.

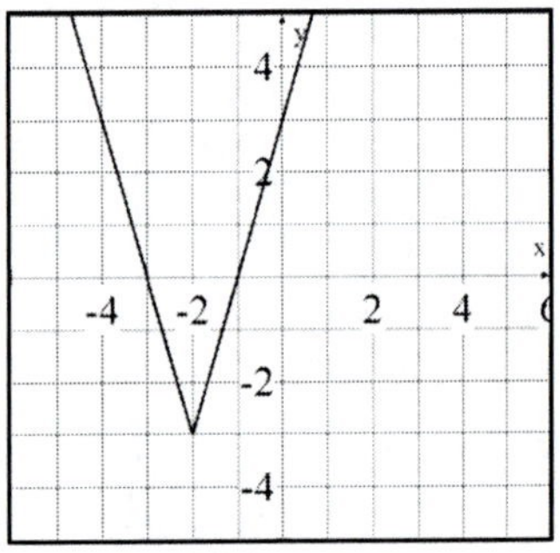

Critical Thinking

33.

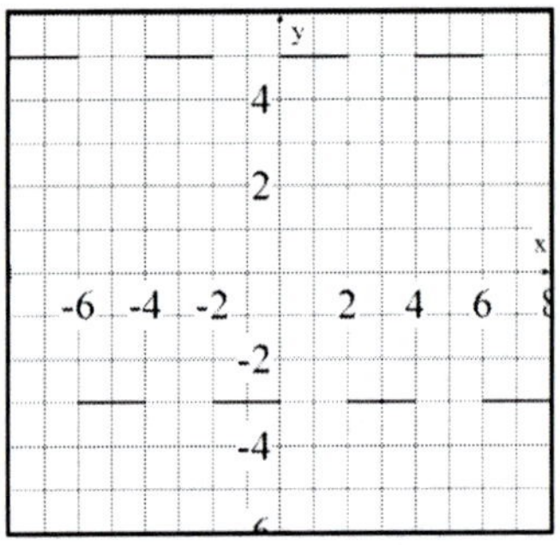

35.

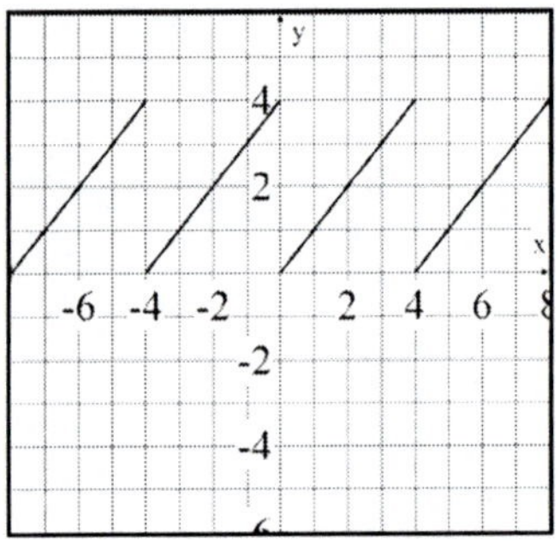

37.

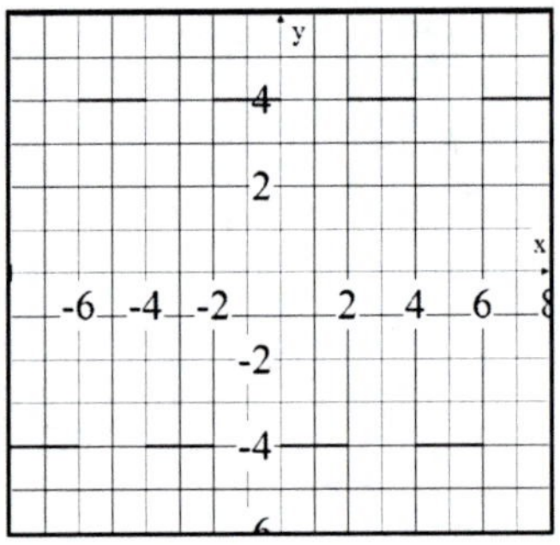

Projects or Group Activities

39.

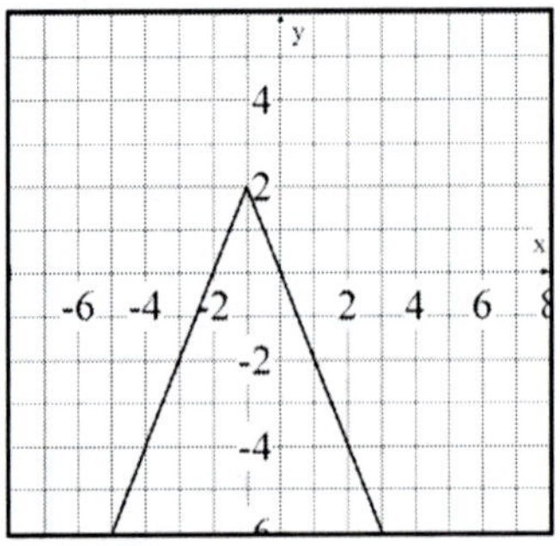

41.

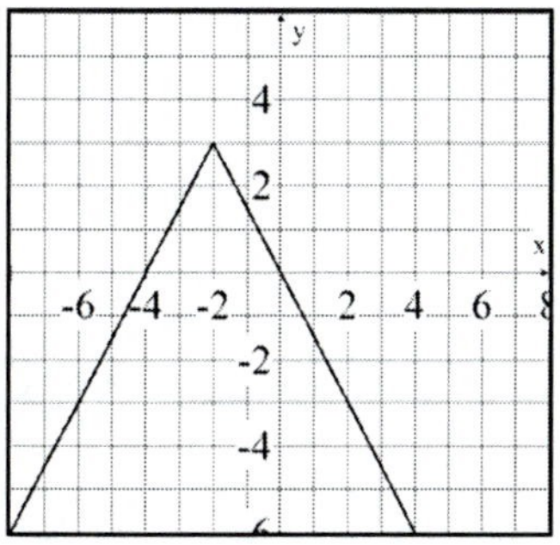

Check Your Progress: Chapter 11

1. $y = x^2 + 6x + 3$

$x = -\frac{b}{2a} = -\frac{6}{2(1)} = -3$

$y = (-3)^2 + 6(-3) + 3 = -6$

Vertex: $(-3,-6)$

Axis of symmetry: $x = -3$

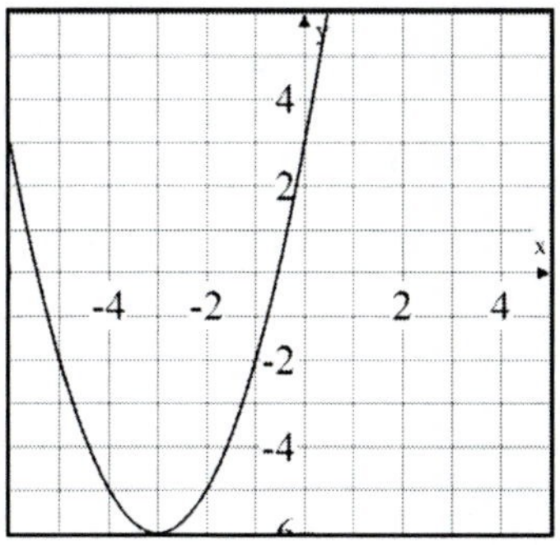

2. $f(x) = -\frac{1}{2}x^2 + 2x + 3$

$x = -\frac{b}{2a} = -\frac{2}{2\left(-\frac{1}{2}\right)} = -\frac{2}{-1} = 2$

$y = \left(-\frac{1}{2}\right)(2)^2 + 2(2) + 3 = 5$

Vertex: $(2,5)$

Axis of symmetry: $x = 2$

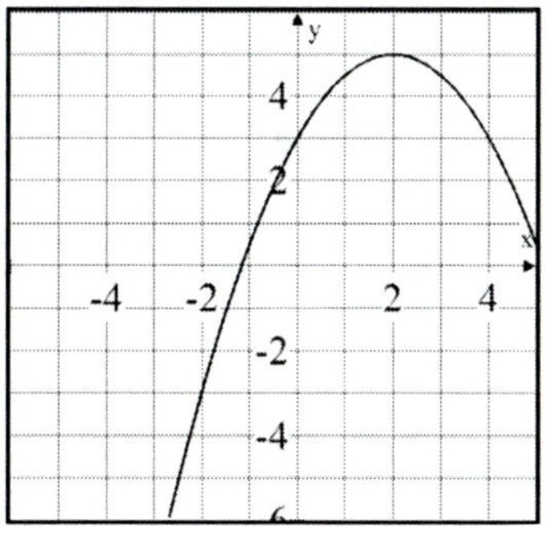

3. $y = 2x^2 + 5x - 3$

$0 = 2x^2 + 5x - 3$

$0 = (x+3)(2x-1)$

$x + 3 = 0 \qquad 2x - 1 = 0$

$x = -3 \qquad 2x = 1$

$x = \frac{1}{2}$

The x-intercepts are (–3, 0) and $\left(\frac{1}{2},0\right)$.

4. $y = x^2 + 2x - 1$

$0 = x^2 + 2x - 1$

$a = 1 \quad b = 2 \quad c = -1$

$x = \frac{-b \pm \sqrt{b^2 - 4ac}}{2a}$

$x = \frac{-2 \pm \sqrt{(2)^2 - 4(1)(-1)}}{2(1)}$

$x = \frac{-2 \pm \sqrt{4+4}}{2} = \frac{-2 \pm \sqrt{8}}{2} = \frac{-2 \pm 2\sqrt{2}}{2}$

$x = -1 \pm \sqrt{2}$

The x-intercepts are $(-1+\sqrt{2},0)$ and $\left(-1-\sqrt{2},0\right)$.

5. $f(x) = x^2 + x - 30$

$x^2 + x - 30 = 0$

$(x+6)(x-5) = 0$

$x + 6 = 0 \quad x - 5 = 0$

$x = -6 \qquad x = 5$

The zeros are –6 and 5.

6. $f(x) = x^2 - 4x + 40$

$x^2 - 4x + 40 = 0$

$a = 1 \quad b = -4 \quad c = 40$

$$x = \frac{-b \pm \sqrt{b^2 - 4ac}}{2a}$$

$$x = \frac{-(-4) \pm \sqrt{(-4)^2 - 4(1)(40)}}{2(1)}$$

$$x = \frac{4 \pm \sqrt{16 - 160}}{2} = \frac{4 \pm \sqrt{-144}}{2} = \frac{4 \pm 12i}{2}$$

$x = 2 \pm 6i$

The zeros are $2 + 6i$ and $2 - 6i$.

7. $f(x) = x^2 - 4x - 32$

$$x = -\frac{b}{2a} = -\frac{-4}{2(1)} = 2$$

$f(x) = x^2 - 4x - 32$

$f(2) = (2)^2 - 4(2) - 32 = 4 - 8 - 32 = -36$

Since $a > 0$, the function has a minimum value. The minimum value of the function is -36.

8. $f(x) = x^2 - 4x + 8$

$$x = -\frac{b}{2a} = -\frac{-4}{2(1)} = 2$$

$f(x) = x^2 - 4x + 8$

$f(2) = (2)^2 - 4(2) + 8 = 4 - 8 + 8 = 4$

Since $a < 0$, the function has a maximum value. The maximum value of the function is 4.

9.

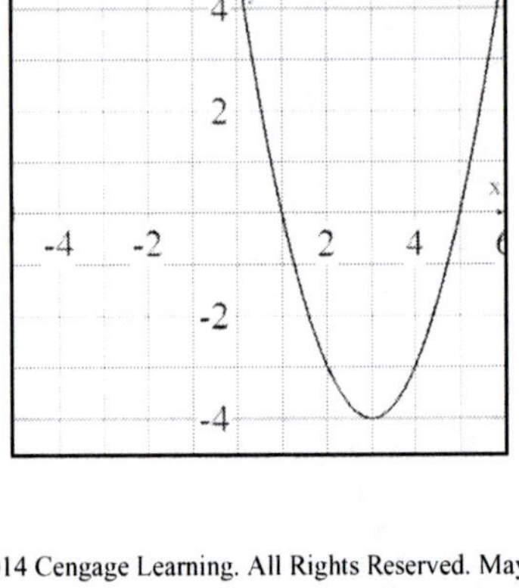

10.

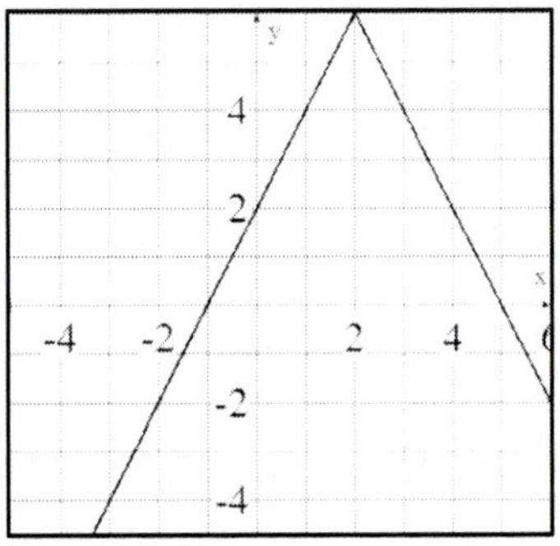

11.

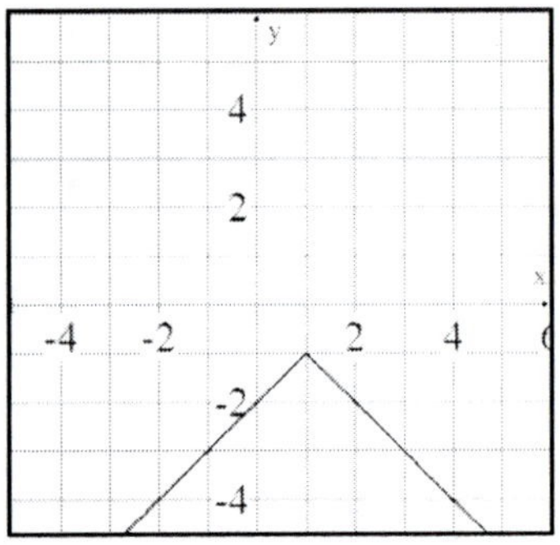

12.

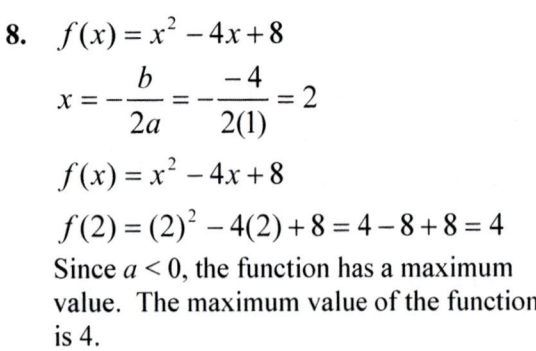

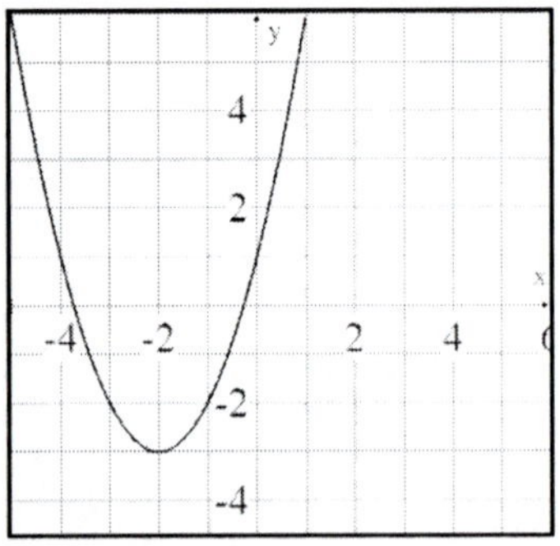

13. **Strategy**: Find the time at which the rock reaches its maximum height by finding the t-coordinate of the vertex. To find the maximum height, evaluate the function at the t-coordinate of the vertex.

Solution:

$t = -\frac{b}{2a} = -\frac{64}{2(-16)} = 2$

$s(t) = -16t^2 + 64t + 76$

$s(2) = -16(2)^2 + 64(2) + 76$

$= -64 + 128 + 76 = 140$

The maximum height of the rock is 140 ft.

Section 11.3

Concept Check

1. $f(x) = x^2 + 4 \qquad g(x) = \sqrt{x+4}$

a) Yes
b) Yes
c) Yes
d) No

3. a) No
b) Yes

Objective A Exercises

5. $f(2) - g(2) = (2\cdot 2^2 - 3) - (-2\cdot 2 + 4)$
$= (2\cdot 4 - 3) - (-4 + 4)$
$= (8 - 3) - (0)$
$= 5$

7. $f(0) + g(0) = (2\cdot 0^2 - 3) + (-2\cdot 0 + 4)$
$= (0 - 3) + (0 + 4)$
$= -3 + 4$
$= 1$

9. $(f\cdot g)(2) = f(2)\cdot g(2)$
$= (2\cdot 2^2 - 3)\cdot(-2\cdot 2 + 4)$
$= (2\cdot 4 - 3)\cdot(-4 + 4)$
$= (8 - 3)\cdot(0)$
$= 0$

11. $\left(\frac{f}{g}\right)(4) = \frac{f(4)}{g(4)}$
$= \frac{2\cdot(4)^2 - 3}{-2\cdot(4) + 4}$
$= \frac{2\cdot 16 - 3}{-8 + 4}$
$= \frac{29}{-4} = -\frac{29}{4}$

13. $f(1) + g(1) = (2\cdot 1^2 + 3\cdot 1 - 1) + (2\cdot 1 - 4)$
$= (2\cdot 1 + 3 - 1) + (2\cdot 1 - 4)$
$= (2 + 3 - 1) + (2 - 4)$
$= 4 - 2$
$= 2$

15. $f(4) - g(4)$
$= (2\cdot(4)^2 + 3\cdot(4) - 1) - (2\cdot(4) - 4)$
$= (2\cdot 16 + 12 - 1) - (2\cdot(4) - 4)$
$= (32 + 12 - 1) - (8 - 4)$
$= 43 - 4$
$= 39$

17. $(f\cdot g)(1) = f(1)\cdot g(1)$
$= (2\cdot(1)^2 + 3\cdot(1) - 1)\cdot(2\cdot(1) - 4)$
$= (2\cdot 1 + 3 - 1)\cdot(2\cdot(1) - 4)$
$= (2 + 3 - 1)\cdot(2 - 4)$
$= 4\cdot(-2)$
$= -8$

19. $\left(\frac{f}{g}\right)(2) = \frac{f(2)}{g(2)}$
$= \frac{2\cdot(2)^2 + 3\cdot(2) - 1}{2\cdot(2) - 4}$
$= \frac{2\cdot 4 + 6 - 1}{4 - 4}$
$= \frac{13}{0}$
Undefined

21. $f(2)-g(2)$
$=(2^2+3\cdot(2)-5)-(2^3-2\cdot(2)+3)$
$=(4+6-5)-(8-4+3)$
$=5-7$
$=-2$

23. $\left(\dfrac{f}{g}\right)(-2)=\dfrac{f(-2)}{g(-2)}$
$=\dfrac{(-2)^2+3\cdot(-2)-5}{(-2)^3-2\cdot(-2)+3}$
$=\dfrac{4-6-5}{-8+4+3}$
$=\dfrac{-7}{-1}$
$=7$

Objective B Exercises

25. $f(x)=2x-3 \qquad g(x)=4x-1$
$g(2x-3)=4(2x-3)-1$
$=8x-12-1$
$=8x-13$
$g[f(x)]=8x-13$

27. $f(x)=2x-3 \qquad g(x)=4x-1$
$f(-2)=2(-2)-3$
$=-4-3=-7$
$g(-7)=4(-7)-1$
$=-28-1=-29$
$g[f(-2)]=-29$

29. $f(x)=2x-3 \qquad g(x)=4x-1$
$g(2x-3)=4(2x-3)-1$
$=8x-12-1$
$=8x-13$
$g[f(x)]=8x-13$

31. $h(x)=2x+4 \qquad f(x)=\dfrac{1}{2}x+2$
$h(0)=2(0)+4$
$=0+4=4$
$f(4)=\dfrac{1}{2}(4)+2$
$=2+2=4$
$f[h(0)]=4$

33. $h(x)=2x+4 \qquad f(x)=\dfrac{1}{2}x+2$
$h(-1)=2(-1)+4$
$=-2+4=2$
$f(2)=\dfrac{1}{2}(2)+2$
$=1+2=3$
$f[h(-1)]=3$

35. $h(x)=2x+4 \qquad f(x)=\dfrac{1}{2}x+2$
$f(2x+4)=\dfrac{1}{2}(2x+4)+2$
$=x+2+2$
$=x+4$
$f[h(x)]=x+4$

37. $g(x)=x^2+3 \qquad h(x)=x-2$
$g(0)=0^2+3=3$
$h(3)=3-2=1$
$h[g(0)]=1$

39. $g(x)=x^2+3 \qquad h(x)=x-2$
$g(-2)=(-2)^2+3=4+3=7$
$h(7)=7-2=5$
$h[g(-2)]=5$

41. $g(x) = x^2 + 3 \qquad h(x) = x - 2$

$h(x^2 + 3) = x^2 + 3 - 2$

$= x^2 + 1$

$h[g(x)] = x^2 + 1$

43. $f(x) = x^2 + x + 1 \qquad h(x) = 3x + 2$

$f(0) = (0)^2 + 0 + 1 = 1$

$h(1) = 3(1) + 2 = 3 + 2 = 5$

$h[f(0)] = 5$

45. $f(x) = x^2 + x + 1 \qquad h(x) = 3x + 2$

$f(-2) = (-2)^2 + (-2) + 1 = 4 - 2 + 1 = 3$

$h(3) = 3(3) + 2 = 9 + 2 = 11$

$h[f(-2)] = 11$

47. $f(x) = x^2 + x + 1 \qquad h(x) = 3x + 2$

$h(x^2 + x + 1) = 3(x^2 + x + 1) + 2$

$= 3x^2 + 3x + 3 + 2$

$= 3x^2 + 3x + 5$

$h[f(x)] = 3x^2 + 3x + 5$

49. $f(x) = x - 2 \qquad g(x) = x^3$

$g(-1) = (-1)^3 = -1$

$f(-1) = -1 - 2 = -3$

$f[g(-1)] = -3$

51. $f(x) = x - 2 \qquad g(x) = x^3$

$f(-1) = -1 - 2 = -3$

$g(-3) = (-3)^3 = -27$

$g[f(-1)] = -27$

53. $f(x) = x - 2 \qquad g(x) = x^3$

$g(x - 2) = (x - 2)^3$

$= (x - 2)(x - 2)(x - 2)$

$= x^3 - 6x^2 + 12x - 8$

$g[f(x)] = x^3 - 6x^2 + 12x - 8$

55. a) **Strategy:** Selling price equals the cost plus the mark-up. If the cost is x and the mark-up is 60% then $S = x + 0.60x$.

If $M(x) = \dfrac{50x + 10{,}000}{x}$ is the cost per camera, then $(S \circ M)(x)$ is the selling price per camera. Find $(S \circ M)(x)$.

Solution:

$(S \circ M)(x) = S(M(x))$

$= M(x) + 0.60(M(x)) = 1.60(M(x))$

$= 1.60\left(\dfrac{50x + 10{,}000}{x}\right)$

$= \dfrac{80x + 16{,}000}{x}$

$S(M(x)) = 80 + \dfrac{16{,}000}{x}$

b) $(S \circ M)(5000) = 80 + \dfrac{16000}{5000}$

$= 80 + 3.2$

$= \$83.20$

c) When 5000 digital cameras are manufactured, the camera store sells each camera for \$83.20.

57. a) $I(n) = 12{,}500n \quad n(m) = 4m$

$(I \circ n)(m) = I(n(m))$

$= 12{,}500(4m)$

$= 50{,}000m$

b) $(I \circ n)(3) = 50{,}000(3) = \$150{,}000$

c) The garage's income from conversions done during a 3 month period is \$150,000.

59. rebate: $r(p) = p - 1500$
discounted price: $d(p) = 0.90p$

a) If the dealer takes the rebate first and then the discount we are finding $d(r(p)) = 0.90(p - 1500) = 0.90p - 1350$.

b) If the dealer takes the discount first and then the rebate we are finding $r(d(p)) = 0.90p - 1500$.

c) As a buyer you would prefer the dealer to use $r(d(p))$ since the cost would be less.

Critical Thinking

61. $f(1) = 2$ and $g(2) = 0$
$g[f(1)] = g(2) = 0$

63. $(f \circ g)(3) = f(g(3))$
$g(3) = 5$ and $f(5) = -2$
$(f \circ g)(3) = f(g(3)) = -2$

65. $g(0) = -4$ and $f(-4) = 7$
$f[g(0)] = f(-4) = 7$

67. $g(x) = x^2 - 1$

$g(3+h) - g(3) = (3+h)^2 - 1 - (3^2 - 1)$

$= 9 + 6h + h^2 - 1 - 8$

$g(3+h) - g(3) = h^2 + 6h$

69. $g(x) = x^2 - 1$

$$\frac{g(1+h) - g(1)}{h} = \frac{(1+h)^2 - 1 - (1^2 - 1)}{h}$$

$$= \frac{1 + 2h + h^2 - 1 - 0}{h}$$

$$= \frac{2h + h^2}{h} = 2 + h$$

$$\frac{g(1+h) - g(1)}{h} = 2 + h$$

71. $g(x) = x^2 - 1$

$$\frac{g(a+h) - g(a)}{h} = \frac{(a+h)^2 - 1 - (a^2 - 1)}{h}$$

$$= \frac{a^2 + 2ah + h^2 - 1 - a^2 + 1}{h}$$

$$= \frac{2ah + h^2}{h} = 2a + h$$

$$\frac{g(a+h) - g(a)}{h} = 2a + h$$

Projects or Group Activities

73. $f(x) = 2x \quad g(x) = 3x - 1 \quad h(x) = x - 2$

$f(1) = 2(1) = 2$

$h(2) = 2 - 2 = 0$

$g(0) = 3(0) - 1 = 0 - 1 = -1$

$g(h[f(1)]) = -1$

75. $f(x) = 2x \quad g(x) = 3x - 1 \quad h(x) = x - 2$

$g(0) = 3(0) - 1 = 0 - 1 = -1$

$h(-1) = -1 - 2 = -3$

$f(-3) = 2(-3) = -6$

$f(h[g(0)]) = -6$

77. $f(x) = 2x \quad g(x) = 3x - 1 \quad h(x) = x - 2$

$h(x) = x - 2$

$f(x-2) = 2(x-2) = 2x - 4$

$g(2x-4) = 3(2x-4) - 1 = 6x - 13$

$g(f[h(x)]) = 6x - 13$

Section 11.4

Concept Check

1. A function is a 1-1 function if, for any a and b in the domain of f, $f(a) = f(b)$ implies that $a = b$.

3. a) Yes
b) No

5. The inverse of a function *f* is the set of ordered pairs formed by reversing the coordinates of each ordered pair of *f*.

7. (ii)

Objective A Exercises

9. Yes, the graph represents a 1-1 function.

11. No, the graph is not a 1-1 function.

13. Yes, the graph represents a 1-1 function.

15. No, the graph is not a 1-1 function.

17. No, the graph is not a 1-1 function.

19. No, the graph is not a 1-1 function.

Objective B Exercises

21. Inverse function: $\{(0,1),(3,2),(8,3),(15,4)\}$

23. No inverse because the numbers 5 and -5 would each be paired with two different values in the range.

25. Inverse function: $\{(-2,0),(5,-1),(3,3),(6,-4)\}$

27. No inverse because the number 3 would be paired with three different values of the range.

29. $f(x) = 4x - 8$

$$y = 4x - 8$$
$$x = 4y - 8$$
$$x + 8 = 4y$$
$$\frac{1}{4}x + 2 = y$$
$$f^{-1}(x) = \frac{1}{4}x + 2$$

31. $f(x) = 2x + 4$

$$y = 2x + 4$$
$$x = 2y + 4$$
$$x - 4 = 2y$$
$$\frac{1}{2}x - 2 = y$$
$$f^{-1}(x) = \frac{1}{2}x - 2$$

33. $f(x) = \frac{1}{2}x - 1$

$$y = \frac{1}{2}x - 1$$
$$x = \frac{1}{2}y - 1$$
$$x + 1 = \frac{1}{2}y$$
$$2x + 2 = y$$
$$f^{-1}(x) = 2x + 2$$

35. $f(x) = -2x + 2$

$$y = -2x + 2$$
$$x = -2y + 2$$
$$x - 2 = -2y$$
$$-\frac{1}{2}x + 1 = y$$
$$f^{-1}(x) = -\frac{1}{2}x + 1$$

37. $f(x)=\frac{2}{3}x+4$

$y=\frac{2}{3}x+4$

$x=\frac{2}{3}y+4$

$x-4=\frac{2}{3}y$

$\frac{3}{2}x-6=y$

$f^{-1}(x)=\frac{3}{2}x-6$

39. $f(x)=-\frac{1}{3}x+1$

$y=-\frac{1}{3}x+1$

$x=-\frac{1}{3}y+1$

$x-1=-\frac{1}{3}y$

$-3x+3=y$

$f^{-1}(x)=-3x+3$

41. $f(x)=2x-5$

$y=2x-5$

$x=2y-5$

$x+5=2y$

$\frac{1}{2}x+\frac{5}{2}=y$

$f^{-1}(x)=\frac{1}{2}x+\frac{5}{2}$

43. $f(x)=5x-2$

$y=5x-2$

$x=5y-2$

$x+2=5y$

$\frac{1}{5}x+\frac{2}{5}=y$

$f^{-1}(x)=\frac{1}{5}x+\frac{2}{5}$

45. $f(x)=6x-3$

$y=6x-3$

$x=6y-3$

$x+3=6y$

$\frac{1}{6}x+\frac{1}{2}=y$

$f^{-1}(x)=\frac{1}{6}x+\frac{1}{2}$

47. $f(x)=3x-5$

$y=3x-5$

$x=3y-5$

$x+5=3y$

$\frac{1}{3}x+\frac{5}{3}=y$

$f^{-1}(x)=\frac{1}{3}x+\frac{5}{3}$

$f^{-1}(0)=\frac{1}{3}(0)+\frac{5}{3}$

$f^{-1}(0)=\frac{5}{3}$

49. $f(x) = 3x - 5$

$y = 3x - 5$

$x = 3y - 5$

$x + 5 = 3y$

$\frac{1}{3}x + \frac{5}{3} = y$

$f^{-1}(x) = \frac{1}{3}x + \frac{5}{3}$

$f^{-1}(4) = \frac{1}{3}(4) + \frac{5}{3}$

$f^{-1}(4) = \frac{9}{3} = 3$

51. Using the vertical line test the graph is a function. Using the horizontal line test the graph is 1-1 and therefore does have an inverse.

53. $f(g(x)) = f\left(\frac{x}{4}\right) = 4\left(\frac{x}{4}\right) = x$

$g(f(x)) = g(4x) = \frac{4x}{4} = x$

Yes, the functions are inverses of each other.

55. $f(h(x)) = f\left(\frac{1}{3x}\right) = 3\left(\frac{1}{3x}\right) = \frac{1}{x}$

$h(f(x)) = h(3x) = \frac{1}{3(3x)} = \frac{1}{9x}$

No, the functions are not inverses of each other.

57. $f(g(x)) = f(3x + 2)$

$= \frac{1}{3}(3x + 2) - \frac{2}{3} = x + \frac{2}{3} - \frac{2}{3} = x$

$g(f(x)) = g\left(\frac{1}{3}x - \frac{2}{3}\right)$

$= 3\left(\frac{1}{3}x - \frac{2}{3}\right) + 2 = x - 2 + 2 = x$

Yes, the functions are inverses of each other.

59. $f(g(x)) = f(2x + 3)$

$= \frac{1}{2}(2x + 3) - \frac{3}{2} = x + \frac{3}{2} - \frac{3}{2} = x$

$g(f(x)) = g\left(\frac{1}{2}x - \frac{3}{2}\right)$

$= 2\left(\frac{1}{2}x - \frac{3}{2}\right) + 3 = x - 3 + 3 = x$

Yes, the functions are inverses of each other.

61. $f(x) = \frac{x}{16}$

$y = \frac{x}{16}$

$x = \frac{y}{16}$

$16x = y$

$f^{-1}(x) = 16x$

The inverse function converts pounds to ounces.

63. $f(x) = x + 30$

$y = x + 30$

$x = y + 30$

$x - 30 = y$

$f^{-1}(x) = x - 30$

The inverse function converts a dress size in France to a dress size in the United States.

65. $f(x) = 90x + 65$

$y = 90x + 65$

$x = 90y + 65$

$x - 65 = 90y$

$\frac{1}{90}x - \frac{13}{18} = y$

$f^{-1}(x) = \frac{1}{90}x - \frac{13}{18}$

The inverse function gives the training intensity percent for a given target heart rate.

67. $f(x) = 120.381x$

$y = 120.381x$

$x = 120.381y$

$\dfrac{x}{120.381} = y$

$0.008307x = y$

$f^{-1}(x) = 0.008307x$

Critical Thinking

69. Inverse of the function:

Grade	Score
A	90-100
B	80-89
C	70-79
D	60-69
F	0-59

No, the inverse of the grading scale is not a function because each grade is paired with more than one score.

71. A constant function is defined as $y = b$ where b is a constant value. The inverse of this function would be $x = a$ where a is a constant. This is not a function.

73.

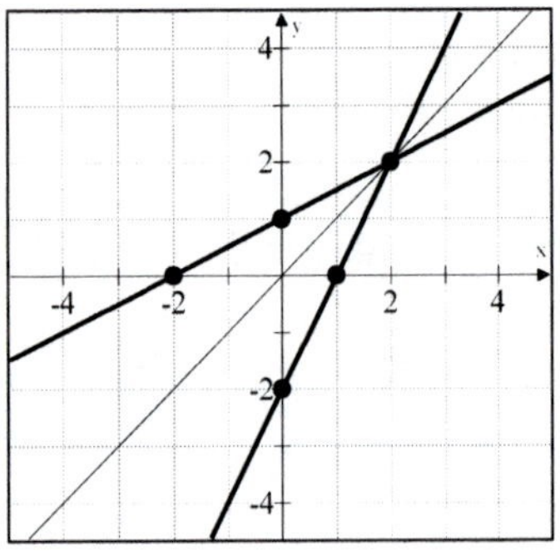

75.

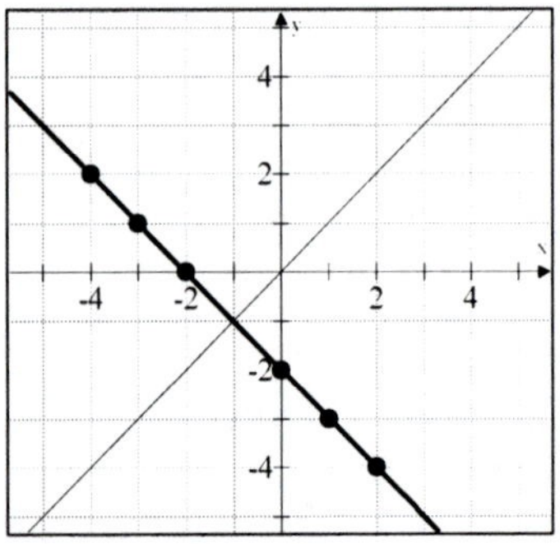

The inverse is the same graph.

77.

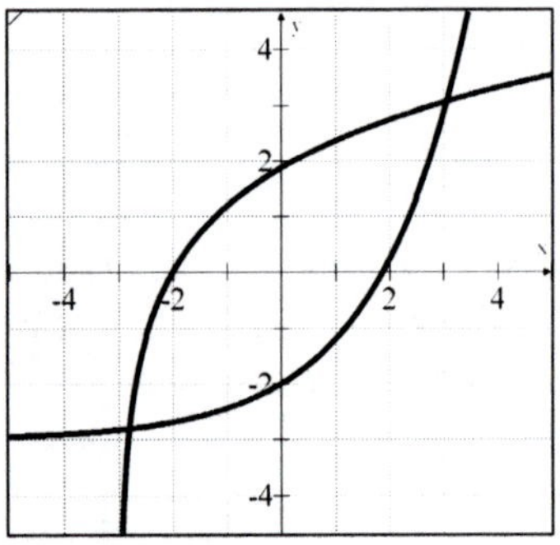

Chapter 11 Review Exercises

1.

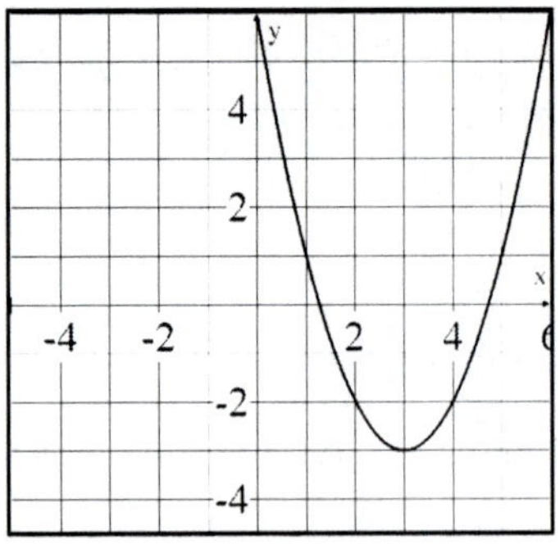

2. Yes, the graph is a function. It passes the vertical line test.

3.

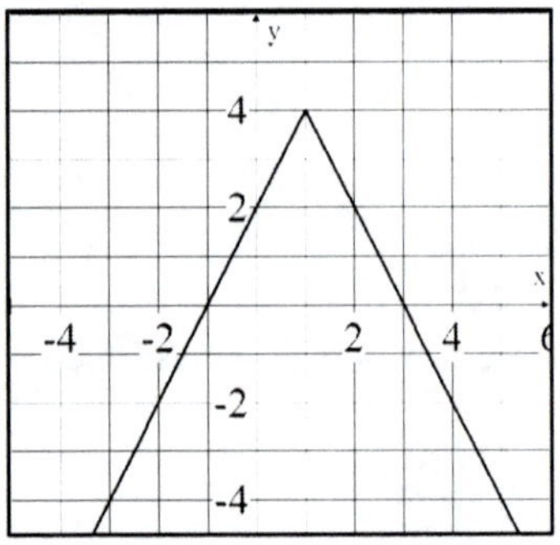

4.

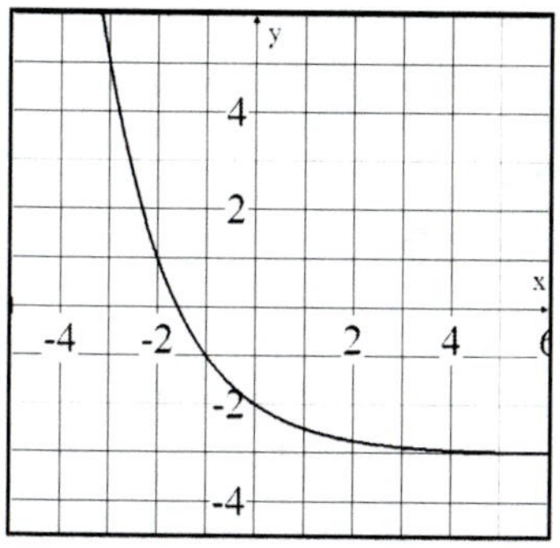

5.

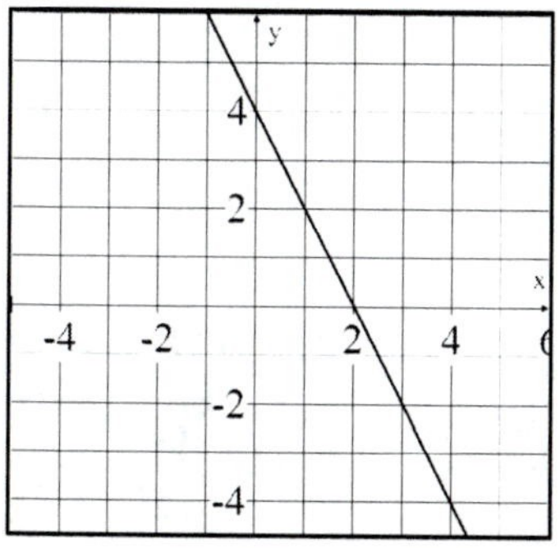

6. $y = x^2 - 2x + 3$

$$-\frac{b}{2a} = -\frac{-2}{2(1)} = -\frac{-2}{2} = -(-1) = 1$$

$$y = 1^2 - 2(1) + 3 = 2$$

Vertex: $(1,2)$

Axis of symmetry: $x = 1$

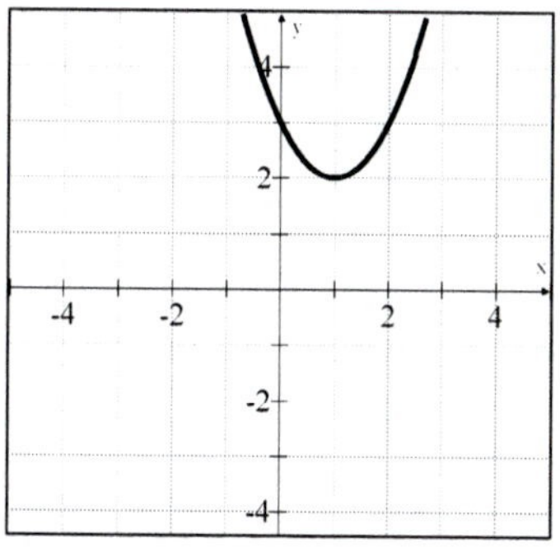

7. $y = -3x^2 + 4x + 6$

$a = -3 \quad b = 4 \quad c = 6$

$b^2 - 4ac$

$(4)^2 - 4(-3)(6) = 16 + 72 = 88$

$88 > 0$

Since the discriminant is greater than zero the parabola has two x-intercepts.

8. $f(x) = 2x^2 + x + 5$

$a = 2 \quad b = 1 \quad c = 5$

$b^2 - 4ac$

$(1)^2 - 4(2)(5) = 1 - 40 = -39$

$-39 < 0$

Since the discriminant is less than zero the parabola has no x-intercepts.

9. $y = 3x^2 + 9x$

$0 = 3x^2 + 9x$

$0 = 3x(x+3)$

$3x = 0 \quad x + 3 = 0$

$x = 0 \qquad x = -3$

The x-intercepts are (0,0) and (−3, 0).

10. $f(x) = x^2 - 6x + 7$

$0 = x^2 - 6x + 7$

$a = 1 \quad b = -6 \quad c = 7$

$$x = \frac{-b \pm \sqrt{b^2 - 4ac}}{2a}$$

$$x = \frac{-(-6) \pm \sqrt{(-6)^2 - 4(1)(7)}}{2(1)}$$

$$x = \frac{6 \pm \sqrt{36 - 28}}{2} = \frac{6 \pm \sqrt{8}}{2} = \frac{6 \pm 2\sqrt{2}}{2}$$

$$x = 3 \pm \sqrt{2}$$

The x-intercepts are $(3 + \sqrt{2}, 0)$ and $(3 - \sqrt{2}, 0)$.

11. $f(x) = 2x^2 - 7x - 15$

$2x^2 - 7x - 15 = 0$

$(2x+3)(x-5) = 0$

$2x + 3 = 0 \qquad x - 5 = 0$

$2x = -3 \qquad x = 5$

$$x = -\frac{3}{2}$$

The zeros are $-\frac{3}{2}$ and 5.

12. $f(x)=x^2-2x+10$

$x^2-2x+10=0$

$a=1 \quad b=-2 \quad c=10$

$$x=\frac{-b\pm\sqrt{b^2-4ac}}{2a}$$

$$x=\frac{-(-2)\pm\sqrt{(-2)^2-4(1)(10)}}{2(1)}$$

$$x=\frac{2\pm\sqrt{4-40}}{2}=\frac{2\pm\sqrt{-36}}{2}=\frac{2\pm 6i}{2}$$

$x=1\pm 3i$

The zeros are $1+3i$ and $1-3i$.

13. $f(x)=-2x^2+4x+1$

$$x=-\frac{b}{2a}=-\frac{4}{2(-2)}=1$$

$f(x)=-2x^2+4x+1$

$f(1)=-2(1)^2+4(1)+1=-2+4+1=3$

The maximum value of the function is 3.

14. $f(x)=x^2-7x+8$

$$x=-\frac{b}{2a}=-\frac{-7}{2(1)}=-\frac{7}{2}$$

$f(x)=x^2-7x+8$

$$f\left(\frac{7}{2}\right)=\left(\frac{7}{2}\right)^2-7\left(\frac{7}{2}\right)+8=\frac{49}{4}-\frac{49}{2}+8=-\frac{17}{4}$$

The minimum value of the function is $-\frac{17}{4}$.

15. $f(x)=x^2+4 \quad g(x)=4x-1$

$g(0)=4(0)-1=-1$

$f(-1)=(-1)^2+4=1+4=5$

$f[g(0)]=5$

16. $f(x)=6x+8 \quad g(x)=4x+2$

$f(-1)=6(-1)+8=-6+8=2$

$g(2)=4(2)+2=8+2=10$

$g[f(-1)]=10$

17. $f(x)=3x^2-4 \quad g(x)=2x+1$

$f(g(x))=f(2x+1)$

$=3(2x+1)^2-4=3(2x+1)(2x+1)-4$

$=3(4x^2+4x+1)-4$

$=12x^2+12x+3-4=12x^2+12x-1$

$f[g(x)]=12x^2+12x-1$

18. $f(x)=2x^2+x-5 \quad g(x)=3x-1$

$g(f(x))=g(2x^2+x-5)$

$=3(2x^2+x-5)-1$

$=6x^2+3x-15-1$

$=6x^2+3x-16$

$g[f(x)]=6x^2+3x-16$

19. $(f+g)(2)=f(2)+g(2)$

$=((2)^2+2(2)-3)+((2)^2-2)$

$=(4+4-3)+(4-2)$

$=5+2$

$=7$

20. $(f-g)(-4)=f(-4)-g(-4)$

$=((-4)^2+2(-4)-3)-((-4)^2-2)$

$=(16-8-3)-(16-2)$

$=5-14$

$=-9$

21. $(f\cdot g)(-4)=f(-4)\cdot g(-4)$

$=((-4)^2+2(-4)-3)\cdot((-4)^2-2)$

$=(16-8-3)\cdot(16-2)$

$=5\cdot 14$

$=70$

22. $\left(\frac{f}{g}\right)(3) = \frac{f(3)}{g(3)}$
$= \frac{(3)^2 + 2(3) - 3}{(3)^2 - 2}$
$= \frac{9 + 6 - 3}{9 - 2}$
$= \frac{12}{7}$

23. $f(x) = -6x + 4$
$y = -6x + 4$
$x = -6y + 4$
$x - 4 = -6y$
$-\frac{1}{6}x + \frac{2}{3} = y$
$f^{-1}(x) = -\frac{1}{6}x + \frac{2}{3}$

24. $f(x) = \frac{2}{3}x - 12$
$y = \frac{2}{3}x - 12$
$x = \frac{2}{3}y - 12$
$x + 12 = \frac{2}{3}y$
$\frac{3}{2}x + 18 = y$
$f^{-1}(x) = \frac{3}{2}x + 18$

25. $f(g(x)) = f(-4x + 5)$
$= -\frac{1}{4}(-4x + 5) + \frac{5}{4} = x - \frac{5}{4} + \frac{5}{4} = x$
$g(f(x)) = g\left(-\frac{1}{4}x + \frac{5}{4}\right)$
$= -4\left(-\frac{1}{4}x + \frac{5}{4}\right) + 5 = x - 5 + 5 = x$
Yes, the functions are inverses of each other.

26. $f(g(x)) = f(2x + 1)$
$= \frac{1}{2}(2x + 1) = x + \frac{1}{2}$
$g(f(x)) = g\left(\frac{1}{2}x\right)$
$= 2\left(\frac{1}{2}x\right) + 1 = x + 1$
No, the functions are not inverses of each other.

27. $p(x) = 0.4x + 15$
$p = 0.4x + 15$
$x = 0.4p + 15$
$x - 15 = 0.4p$
$2.5x - 37.5 = p$
$p^{-1}(x) = 2.5x - 37.5$
The inverse function gives the diver's depth below the surface of the water for a given pressure on the diver.

28. **Strategy**: To find the number of gloves for a maximum profit find the *x*-coordinate of the vertex. To find the maximum profit evaluate the function at the *x*-coordinate of the vertex.

Solution:
$x = -\frac{b}{2a} = -\frac{100}{2(-1)} = -\frac{100}{-2} = -(-50) = 50$
$P(x) = -x^2 + 100x + 2500$
$P(50) = -(50)^2 + 100(50) + 2500$
$= -2500 + 5000 + 2500 = 5000$

The company should make 50 baseball gloves each month to maximize profit. The maximum profit is \$5000.

29. Strategy: Let x represent width of the rectangle.
The length is $14 - x$.
The area is $x(14 - x)$.
To find the width find the x-coordinate of the vertex. To find the length evaluate $14 - x$ at the x-coordinate of the vertex.

Solution:

$x(14 - x) = 14x - x^2$

$x = -\dfrac{b}{2a} = -\dfrac{14}{2(-1)} = -\dfrac{14}{-2} = -(-7) = 7$

$14 - x = 14 - 7 = 7$

The dimensions are 7 ft by 7 ft.

Chapter 11 Test

1. $f(x) = x^2 - 6x + 4$

$-\dfrac{b}{2a} = -\dfrac{-6}{2(1)} = -\dfrac{-6}{2} = -(-3) = 3$

$y = 3^2 - 6(3) + 4 = -5$

Vertex: $(3, -5)$

Axis of symmetry: $x = 3$

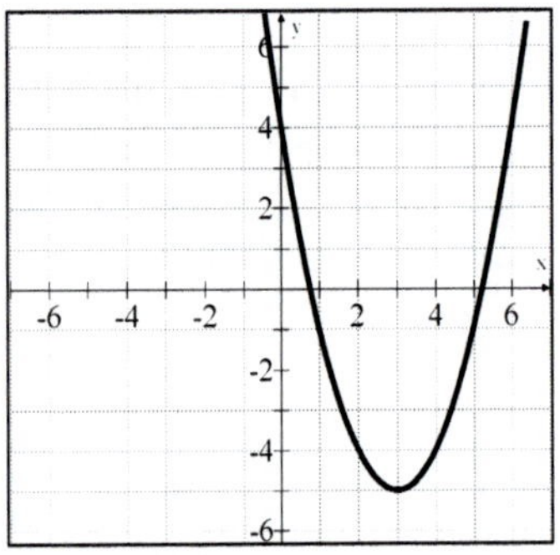

2.

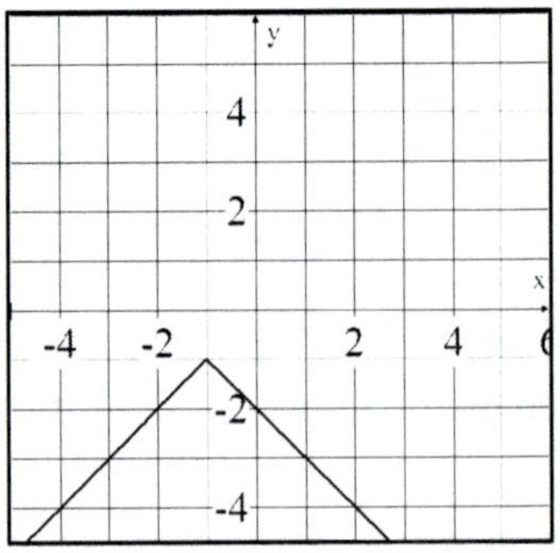

3.

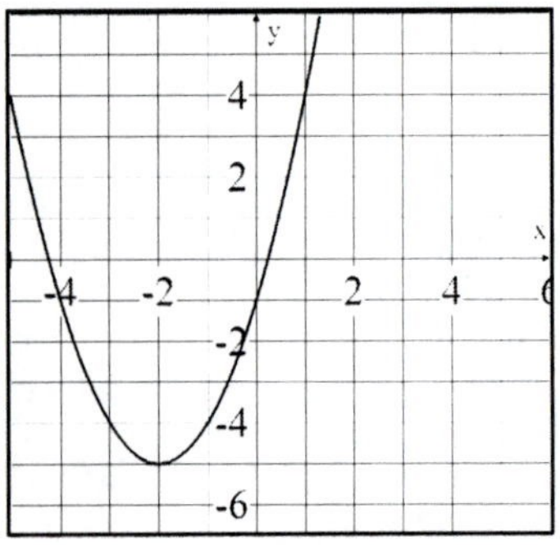

4.

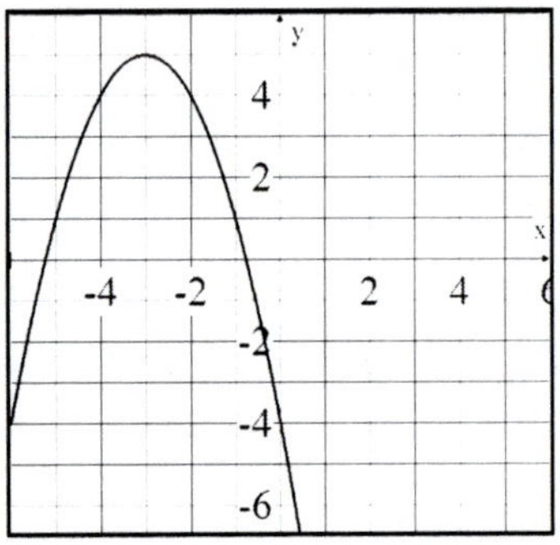

5. $y = 3x^2 - 4x + 6$

$a = 3 \quad b = -4 \quad c = 6$

$b^2 - 4ac$

$(-4)^2 - 4(3)(6) = 16 - 72 = -56$

$-56 < 0$

Since the discriminant is less than zero the parabola has no x-intercepts.

6. $y = 3x^2 - 7x - 6$

$0 = 3x^2 - 7x - 6$

$0 = (3x + 2)(x - 3)$

$3x + 2 = 0 \quad x - 3 = 0$

$3x = -2 \qquad x = 3$

$x = -\dfrac{2}{3}$

The x-intercepts are $(-\dfrac{2}{3}, 0)$ and $(3, 0)$.

7. $f(x) = -x^2 + 8x - 7$

$x = -\frac{b}{2a} = -\frac{8}{2(-1)} = -(-4) = 4$

$f(x) = -x^2 + 8x - 7$

$f(4) = -(4)^2 + 8(4) - 7 = -16 + 32 - 7 = 9$

The maximum value of the function is 9.

8. Domain: $\{x| -\infty < x < \infty\}$
Range: $\{y| y \geq -7\}$

9. $f(x) = x^2 + 2x - 3 \quad g(x) = x^3 - 1$

$(f - g)(2) = f(2) - g(2)$

$= (2^2 + 2(2) - 3) - (2^3 - 1)$

$= (4 + 4 - 3) - (8 - 1)$

$= 5 - 7$

$= -2$

10. $f(x) = x^3 + 1 \quad g(x) = 2x - 3$

$(f \cdot g)(-3) = f(-3) \cdot g(-3)$

$= ((-3)^3 + 1) \cdot (2(-3) - 3)$

$= (-27 + 1) \cdot (-6 - 3)$

$= (-26) \cdot (-9)$

$= 234$

11. $f(x) = 4x - 5 \quad g(x) = x^2 + 3x + 4$

$\left(\frac{f}{g}\right)(-2) = \frac{f(-2)}{g(-2)}$

$= \frac{4(-2) - 5}{(-2)^2 + 3(-2) + 4}$

$= \frac{-8 - 5}{4 - 6 + 4} = \frac{-13}{2}$

$= -\frac{13}{2}$

12. $f(x) = x^2 + 4 \quad g(x) = 2x^2 + 2x + 1$

$(f - g)(-4) = f(-4) - g(-4)$

$= ((-4)^2 + 4) - (2(-4)^2 + 2(-4) + 1)$

$= (16 + 4) - (32 - 8 + 1)$

$= 20 - 25$

$= -5$

13. $f(x) = 2x - 7 \quad g(x) = x^2 - 2x - 5$

$g(2) = 2^2 - 2(2) - 5 = 4 - 4 - 5 = -5$

$f(-5) = 2(-5) - 7 = -10 - 7 = -17$

$f[g(2)] = -17$

14. $f(x) = x^2 + 1 \quad g(x) = x^2 + x + 1$

$f(-2) = (-2)^2 + 1 = 4 + 1 = 5$

$g(5) = (5)^2 + 5 + 1 = 25 + 5 + 1 = 31$

$g[f(-2)] = 31$

15. $f(x) = x^2 - 1 \quad g(x) = 3x + 2$

$g(f(x)) = g(x^2 - 1)$

$= 3(x^2 - 1) + 2 = 3x^2 - 3 + 2$

$= 3x^2 - 1$

$g[f(x)] = 3x^2 - 1$

16. $f(x) = 2x^2 - 7 \quad g(x) = x - 1$

$f(g(x)) = f(x - 1)$

$= 2(x - 1)^2 - 7 = 2(x - 1)(x - 1) - 7$

$= 2(x^2 - 2x + 1) - 7$

$= 2x^2 - 4x + 2 - 7$

$= 2x^2 - 4x - 5$

$f[g(x)] = 2x^2 - 4x - 5$

17. No, inverse because the numbers 4 and 5 would be paired with two different values of the range.

18. Inverse function: $\{(6, 2), (5, 3), (4, 4), (3, 5)\}$

19. $f(x) = 4x - 2$

$y = 4x - 2$

$x = 4y - 2$

$x + 2 = 4y$

$\frac{1}{4}x + \frac{1}{2} = y$

$f^{-1}(x) = \frac{1}{4}x + \frac{1}{2}$

20. $f(x) = \frac{1}{4}x - 4$

$y = \frac{1}{4}x - 4$

$x = \frac{1}{4}y - 4$

$x + 4 = \frac{1}{4}y$

$4x + 16 = y$

$f^{-1}(x) = 4x + 16$

21. $f(g(x)) = f(2x - 4)$

$= \frac{1}{2}(2x - 4) + 2 = x - 2 + 2 = x$

$g(f(x)) = g\left(\frac{1}{2}x + 2\right)$

$= 2\left(\frac{1}{2}x + 2\right) - 4 = x + 4 - 4 = x$

Yes, the functions are inverses of each other.

22. $f(g(x)) = f\left(\frac{3}{2}x + 3\right)$

$= \frac{2}{3}\left(\frac{3}{2}x + 3\right) - 2 = x + 2 - 2 = x$

$g(f(x)) = g\left(\frac{2}{3}x - 2\right)$

$= \frac{3}{2}\left(\frac{2}{3}x - 2\right) + 3 = x - 3 + 3 = x$

Yes, the functions are inverses of each other.

23. No, the graph is not a 1-1 function. It does not pass the horizontal line test.

24. $C(x) = 1.25x + 5$

$C = 1.25x + 5$

$x = 1.25C + 5$

$x - 5 = 1.25C$

$0.8x - 4 = C$

$C^{-1}(x) = 0.8x - 4$

The inverse function gives the number of miles to a certain location for a given cost.

25. **Strategy**: To find the number of speakers for a minimum production cost find the *x*-coordinate of the vertex. To find the minimum cost evaluate the function at the *x*-coordinate of the vertex.

Solution:

$x = -\frac{b}{2a} = -\frac{-50}{2} = -\frac{-50}{2} = -(-25) = 25$

$C(x) = x^2 - 50x + 675$

$C(25) = (25)^2 - 50(25) + 675$

$= 625 - 1250 + 675 = 50$

The company should make 25 speakers each day to minimize production costs.
The minimum daily production cost is \$50.

26. **Strategy**: Let *x* represent one number. The other number is $28 - x$.
Their product is $x(28 - x)$.
To find one number find the *x*-coordinate of the vertex. To find the second number evaluate $28 - x$ at the *x*-coordinate of the vertex.

Solution:

$x(28 - x) = 28x - x^2$

$x = -\frac{b}{2a} = -\frac{28}{2(-1)} = -\frac{28}{-2} = -(-14) = 14$

$28 - x = 28 - 14 = 14$

The two numbers are 14 and 14.

27. Strategy: Let x represent width of the rectangle. The length is 100 - x.
The area is $x(100 - x)$.
To find the width find the x-coordinate of the vertex. To find the length evaluate $100 - x$ at the x-coordinate of the vertex.

Solution:

$$x(100 - x) = 100x - x^2$$

$$x = -\frac{b}{2a} = -\frac{100}{2(-1)} = -\frac{100}{-2} = -(-50) = 50$$

$$100 - x = 100 - 50 = 50$$

$$A = l \cdot w = 50 \cdot 50 = 2500$$

The dimensions of the rectangle are 50 cm by 50 cm. The area is 2500 cm ^{2}.

Cumulative Review Exercises

1. $$-3a + \left|\frac{3b - ab}{3b - c}\right|$$

$$-3(2) + \left|\frac{3(2) - (2)(2)}{3(2) - (-2)}\right|$$

$$= -6 + \left|\frac{6 - 4}{6 + 2}\right| = -6 + \left|\frac{2}{8}\right| = -6 + \left|\frac{1}{4}\right|$$

$$= -6 + \frac{1}{4} = -\frac{23}{4}$$

2. 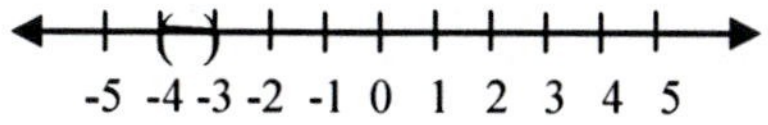

3. $$\frac{3x - 1}{6} - \frac{5 - x}{4} = \frac{5}{6}$$

$$12\left(\frac{3x - 1}{6} - \frac{5 - x}{4}\right) = 12\left(\frac{5}{6}\right)$$

$$2(3x - 1) - 3(5 - x) = 10$$

$$6x - 2 - 15 + 3x = 10$$

$$9x - 17 = 10$$

$$9x = 27$$

$$x = 3$$

The solution is 3.

4. $4x - 2 < -10$ or $3x - 1 > 8$

$4x - 2 < -10 \qquad 3x - 1 > 8$

$4x < -8 \qquad 3x > 9$

$x < -2 \qquad x > 3$

$\{x \mid x < -2\}$ or $\{x \mid x > 3\}$

$\{x \mid x < -2\} \cup \{x \mid x > 3\}$

5. $|8 - 2x| \ge 0$

$8 - 2x \le 0 \qquad 8 - 2x \ge 0$

$-2x \le -8 \qquad -2x \ge -8$

$x \ge 4 \qquad x \le 4$

$\{x \mid x \ge 4\}$ or $\{x \mid x \le 4\}$

$\{x \mid x \ge 4\} \cup \{x \mid x \le 4\}$

$= \{x \mid x \in \text{real numbers}\}$

6. $$\left(\frac{3a^3b}{2a}\right)^2\left(\frac{a^2}{-3b^2}\right)^3 = \left(\frac{3a^2b}{2}\right)^2\left(\frac{a^2}{-3b^2}\right)^3$$

$$= \left(\frac{3^2a^4b^2}{2^2}\right)\left(\frac{a^6}{(-3)^3b^6}\right) = \frac{9a^{10}b^2}{4(-27)b^6}$$

$$= \frac{9a^{10}}{-108b^4} = -\frac{a^{10}}{12b^4}$$

7. $(x - 4)(2x^2 + 4x - 1)$

$= x(2x^2 + 4x - 1) - 4(2x^2 + 4x - 1)$

$= 2x^3 + 4x^2 - x - 8x^2 - 16x + 4$

$= 2x^3 - 4x^2 - 17x + 4$

8. $6x - 2y = -3$

$4x + y = 5$

$6x - 2y = -3$

$\underline{8x + 2y = 10}$

$14x = 7$

$$x = \frac{1}{2}$$

$$6\left(\frac{1}{2}\right) - 2y = -3$$

$3 - 2y = -3$

$-2y = -6$

$y = 3$

The solution is $\left(\frac{1}{2}, 3\right)$.

9. $x^3y + x^2y^2 - 6xy^3 = xy(x^2 + xy - 6y^2)$

$= xy(x + 3y)(x - 2y)$

10. $(b+2)(b-5)=2b+14$
$b^2-3b-10=2b+14$
$b^2-5b-24=0$
$(b-8)(b+3)=0$
$b-8=0 \quad b+3=0$
$b=8 \quad b=-3$
The solutions are −3 and 8.

11. $x^2-2x>15$
$x^2-2x-15>0$
$(x-5)(x+3)>0$
$\{x \mid x<-3\} \cup \{x>5\}$

12. $\dfrac{x^2+4x-5}{2x^2-3x+1}-\dfrac{x}{2x-1}$
$=\dfrac{(x+5)(x-1)}{(2x-1)(x-1)}-\dfrac{x}{2x-1}$
$=\dfrac{x+5}{2x-1}-\dfrac{x}{2x-1}=\dfrac{x+5-x}{2x-1}$
$=\dfrac{5}{2x-1}$

13. $\dfrac{5}{x^2+7x+12}=\dfrac{9}{x+4}-\dfrac{2}{x+3}$
$\dfrac{5}{(x+4)(x+3)}=\dfrac{9}{x+4}\cdot\dfrac{x+3}{x+3}-\dfrac{2}{x+3}\cdot\dfrac{x+4}{x+4}$
$\dfrac{5}{(x+4)(x+3)}=\dfrac{9x+27}{(x+4)(x+3)}-\dfrac{2x+8}{(x+4)(x+3)}$
$5=(9x+27)-(2x+8)$
$5=7x+19$
$-14=7x$
$-2=x$
The solution is −2.

14. $\dfrac{4-6i}{2i}=\dfrac{4-6i}{2i}\cdot\dfrac{i}{i}=\dfrac{4i-6i^2}{2i^2}$
$=\dfrac{4i+6}{-2}=-3-2i$

15. $f(x)=\dfrac{1}{4}x^2$
$-\dfrac{b}{2a}=-\dfrac{0}{2\left(\dfrac{1}{4}\right)}=-\dfrac{0}{\dfrac{1}{4}}=0$
$y=\left(\dfrac{1}{4}\right)0^2=0$

Vertex:
$(0,0)$
Axis of symmetry:
$x=0$

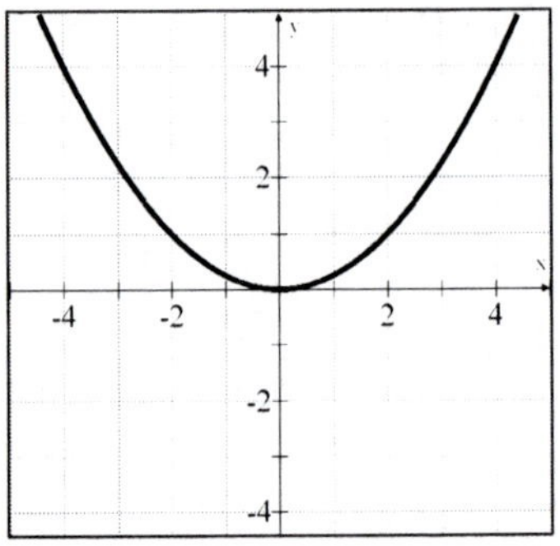

16. $3x-4y\geq 8$
$-4y\geq -3x+8$
$y\leq \dfrac{3}{4}x-2$

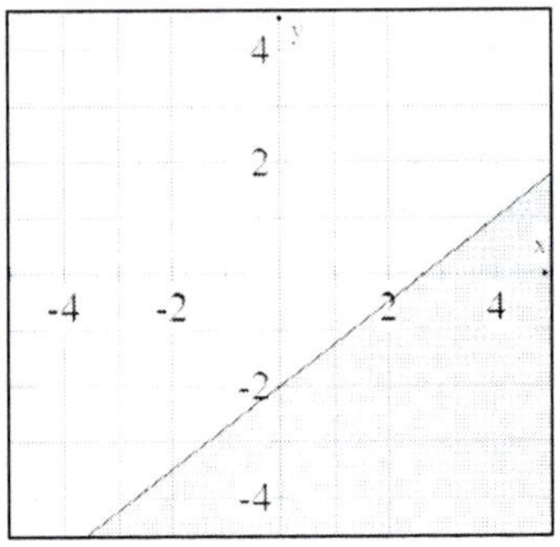

17. $m=\dfrac{y_2-y_1}{x_2-x_1}=\dfrac{-6-4}{2-(-3)}=\dfrac{-10}{5}=-2$
$y-y_1=m(x-x_1)$
$y-4=-2(x-(-3))$
$y-4=-2(x+3)$
$y-4=-2x-6$
$y=-2x-2$

18. The product of the slopes of perpendicular lines is −1.

$$2x - 3y = 6$$
$$-3y = -2x + 6$$
$$y = \frac{2}{3}x - 2$$
$$m_1 \cdot m_2 = -1$$
$$\frac{2}{3} \cdot m_2 = -1$$
$$m_2 = -\frac{3}{2}$$
$$y - y_1 = m(x - x_1)$$
$$y - 1 = -\frac{3}{2}(x - (-3))$$
$$y - 1 = -\frac{3}{2}(x + 3)$$
$$y - 1 = -\frac{3}{2}x - \frac{9}{2}$$
$$y = -\frac{3}{2}x - \frac{7}{2}$$

19. $3x^2 = 3x - 1$

$$3x^2 - 3x + 1 = 0$$
$$a = 3 \quad b = -3 \quad c = 1$$
$$x = \frac{-b \pm \sqrt{b^2 - 4ac}}{2a}$$
$$x = \frac{-(-3) \pm \sqrt{(-3)^2 - 4(3)(1)}}{2(3)}$$
$$x = \frac{3 \pm \sqrt{9 - 12}}{6} = \frac{3 \pm \sqrt{-3}}{6} = \frac{3 \pm i\sqrt{3}}{6}$$
$$x = \frac{1}{2} \pm \frac{i\sqrt{3}}{6}$$

The zeros are $\frac{1}{2} + \frac{i\sqrt{3}}{6}$ and $\frac{1}{2} - \frac{i\sqrt{3}}{6}$.

20. $\sqrt{8x + 1} = 2x - 1$

$$\left(\sqrt{8x + 1}\right)^2 = (2x - 1)^2$$
$$8x + 1 = 4x^2 - 4x + 1$$
$$0 = 4x^2 - 12x$$
$$0 = 4x(x - 3)$$
$$4x = 0 \qquad x - 3 = 0$$
$$x = 0 \qquad x = 3$$

Check both solutions in the original equation:

$$\sqrt{8(0) + 1} = 2(0) - 1$$
$$\sqrt{1} = -1$$
$$1 \neq -1$$
$$\sqrt{8(3) + 1} = 2(3) - 1$$
$$\sqrt{25} = 5$$
$$5 = 5$$

The solution is 3.

21. $f(x) = 2x^2 - 3$

$$x = -\frac{b}{2a} = -\frac{0}{2(2)} = -\frac{0}{4} = 0$$
$$f(x) = 2x^2 - 3$$
$$f(0) = 2(0)^2 - 3 = 0 - 3$$
$$= -3$$

Since $a > 0$, the function has a minimum value. The minimum value of the function is −3.

22. $f(x) = 3x - 4$

$$0 = 3x - 4$$
$$4 = 3x$$
$$x = \frac{4}{3}$$

The zero of the function is $\frac{4}{3}$.

23. Yes

24. $\sqrt[3]{5x-2}=2$

$\left(\sqrt[3]{5x-2}\right)^3=2^3$

$5x-2=8$

$5x=10$

$x=2$

The solution is 2.

25. $g(x)=3x-5 \quad h(x)=\frac{1}{2}x+4$

$h(2)=\frac{1}{2}(2)+4=1+4=5$

$g(5)=3(5)-5=15-5=10$

$g[h(2)]=10$

26. $f(x)=-3x+9$

$y=-3x+9$

$x=-3y+9$

$x-9=-3y$

$-\frac{1}{3}x+3=y$

$f^{-1}(x)=-\frac{1}{3}x+3$

27. Strategy: Let x represent the cost per pound of the mixture.

	Amount	Cost	Value
\$4.50 tea	30	4.50	30(4.50)
\$3.60 tea	45	3.60	45(3.60)
Mixture	75	x	$75x$

The sum of the values before mixing is equal to the value after mixing.

Solution:

$30(4.50)+45(3.60)=75x$

$135+162=75x$

$297=75x$

$x=3.96$

The cost per pound of the mixture is \$3.96.

28. Strategy: Let x represent the number of pounds of 80% copper alloy.

	Amount	Percent	Quantity
80%	x	0.80	$0.80x$
20%	50	0.20	0.20(50)
40%	$50+x$	0.40	$0.40(50+x)$

The sum of the quantities before mixing is equal to the quantity after mixing.

Solution:

$0.80x+0.20(50)=0.40(50+x)$

$0.80x+10=20+0.40x$

$0.40x+10=20$

$0.40x=10$

$x=25$

25 lb of the 80% copper alloy must be used.

29. Strategy: Let x represent the additional amount of insecticide.
The total amount of insecticide is x + 6.
To find the additional amount of insecticide write and solve a proportion.

Solution: $\frac{6}{16}=\frac{x+6}{28}$.

$\frac{3}{8}=\frac{x+6}{28}$

$\frac{3}{8}\cdot 56=\frac{x+6}{28}\cdot 56$

$21=2x+12$

$9=2x$

$4.5=x$

An additional 4.5 oz of insecticide are required.

30. Strategy: Let x represent the time it takes for the smaller pipe to fill the tank.
The time it takes the larger pipe to fill the tank is $x - 8$.

	Rate	Time	Part
Smaller pipe	$\frac{1}{t}$	3	$\frac{3}{t}$
Larger pipe	$\frac{1}{t-8}$	3	$\frac{3}{t-8}$

The sum of the parts of the task completed must equal 1.

Solution:

$$\frac{3}{t} + \frac{3}{t-8} = 1$$

$$t(t-8)\left(\frac{3}{t} + \frac{3}{t-8}\right) = 1(t(t-8))$$

$$3(t-8) + 3t = t^2 - 8t$$

$$3t - 24 + 3t = t^2 - 8t$$

$$6t - 24 = t^2 - 8t$$

$$0 = t^2 - 14t + 24$$

$$0 = (t-2)(t-12)$$

$$t - 2 = 0 \quad t - 12 = 0$$

$$t = 2 \qquad t = 12$$

The solution $t = 2$ is not possible since the time for the larger pipe would then be a negative number.
$t - 8 = 2 - 8 = -6$.
It takes the larger pipe $t - 8 = 12 - 8 = 4$ min to fill the tank.

31. Strategy: Write the basic direct variation equation replacing the variable with the given values. Solve for k.
Write the direct variation equation replacing k with its value. Substitute 40 for f and solve for d.

Solution:

$$d = kf \qquad d = \frac{3}{5}f$$

$$30 = k(50) \qquad = \frac{3}{5}(40)$$

$$\frac{3}{5} = k \qquad = 24$$

A force of 40 lb will stretch the spring 24 in.

32. Strategy: Write the basic inverse variation equation replacing the variable with the given values. Solve for k.
Write the inverse variation equation replacing k with its value. Substitute 1.5 for L and solve for f.

Solution:

$$f = \frac{k}{L} \qquad f = \frac{120}{L}$$

$$60 = \frac{k}{2} \qquad = \frac{120}{1.5}$$

$$120 = k \qquad = 80$$

The frequency is 80 vibrations/min.

Chapter 12: Exponential and Logarithmic Functions

Prep Test

1. $3^{-2} = \dfrac{1}{3^2} = \dfrac{1}{9}$

2. $\left(\dfrac{1}{2}\right)^{-4} = \left(\dfrac{2}{1}\right)^4 = 2^4 = 16$

3. $\dfrac{1}{8} = \dfrac{1}{2^3} = 2^{-3}$

4. $f(x) = x^4 + x^3$
   $f(-1) = (-1)^4 + (-1)^3 = 1 + (-1) = 0$
   $f(3) = (3)^4 + (3)^3 = 81 + 27 = 108$

5. $3x + 7 = x - 5$
   $2x + 7 = -5$
   $2x = -12$
   $x = -6$
   The solution is −6.

6. $16 = x^2 - 6x$
   $0 = x^2 - 6x - 16$
   $0 = (x - 8)(x + 2)$
   $x - 8 = 0 \quad x + 2 = 0$
   $x = 8 \quad x = -2$
   The solutions are −2 and 8.

7. $A(1 + r)^n$
   $5000(1 + 0.04)^6 = 5000(1.04)^6$
   $= 6326.60$

8. $f(x) = x^2 - 1$
   $x = -\dfrac{b}{2a} = \dfrac{0}{2(1)} = \dfrac{0}{2} = 0$
   $f(0) = (0)^2 - 1 = -1$

   Vertex: $(0, -1)$. Axis of symmetry: $x = 0$.

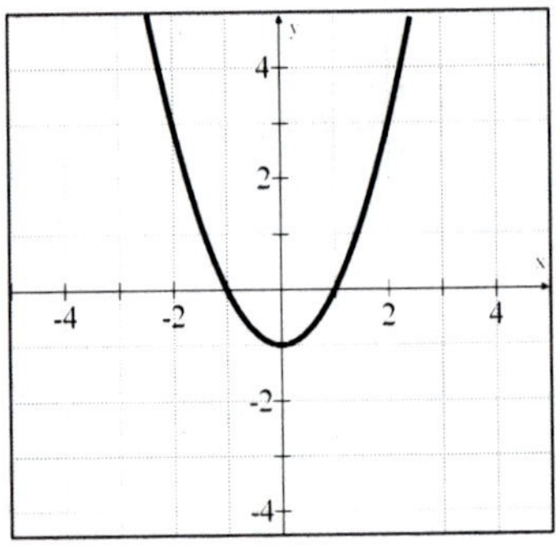

Section 12.1

Concept Check

1. An exponential function with base b is defined by $f(x) = b^x$, $b > 0$, $b \neq 1$, and x is any real number.

3. $f(x) = b^x$, $b > 0$, $b \neq 1$
   (iii) cannot be the base.

Objective A Exercises

5. $f(x) = 3^x$
   a) $f(2) = 3^2 = 9$
   b) $f(0) = 3^0 = 1$
   c) $f(-2) = 3^{-2} = \dfrac{1}{3^2} = \dfrac{1}{9}$

7. $g(x) = 2^{x+1}$
   a) $g(3) = 2^{3+1} = 2^4 = 16$
   b) $g(1) = 2^{1+1} = 2^2 = 4$
   c) $g(-3) = 2^{-3+1} = 2^{-2} = \dfrac{1}{2^2} = \dfrac{1}{4}$

9. $P(x) = \left(\frac{1}{2}\right)^{2x}$

a) $P(0) = \left(\frac{1}{2}\right)^{2(0)} = \left(\frac{1}{2}\right)^{0} = 1$

b) $P\left(\frac{3}{2}\right) = \left(\frac{1}{2}\right)^{2(3/2)} = \left(\frac{1}{2}\right)^{3} = \frac{1}{8}$

c) $P(-2) = \left(\frac{1}{2}\right)^{2(-2)} = \left(\frac{1}{2}\right)^{-4} = 2^4 = 16$

11. $G(x) = e^{x/2}$

a) $G(4) = e^{4/2} = e^2 = 7.3891$

b) $G(-2) = e^{-2/2} = e^{-1} = \frac{1}{e^1} = 0.3679$

c) $G\left(\frac{1}{2}\right) = e^{(1/2)/2} = e^{1/4} = 1.2840$

13. $H(r) = e^{-r+3}$

a) $H(-1) = e^{-(-1)+3} = e^4 = 54.5982$

b) $H(3) = e^{-3+3} = e^0 = 1$

c) $H(5) = e^{-5+3} = e^{-2} = \frac{1}{e^2} = 0.1353$

15. $F(x) = 2^{x^2}$

a) $F(2) = 2^{2^2} = 2^4 = 16$

b) $F(-2) = 2^{(-2)^2} = 2^4 = 16$

c) $F\left(\frac{3}{4}\right) = 2^{(3/4)^2} = 2^{9/16} = 1.4768$

17. $f(x) = e^{-x^2/2}$

a) $f(-2) = e^{-(-2)^2/2} = e^{-2} = \frac{1}{e^2} = 0.1353$

b) $f(2) = e^{-2^2/2} = e^{-2} = \frac{1}{e^2} = 0.1353$

c)

$f(-3) = e^{-(-3)^2/2} = e^{-9/2} = \frac{1}{e^{9/2}} = 0.0111$

19. $f(a) > f(b)$

Objective B Exercises

21. $f(x) = 3^x$

x	y
0	1
-1	⅓
1	3

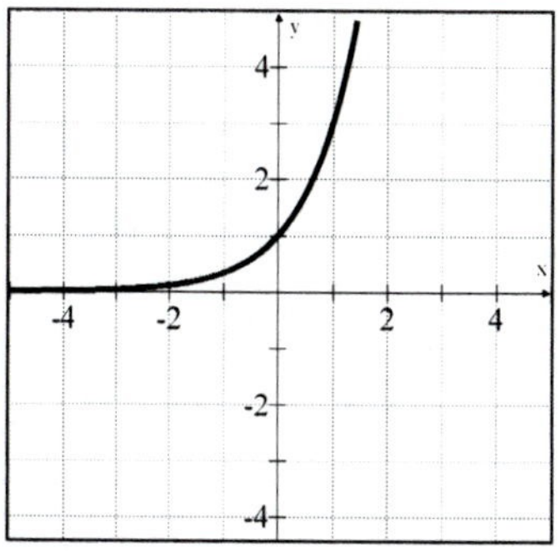

23. $f(x) = 2^{x+1}$

x	y
0	2
-1	1
1	4

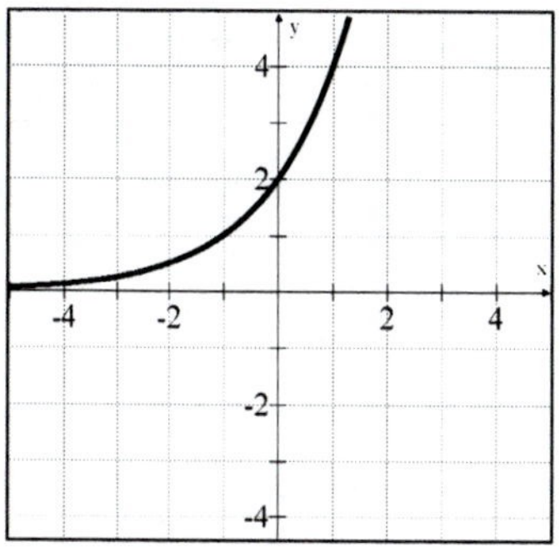

25. $f(x) = \left(\frac{1}{3}\right)^x$

x	y
0	1
-1	3
1	⅓

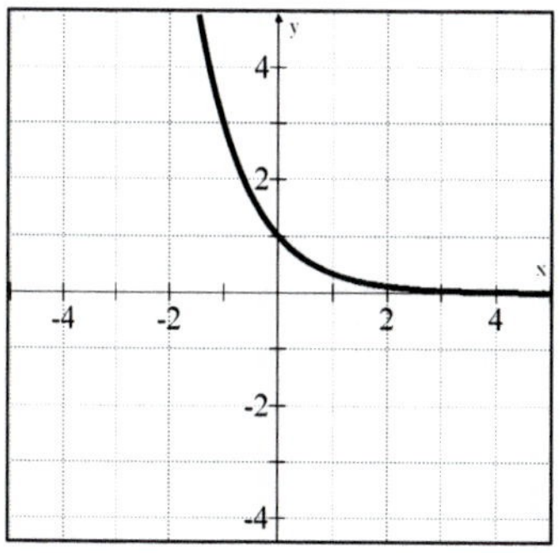

27. $f(x) = 2^{-x} + 1$

x	y
0	2
-1	3
1	3/2

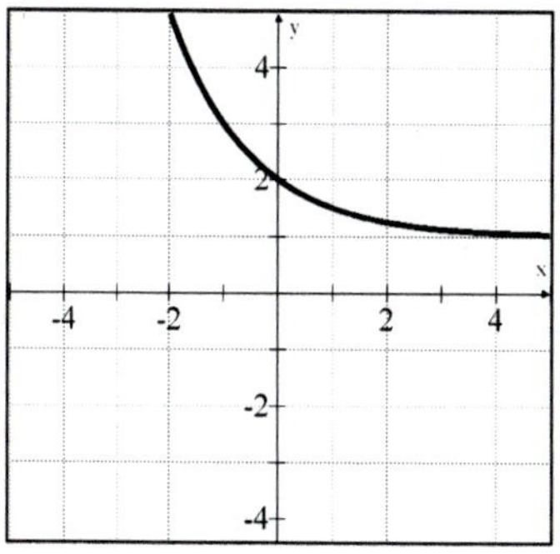

29. $f(x) = \left(\frac{1}{3}\right)^{-x}$

x	y
0	1
-1	$\frac{1}{3}$
1	3

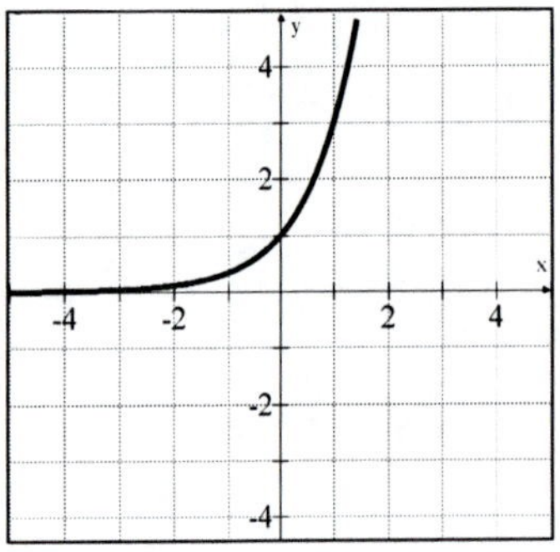

31. $f(x) = e^{x+1} - 1$

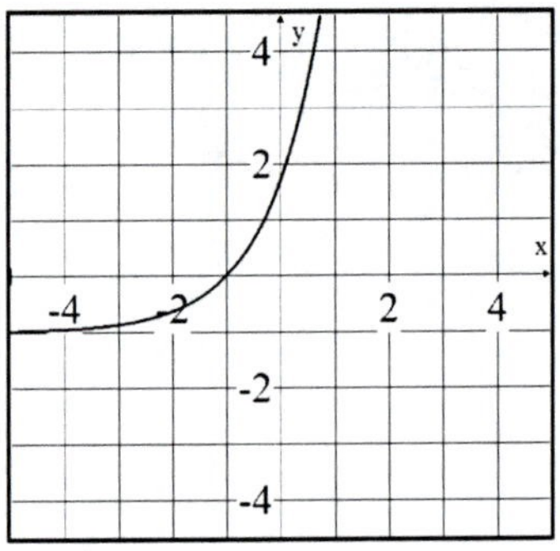

33. $v(t) = 32(1 - e^{-t})$

$v(4) = 32(1 - e^{-4}) \approx 31.41$

The speed of the object after 4 s is 31.41 ft/sec.

35. (i) and (iii) have the same graphs.
(ii) and (iv) have the same graphs.

Critical Thinking

37. $P(x) = (\sqrt{3})^x$

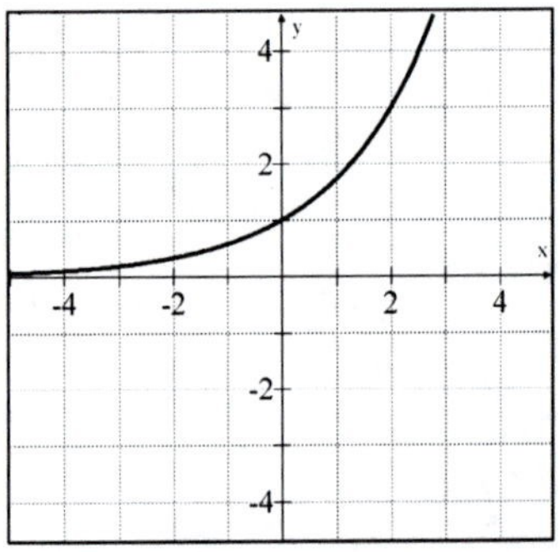

39. $f(x) = \pi^x$

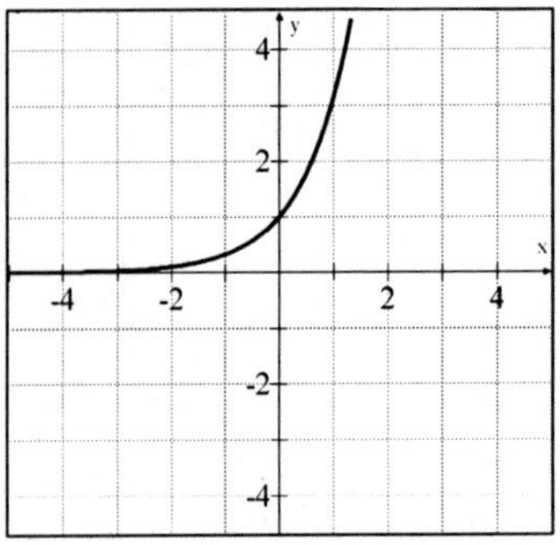

Projects or Group Activities

41.

n	$(1+n)^{1/n}$
0.01	2.704814
0.001	2.716924
0.0001	2.718146
0.00001	2.718268

As n decreases, $(1+n)^{1/n}$ becomes closer to e.

Section 12.2

Objective A Exercises

1. A common logarithm is a logarithm with a base of 10.

3. $\log_5 25 = 2$

5. $\log_4 \frac{1}{16} = -2$

7. $3^4 = 81$

9. $e^q = p$

11. False

13. True

15. True

Objective A Exercises

17. $\log_3 81 = x$

$3^x = 81$

$x = 4$

$\log_3 81 = 4$

19. $\log_2 128 = x$

$2^x = 128$

$x = 7$

$\log_2 128 = 7$

21. $\log 100 = x$

$10^x = 100$

$x = 2$

$\log 100 = 2$

23. $\ln e^3 = x$

$3 \ln e = x$

$3(1) = x$

$x = 3$

$\ln e^3 = 3$

25. $\log_8 1 = x$

$8^x = 1$

$x = 0$

$\log_8 1 = 0$

27. $\log_5 625 = x$

$5^x = 625$

$x = 4$

$\log_5 625 = 4$

29. $\log_3 x = 2$

$3^2 = x$

$x = 9$

31. $\log_4 x = 3$

$4^3 = x$

$x = 64$

33. $\log_7 x = -1$

$7^{-1} = x$

$x = \frac{1}{7}$

35. $\log_6 x = 0$

$6^0 = x$

$x = 1$

37. $\log x = 2.5$

$10^{2.5} = x$

$x = 316.23$

39. $\log x = -1.75$

$10^{-1.75} = x$

$x = 0.02$

41. $\ln x = 2$

$e^2 = x$

$x = 7.39$

43. $\ln x = -\frac{1}{2}$

$e^{-1/2} = x$

$x = 0.61$

45. $x > 1$

Objective B Exercises

47. $\log_b(xy) = \log_b(x) + \log_b(y)$

49. False

51. $\log_{12} 1 = 0$

53. $\ln e = 1$

55. $\log_3 3^x = x$

57. $e^{\ln v} = v$

59. $2^{\log_2(x^2+1)} = x^2 + 1$

61. $\log_5 5^{x^2-x-1} = x^2 - x - 1$

63. $\log_8(xz) = \log_8 x + \log_8 z$

65. $\log_3 x^5 = 5\log_3 x$

67. $\log_b\left(\frac{r}{s}\right) = \log_b r - \log_b s$

69. $\log_3\left(x^2y^6\right) = \log_3 x^2 + \log_3 y^6$
$= 2\log_3 x + 6\log_3 y$

71. $\log_7\left(\frac{u^3}{v^4}\right) = \log_7 u^3 - \log_7 v^4$
$= 3\log_7 u - 4\log_7 v$

73. $\log_2(rs)^2 = 2\log_2(rs) = 2[\log_2 r + \log_2 s]$

75. $\ln\left(x^2yz\right) = \ln x^2 + \ln y + \ln z$
$= 2\ln x + \ln y + \ln z$

77. $\log_5\left(\frac{xy^2}{z^4}\right) = \log_5 xy^2 - \log_5 z^4$
$= \log_5 x + \log_5 y^2 - \log_5 z^4$
$= \log_5 x + 2\log_5 y - 4\log_5 z$

79. $\log_8\left(\frac{x^2}{yz^2}\right) = \log_8 x^2 - \log_8 yz^2$
$= \log_8 x^2 - (\log_8 y + \log_8 z^2)$
$= \log_8 x^2 - \log_8 y - \log_8 z^2$
$= 2\log_8 x - \log_8 y - 2\log_8 z$

81. $\log_4 \sqrt{x^3y} = \log_4(x^3y)^{1/2} = \frac{1}{2}\log_4(x^3y)$
$= \frac{1}{2}[\log_4 x^3 + \log_4 y]$
$= \frac{1}{2}[3\log_4 x + \log_4 y]$
$= \frac{3}{2}\log_4 x + \frac{1}{2}\log_4 y$

83. $\log_7 \sqrt{\frac{x^3}{y}} = \log_7\left(\frac{x^3}{y}\right)^{1/2} = \frac{1}{2}\log_7 \frac{x^3}{y}$
$= \frac{1}{2}[\log_7 x^3 - \log_7 y]$
$= \frac{1}{2}[3\log_7 x - \log_7 y]$
$= \frac{3}{2}\log_7 x - \frac{1}{2}\log_7 y$

85. $\log_3\left(\frac{t}{\sqrt{x}}\right) = \log_3\left(\frac{t}{x^{1/2}}\right)$
$= \log_3 t - \log_3 x^{1/2}$
$= \log_3 t - \frac{1}{2}\log_3 x$

87. $\log_3 x^3 + \log_3 y^2 = \log_3(x^3y^2)$

89. $\ln x^4 - \ln y^2 = \ln\left(\frac{x^4}{y^2}\right)$

91. $3\log_7 x = \log_7 x^3$

93. $3\ln x + 4\ln y = \ln x^3 + \ln y^4 = \ln(x^3y^4)$

95. $2(\log_4 x + \log_4 y) = 2\log_4(xy)$
$= \log_4(xy)^2$
$= \log_4(x^2y^2)$

97. $2\log_3 x - \log_3 y + 2\log_3 z$
$= \log_3 x^2 - \log_3 y + \log_3 z^2$
$= \log_3\left(\frac{x^2}{y}\right) + \log_3 z^2$
$= \log_3\left(\frac{x^2z^2}{y}\right)$

99. $\ln x - (2\ln y + \ln z) = \ln x - (\ln y^2 + \ln z)$
$= \ln x - \ln(y^2z)$
$= \ln\left(\frac{x}{y^2z}\right)$

101. $\frac{1}{2}(\log_6 x - \log_6 y) = \frac{1}{2}\log_6\left(\frac{x}{y}\right)$

$= \log_6\left(\frac{x}{y}\right)^{1/2}$

$= \log_6 \sqrt{\frac{x}{y}}$

103. $2(\log_4 s - 2\log_4 t + \log_4 r)$

$= 2(\log_4 s - \log_4 t^2 + \log_4 r)$

$= 2(\log_4 \frac{s}{t^2} + \log_4 r)$

$= 2\log_4\left(\frac{sr}{t^2}\right)$

$= \log_4\left(\frac{sr}{t^2}\right)^2$

$= \log_4 \frac{s^2r^2}{t^4}$

105. $\ln x - 2(\ln y + \ln z)$

$= \ln x - 2\ln(yz)$

$= \ln x - \ln(yz)^2$

$= \ln\left(\frac{x}{(yz)^2}\right)$

$= \ln \frac{x}{y^2z^2}$

107. $\frac{1}{2}(3\log_4 x - 2\log_4 y + \log_4 z)$

$= \frac{1}{2}(\log_4 x^3 - \log_4 y^2 + \log_4 z)$

$= \frac{1}{2}(\log_4\left(\frac{x^3}{y^2}\right) + \log_4 z)$

$= \log_4\left(\frac{x^3z}{y^2}\right)^{1/2}$

$= \log_4 \sqrt{\frac{x^3z}{y^2}}$

109. $\frac{1}{2}\log_2 x - \frac{2}{3}\log_2 y + \frac{1}{2}\log_2 z$

$= \log_2 x^{1/2} - \log_2 y^{2/3} + \log_2 z^{1/2}$

$= \log_2\left(\frac{x^{1/2}}{y^{2/3}}\right) + \log_5 z^{1/2}$

$= \log_2\left(\frac{x^{1/2}z^{1/2}}{y^{2/3}}\right)$

$= \log_2 \frac{\sqrt{xy}}{\sqrt[3]{y^2}}$

Objective C Exercises

111. $\log_8 6 = \frac{\log_{10} 6}{\log_{10} 8} = 0.8617$

113. $\log_5 30 = \frac{\log_{10} 30}{\log_{10} 5} = 2.1133$

115. $\log_3 0.5 = \frac{\log_{10} 0.5}{\log_{10} 3} = -0.6309$

117. $\log_7 1.7 = \frac{\log_{10} 1.7}{\log_{10} 7} = 0.2727$

119. $\log_5 15 = \frac{\log_{10} 15}{\log_{10} 5} = 1.6826$

121. $\log_{12} 120 = \frac{\log_{10} 120}{\log_{10} 12} = 1.9266$

123. $\log_4 2.55 = \frac{\log_{10} 2.55}{\log_{10} 4} = 0.6752$

125. $\log_5 67 = \frac{\log_{10} 67}{\log_{10} 5} = 2.6125$

127. $\log_5 x = \frac{\log_{10} x}{\log_{10} 5}$

Critical Thinking

129. $\log_3(\log_3 x) = 2$

Let $y = \log_3(x)$

$\log_3(\log_3(x)) = 2$

$\log_3(y) = 2$

$3^2 = y$

$y = 9$

$y = \log_3(x)$

$9 = \log_3(x)$

$3^9 = x$

$x = 19{,}683$

131. $\log_2(\log_2 256) = x$

Let $y = \log_2(256)$

$y = \log_2(256) = 6$

$\log_2(8) = x$

$x = 3$

133. Because $x = 4$, $x - 5 = -1$. The logarithm of a negative number is undefined.

Projects or Groups Activities

135. a) $D = -(p_1 \log_2 p_1 + p_2 \log_2 p_2 + p_3 \log_2 p_3 + p_4 \log_2 p_4 + p_5 \log_2 p_5)$

$$D = -\left(\frac{1}{5}\log_2\frac{1}{5} + \frac{1}{5}\log_2\frac{1}{5} + \frac{1}{5}\log_2\frac{1}{5} + \frac{1}{5}\log_2\frac{1}{5} + \frac{1}{5}\log_2\frac{1}{5}\right)$$

$$D = -5\left(\frac{1}{5}\log_2 5\right) = \log_2 5 = 2.3219281$$

b) $D = -(p_1 \log_2 p_1 + p_2 \log_2 p_2 + p_3 \log_2 p_3 + p_4 \log_2 p_4 + p_5 \log_2 p_5)$

$$D = -\left(\frac{1}{8}\log_2\frac{1}{8} + \frac{3}{8}\log_2\frac{3}{8} + \frac{1}{16}\log_2\frac{1}{16} + \frac{1}{8}\log_2\frac{1}{8} + \frac{5}{16}\log_2\frac{5}{16}\right)$$

$D = 2.055036$

Less diversity

c) $D = -(p_1 \log_2 p_1 + p_2 \log_2 p_2 + p_3 \log_2 p_3 + p_4 \log_2 p_4 + p_5 \log_2 p_5)$

$$D = -\left(0\log_2 0 + \frac{1}{4}\log_2\frac{1}{4} + 0\log_2 0 + 0\log_2 0 + \frac{3}{3}\log_2\frac{3}{4}\right)$$

$$D = -\left(0 + \frac{1}{4}\log_2\frac{1}{4} + 0 + 0 + \frac{3}{3}\log_2\frac{3}{4}\right) = 0.82600$$

Less diversity

d) $D = -(p_1 \log_2 p_1 + p_2 \log_2 p_2 + p_3 \log_2 p_3 + p_4 \log_2 p_4 + p_5 \log_2 p_5)$

$D = -(0\log_2 0 + 0\log_2 0 + 0\log_2 0 + 0\log_2 0 + 1\log_2 1)$

$D = -(0 + 0 + 0 + 0 + 0) = 0$

Because this system has only one species, there is no diversity in the system.

Section 12.3

Objective A Exercises

1. They have the same graph.

3. $x = 2^{\frac{y}{3}}$

5. $x = e^{\frac{y-2}{3}}$

Objective A Exercises

7. $f(x) = \log_4 x$

$y = \log_4 x$ is equivalent to $x = 4^y$.

x	y
$\frac{1}{16}$	-2
$\frac{1}{4}$	-1
1	0
4	1

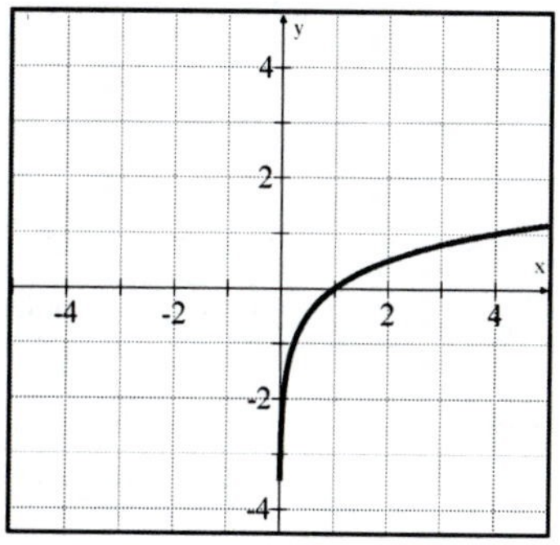

9. $f(x) = \log_3(2x-1)$

$y = \log_3(2x-1)$ is equivalent to

$2x - 1 = 3^y$ or $x = \frac{1}{2}(3^y + 1)$

x	y
$\frac{2}{3}$	-1
1	0
2	1
5	2

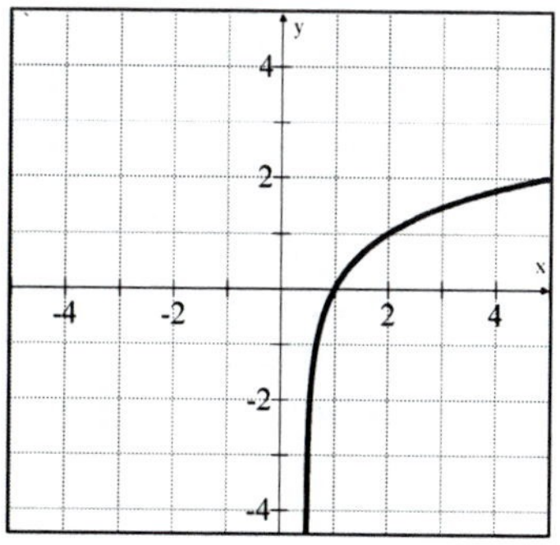

11. $f(x) = 3\log_2 x$

$y = 3\log_2 x$ is equivalent to $x = 2^{y/3}$

x	y
$\frac{1}{2}$	-3
1	0
2	3
4	6

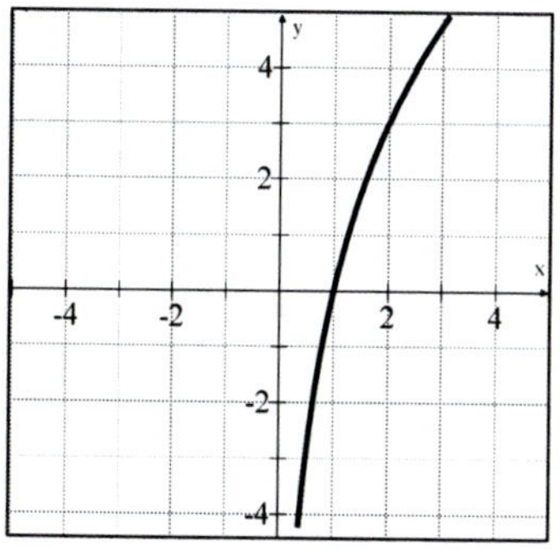

13. $f(x) = -\log_2 x$

$y = -\log_2 x$ is equivalent to $x = 2^{-y}$

x	y
2	-1
1	0
$\frac{1}{2}$	1
$\frac{1}{4}$	2

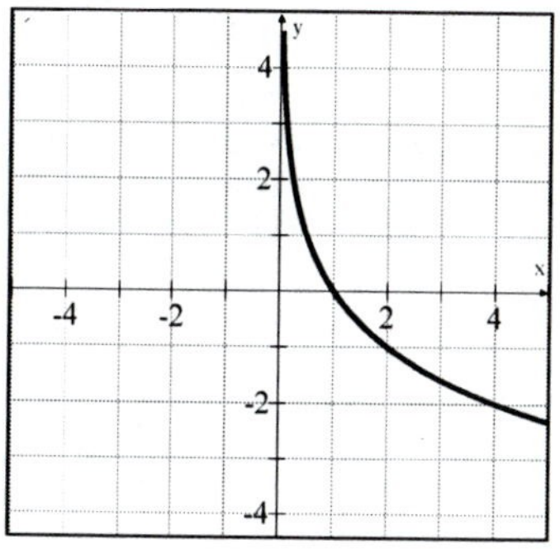

15. $f(x) = \log_2(x-1)$

$y = \log_2(x-1)$ is equivalent to $x - 1 = 2^y$

or $x = 2^y + 1$

x	y
$\frac{3}{2}$	-1
2	0
3	1
5	2

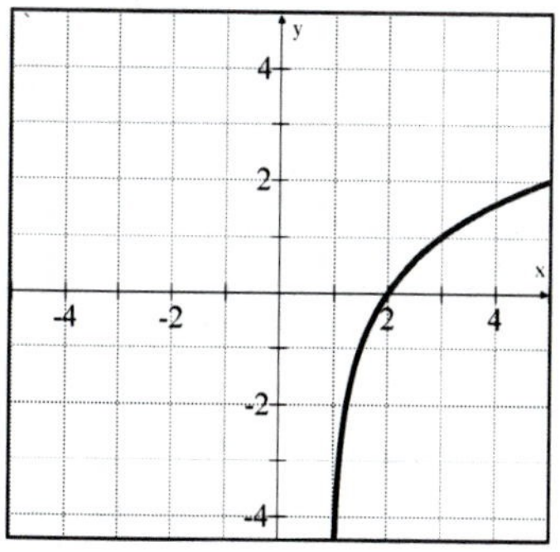

17. $f(x) = -\log_2(x-1)$

$y = -\log_2(x-1)$ is equivalent to

$x - 1 = 2^{-y}$ or $x = 2^{-y} + 1$

x	y
3	-1
2	0
$\frac{3}{2}$	1
$\frac{5}{4}$	2

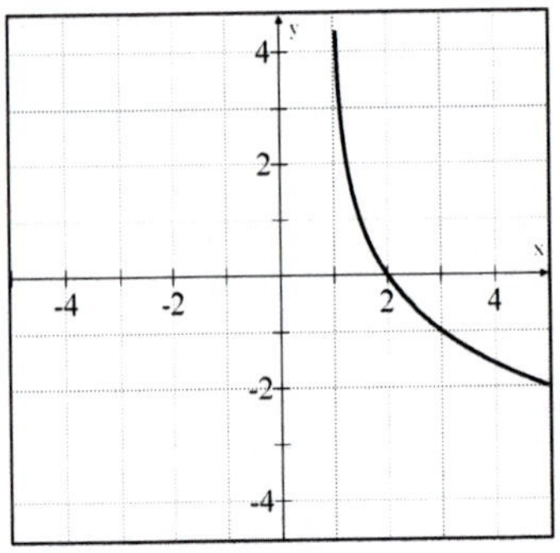

19. They intersect at the point (1,0).

Critical Thinking

21. $f(x) = x - \log_2(1-x)$

$y = x - \log_2(1-x)$

$y = x - \dfrac{\log(1-x)}{\log 2}$

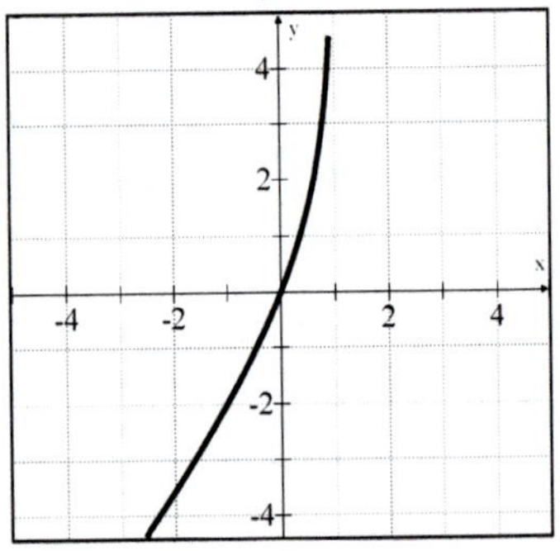

23. $f(x) = \dfrac{x}{2} - 2\log_2(x+1)$

$y = \dfrac{x}{2} - 2\log_2(x+1)$

$y = \dfrac{x}{2} - \log_2(x+1)^2$

$y = \dfrac{x}{2} - \dfrac{\log(x+1)^2}{\log 2}$

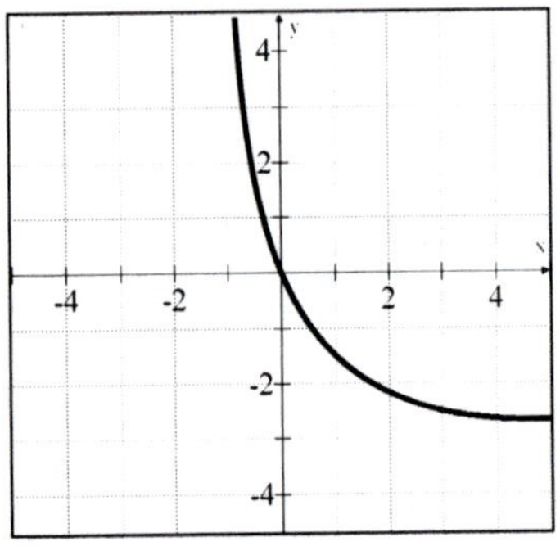

Projects or Group Activities

25. a) $M = 5\log s - 5$

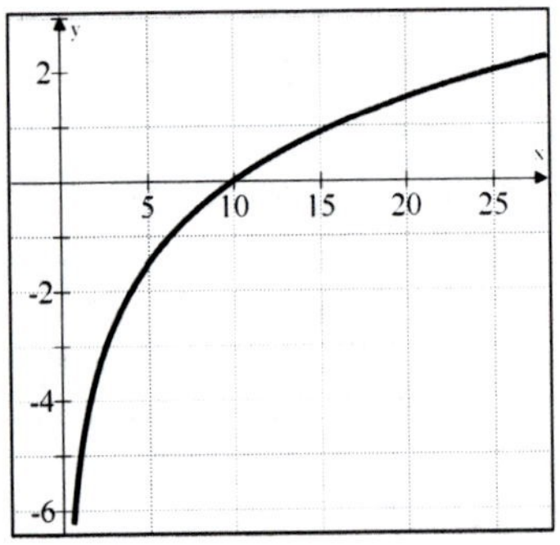

b) The point (25.1, 2) means that a star that is 25.1 parsecs from Earth has a distance modulus of 2.

c) 6.3 parsecs

Check Your Progress: Chapter 12

1. $f(x) = 3^x$

$f(4) = 3^4 = 81$

2. $f(x) = 2^{x-5}$

$f(2) = 2^{2-5} = 2^{-3} = \dfrac{1}{2^3} = \dfrac{1}{8}$

3. $f(x) = 4^{2x+3}$

$f(-2) = 4^{2(-2)+3} = 4^{-1} = \dfrac{1}{4^1} = \dfrac{1}{4}$

4.

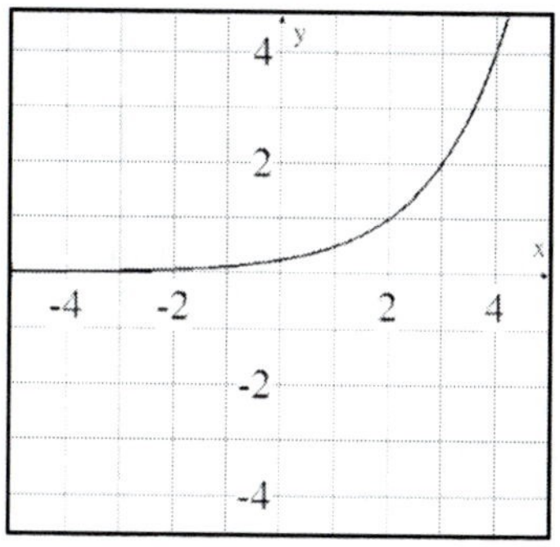

5.

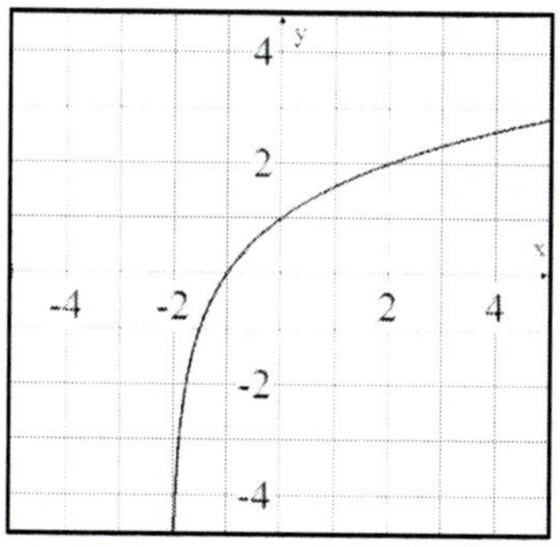

6.

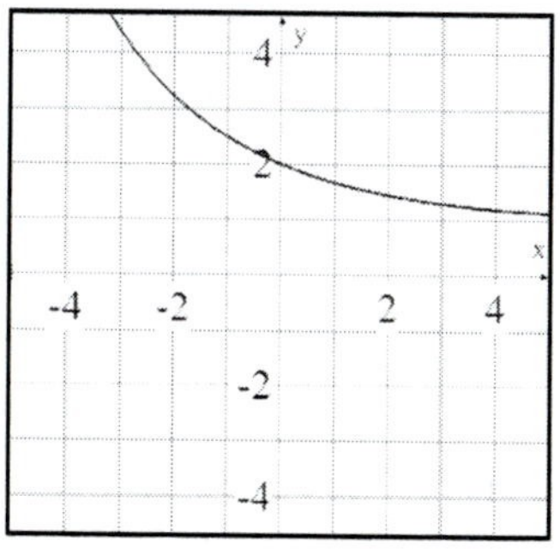

7. $\log_3 81 = x$

$3^x = 81$

$x = 4$

$\log_3 81 = 4$

8. $\log_4\left(\dfrac{1}{64}\right) = x$

$4^x = \dfrac{1}{64}$

$x = -3$

$\log_4\left(\dfrac{1}{64}\right) = -3$

9. $\log_5\left(\dfrac{1}{5}\right) = x$

$5^x = \dfrac{1}{5}$

$x = -1$

$\log_5\left(\dfrac{1}{5}\right) = -1$

10. $\log_7 7^{33} = x$

$7^x = 7^{33}$

$x = 33$

$\log_7 7^{33} = 33$

11. $\log_5 x = 4$

$5^4 = x$

$x = 625$

12. $\log_3 x = -3$

$3^{-3} = x$

$x = \dfrac{1}{27}$

13. $\log_7 x = 1$

$7^1 = x$

$x = 7$

14. $\log x = -4$

$10^{-4} = x$

$x = 0.0001$

15. $\log_7(x^2y^5) = \log_7 x^2 + \log_7 y^5$

$= 2\log_7 x + 5\log_7 y$

16. $\log_8\left(\frac{x}{y^3}\right) = \log_8 x - \log_8 y^3$

$= \log_8 x - 3\log_8 y$

17. $\log_3\left(\frac{x^2}{\sqrt{yz}}\right) = \log_3 x^2 - \log_3(yz)^{1/2}$

$= 2\log_3 x - \frac{1}{2}\log_3(yz)$

$= 2\log_3 x - \frac{1}{2}\log_3 y - \frac{1}{2}\log_3 z$

18. $3\log_3 x - 4\log_3 y = \log_3 x^3 - \log_3 y^4$

$= \log_3\left(\frac{x^3}{y^4}\right)$

19. $\ln x - (2\ln y - 5\ln z) = \ln x - (\ln y^2 - \ln z^5)$

$= \ln x - \ln\frac{y^2}{z^5} = \ln\frac{x}{\frac{y^2}{z^5}} = \ln\frac{xz^5}{y^2}$

20. $\frac{1}{2}(\log x + \log y) = \frac{1}{2}\log(xy) = \log(xy)^{1/2}$

$= \log\sqrt{xy}$

21. $\log_3 12 = \frac{\log_{10} 12}{\log_{10} 3} \approx 2.2619$

22. $\log_5 0.1 = \frac{\log_{10} 0.1}{\log_{10} 5} \approx -1.4307$

23. $\log_7 5 = \frac{\log_{10} 5}{\log_{10} 7} \approx 0.8271$

Section 12.4

Concept Check

1. The 1-1 Property of Exponential Functions states that for $b > 0$, $b \neq 1$, if $b^x = b^y$, then $x = y$.

3. $x < 0$

Objective A Exercises

5. $5^{4x-1} = 5^{x-2}$

$4x - 1 = x - 2$

$3x - 1 = -2$

$3x = -1$

$x = -\frac{1}{3}$

The solution is $-\frac{1}{3}$.

7. $8^{x-4} = 8^{5x+8}$

$x - 4 = 5x + 8$

$-4 = 4x + 8$

$4x = -12$

$x = -3$

The solution is −3.

9. $9^x = 3^{x+1}$

$3^{2x} = 3^{x+1}$

$2x = x + 1$

$x = 1$

The solution is 1.

11. $8^{x+2} = 16^x$

$(2^3)^{x+2} = 2^{4x}$

$2^{3x+6} = 2^{4x}$

$3x + 6 = 4x$

$x = 6$

The solution is 6.

13. $16^{2-x} = 32^{2x}$

$(2^4)^{2-x} = (2^5)^{2x}$

$2^{8-4x} = 2^{10x}$

$8 - 4x = 10x$

$8 = 14x$

$x = \frac{8}{14} = \frac{4}{7}$

The solution is $\frac{4}{7}$.

15. $25^{3-x} = 125^{2x-1}$

$(5^2)^{3-x} = (5^3)^{2x-1}$

$5^{6-2x} = 5^{6x-3}$

$6 - 2x = 6x - 3$

$6 = 8x - 3$

$9 = 8x$

$x = \frac{9}{8}$

The solution is $\frac{9}{8}$.

17. $5^x = 6$

$\log 5^x = \log 6$

$x \log 5 = \log 6$

$x = \frac{\log 6}{\log 5}$

$x = 1.1133$

The solution is 1.1133.

19. $8^{x/4} = 0.4$

$\log 8^{x/4} = \log 0.4$

$\frac{x}{4} \log 8 = \log 0.4$

$\frac{x}{4} = \frac{\log 0.4}{\log 8}$

$x = 4 \cdot \frac{\log 0.4}{\log 8}$

$x = -1.7626$

The solution is -1.7626.

21. $2^{3x} = 5$

$\log 2^{3x} = \log 5$

$3x \log 2 = \log 5$

$3x = \frac{\log 5}{\log 2}$

$3x = 2.3219$

$x = 0.7740$

The solution is 0.7740.

23. $2^{-x} = 7$

$\log 2^{-x} = \log 7$

$-x \log 2 = \log 7$

$-x = \frac{\log 7}{\log 2}$

$-x = 2.8074$

$x = -2.8074$

The solution is -2.8074.

25. $2^{x-1} = 6$

$\log 2^{x-1} = \log 6$

$(x-1)\log 2 = \log 6$

$x - 1 = \frac{\log 6}{\log 2}$

$x = \frac{\log 6}{\log 2} + 1$

$x = 3.5850$

The solution is 3.5850.

27. $3^{2x-1} = 4$

$\log 3^{2x-1} = \log 4$

$(2x-1)\log 3 = \log 4$

$2x - 1 = \frac{\log 4}{\log 3}$

$2x - 1 = 1.2619$

$2x = 2.2619$

$x = 1.1309$

The solution is 1.1309.

29. $\left(\frac{1}{2}\right)^{x+1} = 3$

$\log\left(\frac{1}{2}\right)^{x+1} = \log 3$

$(x+1)\log\left(\frac{1}{2}\right) = \log 3$

$x+1 = \dfrac{\log 3}{\log \frac{1}{2}}$

$x = \dfrac{\log 3}{\log \frac{1}{2}} - 1$

$x = -2.5850$

The solution is −2.5850.

31. $3 \cdot 2^x = 7$

$\log(3 \cdot 2^x) = \log 7$

$\log 3 + \log 2^x = \log 7$

$\log 3 + x \log 2 = \log 7$

$x \log 2 = \log 7 - \log 3$

$x = \dfrac{\log 7 - \log 3}{\log 2}$

$x = 1.2224$

The solution is 1.2224.

33. $7 = 10\left(\frac{1}{2}\right)^{x/8}$

$\log 7 = \log 10\left(\frac{1}{2}\right)^{x/8}$

$\log 7 = \log 10 + \log\left(\frac{1}{2}\right)^{x/8}$

$\log 7 = \log 10 + \frac{x}{8}\log\frac{1}{2}$

$\log 7 - \log 10 = \frac{x}{8}\log\frac{1}{2}$

$\dfrac{\log 7 - \log 10}{\log \frac{1}{2}} = \dfrac{x}{8}$

$0.5146 = \dfrac{x}{8}$

$4.1166 = x$

The solution is 4.1166.

35. $15 = 12(e)^{0.05x}$

$\ln 15 = \ln 12(e)^{0.05x}$

$\ln 15 = \ln 12 + \ln(e)^{0.05x}$

$\ln 15 = \ln 12 + 0.05x$

$\ln 15 - \ln 12 = 0.05x$

$0.2231 = 0.05x$

$4.4629 = x$

The solution is 4.4629.

Objective B Exercises

37. $\log x = \log(1-x)$

$x = 1 - x$

$2x = 1$

$x = \dfrac{1}{2}$

The solution is $\dfrac{1}{2}$.

39. $\ln(3x+2) = \ln(5x+4)$

$3x + 2 = 5x + 4$

$-2x = 2$

$x = -1$

When we substitute $x = -1$ in either side of the equation we get a logarithm of a negative number.

Because the logarithm of a negative number is not a real number there is no solution.

41. $\log_2(8x) - \log_2(x^2 - 1) = \log_2 3$

$\log_2 \dfrac{8x}{x^x - 1} = \log_2 3$

$\dfrac{8x}{x^2 - 1} = 3$

$(x^2 - 1)\dfrac{8x}{x^2 - 1} = 3(x^2 - 1)$

$8x = 3x^2 - 3$

$0 = 3x^2 - 8x - 3$

$0 = (3x + 1)(x - 3)$

$3x + 1 = 0 \quad x - 3 = 0$

$3x = -1 \quad x = 3$

$x = -\dfrac{1}{3}$

$-\dfrac{1}{3}$ does not check as a solution.
The solution is 3.

43. $\log_9 x + \log_9(2x - 3) = \log_9 2$

$\log_9(x(2x - 3)) = \log_9 2$

$x(2x - 3) = 2$

$2x^2 - 3x = 2$

$2x^2 - 3x - 2 = 0$

$(2x + 1)(x - 2) = 0$

$2x + 1 = 0 \quad x - 2 = 0$

$2x = -1 \quad x = 2$

$x = -\dfrac{1}{2}$

$-\dfrac{1}{2}$ does not check as a solution.
The solution is 2.

45. $\log_2(2x - 3) = 3$

$2x - 3 = 2^3$

$2x - 3 = 8$

$2x = 11$

$x = \dfrac{11}{2}$

The solution is $\dfrac{11}{2}$.

47. $\ln(3x + 2) = 4$

$3x + 2 = e^4$

$3x + 2 = 54.5982$

$3x = 52.5982$

$x = 17.5327$
The solution is 17.5327.

49. $\log_2(x + 1) + \log_2(x + 3) = 3$

$\log_2((x + 1)(x + 3)) = 3$

$\log_2(x^2 + 4x + 3) = 3$

$x^2 + 4x + 3 = 2^3$

$x^2 + 4x + 3 = 8$

$x^2 + 4x - 5 = 0$

$(x + 5)(x - 1) = 0$

$x + 5 = 0 \quad x - 1 = 0$

$x = -5 \quad x = 1$

-5 does not check as a solution.
The solution is 1.

51. $\log_5(2x) - \log_5(x-1) = 1$

$\log_5 \dfrac{2x}{x-1} = 1$

$\dfrac{2x}{x-1} = 5^1$

$(x-1)\dfrac{2x}{x-1} = 5(x-1)$

$2x = 5x - 5$

$-3x = -5$

$x = \dfrac{-5}{-3} = \dfrac{5}{3}$

The solution is $\dfrac{5}{3}$.

53. $\log_8(6x) = \log_8 2 + \log_8(x-4)$

$\log_8(6x) = \log_8(2(x-4))$

$6x = 2x - 8$

$4x = -8$

$x = -2$

-2 does not check as a solution. The equation has no solution.

55. $x - 2 < x$ and therefore $\log(x-2) < \log x$. This means that $\log(x-2) - \log x < 0$ and could not equal the positive number 3.

Critical Thinking

57. $3^{x+1} = 2^{x-2}$

$\log 3^{x+1} = \log 2^{x-2}$

$(x+1)\log 3 = (x-2)\log 2$

$x + 1 = \dfrac{\log 2}{\log 3}(x-2)$

$x + 1 = 0.6309297536(x-2)$

$x + 1 = 0.6309297536x - 1.261859507$

$0.3690702464x = -2.261859507$

$x = -6.1285$

The solution is -6.1285.

59. $7^{2x-1} = 3^{2x+3}$

$\log 7^{2x-1} = \log 3^{2x+3}$

$(2x-1)\log 7 = (2x+3)\log 3$

$2x - 1 = \dfrac{\log 3}{\log 7}(2x+3)$

$2x - 1 = 0.5645750341(2x+3)$

$2x - 1 = 1.129150068x + 1.693725102$

$0.870849932x = 2.693725102$

$x = 3.0932$

The solution is 3.09352.

Projects or Group Activities

61. a) $s = 312.5 \ln \dfrac{e^{0.32t} + e^{-0.32t}}{2}$

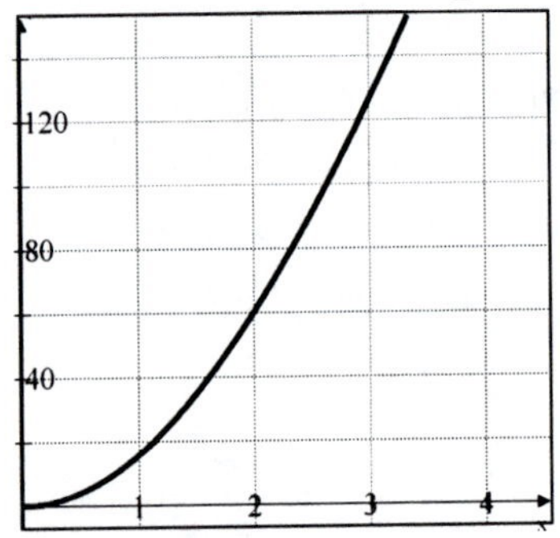

b) Use the graphing calculator to find t when $s = 100$.

$t = 2.64$

It will take 2.64s for the object to fall 100 ft.

Section 12.5

Concept Check

1. Compound interest is interest that is compounded not only on the original principal, but also on the interest already earned.

Objective A Exercises

3. **Strategy**: To find the value of the investment use the compound interest formula.

$P = 1000, n = 8, i = \frac{8\%}{4} = \frac{0.08}{4} = 0.02$

Solution: $A = P(1+i)^n$

$A = 1000(1+0.02)^8$

$A = 1000(1.02)^8$

$A = 1171.66$

The value of the investment after 2 years is \$1172.

5. **Strategy**: To find how many years it will take for the investment to be worth \$15,000 solve the compound interest formula for n.

$A = 15{,}000, P = 5000, i = \frac{6\%}{12} = \frac{0.06}{12} = 0.005$

Solution: $A = P(1+i)^n$

$15000 = 5000(1+0.005)^n$

$3 = (1.005)^n$

$\log 3 = \log(1.005)^n$

$\log 3 = n \log 1.005$

$\frac{\log 3}{\log 1.005} = n$

$n \approx 220$

$\frac{n}{12} = \frac{220}{12} \approx 18$

The investment will be worth \$15,000 in approximately 18 years.

7. a) **Strategy**: To find the technetium level use the exponential decay formula.

$A_0 = 30, k = 6, t = 3$

Solution: $A = A_0(0.5)^{t/k}$

$A = 30(0.5)^{3/6}$

$A = 21.2$

The technetium level is 21.2 mg after 3 h.

b) **Strategy**: To find out how long it will take the technetium level to reach 20 mg use the exponential decay formula.

$A_0 = 30, A = 20, k = 6$

Solution: $A = A_0(0.5)^{t/k}$

$20 = 30(0.5)^{t/6}$

$\frac{2}{3} = 0.5^{t/6}$

$\log \frac{2}{3} = \log 0.5^{t/6}$

$\log \frac{2}{3} = \frac{t}{6} \log 0.5$

$\frac{6 \log \frac{2}{3}}{\log 0.5} = t$

$t = 3.5$

The technetium level is 20 mg after 3.5 h.

9. **Strategy**: To find the half life use the exponential decay formula.

$A_0 = 25, A = 18.95, t = 1$

Solution: $A = A_0(0.5)^{t/k}$

$18.95 = 25(0.5)^{1/k}$

$0.758 = 0.5^{1/k}$

$\log 0.758 = \log 0.5^{1/k}$

$\log 0.758 = \frac{1}{k} \log 0.5$

$k \log 0.758 = \log 0.5$

$k = \frac{\log 0.5}{\log 0.758}$

$k = 2.5$

The half life is 2.5 years.

11. Strategy: To determine the intensity of the earthquake use the Richter scale equation.
$M = 8.9$

Solution: $M = \log \frac{I}{I_0}$

$8.9 = \log \frac{I}{I_0}$

$10^{8.9} = \frac{I}{I_0}$

$I = 10^{8.9} I_0$

$I = 794{,}328{,}235 I_0$

The intensity of the earthquake was $794{,}328{,}235I_0$.

13. Strategy: To determine the how many times stronger the Honshu earthquake was use the Richter scale equation.
$M_1 = 6.9 \quad M_2 = 6.4$

Solution: $M = \log \frac{I}{I_0} = \log I - \log I_0$

$6.9 = \log I_1 - \log I_0$

$6.4 = \log I_2 - \log I_0$

Subtract the equations.

$0.5 = \log I_1 - \log I_2$

$0.5 = \log \frac{I_1}{I_2}$

$10^{0.5} = \frac{I_1}{I_2}$

$I_1 = 10^{0.5} I_2$

$I_1 = 3.16 I_2$

The Honshu earthquake was 3.2 times stronger than the Quetta earthquake.

15. Strategy: To determine the magnitude of the earthquake for the seismogram given use the given equation.
$A = 23 \quad t = 24$

Solution: $M = \log A + 3\log 8t - 2.92$

$M = \log 23 + 3\log 8(24) - 2.92$

$M = \log 23 + 3\log 192 - 2.92$

$M = 5.29$

The magnitude of the earthquake was 5.3.

17. Strategy: To determine the magnitude of the earthquake for the seismogram given use the given equation.
$A = 28 \quad t = 28$

Solution: $M = \log A + 3\log 8t - 2.92$

$M = \log 28 + 3\log 8(28) - 2.92$

$M = \log 28 + 3\log 224 - 2.92$

$M = 5.58$

The magnitude of the earthquake was 5.6.

19. Strategy: To find the pH replace H^+ with its given value and solve for pH.

Solution: $pH = -\log(H^+)$

$pH = -\log(3.98 \times 10^{-9})$

$pH = 8.4$

The pH of baking soda is 8.4.

21. Strategy: To find the hydrogen ion concentration replace pH with its given value and solve for H^+.

Solution: $pH = -\log(H^+)$

$5.3 < -\log(H^+) < 6.6$

$-5.3 > \log(H^+) > -6.6$

$10^{-5.3} > H^+ > 10^{-6.6}$

$5 \times 10^{-6} > H^+ > 2.5 \times 10^{-7}$

The range of hydrogen ion concentration for peanuts is 2.5×10^{-7} to 5.0×10^{-6}.

23. **Strategy**: To find the number of decibels replace I with its given value in the equation and solve for D.

Solution: $D = 10(\log I + 16)$

$D = 10(\log(630) + 16)$

$D = 10(18.7993)$

$D = 187.993$

The blue whale sounds emit 188 decibels.

25. **Strategy**: To find the intensity replace D with its given value in the equation and solve for I.

Solution: $D = 10(\log I + 16)$

$25 = 10(\log(I) + 16)$

$25 = 10\log I + 160$

$-135 = 10\log I$

$-13.5 = \log I$

$10^{-13.5} = I$

$I = 3.16 \times 10^{-14}$

The intensity is 3.16×10^{-14} watts/cm^2.

27. **Strategy**: To find the percent solve the equation for P.

$d = 0.005 \quad k = 20$

Solution: $\log P = -kd$

$\log P = -20(0.005)$

$\log P = -0.1$

$P = 10^{-.1}$

$P = 0.7943$

79.4% of the light will pass through the glass.

29. **Strategy**: To find the thickness of copper needed replace I and I_0 with the given values then solve for x. $I = 0.25 \quad I_0 = 1$

Solution: $I = I_0 e^{-3.2x}$

$0.25 = e^{-3.2x}$

$\ln 0.25 = \ln e^{-3.2x}$

$-1.39 = -3.2x$

$x = 0.43$

The thickness of the copper is 0.4 cm.

31. a) decay
b) growth
c) decay
d) growth

Critical Thinking

33. a) **Strategy:** To find the value of the investments after 3 years, use the given equation.

Solution: $A = A_0 e^{rt}$

$A = 5000e^{(0.06)(3)}$

$A = 5000e^{0.18}$

$A \approx 5986.09$

The value of the investment will be worth $5986.09.

b) **Strategy:** To find the interest rate needed to grow an investment from $1000 to $1250 in 2 years, use the given equation.

Solution: $A = A_0 e^{rt}$

$1250 = 1000e^{2r}$

$1.25 = e^{2r}$

$\ln 1.25 = \ln e^{2r}$

$0.2231 = 2r$

$r \approx 0.112$

The rate must be 11.2%.

Projects or Group Activities

35. a) **Strategy**: Evaluate the given function at $x = 375$ ft.

$$f(x)=\left(\frac{0.5774v+155.3}{v}\right)x+565.3\ln\left(\frac{v-0.2747x}{v}\right)+3.5$$

Solution:

$$f(375)=\left(\frac{0.5774(160)+155.3}{160}\right)375+565.3\ln\left(\frac{160-0.2747(375)}{160}\right)+3.5$$

$$f(375)=\left(\frac{247.684}{160}\right)375+565.3\ln\left(\frac{56.9875}{160}\right)+3.5$$

$$f(375)=580.5094-583.5829+3.5$$

$$f(375)=0.4265$$

The ball will hit 0.43ft from the bottom of the fence.

b) **Strategy**: Increase the speed by 4% so that $v = 166.4$ ft/s.

$$f(375)=\left(\frac{0.5774(166.4)+155.3}{166.4}\right)375+565.3\ln\left(\frac{166.4-0.2747(375)}{166.4}\right)+3.5$$

$$f(375)=566.5100-545.5868+3.5$$

$$f(375)=24.4$$

The height of the ball is 24.4 feet so it will clear the 15 ft fence by approximately 9 ft.

c) **Strategy**: Determine the value for x for which $f(x)$ is greater than 15.

$$f(x)=\left(\frac{0.5774v+155.3}{v}\right)x+565.3\ln\left(\frac{v-0.2747x}{v}\right)+3.5$$

Solution:

$$f(x)=\left(\frac{0.5774(166.4)+155.3}{166.4}\right)x+565.3\ln\left(\frac{166.4-0.2747x}{166.4}\right)+3.5$$

$$x=385$$

Use the graphing calculator to determine if $x = 385$ ft. the height of the ball will be 15.029 ft and will clear the fence.

Chapter 12 Review Exercises

1. $f(2)=e^{2-2}=e^{0}=1$

2. $5^2 = 25$

3. $f(x)=3^{-x}+2$

x	y
0	3
-1	5
1	$7/3$
2	$19/9$

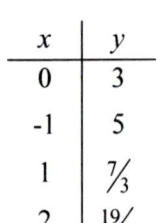

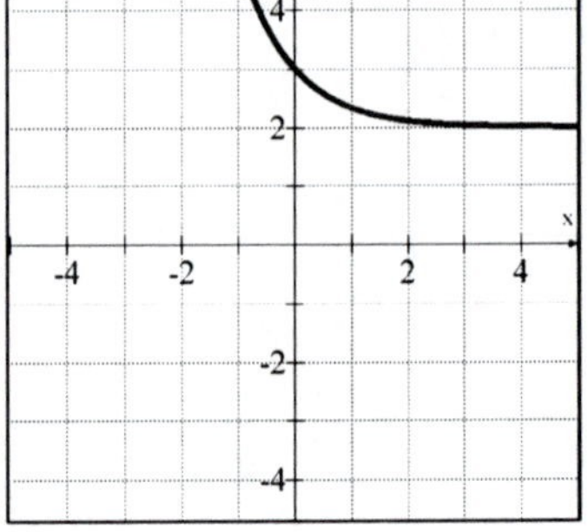

4. $f(x) = \log_3(x-1)$

$y = \log_3(x-1)$ is equivalent to

$x - 1 = 3^y$ or $x = 3^y + 1$

x	y
$4/3$	-1
2	0
4	1

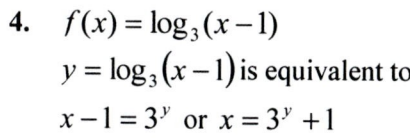

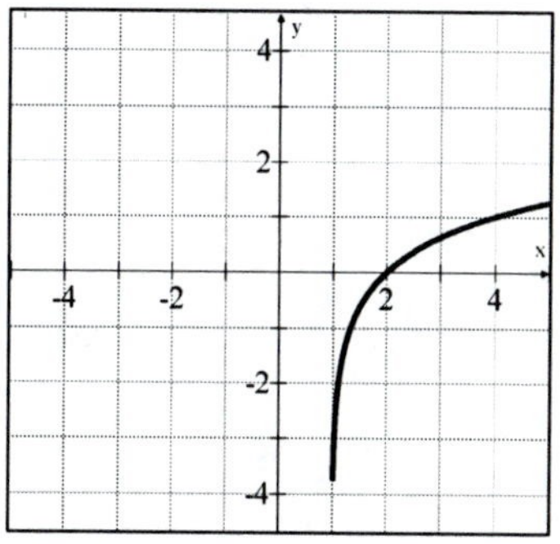

5. $\log_3 \sqrt[5]{x^2y^4} = \log_3(x^2y^4)^{1/5} = \frac{1}{5}\log_3(x^2y^4)$

$= \frac{1}{5}[\log_3 x^2 + \log_3 y^4]$

$= \frac{1}{5}[2\log_3 x + 4\log_3 y]$

$= \frac{2}{5}\log_3 x + \frac{4}{5}\log_3 y$

6. $2\log_3 x - 5\log_3 y = \log_3 x^2 - \log_3 y^5$

$= \log_3\left(\frac{x^2}{y^5}\right)$

7. $27^{2x+4} = 81^{x-3}$

$(3^3)^{2x+4} = (3^4)^{x-3}$

$6x + 12 = 4x - 12$

$2x = -24$

$x = -12$

The solution is -12.

8. $\log_5 \frac{7x+2}{3x} = 1$

Rewrite in exponential form.

$5^1 = \frac{7x+2}{3x}$

$15x = 7x + 2$

$8x = 2$

$x = \frac{1}{4}$

The solution is $\frac{1}{4}$.

9. $\log_6 22 = \frac{\log_{10} 22}{\log_{10} 6} = 1.7251$

10. $\log_2 x = 5$

Rewrite in exponential form.

$2^5 = x$

$x = 32$

The solution is 32.

11. $\log_3(x+2) = 4$

Rewrite in exponential form.

$3^4 = x + 2$

$81 = x + 2$

$x = 79$

The solution is 79.

12. $\log_{10} x = 3$

Rewrite in exponential form.

$10^3 = x$

$1x = 1000$

The solution is 1000.

13. $\frac{1}{3}(\log_7 x + 4\log_7 y) = \frac{1}{3}(\log_7 x + \log_7 y^4)$

$\frac{1}{3}(\log_7 (xy^4)) = \log_7 (xy^4)^{1/3}$

$= \log_7 \sqrt[3]{xy^4}$

14. $\log_8 \sqrt{\frac{x^5}{y^3}} = \log_8 \left(\frac{x^5}{y^3}\right)^{1/2}$

$= \frac{1}{2}(\log_8 x^5 - \log_8 y^3)$

$= \frac{5}{2}\log_8 x - \frac{3}{2}\log_8 y$

15. $\log_2 32 = 5$

16. $\log_3 1.6 = \frac{\log_{10} 1.6}{\log_{10} 3} = 0.4278$

17. $3^{x+2} = 5$

$\log 3^{x+2} = \log 5$

$(x+2)\log 3 = \log 5$

$x + 2 = \frac{\log 5}{\log 3}$

$x = \frac{\log 5}{\log 3} - 2$

$x = -0.535$

The solution is −0.535.

18. $f(-3) = \left(\frac{2}{3}\right)^{-3+2} = \left(\frac{2}{3}\right)^{-1} = \frac{3}{2}$

19. $\log_2 (x+3) - \log_2 (x-1) = 3$

$\log_2 \frac{x+3}{x-1} = 3$

$\frac{x+3}{x-1} = 2^3$

$\frac{x+3}{x-1} = 8$

$(x-1)\frac{x+3}{x-1} = 8(x-1)$

$x + 3 = 8x - 8$

$3 = 7x - 8$

$11 = 7x$

$x = \frac{11}{7}$

The solution is $\frac{11}{7}$.

20. $\log_3 (2x+3) + \log_3 (x-2) = 2$

$\log_3 ((2x+3)(x-2)) = 2$

$2x^2 - x - 6 = 3^2$

$2x^2 - x - 6 = 9$

$2x^2 - x - 15 = 0$

$(2x+5)(x-3) = 0$

$2x + 5 = 0 \quad x - 3 = 0$

$2x = -5 \quad\quad x = 3$

$x = -\frac{5}{2}$

$-\frac{5}{2}$ does not check as a solution.

The solution is 3.

21. $f(x)=\left(\frac{2}{3}\right)^{x+1}$

x	y
-1	1
-2	$3/2$
0	$2/3$
1	$4/9$

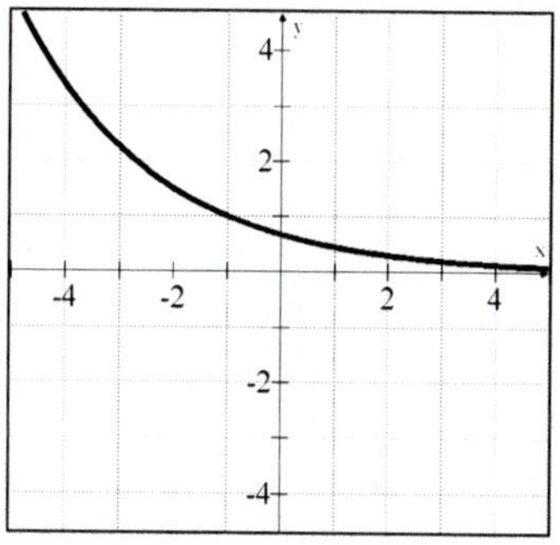

22. $f(x)=\log_2(2x-1)$

$y=\log_2(2x-1)$ is equivalent to

$2x-1=2^y$ or $x=\frac{2^y+1}{2}$

x	y
$3/4$	-1
1	0
$3/2$	1

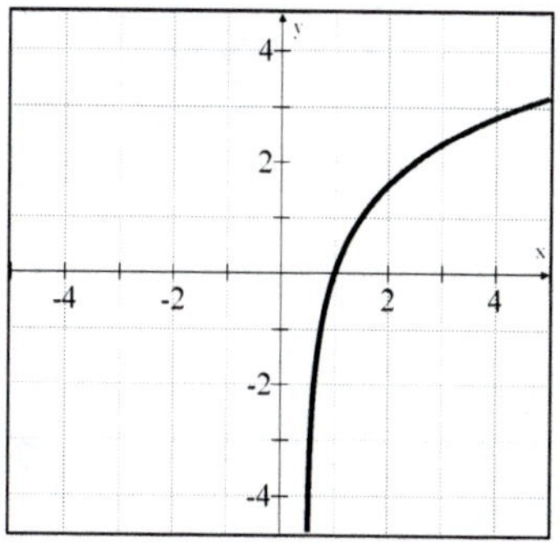

23. $\log_6 36=x$

$6^x=36$

$x=2$

24. $\frac{1}{3}(\log_2 x-\log_2 y)=\frac{1}{3}\log_2\left(\frac{x}{y}\right)$

$=\log_2\left(\frac{x}{y}\right)^{1/3} == \log_2\sqrt[3]{\frac{x}{y}}$

25. $9^{2x}=3^{x+3}$

$(3^2)^{2x}=3^{x+3}$

$3^{4x}=3^{x+3}$

$4x=x+3$

$3x=3$

$x=1$

The solution is 1.

26. $5\cdot 3^{x/2}=12$

$3^{x/2}=2.4$

$\log 3^{x/2}=\log 2.4$

$\frac{x}{2}\log 3=\log 2.4$

$\frac{x}{2}=\frac{\log 2.4}{\log 3}$

$\frac{x}{2}=0.7969$

$x=1.5938$

The solution is 1.5938.

27. $\log_5 x=-1$

Rewrite in exponential form.

$5^{-1}=x$

$\frac{1}{5}=x$

The solution is $\frac{1}{5}$.

28. $\log_3 81=4$

29. $\log x+\log(x-2)=\log 15$

$\log(x(x-2))=\log 15$

$x(x-2)=15$

$x^2-2x-15=0$

$(x-5)(x+3)=0$

$x-5=0 \quad x+3=0$

$x=5 \qquad x=-3$

-3 does not check as a solution.

The solution is 5.

30. $\log_5 \sqrt[3]{x^2 y} = \log_3 (x^2 y)^{1/3} = \frac{1}{3}\log_3 (x^2 y)$

$= \frac{1}{3}[\log_3 x^2 + \log_3 y]$

$= \frac{1}{3}[2\log_3 x + \log_3 y]$

$= \frac{2}{3}\log_3 x + \frac{1}{3}\log_3 y$

31. $6e^{-2x} = 17$

$\ln 6e^{-2x} = \ln 17$

$\ln 6 + \ln e^{-2x} = \ln 17$

$-2x = \ln 17 - \ln 6$

$-2x = 1.0415$

$x = -0.5207$

32. $f(-3) = 7^{-3+2} = 7^{-1} = \frac{1}{7}$

33. $\log_2 16 = x$

$2^x = 16$

$x = 4$

34. $\log_6 x = \log_6 2 + \log_6 (2x - 3)$

$\log_6 x = \log_6 (2(2x-3))$

$x = 2(2x - 3)$

$x = 4x - 6$

$-3x = -6$

$x = 2$

The solution is 2.

35. $\log_2 5 = x$

$x = \frac{\log 5}{\log 2} = 2.3219$

36. $4^x = 8^{x-1}$

$(2^2)^x = (2^3)^{x-1}$

$2^{2x} = 2^{3x-3}$

$2x = 3x - 3$

$-x = -3$

$x = 3$

The solution is 3.

37. $\log_5 x = 4$

Rewrite in exponential form

$x = 5^4 = 625$

38. $3\log_b x - 7\log_b y = \log_b x^3 - \log_b y^7$

$= \log_b\left(\frac{x^3}{y^7}\right)$

39. $f(x) = 5^{-x-1}$

$f(-2) = 5^{-(-2)-1} = 5^{2-1} = 5^1 = 5$

40. $5^{x-2} = 7$

$\log 5^{x-2} = \log 7$

$(x-2)\log 5 = \log 7$

$x - 2 = \frac{\log 7}{\log 5}$

$x = \frac{\log 7}{\log 5} + 2$

$x = 3.2091$

41. **Strategy**: To find the value of the investment use the compound interest formula.

$P = 4000,\ n = 24,\ i = \frac{8\%}{12} = \frac{0.08}{12} = 0.00\overline{6}$

Solution: $A = P(1+i)^n$

$A = 4000(1 + 0.00\overline{6})^{24}$

$A = 4000(1.00\overline{6})^{24}$

$A = 4691.55$

The value of the investment after 2 years is \$4692.

42. Strategy: To determine the magnitude of the earthquake, use the Richter scale equation.

Solution: $M = \log\frac{I}{I_0}$

$M = \log\frac{1{,}584{,}893{,}192I_0}{I_0}$

$M = \log 1584893192$

$M = 9.2$

The magnitude of the earthquake was 9.2.

43. Strategy: To find the half life use the exponential decay formula.

$A_0 = 25,\ A = 15,\ t = 20$

Solution: $A = A_0(0.5)^{t/k}$

$15 = 25(0.5)^{20/k}$

$0.6 = 0.5^{20/k}$

$\log 0.6 = \log 0.5^{20/k}$

$\log 0.6 = \frac{20}{k}\log 0.5$

$k\log 0.6 = 20\log 0.5$

$k = \frac{20\log 0.5}{\log 0.6}$

$k = 27.14$

The half life is 27 days.

44. Strategy: To find the number of decibels replace I with its given value in the equation and solve for D.

Solution: $D = 10(\log I + 16)$

$D = 10(\log(5\times10^{-6}) + 16)$

$D = 10(10.6990)$

$D = 106.99$

The sound emitted from a busy street corner is 107 decibels.

Chapter 12 Test

1. $f(0) = \left(\frac{2}{3}\right)^0 = 1$

2. $f(-2) = 3^{-2+1} = 3^{-1} = \frac{1}{3}$

3. $f(x) = 2^x - 3$

x	y
-2	$-1\frac{1}{4}$
0	-2
2	1

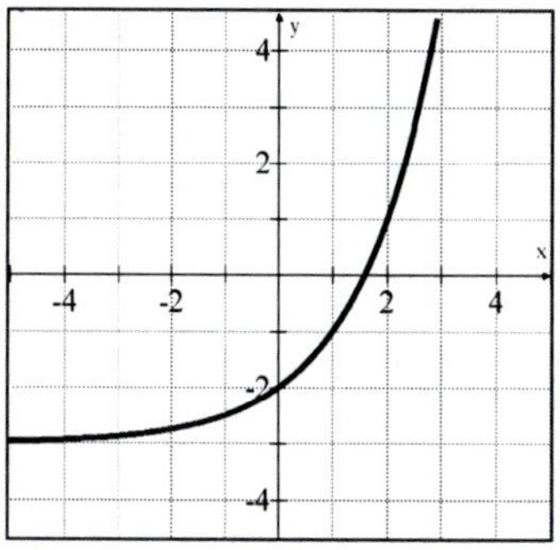

4. $f(x) = 2^x + 2$

x	y
-2	$\frac{9}{4}$
0	3
1	4

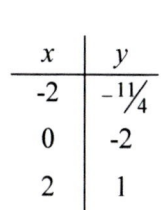

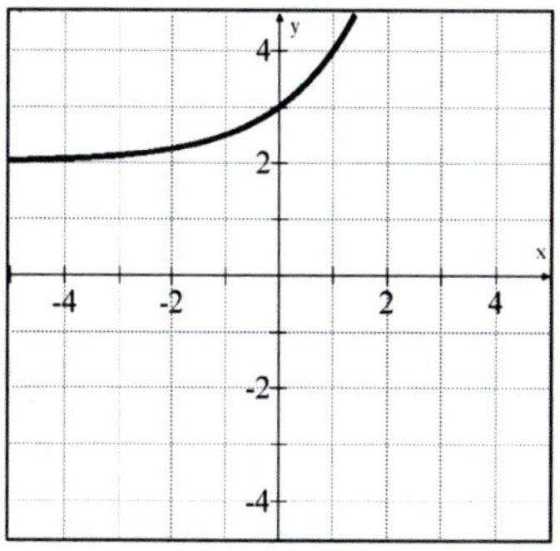

5. $\log_4 16 = x$

$4^x = 16$

$x = 2$

6. $\log_3 x = -2$

$x = 3^{-2} = \frac{1}{3^2} = \frac{1}{9}$

7. $f(x) = \log_2(2x)$

$y = \log_2(2x)$ is equivalent to

$2x = 2^y$ or $x = \dfrac{2^y}{2}$

x	y
$\frac{1}{4}$	-1
$\frac{1}{2}$	0
1	1
2	2

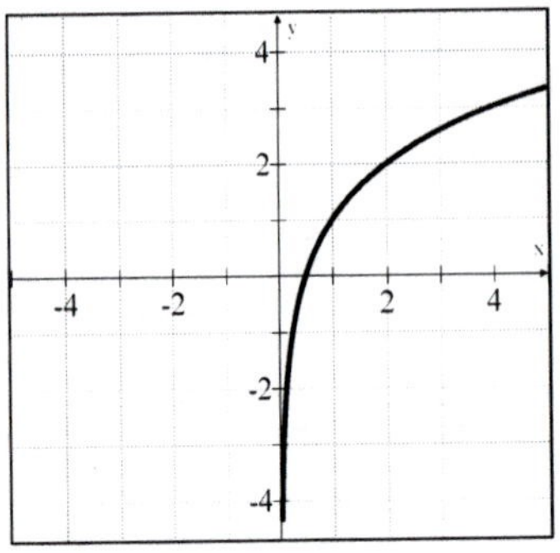

8. $f(x) = \log_3(x+1)$

$y = \log_3(x+1)$ is equivalent to

$x+1 = 3^y$ or $x = 3^y - 1$

x	y
$-\frac{2}{3}$	-1
0	0
2	1

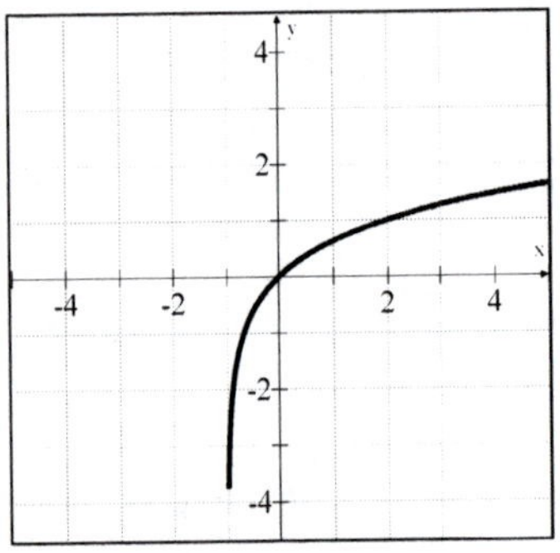

9. $\log_6 \sqrt{xy^3} = \log_6 (xy^3)^{1/2} = \dfrac{1}{2}\log_3(xy^3)$

$= \dfrac{1}{2}[\log_6 x + \log_6 y^3]$

$= \dfrac{1}{2}[\log_6 x + 3\log_6 y]$

$= \dfrac{1}{2}\log_6 x + \dfrac{3}{2}\log_6 y$

10. $\dfrac{1}{2}(\log_3 x - \log_3 y) = \dfrac{1}{2}\log_3\left(\dfrac{x}{y}\right)$

$= \log_3 \sqrt{\dfrac{x}{y}}$

11. $\ln\dfrac{x}{\sqrt{z}} = \ln x - \ln\sqrt{z} = \ln x - \ln z^{1/2}$

$= \ln x - \dfrac{1}{2}\ln z$

12. $3\ln x - \ln y - \dfrac{1}{2}\ln z = \ln x^3 - \ln y - \ln z^{1/2}$

$= \ln\dfrac{x^3}{y} - \ln z^{1/2} = \ln\dfrac{x^3}{yz^{1/2}}$

$= \ln\dfrac{x^3}{y\sqrt{z}}$

13. $3^{7x+1} = 3^{4x-5}$

$7x + 1 = 4x - 5$

$3x = -6$

$x = -2$

The solution is -2.

14. $8^x = 2^{x-6}$

$(2^3)^x = 2^{x-6}$

$3x = x - 6$

$2x = -6$

$x = -3$

The solution is -3.

15. $3^x = 17$

$\log 3^x = \log 17$

$x\log 3 = \log 17$

$x = \dfrac{\log 17}{\log 3}$

$x = 2.5789$

The solution is 2.5789.

16. $\log x + \log(x-4) = \log 12$

$\log(x(x-4)) = \log 12$

$x(x-4) = 12$

$x^2 - 4x - 12 = 0$

$(x-6)(x+2) = 0$

$x-6=0 \quad x+2=0$

$x = 6 \qquad x = -2$

-2 does not check as a solution.
The solution is 6.

17. $\log_6 x + \log_6(x-1) = 1$

$\log_6(x(x-1)) = 1$

$x(x-1) = 6^1$

$x^2 - x - 6 = 0$

$(x-3)(x+2) = 0$

$x-3=0 \quad x+2=0$

$x = 3 \qquad x = -2$

-2 does not check as a solution.
The solution is 3.

18. $\log_5 9 = x$

$x = \dfrac{\log 9}{\log 5} = 1.3652$

19. $\log_3 19 = x$

$x = \dfrac{\log 19}{\log 3} = 2.6801$

20. $5^{2x-5} = 9$

$\log 5^{2x-5} = \log 9$

$(2x-5)\log 5 = \log 9$

$2x - 5 = \dfrac{\log 9}{\log 5}$

$2x - 5 = 1.3652$

$2x = 6.3652$

$x = 3.1826$

The solution is 3.1826.

21. $2e^{x/4} = 9$

$\ln 2e^{x/4} = \ln 9$

$\ln 2 + \ln e^{x/4} = \ln 9$

$\dfrac{x}{4} = \ln 9 - \ln 2$

$\dfrac{x}{4} = 1.5041$

$x = 6.0163$

22. $\log_5(30x) - \log_5(x+1) = 2$

$\log_5 \dfrac{30x}{x+1} = 2$

$\dfrac{30x}{x+1} = 5^2$

$\dfrac{30x}{x+1} = 25$

$(x+1)\dfrac{30x}{x+1} = 25(x+1)$

$30x = 25x + 25$

$5x = 25$

$x = 5$

The solution is 5.

23. **Strategy**: To find the approximate age of the shard use the given equation.

$A_0 = 250, A = 170$

Solution: $A = A_0(0.5)^{t/5570}$

$170 = 250(0.5)^{t/5570}$

$0.68 = 0.5^{t/5570}$

$\log 0.68 = \log 0.5^{t/5570}$

$\log 0.68 = \dfrac{t}{5570}\log 0.5$

$\dfrac{5570 \log 0.68}{\log 0.5} = t$

$t = 3099$

The shard is approximately 3099 years old.

24. **Strategy**: To find the intensity replace D with its given value in the equation and solve for I.

Solution: $D = 10(\log I + 16)$

$75 = 10(\log(I) + 16)$

$75 = 10\log I + 160$

$-85 = 10\log I$

$-8.5 = \log I$

$10^{-8.5} = I$

$I = 3.16 \times 10^{-9}$

The intensity is 3.16×10^{-9} watts/cm^2.

25. **Strategy**: To find out the half life of a radioactive material use the exponential decay formula.

$A_0 = 10, A = 9, t = 5$

Solution: $A = A_0(0.5)^{t/k}$

$9 = 10(0.5)^{5/k}$

$0.9 = 0.5^{5/k}$

$\log 0.9 = \log 0.5^{5/k}$

$\log 0.9 = \frac{5}{k}\log 0.5$

$k = \frac{5\log 0.5}{\log 0.9}$

$k = 32.9$

The half life is 33 h.

Cumulative Review Exercises

1. $4 - 2[x - 3(2 - 3x) - 4x] = 2x$

$4 - 2[x - 6 + 9x - 4x] = 2x$

$4 - 2[6x - 6] = 2x$

$4 - 12x + 12 = 2x$

$16 = 14x$

$x = \frac{16}{14} = \frac{8}{7}$

The solution is $\frac{8}{7}$.

2. $2x - y = 5$

$-y = -2x + 5$

$y = 2x - 5$

$m = 2$ and $(2, -2)$

$y - y_1 = m(x - x_1)$

$y - (-2) = 2(x - 2)$

$y + 2 = 2x - 4$

$y = 2x - 6$

3. $4x^{2n} + 7x^n + 3 = (4x^n + 3)(x^n + 1)$

4. $\dfrac{1 - \frac{5}{x} + \frac{6}{x^2}}{1 + \frac{1}{x} - \frac{6}{x^2}} = \dfrac{1 - \frac{5}{x} + \frac{6}{x^2}}{1 + \frac{1}{x} - \frac{6}{x^2}} \cdot \dfrac{x^2}{x^2}$

$= \dfrac{x^2 - 5x + 6}{x^2 + x - 6} = \dfrac{(x-2)(x-3)}{(x-2)(x+3)}$

$= \dfrac{x-3}{x+3}$

5. $\dfrac{\sqrt{xy}}{\sqrt{x} - \sqrt{y}} = \dfrac{\sqrt{xy}}{\sqrt{x} - \sqrt{y}} \cdot \dfrac{\sqrt{x} + \sqrt{y}}{\sqrt{x} + \sqrt{y}}$

$= \dfrac{\sqrt{x^2 y} + \sqrt{xy^2}}{\sqrt{x^2} - \sqrt{y^2}}$

$= \dfrac{x\sqrt{y} - y\sqrt{x}}{x - y}$

6. $x^2 - 4x - 6 = 0$

$x^2 - 4x + 4 = 6 + 4$

$(x-2)^2 = 10$

$\sqrt{(x-2)^2} = \sqrt{10}$

$x - 2 = \pm\sqrt{10}$

$x = 2 \pm \sqrt{10}$

The solutions are $2 + \sqrt{10}$ and $2 - \sqrt{10}$.

7. $(x-r_1)(x-r_2)=0$

$\left(x-\frac{1}{3}\right)(x-(-3))=0$

$\left(x-\frac{1}{3}\right)(x+3)=0$

$x^2+\frac{8}{3}x-1=0$

$3x^2+8x-3=0$

8. $2x-y<3$ $\quad$ $x+y<1$

$-y<-2x+3$ $\quad$ $y<-x+1$

$y>2x-3$

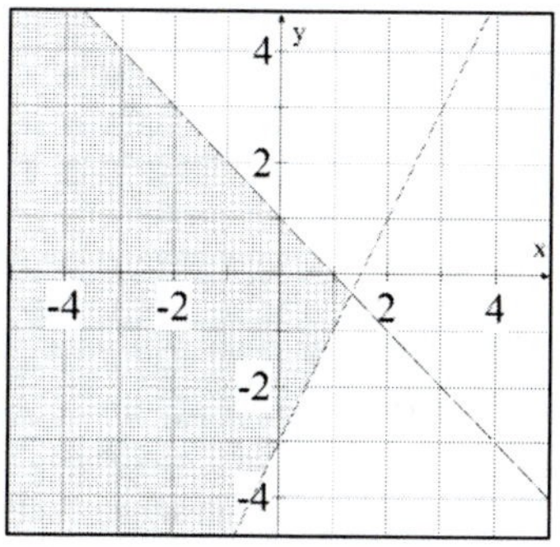

9. (1) $3x-y+z=3$

(2) $x+y+4z=7$

(3 $3x-2y+3z=8$

Eliminate y. Add equations (1) and (2).

$3x-y+z=3$

$x+y+4z=7$

(4) $4x+5z=10$

Multiply equation (2) by 2 and add to equation (3).

$2(x+y+4z)=2(7)$

$3x-2y+3z=8$

$2x+2y+8z=14$

$3x-2y+3z=8$

(5) $5x+11z=22$

Multiply equation (4) by 5 and equation (5) by −4, and add.

$5(4x+5z)=5(10)$

$-4(5x+11z)=-4(22)$

$20x+25z=50$

$-20x-44z=-88$

$-19z=-38$

$z=2$

Replace z with 2 in equation (4).

$4x+5(2)=10$

$4x=0$

$x=0$

Replace x with 0 and z with 2 in equation (1).

$3(0)-y+2=3$

$-y+2=3$

$-y=1$

$y=-1$

The solution is (0, −1, 2).

10. $\frac{x-4}{2-x}-\frac{1-6x}{2x^2-7x+6}$

$=\frac{x-4}{2-x}-\frac{1-6x}{(2x-3)(x-2)}$

$=\frac{x-4}{2-x}+\frac{1-6x}{(2x-3)(2-x)}$

$=\frac{(x-4)}{(2-x)}\cdot\frac{(2x-3)}{(2x-3)}+\frac{1-6x}{(2x-3)(2-x)}$

$=\frac{2x^2-11x+12+1-6x}{(2-x)(2x-3)}=\frac{2x^2-17x+13}{(2-x)(2x-3)}$

$=-\frac{2x^2-17x+13}{(x-2)(2x-3)}$

11. $x^2+4x-5\leq 0$

$(x+5)(x-1)\leq 0$

$\{x\mid -5\leq x\leq 1\}$

12. $|2x-5|\leq 3$

$-3\leq 2x-5\leq 3$

$-3+5\leq 2x-5+5\leq 3+5$

$2\leq 2x\leq 8$

$1\leq x\leq 4$

$\{x\mid 1\leq x\leq 4\}$

13. $f(x) = \left(\frac{1}{2}\right)^x + 1$

x	y
-1	3
0	2
2	$\frac{5}{4}$

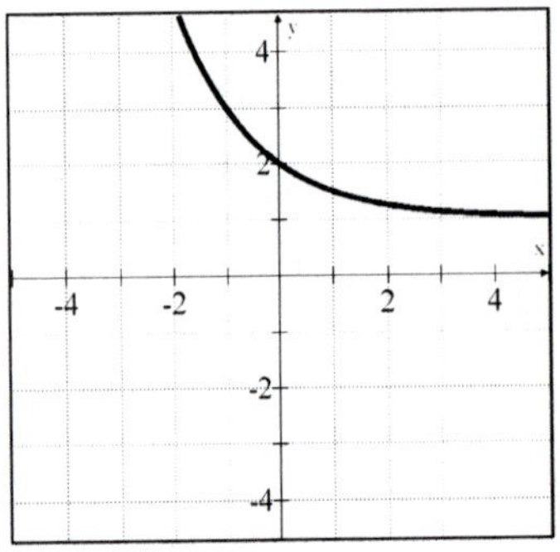

14. $f(x) = \log_2 x - 1$

$y + 1 = \log_2 x$

$x = 2^{y+1}$

x	y
1	-1
2	0
4	1

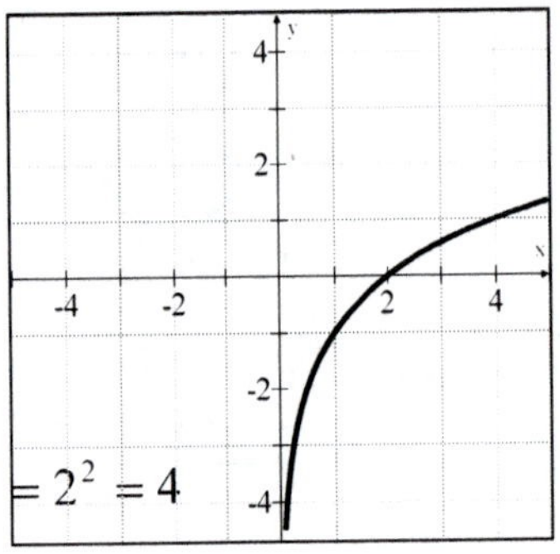

15. $f(-3) = 2^{-(-3)-1} = 2^2 = 4$

16. $\log_5 x = 3$

$x = 5^3 = 125$

17. $3\log_b x - 5\log_b y = \log_b x^3 - \log_b y^5$

$= \log_b \frac{x^3}{y^5}$

18. $\log_3 7 = x$

$x = \frac{\log 7}{\log 3} = 1.7712$

19. $4^{5x-2} = 4^{3x+2}$

$5x - 2 = 3x + 2$

$2x = 4$

$x = 2$

The solution is 2.

20. $\log x + \log(2x+3) = \log 2$

$\log(x(2x+3)) = \log 2$

$x(2x+3) = 2$

$2x^2 + 3x - 2 = 0$

$(2x-1)(x+2) = 0$

$2x - 1 = 0 \quad x + 2 = 0$

$2x = 1 \qquad x = -2$

$x = \frac{1}{2}$

-2 does not check as a solution.

The solution is $\frac{1}{2}$.

21. Strategy: Let c represent the number of checks. To find the number of checks, write and solve an inequality.

Solution: $5.00 + 0.02c > 2.00 + 0.08c$

$5 > 2 + 0.06c$

$3 > 0.06c$

$50 > c$

The customer can write at most 49 checks.

22. Strategy: Let x represent the cost per pound of the mixture.

	Amount	Cost	Value
\$4.00	16	4.00	16(4.00)
\$2.50	24	2.50	24(2.50)
Mixture	40	x	$40x$

The sum of the values before mixing is equal to the value after mixing.

Solution:

$16(4.00) + 24(2.50) = 40x$

$64 + 60 = 40x$

$124 = 40x$

$x = 3.1$

The cost per pound of the mixture is \$3.10.

23. Strategy: Let x represent the rate of the wind.

	Distance	Rate	Time
With wind	1000	$225+x$	$\frac{1000}{225+x}$
Against wind	800	$225-x$	$\frac{800}{225-x}$

The flying time with the wind is the same as the flying time against the wind.

Solution:

$$\frac{1000}{225+x}=\frac{800}{225-x}$$

$$(225+x)(225-x)\frac{1000}{225+x}=\frac{800}{225-x}(225+x)(225-x)$$

$$(225-x)1000=800(225+x)$$

$$225000-1000x=180000+800x$$

$$45000=1800x$$

$$25=x$$

The rate of the wind is 25 mph.

24. Strategy: Write the basic direct variation equation replacing the variable with the given values. Solve for k.
Write the direct variation equation replacing k with its value. Substitute 34 for f and solve for d.

Solution:

$d=kf \qquad d=0.3f$

$6=k(20) \qquad =0.3(34)$

$0.3=k \qquad =10.2$

The string will stretch 10.2 in.

25. Strategy: Let x represent the cost of redwood. The cost of fir is y.
First purchase:

	Amount	Cost	Value
Redwood	80	x	$80x$
Fir	140	y	$140y$

Second purchase:

	Amount	Cost	Value
Redwood	140	x	$140x$
Fir	100	y	$100y$

The total cost of the first purchase is \$67.
The total cost of the second purchase is \$81.

Solution:

$80x+140y=67$
$140x+100y=81$

$-5(80x+140y)=-5(67)$
$7(140x+100y)=7(81)$

$-400x-700y=-335$
$980x+700y=567$

$580x=232$
$x=0.40$

$80(0.40)+140y=67$
$32+140y=67$
$140y=35$
$y=0.25$

The cost of the redwood is \$0.40 per foot.
The cost of the fir is \$0.25 per foot.

26. Strategy: To find how many years it will take for the investment to double in value.

$$A=10{,}000,\ P=5000,\ i=\frac{7\%}{2}=\frac{0.07}{2}=0.035$$

Solution: $A=P(1+i)^n$

$$10000=5000(1+0.035)^n$$

$$2=(1.035)^n$$

$$\log 2=\log(1.035)^n$$

$$\log 2=n\log 1.035$$

$$\frac{\log 2}{\log 1.035}=n$$

$$n=20$$

$$\frac{n}{2}=\frac{20}{2}=10$$

The investment will take 10 years to double in value.

Final Exam

1. $12-8[3-(-2)]^2 \div 5-3$

$=12-8(3+2)^2 \div 5-3$

$=12-8(5)^2 \div 5-3$

$=12-8(25)\div 5-3$

$=12-200\div 5-3$

$=12-40-3$

$=-31$

2. $\dfrac{a^2-b^2}{a-b}$

$\dfrac{3^2-(-4)^2}{3-(-4)}=\dfrac{9-16}{3+4}=\dfrac{-7}{7}$

$=-1$

3. $5-2[3x-7(2-x)-5x]$

$=5-2[3x-14+7x-5x]$

$=5-2[5x-14]$

$=5-10x+28$

$=-10x+33$

4. $\dfrac{3}{4}x-2=4$

$\dfrac{3}{4}x=6$

$\dfrac{4}{3}\cdot\dfrac{3}{4}x=6\cdot\dfrac{4}{3}$

$x=8$

The solution is 8.

5. $8-|5-3x|=1$

$-|5-3x|=-7$

$|5-3x|=7$

$5-3x=7 \quad 5-3x=-7$

$-3x=2 \quad -3x=-12$

$x=-\dfrac{2}{3} \quad x=4$

The solutions are $-\dfrac{2}{3}$ and 4.

6. $V=\dfrac{4}{3}\pi r^3=\dfrac{4}{3}\pi(4)^3=268.1\ ft^3$

7. $2x-3y=9$

$2x-3(0)=9$

$2x=9$

$x=\dfrac{9}{2}$

The x-intercept is $\left(\dfrac{9}{2},0\right)$.

$2(0)-3y=9$

$-3y=9$

$y=-3$

The y-intercept is $(0,-3)$.

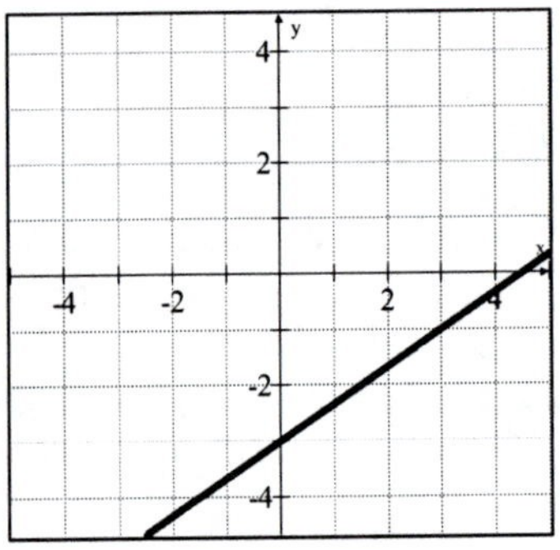

8. $(3,-2)$ and $(1,4)$

$m=\dfrac{y_2-y_1}{x_2-x_1}=\dfrac{4-(-2)}{1-3}=\dfrac{4+2}{-2}=\dfrac{6}{-2}$

$m=-3$

$y-y_1=m(x-x_1)$

$y-4=-3(x-1)$

$y-4=-3x+3$

$y=-3x+7$

9. $3x - 2y = 6$

$-2y = -3x + 6$

$y = \frac{3}{2}x - 3$

$m_1 = \frac{3}{2}$

$m_1 \cdot m_2 = -1$

$\frac{3}{2}m_2 = -1$

$m_2 = -\frac{2}{3} \quad (-2,1)$

$y - y_1 = m(x - x_1)$

$y - 1 = -\frac{2}{3}(x - (-2))$

$y - 1 = -\frac{2}{3}(x + 2)$

$y - 1 = -\frac{2}{3}x - \frac{4}{3}$

$y = -\frac{2}{3}x - \frac{1}{3}$

10. $2a[5 - a(2 - 3a) - 2a] + 3a^2$
$= 2a[5 - 2a + 3a^2 - 2a] + 3a^2$
$= 2a[5 - 4a + 3a^2] + 3a^2$
$= 10a - 8a^2 + 6a^3 + 3a^2$
$= 6a^3 - 5a^2 + 10a$

11. $8 - x^3y^3 = 2^3 - (xy)^3$
$= (2 - xy)(4 + 2xy + x^2y^2)$

12. $x - y - x^3 + x^2y = x - y - x^2(x - y)$
$= 1(x - y) - x^2(x - y)$
$= (x - y)(1 - x^2)$
$= (x - y)(1 - x)(1 + x)$

13.

$$\begin{array}{r} x^2 - 2x - 3 \\ 2x - 3\overline{\smash{\big)}\,2x^3 - 7x^2 + 0x + 4} \\ \underline{2x^3 - 3x^2} \qquad\qquad \\ -4x^2 + 0x \qquad \\ \underline{-4x^2 + 6x} \qquad \\ -6x + 4 \\ \underline{-6x + 9} \\ -5 \end{array}$$

$(2x^3 - 7x^2 + 4) \div (2x - 3) = x^2 - 2x - 3 + \frac{-5}{2x - 3}$

14. $\frac{x^2 - 3x}{2x^2 - 3x - 5} \div \frac{4x - 12}{4x^2 - 4}$

$= \frac{x^2 - 3x}{2x^2 - 3x - 5} \cdot \frac{4x^2 - 4}{4x - 12}$

$= \frac{x(x - 3)}{(2x - 5)(x + 1)} \cdot \frac{4(x + 1)(x - 1)}{4(x - 3)}$

$= \frac{x(x - 3) \cdot 4(x + 1)(x - 1)}{(2x - 5)(x + 1) \cdot 4(x - 3)}$

$= \frac{x(x - 1)}{2x - 5}$

15. The LCM is $(x - 3)(x + 2)$.

$\frac{x - 2}{x + 2} - \frac{x + 3}{x - 3} = \frac{x - 2}{x + 2} \cdot \frac{x - 3}{x - 3} - \frac{x + 3}{x - 3} \cdot \frac{x + 2}{x + 2}$

$= \frac{(x - 2)(x - 3) - (x + 3)(x + 2)}{(x - 3)(x + 2)}$

$= \frac{x^2 - 5x + 6 - x^2 - 5x - 6}{(x - 3)(x + 2)}$

$= \frac{-10x}{(x - 3)(x + 2)}$

16. The LCM is $x(x+4)$.

$$\frac{\frac{3}{x}+\frac{1}{x+4}}{\frac{1}{x}+\frac{3}{x+4}}=\frac{\frac{3}{x}+\frac{1}{x+4}}{\frac{1}{x}+\frac{3}{x+4}}\cdot\frac{x(x+4)}{x(x+4)}$$
$$=\frac{3x+12+x}{x+4+3x}=\frac{4x+12}{4x+4}$$
$$=\frac{4(x+3)}{4(x+1)}=\frac{x+3}{x+1}$$

17. $\frac{5}{x-2}-\frac{5}{x^2-4}=\frac{1}{x+2}$

$$\frac{5}{x-2}-\frac{5}{(x+2)(x-2)}=\frac{1}{x+2}$$
$$(x-2)(x+2)\left(\frac{5}{x-2}-\frac{5}{(x+2)(x-2)}\right)$$
$$=(x-2)(x+2)\left(\frac{1}{x+2}\right)$$
$$5(x+2)-5=x-2$$
$$5x+10-5=x-2$$
$$4x+5=-2$$
$$4x=-7$$
$$x=-\frac{7}{4}$$

The solution is $-\frac{7}{4}$.

18. $a_n=a_1+(n-1)d$

$$a_n-a_1=(n-1)d$$
$$d=\frac{a_n-a_1}{n-1}$$

19. $\left(\frac{4x^2y^{-1}}{3x^{-1}y}\right)^{-2}\left(\frac{2x^{-1}y^2}{9x^{-2}y^2}\right)^3$

$$\frac{4^{-2}x^{-4}y^2}{3^{-2}x^2y^{-2}}\cdot\frac{2^3x^{-3}y^6}{9^3x^{-6}y^6}$$
$$=\frac{2^3\cdot 3^2y^4x^3}{4^2\cdot 9^3x^6}$$
$$=\frac{y^4}{162x^3}$$

20. $\left(\frac{3x^{2/3}y^{1/2}}{6x^2y^{4/3}}\right)^6=\frac{3^6x^4y^3}{6^6x^{12}y^8}=\frac{1}{64x^8y^5}$

21. $x\sqrt{18x^2y^3}-y\sqrt{50x^4y}$

$$=x\sqrt{3^2x^2y^2(2y)}-y\sqrt{5^2x^4(2y)}$$
$$=3x^2y\sqrt{2y}-5x^2y\sqrt{2y}$$
$$=-2x^2y\sqrt{2y}$$

22. $\frac{\sqrt{16x^5y^4}}{\sqrt{32xy^7}}=\sqrt{\frac{16x^5y^4}{32xy^7}}=\sqrt{\frac{x^4}{2y^3}}$

$$=\sqrt{\frac{x^4}{y^2(2y)}}=\frac{x^2}{y}\sqrt{\frac{1}{2y}}\cdot\sqrt{\frac{2y}{2y}}$$
$$=\frac{x^2}{y}\sqrt{\frac{2y}{(2y)^2}}$$
$$=\frac{x^2\sqrt{2y}}{2y^2}$$

23. $\frac{3}{2+i}\cdot\frac{2-i}{2-i}=\frac{6-3i}{4-i^2}=\frac{6-3i}{4+1}$

$$=\frac{6-3i}{5}=\frac{6}{5}-\frac{3}{5}i$$

24. $(x - r_1)(x - r_2) = 0$

$$\left(x - \left(-\frac{1}{2}\right)\right)(x - 2) = 0$$

$$\left(x + \frac{1}{2}\right)(x - 2) = 0$$

$$x^2 - \frac{3}{2}x - 1 = 0$$

$$2(x^2 - \frac{3}{2}x - 1) = 0(2)$$

$$2x^2 - 3x - 2 = 0$$

25. $2x^2 - 3x - 1 = 0$

$a = 2, b = -3, c = -1$

$$x = \frac{-b \pm \sqrt{b^2 - 4ac}}{2a}$$

$$x = \frac{-(-3) \pm \sqrt{(-3)^2 - 4(2)(-1)}}{2(2)}$$

$$x = \frac{3 \pm \sqrt{9 + 8}}{4}$$

$$x = \frac{3 \pm \sqrt{17}}{4}$$

The solutions are $\frac{3 + \sqrt{17}}{4}$ and $\frac{3 - \sqrt{17}}{4}$.

26. $x^{2/3} - x^{1/3} - 6 = 0$

$$\left(x^{1/3}\right)^2 - x^{1/3} - 6 = 0$$

$$u^2 - u - 6 = 0$$

$$(u - 3)(u + 2) = 0$$

$$u - 3 = 0 \quad u + 2 = 0$$

$$u = 3 \qquad u = -2$$

Replace u with $x^{1/3}$.

$$x^{1/3} = 3 \qquad x^{1/3} = -2$$

$$\left(x^{1/3}\right)^3 = (3)^3 \quad \left(x^{1/3}\right)^3 = (-2)^3$$

$$x = 27 \qquad x = -8$$

The solutions are −8 and 27.

27. $f(x) = -x^2 + 4$

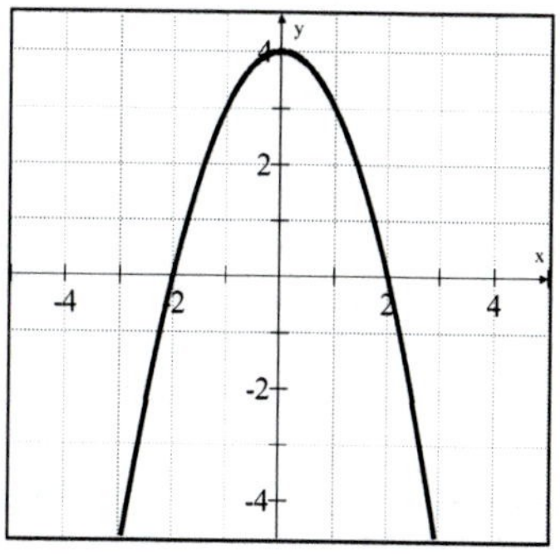

28. $f(x) = -\frac{1}{2}x - 3$

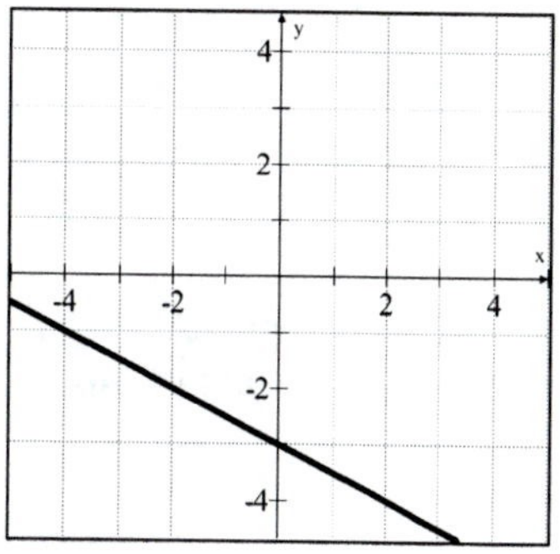

29. $\frac{2}{x} - \frac{2}{2x + 3} = 1$

$$x(2x + 3)\left(\frac{2}{x} - \frac{2}{2x + 3}\right) = 1x(2x + 3)$$

$$2(2x + 3) - 2x = 2x^2 + 3x$$

$$4x + 6 - 2x = 2x^2 + 3x$$

$$0 = 2x^2 + x - 6$$

$$(2x - 3)(x + 2) = 0$$

$$2x - 3 = 0 \quad x + 2 = 0$$

$$2x = 3 \qquad x = -2$$

$$x = \frac{3}{2}$$

The solutions are −2 and $\frac{3}{2}$.

30. $f(x) = \frac{2}{3}x - 4$

$y = \frac{2}{3}x - 4$

$x = \frac{2}{3}y - 4$

$x + 4 = \frac{2}{3}y$

$\frac{3}{2}(x+4) = \frac{3}{2}\left(\frac{2}{3}y\right)$

$y = \frac{3}{2}x + 6$

$f^{-1}(x) = \frac{3}{2}x + 6$

31. (1) $3x - 2y = 1$
(2) $5x - 3y = 3$
Eliminate y. Multiply equation (1) by -3 and equation (2) by 2. Add the two equations.

$-3(3x - 2y) = -3(1)$
$2(5x - 3y) = 2(3)$

$-9x + 6y = -3$
$10x - 6y = 6$
$x = 3$

Substitute 3 for x in equation (1).
$3(3) - 2y = 1$
$9 - 2y = 1$
$-2y = -8$
$y = 4$
The solution is (3, 4).

32. $\sqrt{49x^6} = \sqrt{7^2x^6} = 7x^3$

33. $2 - 3x < 6$ and $2x + 1 > 4$
$-3x < 4$ $\quad 2x > 3$
$x > -\frac{4}{3}$ $\quad x > \frac{3}{2}$
The solution is $\left\{x \mid x > \frac{3}{2}\right\}$.

34. $|2x + 5| < 3$
$-3 < 2x + 5 < 3$
$-3 - 5 < 2x + 5 - 5 < 3 - 5$
$-8 < 2x < -2$
$-4 < x < -1$
$\{x \mid -4 < x < -1\}$

35. $3x + 2y > 6$
$2y > -3x + 6$
$y > -\frac{3}{2}x + 3$

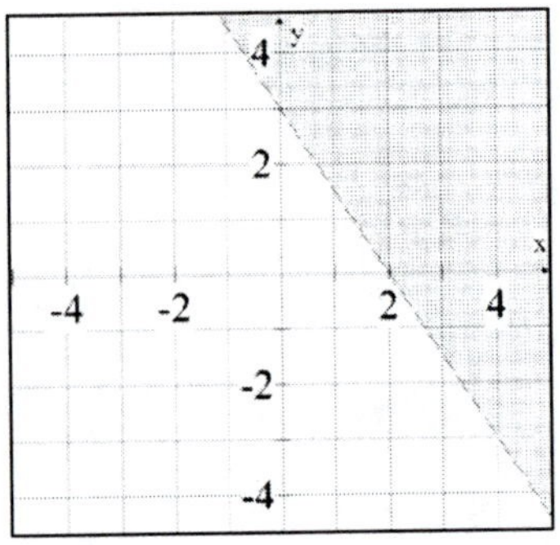

36. $f(x) = 3^{-x} - 2$

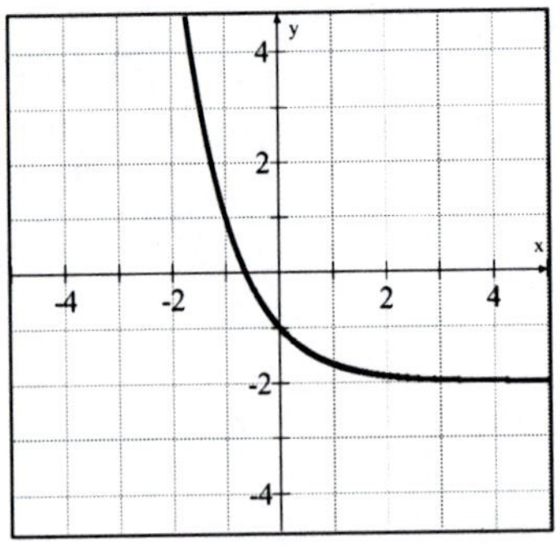

37. $f(x) = \log_2(x+1)$

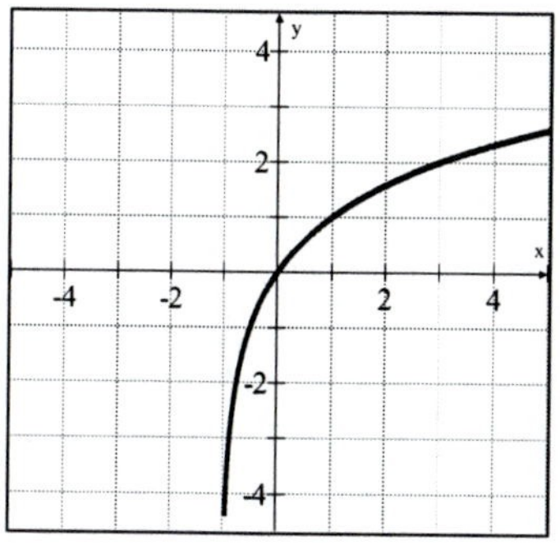

38. $2(\log_2 a - \log_2 b) = 2\log_2 \frac{a}{b} = \log_2 \frac{a^2}{b^2}$

39. $\log_3(x) - \log_3(x-3) = \log_3 2$

$\log_3 \frac{x}{x-3} = \log_3 2$

$\frac{x}{x-3} = 2$

$(x-3)\frac{x}{x-3} = 2(x-3)$

$x = 2x - 6$

$-x = -6$

$x = \frac{-6}{-1} = 6$

The solution is 6.

40. Strategy: Let x represent the score on the last test.
To find the range of scores solve the inequality.

Solution:

$70 \le \frac{64+58+82+77+x}{5} \le 79$

$70 \le \frac{281+x}{5} \le 79$

$350 \le 281 + x \le 395$

$69 \le x \le 114$

Since 100 is the maximum score, the range of scores is $69 \le x \le 100$.

41. Strategy: Let x represent the average speed of the jogger.
The average speed of the cyclist is $2.5x$.

	Rate	Time	Distance
Jogger	x	2	$2(x)$
Cyclist	$2.5x$	2	$2(2.5x)$

The distance traveled by the cyclist is 24 miles more than the distance traveled by th2 jogger.

Solution: $2x + 24 = 2(2.5x)$

$2x + 24 = 5x$

$24 = 3x$

$x = 8$

$2(2.5x) = 5(8) = 40$

The cyclist traveled 40 mi.

42. Strategy: Let x represent the amount invested at 8.5%.
The amount invested at 6.4% is $12000 - x$.

	Principal	Rate	Interest
8.5%	x	0.085	$0.085x$
6.4%	$12000 - x$	0.064	$0.064(12000 - x)$

The sum of the interest earned from the two investments is $936.

Solution:

$0.085x + 0.064(12000 - x) = 936$

$0.085x + 768 - 0.064x = 936$

$0.021x + 768 = 936$

$0.021x = 168$

$x = 8000$

$12000 - x = 12000 - 8000 = 4000$

$8,000 is invested at 8.5% and $4,000 is invested at 6.4%.

43. Strategy: Let x represent the width of the rectangle.
The length of the rectangle is $3x - 1$.
The area of the rectangle is 140 ft^2.

Solution: $A = LW$

$$140 = x(3x-1)$$
$$140 = 3x^2 - x$$
$$0 = 3x^2 - x - 140$$
$$0 = (3x+20)(x-7)$$
$$3x+20=0 \quad x-7=0$$
$$3x=-20 \quad x=7$$
$$x=-\frac{20}{3}$$

The width cannot be a negative number.
$3x - 1 = 3(7) - 1 = 20$
The width is 7 ft.
The length is 20 ft.

44. Strategy: Let x represent the number of additional shares. Write and solve a proportion.

Solution: $\frac{300}{486} = \frac{300+x}{810}$

$$(810 \cdot 486)\frac{300}{486} = \frac{300+x}{810}(810 \cdot 486)$$
$$300(810) = 486(300+x)$$
$$243{,}000 = 145{,}800 + 486x$$
$$97200 = 486x$$
$$x = 200$$

200 additional shares would need to be purchased.

45. Strategy: Let x represent the rate of the car.
The rate of the plane is $7x$.

	Distance	Rate	Time
Car	45	x	$\frac{45}{x}$
Plane	1050	$7x$	$\frac{1050}{7x}$

The total time traveled is $3\frac{1}{4}$ h.

Solution: $\frac{45}{x} + \frac{1050}{7x} = 3\frac{1}{4}$

$$\frac{45}{x} + \frac{150}{x} = \frac{13}{4}$$
$$4x\left(\frac{45}{x} + \frac{150}{x}\right) = \left(\frac{13}{4}\right)4x$$
$$180 + 600 = 13x$$
$$780 = 13x$$
$$x = 60$$
$$7x = 7(60) = 420$$

The rate of the plane is 420 mph.

46. Strategy: To find the distance the object has fallen, substitute 75 ft/s for v and solve for d.

Solution: $v = \sqrt{64d}$

$$75 = \sqrt{64d}$$
$$75^2 = \left(\sqrt{64d}\right)^2$$
$$5625 = 64d$$
$$d = 87.89$$

The distance traveled is 88 ft.

47. Strategy: Let x represent the rate traveled during the first 360 mi.
The rate traveled during the next 300 mi is $x + 30$.

	Distance	Rate	Time
First part of the trip	360	x	$\frac{360}{x}$
Second part of the trip	300	$x + 30$	$\frac{300}{x+30}$

The total time traveled was 5 h.

Solution: $\frac{360}{x} + \frac{300}{x+30} = 5$

$$x(x+30)\left(\frac{360}{x} + \frac{300}{x+30}\right) = 5(x)(x+30)$$
$$360(x+30) + 300x = 5x^2 + 150x$$
$$360x + 10800 + 300x = 5x^2 + 150x$$

$0 = 5x^2 - 510x - 10800$

$0 = 5(x^2 + 102x - 2160)$

$0 = 5(x + 18)(x - 120)$

$x + 18 = 0 \quad x - 120 = 0$

$x = -18 \qquad x = 120$

The rate cannot be a negative number.
The rate of the plane for the first 360 mi is 120 mph.

48. Strategy: Write the basic inverse variation equation replacing the variable with the given values. Solve for *k*.
Write the inverse variation equation replacing *k* with its value. Substitute 4 for *d* and solve for *I*.

Solution:

$I = \frac{k}{d^2}$

$8 = \frac{k}{20^2}$

$8 = \frac{k}{400}$

$3200 = k$

$I = \frac{3200}{d^2} = \frac{3200}{4^2} = \frac{3200}{16} = 200$

The intensity is 200 foot-candles.

49. Strategy: Let *x* represent the rate of the boat in calm water.
The rate of the current is *y*.

	Rate	Time	Distance
With current	$x + y$	2	$2(x + y)$
Against current	$x - y$	3	$3(x - y)$

The distance traveled with the current is 30 miles. The distance traveled against the current is 30 miles.

$2(x + y) = 30$

$3(x - y) = 30$

Solution:

$2(x + y) = 30$

$3(x - y) = 30$

$\frac{1}{2} \cdot 2(x + y) = \frac{1}{2} \cdot 30$

$\frac{1}{3} \cdot 3(x - y) = \frac{1}{3} \cdot 30$

$x + y = 15$

$x - y = 10$

$2x = 25$

$x = 12.5$

$x + y = 15$

$12.5 + y = 15$

$y = 2.5$

The rate of the boat in calm water is 12.5 mph. The rate of the current is 2.5 mph.

50. Strategy: To find the value of the investment after two years use the compound interest formula.

$A = 4000,\ n = 24,\ i = \frac{9\%}{12} = \frac{0.09}{12} = 0.0075$

Solution: $P = A(1 + i)^n$

$P = 4000(1 + 0.0075)^{24}$

$P = 4000(1.0075)^{24}$

$P = 4785.65$

The value of the investment after 2 years is \$4785.65.

CHAPTER T: Transitioning To Intermediate Algebra

Section T.1

Objective A Exercises

1. $a-2c$

$2-2(-4)=2+8=10$

3. $3b-3c$

$3(3)-3(-4)=9+12=21$

5. $16\div(2c)$

$16\div[(2)(-4)]=16\div(-8)=-2$

7. $3b-(a+c)^2$

$3(3)-(2+(-4))^2=9-(-2)^2=9-4=5$

9. $(b-3a)^2+bc$

$(3-3(2))^2+3(-4)=(3-6)^2-12$

$=(-3)^2-12=9-12=-3$

11. $\dfrac{d-b}{a}$

$=\dfrac{4-3}{-1}=\dfrac{1}{-1}=-1$

13. $\dfrac{b-d}{c-a}$

$\dfrac{3-4}{-2-(-1)}=\dfrac{-1}{-2+1}=\dfrac{-1}{-1}=1$

15. $3(b-a)-bc$

$3[3-(-1)]-3(-2)=3(3+1)+6$

$=3(4)+6=12+6=18$

17. $\dfrac{abc}{b-d}$

$\dfrac{(-1)(3)(-2)}{3-4}=\dfrac{-3(-2)}{-1}=\dfrac{6}{-1}=-6$

19. $(-b+d)^2+(-a+c)^2$

$(-3+4)^2+(-(-1)+(-2))^2=(1)^2+(1-2)^2$

$=1+(-1)^2=1+1=2$

21. $3cd-(4a)^2$

$3(-2)(4)-[4(-1)]^2=-6(4)-(-4)^2$

$=-24-16=-40$

$(a+b)^2-c$

23. $[2.7+(-1.6)]^2-(-0.8)=(1.1)^2+0.8$

$=1.21+0.8=2.01$

Objective B Exercises

25. $x+7x=8x$

27. $8b-5b=3b$

29. $-12a+17a=5a$

31. $4x+5x+2x=9x+2x=11x$

33. $6x-2y+9x=(6x+9x)-2y=15x-2y$

35. $5a+6a-2a=11a-2a=9a$

37. $12y^2+10y^2=22y^2$

39. $\dfrac{3}{4}x-\dfrac{1}{4}x=\dfrac{2}{4}x=\dfrac{1}{2}x$

41. $-4(5x)=-20x$

43. $(6a)(-4)=-24a$

45. $\dfrac{1}{4}(4x)=\dfrac{4}{4}x=x$

47. $\dfrac{1}{3}(21x)=7x$

49. $(36y)\left(\dfrac{1}{12}\right)=\dfrac{36y}{12}=3y$

51. $-3(a+5)=-3a-15$

53. $(-2x-6)8=-16x-48$

55. $-5(2y^2-1)=-10y^2+5$

57. $6(3x^2-2xy-y^2)=18x^2-12xy-6y^2$

59. $3-(10+8y)=3-10-8y=-8y-7$

61. $-5[2x+3(5-x]=-5[2x+15-3x]$

$=-5[-x+15]=5x-75$

63. $-5a-2[2a-4(a+7)]=-5a-2[2a-4a-28]$

$=-5a-2[-2a-28]$

$=-5a+4a+56=-a+56$

Section T.2

Objective A Exercises

1.
$$\begin{aligned} x+7&=-5 \\ x+7-7&=-5-7 \\ x&=-12 \end{aligned}$$
The solution is –12.

3.
$$\begin{aligned} -9&=z-8 \\ -9+8&=z-8+8 \\ -1&=z \end{aligned}$$
The solution is –1.

5.
$$\begin{aligned} -48&=6z \\ \frac{-48}{6}&=\frac{6z}{6} \\ -8&=z \end{aligned}$$
The solution is –8

7.
$$\begin{aligned} -\frac{3}{4}x&=15 \\ -\frac{4}{3}\left(-\frac{3}{4}x\right)&=-\frac{4}{3}(15) \\ x&=-20 \end{aligned}$$
The solution is –20.

9.
$$\begin{aligned} -\frac{x}{4}&=-2 \\ -4\left(-\frac{x}{4}\right)&=-4(-2) \\ x&=8 \end{aligned}$$
The solution is 8.

11.
$$\begin{aligned} 4-2b&=-2-4b \\ 4-2b+4b&=2-4b+4b \\ 4+2b&=2 \\ 4-4+2b&=2-4 \\ 2b&=-2 \\ \frac{2b}{2}&=\frac{-2}{2} \\ b&=-1 \end{aligned}$$
The solution is –1.

13.
$$\begin{aligned} 5x-3&=9x-7 \\ 5x-9x-3&=9x-9x-7 \\ -4x-3&=-7 \\ -4x-3+3&=-7+3 \\ -4x&=-4 \\ \frac{-4x}{-4}&=\frac{-4}{-4} \\ x&=1 \end{aligned}$$
The solution is 1.

15.
$$\begin{aligned} 6a-1&=2+2a \\ 6a-2a-1&=2+2a-2a \\ 4a-1&=2 \\ 4a-1+1&=2+1 \\ 4a&=3 \\ \frac{4a}{4}&=\frac{3}{4} \\ a&=\frac{3}{4} \end{aligned}$$
The solution is $\frac{3}{4}$.

17.
$$\begin{aligned} 2-6y&=5-7y \\ 2-6y+7y&=5-7y+7y \\ 2+y&=5 \\ 2-2+y&=5-2 \\ y&=3 \end{aligned}$$
The solution is 3.

19.
$$\begin{aligned} 2(x+1)+5x&=23 \\ 2x+2+5x&=23 \\ 7x+2&=23 \\ 7x+2-2&=23-2 \\ 7x&=21 \\ \frac{7x}{7}&=\frac{21}{7} \\ x&=3 \end{aligned}$$
The solution is 3.

21.
$$\begin{aligned} 7a-(3a-4)&=12 \\ 7a-3a+4&=12 \\ 4a+4&=12 \\ 4a+4-4&=12-4 \\ 4a&=8 \\ \frac{4a}{4}&=\frac{8}{4} \\ a&=2 \end{aligned}$$
The solution is 2.

23.
$$\begin{aligned} 9-7x&=4(1-3x) \\ 9-7x&=4-12x \\ 9-7x+12x&=4-12x+12x \\ 9+5x&=4 \\ 9-9+5x&=4-9 \\ 5x&=-5 \\ \frac{5x}{5}&=\frac{-5}{5} \\ x&=-1 \end{aligned}$$
The solution is –1.

25.
$$\begin{aligned} 2z-2&=5-(9-6z)\\ 2z-2&=5-9+6z\\ 2z-2&=-4+6z\\ 2z-2z-2&=-4+6z-2z\\ -2&=-4+4z\\ -2+4&=-4+4+4z\\ 2&=4z\\ \frac{2}{4}&=\frac{4z}{4}\\ \frac{1}{2}&=z \end{aligned}$$
The solution is $\frac{1}{2}$.

27.
$$\begin{aligned} 5(6-2x)&=2(5-3x)\\ 30-10x&=10-6x\\ 30-10x+10x&=10-6x+10x\\ 30&=10+4x\\ 30-10&=10-10+4x\\ 20&=4x\\ \frac{20}{4}&=\frac{4x}{4}\\ 5&=x \end{aligned}$$
The solution is 5.

29.
$$\begin{aligned} 2(3b-5)&=4(6b-2)\\ 6b-10&=24b-8\\ 6b-6b-10&=26b-6b-8\\ -10&=18b-8\\ -10+8&=18b-8+8\\ -2&=18b\\ \frac{-2}{18}&=\frac{18b}{18}\\ -\frac{1}{9}&=b \end{aligned}$$
The solution is $-\frac{1}{9}$.

Objective B Exercises

31.
$$\begin{aligned} x-5&>-2\\ x-5+5&>-2+5\\ x&>3 \end{aligned}$$
The solution is set is $\{x\,|\,x>3\}$.

33.
$$\begin{aligned} -2+n&\ge 0\\ -2+2+n&\ge 0+2\\ n&\ge 2 \end{aligned}$$
The solution is set is $\{n\,|\,n\ge 2\}$.

35.
$$\begin{aligned} 8x&\le -24\\ \frac{8x}{8}&\le\frac{-24}{8}\\ x&\le -3 \end{aligned}$$
The solution is set is $\{x\,|\,x\le -3\}$.

37.
$$\begin{aligned} 3n&>0\\ \frac{3n}{3}&>\frac{0}{3}\\ n&>0 \end{aligned}$$
The solution is set is $\{n\,|\,n>0\}$.

39.
$$\begin{aligned} 2x-1&>7\\ 2x-1+1&>7+1\\ 2x&>8\\ \frac{2x}{2}&>\frac{8}{2}\\ x&>4 \end{aligned}$$
The solution is set is $\{x\,|\,x>4\}$.

41.
$$\begin{aligned} 4-3x&<10\\ 4-4-3x&<10=4\\ -3x&<6\\ \frac{-3x}{-3}&>\frac{6}{-3}\\ x&>-2 \end{aligned}$$
The solution is set is $\{x\,|\,x>-2\}$.

43.
$$\begin{aligned} 3x-1&>2x+2\\ 3x-2x-1&>2x-2x+2\\ x-1&>2\\ x-1+1&>2+1\\ x&>3 \end{aligned}$$
The solution is set is $\{x\,|\,x>3\}$.

45.
$$\begin{aligned} 8x+1&\ge 2x+13\\ 8x-2x+1&\ge 2x-2x+13\\ 6x+1&\ge 13\\ 6x+1-1&\ge 13-1\\ 6x&\ge 12\\ \frac{6x}{6}&\ge\frac{12}{6}\\ x&\ge 2 \end{aligned}$$
The solution is set is $\{x\,|\,x\ge 2\}$.

47.
$$\begin{aligned} -3-4x&>-11\\ -3+3-4x&>-11+3\\ -4x&>-8\\ \frac{-4x}{-4}&<\frac{-8}{-4}\\ x&<2 \end{aligned}$$

The solution is set is $\{x \mid x < 2\}$.

49.
$$\begin{aligned} 4x - 2 &> 3x + 1 \\ 4x - 3x - 2 &> 3x - 3x + 1 \\ x - 2 &> 1 \\ x - 2 + 2 &> 1 + 2 \\ x &> 3 \end{aligned}$$
The solution is set is $\{x \mid x > 3\}$.

51.
$$\begin{aligned} 9x + 2 &\geq 3x + 14 \\ 9x - 3x + 2 &\geq 3x - 3x + 14 \\ 6x + 2 &\geq 14 \\ 6x + 2 - 2 &\geq 14 - 2 \\ 6x &\geq 12 \\ \frac{6x}{6} &\geq \frac{12}{6} \\ x &\geq 2 \end{aligned}$$
The solution is set is $\{x \mid x \geq 2\}$.

53.
$$\begin{aligned} -5 - 2x &> -13 \\ -5 + 5 - 2x &> -13 + 5 \\ -2x &> -8 \\ \frac{-2x}{-2} &< \frac{-8}{-2} \\ x &< 4 \end{aligned}$$
The solution is set is $\{x \mid x < 4\}$.

55.
$$\begin{aligned} 4(2x - 1) &> 3x - 2(3x - 5) \\ 8x - 4 &> 3x - 6x + 10 \\ 8x - 4 &> -3x + 10 \\ 8x + 3x - 4 &> -3x + 3x + 10 \\ 11x - 4 &> 10 \\ 11x - 4 + 4 &> 10 + 4 \\ 11x &> 14 \\ \frac{11x}{11} &> \frac{14}{11} \end{aligned}$$
The solution is set is $\left\{x \mid x > \frac{14}{11}\right\}$.

57.
$$\begin{aligned} 3(4x + 3) &\leq 7 - 4(x - 2) \\ 12x + 9 &\leq 7 - 4x + 8 \\ 12x + 9 &\leq 15 - 4x \\ 12x + 4x + 9 &\leq 15 - 4x + 4x \\ 16x + 9 &\leq 15 \\ 16x + 9 - 9 &\leq 15 - 9 \\ 16x &\leq 6 \\ \frac{16x}{16} &\leq \frac{6}{16} \\ x &\leq \frac{3}{8} \end{aligned}$$
The solution is set is $\left\{x \mid x \leq \frac{3}{8}\right\}$.

59.
$$\begin{aligned} 3 - 4(x + 2) &\leq 6 + 4(2x + 1) \\ 3 - 4x - 8 &\leq 6 + 8x + 4 \\ -4x - 5 &\leq 8x + 10 \\ -4x - 8x - 5 &\leq 8x - 8x + 10 \\ -12x - 5 &\leq 10 \\ -12x - 5 + 5 &\leq 10 + 5 \\ -12x &\leq 15 \\ \frac{-12x}{-12} &\geq \frac{15}{-12} \\ x &\geq -\frac{5}{4} \end{aligned}$$
The solution is set is $\left\{x \mid x \geq -\frac{5}{4}\right\}$.

Section T.3

Objective A Exercises

1. Graphing the points (2, 3), (4, 0), (–4, 1), and (–2, –2).

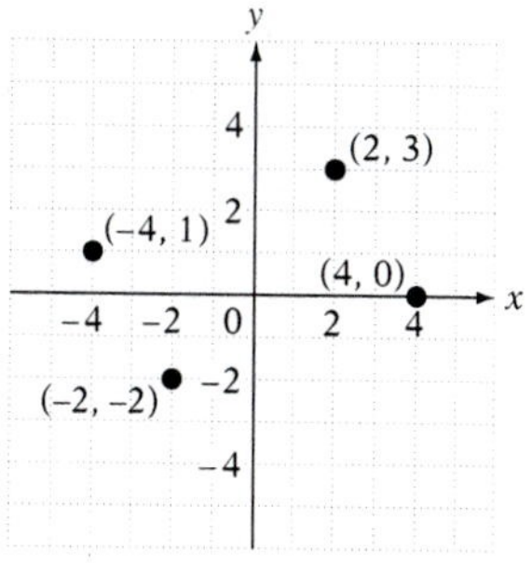

3. Graphing the points (–2, 5), (3, 4), (0, 0), and (–3, –2).

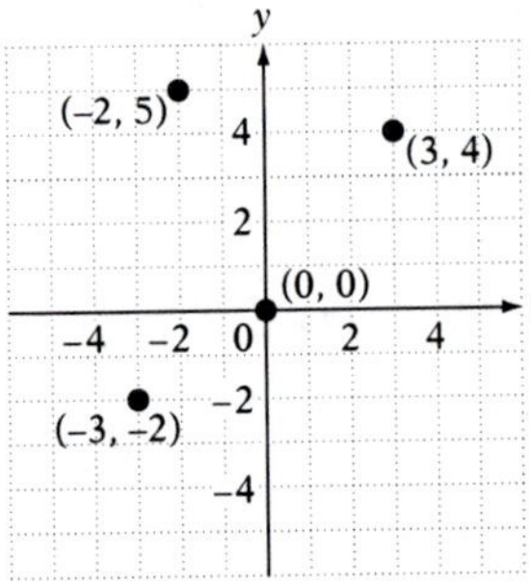

Objective B Exercises

5. $y = 2x + 1$

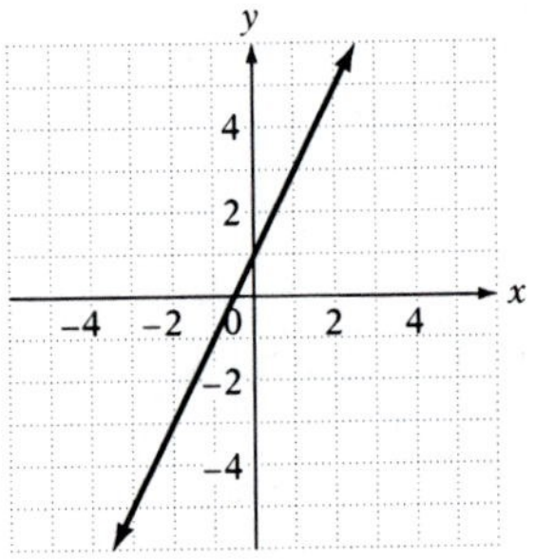

7. $y = -3x + 4$

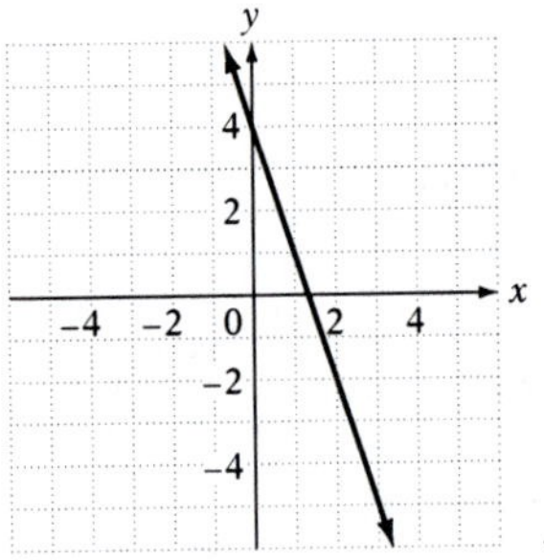

9. $y = 3x$

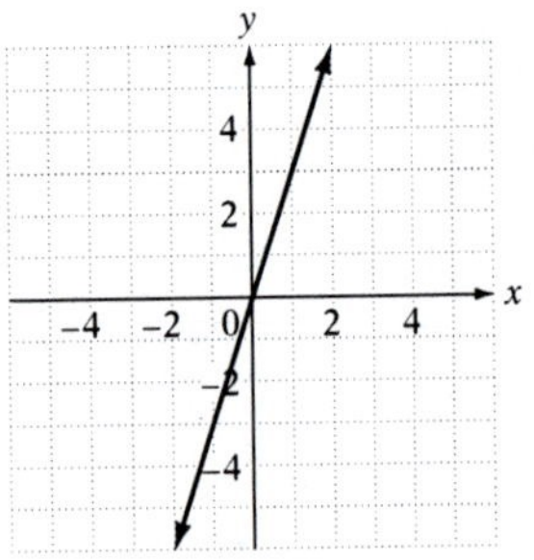

11. $y = -\frac{4}{3}x$

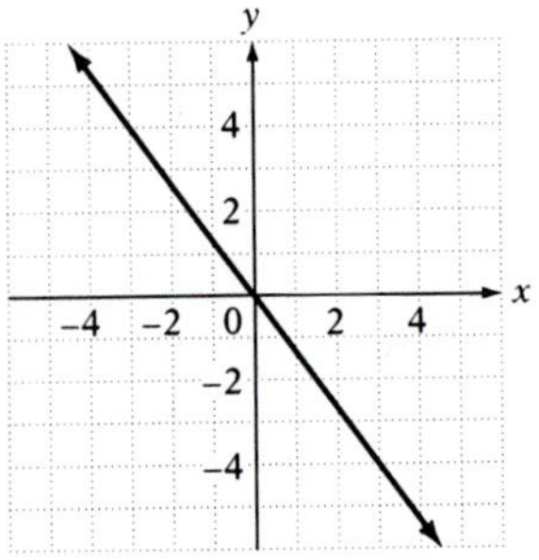

13. $y = \frac{3}{2}x - 1$

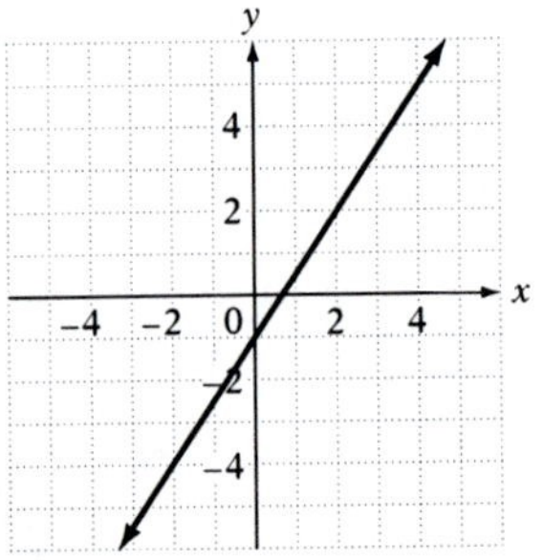

15. $y = -\frac{2}{3}x + 1$

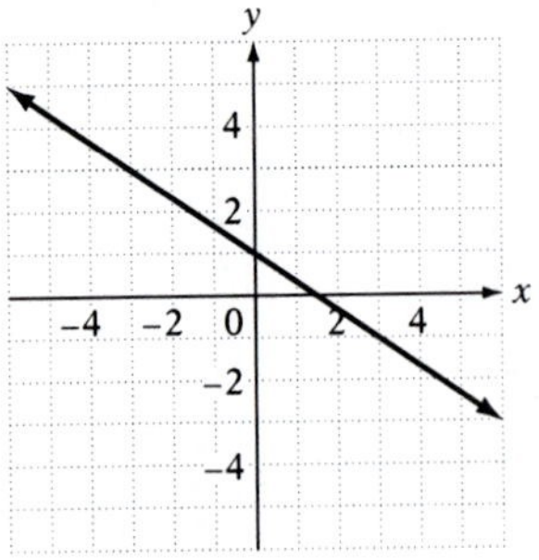

17. $2x + y = -3$

$y = -2x - 3$

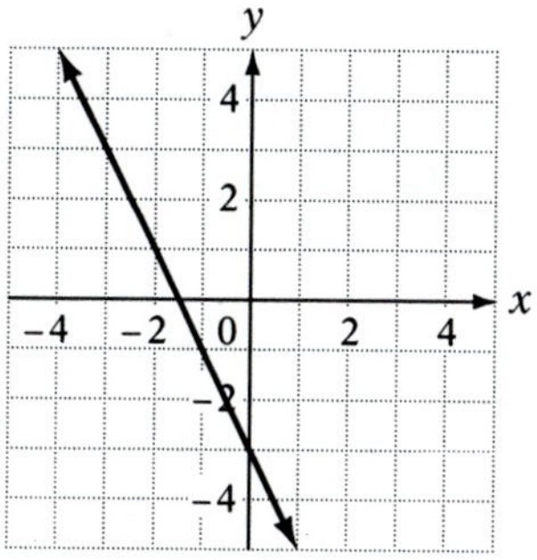

19. $x - 4y = 8$

$$-4y = -x + 8$$

$$y = \frac{1}{4}x - 2$$

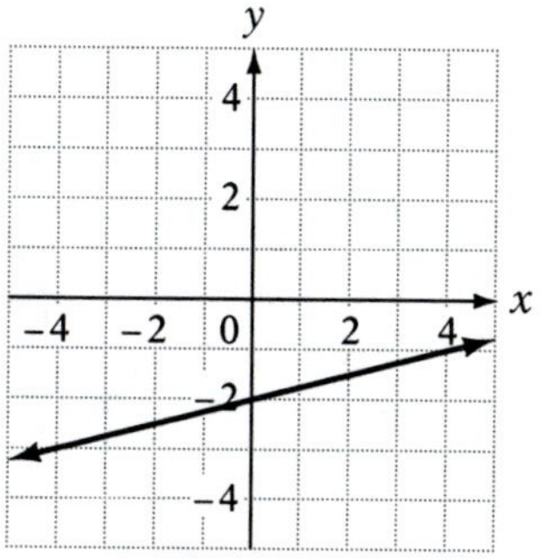

21. $3x - 2y = 8$

$$-2y = -3x + 8$$

$$y = \frac{3}{2}x - 4$$

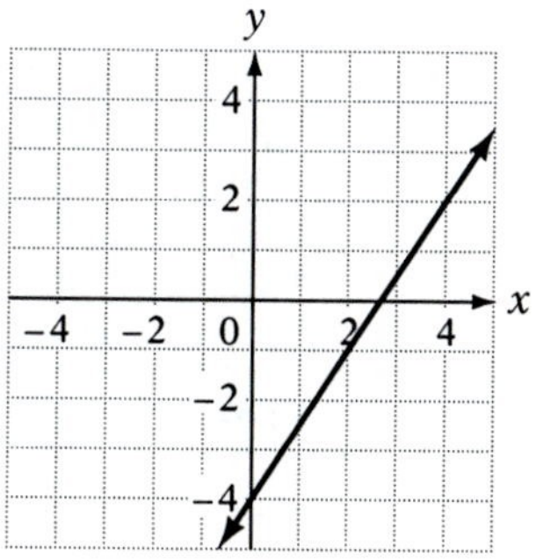

23. $$m = \frac{y_2 - y_1}{x_2 - x_1}$$

$$m = \frac{4-1}{3-2} = \frac{3}{1} = 3$$

25. $$m = \frac{y_2 - y_1}{x_2 - x_1}$$

$$m = \frac{2-1}{2-(-2)} = \frac{1}{2+2} = \frac{1}{4}$$

27. $$m = \frac{y_2 - y_1}{x_2 - x_1}$$

$$m = \frac{-3-3}{5-1} = \frac{-6}{4} = -\frac{3}{2}$$

29. $$m = \frac{y_2 - y_1}{x_2 - x_1}$$

$$m = \frac{3-2}{-1-(-1)} = \frac{1}{-1+1} = \frac{1}{0}$$

The slope is undefined.

31. $$m = \frac{y_2 - y_1}{x_2 - x_1}$$

$$m = \frac{1-1}{-2-5} = \frac{0}{-7} = 0$$

33. $$m = \frac{y_2 - y_1}{x_2 - x_1}$$

$$m = \frac{-1-0}{2-3} = \frac{-1}{-1} = 1$$

35. For $y = \frac{5}{2}x - 4$, $m = \frac{5}{2}$, and $b = (0, -4)$.

37. For $y = x$, $m = 1$, and $b = (0, 0)$.

39. $$y = \frac{1}{2}x + 2$$

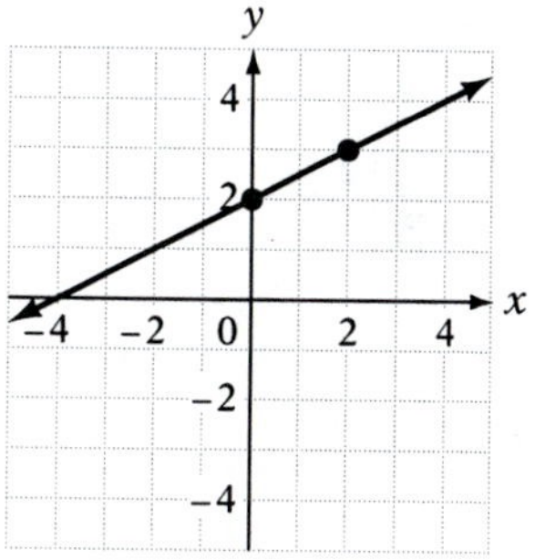

41. $$y = -\frac{3}{2}x$$

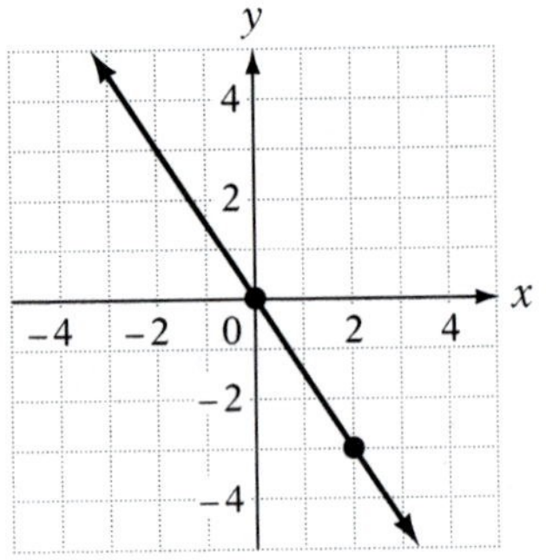

43. $y = -\frac{1}{2}x + 2$

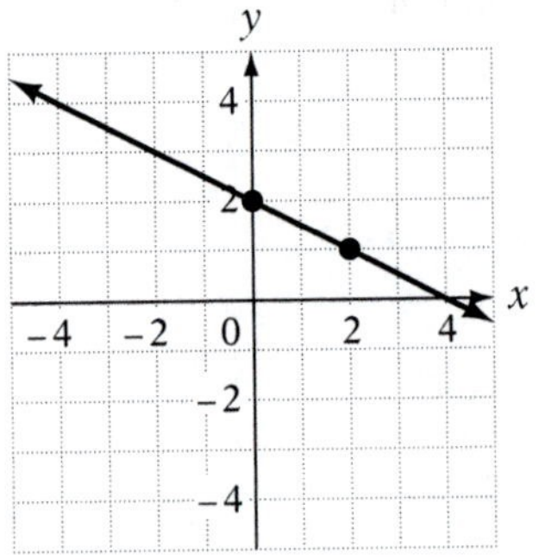

45.
$$x - 3y = 3$$
$$-3y = -x + 3$$
$$y = \frac{1}{3}x - 1$$

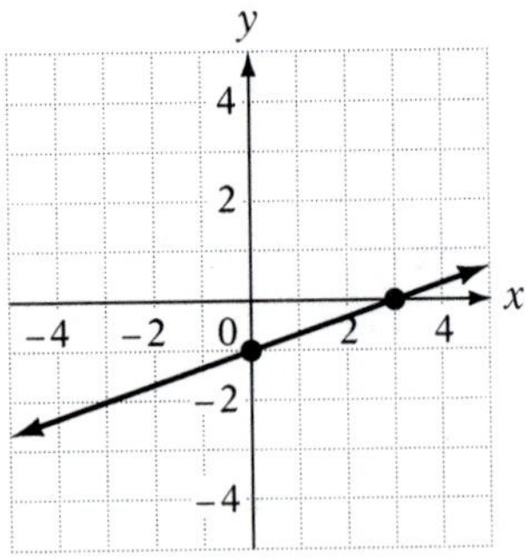

Objective C Exercises

47.
$$g(x) = -3x + 1$$
$$g(-4) = -3(-4) + 1$$
$$g(-4) = 12 + 1$$
$$g(-4) = 13$$
(–4, 13)

49.
$$p(x) = 6 - 8x$$
$$p(-1) = 6 - 8(-1)$$
$$p(-1) = 6 + 8$$
$$p(-1) = 14$$
(–1, 14)

51.
$$f(t) = t^2 - t - 3$$
$$f(2) = 2^2 - 2 - 3$$
$$f(2) = 4 - 2 - 3$$
$$f(2) = -1$$
(2, –1)

53.
$$h(x) = -3x^2 + x - 1$$
$$h(-2) = -3(-2)^2 + (-2) - 1$$
$$h(-2) = -3(4) - 2 - 1$$
$$h(-2) = -12 - 2 - 1$$
$$h(-2) = -15$$
(–2, –15)

55.
$$g(t) = 4t^3 - 2t$$
$$g(-1) = 4(-1)^3 - 2(-1)$$
$$g(-1) = 4(-1) + 2$$
$$g(-1) = -4 + 2$$
$$g(-1) = -2$$
(–1, –2)

Objective D Exercises

57.
$$y - y_1 = m(x - x_1)$$
$$y - 2 = -3(x - (-1))$$
$$y - 2 = -3(x + 1)$$
$$y - 2 = -3x - 3$$
$$y = -3x - 1$$

59.
$$y - y_1 = m(x - x_1)$$
$$y - (-5) = -2(x - 4)$$
$$y + 5 = -2x + 8$$
$$y = -2x + 3$$

61.
$$y - y_1 = m(x - x_1)$$
$$y - (-3) = -\frac{3}{5}(x - 5)$$
$$y + 3 = -\frac{3}{5}x + 3$$
$$y = -\frac{3}{5}x$$

63.
$$y - y_1 = m(x - x_1)$$
$$y - (-2) = -\frac{2}{3}(x - (-3))$$
$$y + 2 = -\frac{2}{3}x(x + 3)$$
$$y + 2 = -\frac{2}{3}x - 2$$
$$y = -\frac{2}{3}x - 4$$

Section T.4

Objective A Exercises

1. $z^3 \cdot z \cdot z^4 = z^{3+1+4} = z^8$

3. $(x^3)^5 = x^{3\cdot 5} = x^{15}$

5. $(x^2y^3)^6 = x^{2\cdot 6}y^{3\cdot 6} = x^{12}y^{18}$

7. $\frac{a^8}{a^2} = a^{8-2} = a^6$

9. $(-m^3n)(m^6n^2) = -m^{3+6}n^{1+2} = -m^9n^3$

11. $(-2a^3bc^2)^3 = (-2)^3 a^{3\cdot 3}b^{1\cdot 3}c^{2\cdot 3} = -8a^9b^3c^6$

13. $\frac{m^4n^7}{m^3n^5} = m^{4-3}n^{7-5} = mn^2$

15. $\frac{-16a^7}{24a^6} = -\frac{2a^{7-6}}{3} = -\frac{2a}{3}$

17. $(9mn^4p)(-3mp^2) = -27m^{1+1}n^4p^{1+2} = -27m^2n^4p^3$

19.
$$(-2n^2)(-3n^4)^3 = (-2n^2)((-3)^3 n^{4\cdot 3})$$
$$= (-2n^2)(-27n^{12})$$
$$= 54n^{2+12} = 54n^{14}$$

21. $\frac{14x^4y^6z^2}{16x^3y^9z} = \frac{7x^{4-3}y^{6-9}z^{2-1}}{8} = \frac{7xy^{-3}z}{8} = \frac{7xz}{8y^3}$

23.
$$(-2x^3y^2)^3(-xy^2)^4 = ((-2)^3 x^{3\cdot 3}y^{2\cdot 3})((-1)^4 x^{1\cdot 4}y^{2\cdot 4})$$
$$= (-8x^9y^6)(x^4y^8)$$
$$= -8x^{9+4}y^{6+8} = -8x^{13}y^{14}$$

25. $4x^{-7} = 4\cdot\frac{1}{x^7} = \frac{4}{x^7}$

27. $d^{-4}d^{-6} = d^{-10} = \frac{1}{d^{10}}$

29. $\frac{x^{-3}}{x^2} = \frac{1}{x^{3+2}} = \frac{1}{x^5}$

31. $\frac{1}{3x^{-2}} = \frac{1}{3}\cdot x^2 = \frac{x^2}{3}$

33. $(x^2y^{-4})^3 = x^6y^{-12} = \frac{x^6}{y^{12}}$

35. $(3x^{-1}y^{-2})^2 = 3^2x^{-2}y^{-4} = \frac{9}{x^2y^4}$

37. $(2x^{-1})(x^{-3}) = 2x^{-4} = \frac{2}{x^4}$

39. $\frac{3x^{-2}y^2}{6xy^2} = \frac{1}{2x^3}$

41. $\frac{2x^{-1}y^{-4}}{4xy^2} = \frac{1}{2x^2y^6}$

43. $(x^{-2}y)^2(xy)^{-2} = (x^{-4}y^2)(x^{-2}y^{-2}) = x^{-6} = \frac{1}{x^6}$

45. $\left(\frac{x^2y^{-1}}{xy}\right)^{-4} = \frac{x^{-8}y^4}{x^{-4}y^{-4}} = x^{-8+4}y^{4+4} = x^{-4}y^8 = \frac{y^8}{x^4}$

47. $\left(\frac{4a^{-2}b}{8a^3b^{-4}}\right)^2 = \left(\frac{b^5}{2a^5}\right)^2 = \frac{b^{10}}{4a^{10}}$

Objective B Exercises

49.
$$(4b^2 - 5b) + (3b^2 + 6b - 4)$$
$$= (4b^2 + 3b^2) + (-5b + 6b) - 4$$
$$= 7b^2 + b - 4$$

51.
$$(2a^2 - 7a + 10) + (a^2 + 4a + 7)$$
$$= (2a^2 + a^2) + (-7a + 4a) + (10 + 7)$$
$$= 3a^2 - 3a + 17$$

53.
$$(x^2 - 2x + 1) - (x^2 + 5x + 8)$$
$$= x^2 - 2x + 1 - x^2 - 5x - 8$$
$$= -7x - 7$$

55.
$$(-2x^3 + x - 1) - (-x^2 + x - 3)$$
$$-2x^3 + x - 1 + x^2 - x + 3$$
$$= -2x^3 + x^2 + 2$$

57. $\left(x^3-7x+4\right)+\left(2x^2+x-10\right)$
$=x^3+2x^2-6x-6$

59. $\left(5x^3+7x-7\right)+\left(10x^2-8x+3\right)$
$=5x^3+10x^2-x-4$

61. $\left(2y^3+6y-2\right)-\left(y^3+y^2+4\right)$
$=2y^3+6y-2-y^3-y^2-4$
$=y^3-y^2+6y-6$

63. $\left(4y^3-y-1\right)-\left(2y^2-3y+3\right)$
$=4y^3-y-1-2y^2+3y-3$
$=4y^3-2y^2+2y-4$

Objective C Exercises

65. $4b\left(3b^3-12b^2-6\right)=12b^4-48b^3-24b$

67. $3b\left(3b^4-3b^2+8\right)=9b^5-9b^3+24b$

69. $-2x^2y\left(x^2-3xy+2y^2\right)=-2x^4y+6x^3y^2-4x^2y^3$

71.
$$\begin{array}{r} x^2+3x+2 \\ \times \qquad x+1 \\ \hline x^2+3x+2 \\ x^3+3x^2+2x \qquad \\ \hline x^3+4x^2+5x+2 \end{array}$$

73.
$$\begin{array}{r} a^2-3a+4 \\ \times \qquad a-3 \\ \hline -3a^2+9a-12 \\ a^3-3a^2+4a \qquad \\ \hline a^3-6a^2+13a-12 \end{array}$$

75.
$$\begin{array}{r} -2b^2-3b+4 \\ \times \qquad b-5 \\ \hline 10b^2+15b-20 \\ -2b^3-3b^2+4b \qquad \\ \hline -2b^3+7b^2+19b-20 \end{array}$$

77.
$$\begin{array}{r} x^3 \qquad -3x+2 \\ \times\, x-4 \\ \hline -4x^3 \qquad +12x-8 \\ x^4 \qquad -3x^2+2x \qquad \\ \hline x^4-4x^3-3x^2+14x-8 \end{array}$$

79.
$$\begin{array}{r} y^3+2y^2-3y+1 \\ \times \qquad y+2 \\ \hline 2y^3+4y^2-6y+2 \\ y^4+2y^3-3y^2+y \qquad \\ \hline y^4+4y^3+y^2-5y+2 \end{array}$$

81. $(a-3)(a+4)=a^2+4a-3a-12$
$=a^2+a-12$

83. $(y-7)(y-3)=y^2-3y-7y+21$
$=y^2-10y+21$

85. $(2x+1)(x+7)=2x^2+14x+x+7$
$=2x^2+15x+7$

87. $(3x-1)(x+4)=3x^2+12x-x-4$
$=3x^2+11x-4$

89. $(4x-3)(x-7)=4x^2-28x-3x+21$
$=4x^2-31x+21$

91. $(3y-8)(y+2)=3y^2+6y-8y-16$
$=3y^2-2y-16$

93. $(7a-16)(3a-5)=21a^2-35a-48a+80$
$=21a^2-83a+80$

95. $(x+y)(2x+y)=2x^2+xy+2xy+y^2$
$=2x^2+3xy+y^2$

97. $(3x-4y)(x-2y)=3x^2-6xy-4xy+8y^2$
$=3x^2-10xy+8y^2$

99. $(5a-3b)(2a+4b)=10a^2+20ab-6ab-12b^2$
$=10a^2+14ab-12b^2$

101. $(4x-7)(4x+7)=16x^2-49$

Objective D Exercises

103.
$$\begin{array}{r} x+2 \\ x-3\overline{)x^2-x-6} \\ \underline{x^2-3x} \\ 2x-6 \\ \underline{2x-6} \\ 0 \end{array}$$

$\left(x^2-x-6\right)\div(x-3)=x+2$

105.
$$\begin{array}{r} 2y-7 \\ y-3\overline{)2y^2-13y+21} \\ \underline{2y^2-6y} \\ -7y+21 \\ \underline{-7y+21} \\ 0 \end{array}$$
$$\left(2y^2-13y+21\right)\div\left(y-3\right)=2y-7$$

107.
$$\begin{array}{r} x-2 \\ x+2\overline{)x^2+0+4} \\ \underline{x^2+2x} \\ -2x+4 \\ \underline{-2x-4} \\ 8 \end{array}$$
$$\left(x^2+4\right)\div\left(x+2\right)=x-2+\frac{8}{x+2}$$

109.
$$\begin{array}{r} 3y-5 \\ 2y+4\overline{)6y^2+2y+0} \\ \underline{2y^2+12y} \\ -10y+0 \\ \underline{-10y-20} \\ 20 \end{array}$$
$$\left(6y^2+2y\right)\div\left(2y+4\right)=3y-5+\frac{20}{2y+4}$$

111.
$$\begin{array}{r} b-5 \\ b-3\overline{)b^2-8b-9} \\ \underline{b^2-3b} \\ -5b-9 \\ \underline{-5b+15} \\ -24 \end{array}$$
$$\left(b^2-8b-9\right)\div\left(b-3\right)=b-5-\frac{24}{b-3}$$

113.
$$\begin{array}{r} 3x+17 \\ x-4\overline{)3x^2+5x-4} \\ \underline{3x^2-12x} \\ 17x-4 \\ \underline{17x-68} \\ 64 \end{array}$$
$$\left(3x^2+5x-4\right)\div\left(x-4\right)=3x+17+\frac{64}{x-4}$$

115.
$$\begin{array}{r} 5y+3 \\ 2y+3\overline{)10y^2+21y+10} \\ \underline{10y^2+15y} \\ 6y+10 \\ \underline{6y+9} \\ 1 \end{array}$$
$$\left(10y^2+21y+10\right)\div\left(2y+3\right)=5y+3+\frac{1}{2y+3}$$

117.
$$\begin{array}{r} x^2-5x+2 \\ x-1\overline{)x^3-6x^2+7x-2} \\ \underline{x^3-x^2} \\ -5x^2+7x \\ \underline{-5x^2+5x} \\ 2x-2 \\ \underline{2x-2} \\ 0 \end{array}$$
$$\left(x^3-6x^2+7x-2\right)\div\left(x-1\right)=x^2-5x+2$$

119.
$$\begin{array}{r} x^2+5 \\ x^2+0-2\overline{)x^4+0+3x^2+0-10} \\ \underline{x^4+0-2x^2} \\ 5x^2+0-10 \\ \underline{5x^2+0-10} \\ 0 \end{array}$$
$$\left(x^2+3x^2-10\right)\div\left(x^2-2\right)=x^2+5$$

Objective E Exercises

121. $12y^2=2\cdot 2\cdot 3y^2$

$5y=5y$

The GCF is y.

$$\begin{aligned} 12y^2-5y &= y(12y)-y(5) \\ &= y(12y-5) \end{aligned}$$

123. $10x^2yz^2=2\cdot 5x^2yz^2$

$15xy^3z=3\cdot 5xy^3z$

The GCF is $5xyz$.

$$\begin{aligned} 10x^2yz^2+15xy^3z &= 5xyz(2xz)+5xyz\left(3y^2\right) \\ &= 5xyz\left(2xz+3y^2\right) \end{aligned}$$

125. $5x^2=5x^2$

$15x=3\cdot 5x$

$35=5\cdot 7$

The GCF is 5.

$$5x^2-15x+35=5(x^2)+5(-3x)+5(7)$$
$$=5(x^2-3x+7)$$

127. The GCF is $3y^2$.

$$3y^4-9y^3-6y^2=3y^2(y^2)+3y^2(-3y)+3y^2(-2)$$
$$=3y^2(y^2-3y-2)$$

129. The GCF is xyz.

$$x^4y^4-3x^3y^3+6x^2y^2$$
$$=x^2y^2(x^2y^2)+x^2y^2(-3xy)+x^2y^2(6)$$
$$=x^2y^2(x^2y^2-3xy+6)$$

131. $16x^2y=2\cdot2\cdot2\cdot2x^2y$

$8x^3y^4=2\cdot2\cdot2x^3y^4$

$48x^2y^2=2\cdot2\cdot2\cdot2\cdot3x^2y^2$

The GCF is $8x^2y$.

$$16x^2y-8x^3y^4-48x^2y^2$$
$$=8x^2y(2)+8x^2y(-xy^3)+8x^2y(-6y)$$
$$=8x^2(2-xy^3-6y)$$

133.

Factors	Sum
+1, −2	−1
−1, +2	1

$x^2+x-2=(x+2)(x-1)$

135.

Factors	Sum
−1,+12	11
+1, −12	−11
−2, +6	4
+2, −6	−4
−3, +4	1
+3, −4	−1

$a^2+a-12=(a+4)(a-3)$

137.

Factors	Sum
−1, −2	−3

$a^2-3a+2=(a-1)(a-2)$

139.

Factors	Sum
−1, +8	7
+1, −8	−7
−2, +4	2
+2, −4	−2

$b^2+7b-8=(b+8)(b-1)$

141.

Factors	Sum
−1, +45	44
+1, −45	−44
−3, +15	12
+3, −15	−12
−5, +9	4
+5, −9	−4

$z^2-4z-45=(z+5)(z-9)$

143.

Factors	Sum
−1, −45	−46
−3, −15	−18
−5, −9	−14

$z^2-14z+45=(z-5)(z-9)$

145.

Factors	Sum
+1, +20	21
+2, +10	12
+4, +5	9

$b^2+9b+20=(b+4)(b+5)$

147.

Factors	Sum
−1, −81	7
−3, −27	−30
−9, −9	−18

$y^2-9y+81$ is nonfactorable over the integers.

149.

Factors	Sum
−1, −56	−57
−2, −28	−30
−4, −14	−16
−7, −8	−15

$x^2-15x+56=(x-7)(x-8)$

151. Factors of 2: 1, 2 Factors of 3: 1, 3

Trial Factors	Middle Term
$(1y+1)(2y+3)$	$3y+2y=5y$
$(1y+3)(2y+1)$	$y+6y=7y$

$2y^2+7y+3=(y+3)(2y+1)$

153. Factors of 3: 1, 3 Factors of 1: −1, −1

Trial Factors	Middle Term
$(1a-1)(3a-1)$	$-1-3a=-4a$

$3a^2-4a+1=(a-1)(3a-1)$

155. Factors of 2: 1, 2

Factors of −3: −1, +3 or +1, −3

Trial Factors	Middle Term
$(1x-1)(2x+3)$	$3x-2x=x$
$(1x+3)(2x-1)$	$-x+6x=5x$
$(1x+1)(2x-3)$	$-3x+2x=-x$
$(1x-3)(2x+1)$	$x-6x=-5x$

$2x^2-5x-3=(x-3)(2x+1)$

157. Factors of 10: 1, 10 or 2, 5
Factors of 3: +1, +3

Trial Factors	Middle Term
$(1t+1)(10t+3)$	$3t+10t=13t$
$(1t+3)(10t+1)$	$t+30t=31t$
$(2t+1)(5t+3)$	$6t+5t=11t$
$(2t+3)(5t+1)$	$2t+15t=17t$

$10t^2+11t+3=(2t+1)(5t+3)$

159. Factors of 10: 1, 10 or 2, 5
Factors of –4:–1,+4 or +1,–4 or +2,–2

Trial Factors	Middle Term
$(1z-1)(10z+4)$	Common factor
$(1x+4)(10z-1)$	$-z+40z=39z$
$(1z+1)(10z-4)$	Common factor
$(1z-4)(10z+1)$	$z-40z=-39z$
$(1z-2)(10z+2)$	Common factor
$(1z+2)(10z-2)$	Common factor
$(2z-1)(5x+4)$	$8z-5z=3z$
$(2z+4)(5z-1)$	Common factor
$(2x+1)(5z-4)$	$-8z+5z=-3z$
$(2z-2)(5z+2)$	Common factor
$(2z-2)(5z+2)$	Common factor
$(2z+2)(5z-2)$	Common factor

$10z^2+3z-4=(2z-1)(5z+4)$

161. Factors of 3: 1, 3
Factors of 10: +1, +10 or +2, +5

Trial Factors	Middle Term
$(1z+1)(3z+10)$	$10z+3z=13z$
$(1z+10)(3z+1)$	$z+30z=31z$
$(1z+2)(3z+5)$	$5z+6z=11z$
$(1z+5)(3z+2)$	$2z+15z=17z$

$3z^2+95z+10$ is nonfactorable over the integers.

163. $2t^2-t-10$

$2(-10)=-20$

The factors of –20 whose sum if –1: 4 and –5

$$\begin{aligned}2t^2-t-10&=2t^2+4t-5t-10\\&=(2t^2+4t)+(-5t-10)\\&=2t(t+2)-5(t+2)\\&=(t+2)(2t-5)\end{aligned}$$

165. $12y^2+19y+5$

$12(5)=60$

The factors of 60 whose sum is 19: 4 and 15

$$\begin{aligned}12y^2+19y+5&=12y^2+4y+15y+5\\&=(12y^2+4y)+(15y+5)\\&=4y(3y+1)+5(3y+1)\\&=(3y+1)(4y+5)\end{aligned}$$

167. $11a^2-54a-5$

$11(-5)=-55$

The factors –55 whose sum is –54: –55 and 1

$$\begin{aligned}11a^2-54a-5&=11a^2-55a+a-5\\&=(11a^2-55a)+(a-5)\\&=11a(a-5)+1(a-5)\\&=(a-5)(11a+1)\end{aligned}$$

169. $6b^2-13b+6$

$6(6)=36$

The factors of 36 whose sum is –13: –9 and –4

$$\begin{aligned}6b^2-13b+6&=6b^2-9b-4b+6\\&=(6b^2-9b)+(-4b+6)\\&=3b(2b-3)-2(2b-3)\\&=(2b-3)(3b-2)\end{aligned}$$

171. The GCF is 3.

$3x^2+15x+18=3(x^2+5x+6)$

Factor the trinomial.

Factors	Sum
+1, +6	7
+2, +3	5

$3x^2+15x+18=3(x+2)(x+3)$

The GCF is a.

173. $ab^2+7ab-8a=a(b^2+7b-8)$

Factor the trinomial.

Factors	Sum
−1, +8	7
+1, −8	−7
−2, +4	2
+2, −4	−2

$ab^2+7ab-8a=a(b+8)(b-1)$

175. The GCF is $2y^2$.

$2y^4-26y^3-96y^2=2y^2(y^2-13y-48)$

Factor the trinomial.

Factors	Sum
−1, +48	47
+1, −48	−47
−2, +24	22
+2, −24	−22
−3, +16	13
+3, −16	−13
−4, +12	8
+4, −12	−8
−6, +8	2
+6, −8	−2

$2y^4-26y^3-96y^2=2y^2(y+3)(y-16)$

177. $2x^3-11x^2+5x$

The GCF is x.

$2x^3-11x^2+5x=x(2x^2-11x+5)$

$2(5)=10$

The factors of 10 whose sum is −11: −1 and −10

$$\begin{aligned}2x^3-11x^2+5x&=x(2x^2-11x+5)\\&=x[2x^2-x-10x+5]\\&=x[(2x^2-x)+(-10x+5)]\\&=x[x(2x-1)-5(2x-1)]\\&=x(2x-1)(x-5)\end{aligned}$$

179. $10t^2-5t-50$

The GCF is 5.

$10t^2-5t-50=5(2t^2-t-10)$

$2(-10)=-20$

The factors of −20 whose sum is −1: −5 and 4

$$\begin{aligned}10t^2-5t-50&=5(2t^2-t-10)\\&=5[2t^2+4t-5t-10]\\&=5[(2t^2+4t)+(-5t-10)]\\&=5[2t(t+2)-5(t+2)]\\&=5(t+2)(2t-5)\end{aligned}$$

181. $6p^3+5p^2+p$

The GCF is p.

$6p^3+5p^2+p=p(6p^2+5p+1)$

$6(1)=6$

The factors of 6 whose sum is 5: 3 and 2

$$\begin{aligned}6p^3+5p^2+p&=p(6p^2+5p+1)\\&=p(6p^2+2p+3p+1)\\&=p[(6p^2+2p)+(3p+1)]\\&=p[2p(3p+1)+1(3p+1)]\\&=p(3p+1)(2p+1)\end{aligned}$$

CPSIA information can be obtained
at www.ICGtesting.com
Printed in the USA
FFOW02n0105061215
19324FF